Fuel Systems and Emission Controls

Shop Manual

Fifth Edition

Chek-Chart

Warren M. Farnell
Revision Author

James D. Halderman
Series Advisor

PEARSON
Prentice Hall

Upper Saddle River, New Jersey
Columbus, Ohio

Library of Congress Cataloging-in-Publication Data
Farnell, Warren.
Fuel systems and emission controls. Shop manual / Warren M. Farnell, revision author.—5th ed.
p. cm.
ISBN 0-13-140785-6
1. Automobiles—Fuel systems—Maintenance and repair—Handbooks, manuals, etc. 2. Automobiles—Pollution control devices—Maintenance and repair—Handbooks, manuals, etc. I. Title.
TL214.F8F37 2006
629.25′3—dc22

2005020026

Acquisitions Editor: Tim Peyton
Assistant Editor: Linda Cupp
Editorial Assistant: Nancy Kesterson
Production Coordination: Carlisle Publishers Services
Production Editor: Holly Shufeldt
Design Coordinator: Diane Ernsberger
Cover Designer: Jeff Vanik
Cover art: Corel
Production Manager: Deidra Schwartz
Marketing Manager: Ben Leonard
Senior Marketing Coordinator: Liz Farrell
Marketing Assistant: Les Roberts

This book was set in Times by Carlisle Communications, Ltd. It was printed and bound by Bind-Rite Graphics. The cover was printed by The Lehigh Press, Inc.

Portions of materials contained herein have been reprinted with permission of General Motors Corporation, Service and Parts Operations. License Agreement 0510867.

Pearson Education Ltd.
Pearson Education Singapore Pte. Ltd.
Pearson Education Canada, Ltd.
Pearson Education—Japan
Pearson Education Australia Pty. Limited
Pearson Education North Asia Ltd.
Pearson Educación de Mexico, S.A. de. C.V.
Pearson Education Malaysia Pte. Ltd.

PEARSON
Prentice
Hall

10 9 8 7 6 5 4 3
0-13-140785-6

Introduction

Fuel Systems and Emission Controls is part of the Chek-Chart automotive series. The entire series is job-oriented and designed especially for students who intend to work in the automotive service profession. The package for each course consists of two volumes, a *Classroom Manual* and a *Shop Manual.*

The fifth edition of *Fuel Systems and Emission Controls* has been completely revised to include in-depth coverage of the latest developments in automotive emission controls and fuel systems. Students will be able to use the knowledge gained from these books and from their instructor to diagnose and repair automotive emission controls and fuel systems used on today's automobiles.

This package retains the traditional thoroughness and readability of the Chek-Chart automotive series. Furthermore, both the *Classroom Manual* and the *Shop Manual,* as well as the *Instructor's Manual,* have been greatly enhanced.

CLASSROOM MANUAL

New features in the *Classroom Manual* include:

- New chapters on computer input devices, computer output devices, and emissions.
- Three new chapters covering ignition systems.
- Objectives in each chapter that alert students to the important themes and learning goals.
- Over 65 new illustrations.
- An added appendix that includes OBD II diagnostic trouble codes, major elements of operating I/M programs, vehicle manufacturer service information websites, and links to state emission programs.
- Obsolete material has been deleted.

SHOP MANUAL

Each chapter of the completely revised *Shop Manual* correlates with the *Classroom Manual.* Like the *Classroom Manual,* the *Shop Manual* features an overhauled illustration program. It includes over 135 new or revised figures and extensive photo sequences showing step-by-step repair procedures.

INSTRUCTOR'S MANUAL

The *Instructor's Manual* includes task sheets that cover many of the NATEF tasks for A8 Engine Performance. Instructors may reproduce these task sheets for use by the students in the lab or during an internship. The *Instructor's Manual* also includes a test bank and answers to end-of-chapter questions in the *Classroom Manual.*

The *Instructor's Resource* CD that accompanies the *Instructor's Manual* includes Microsoft® PowerPoint® presentations and photographs that appear in the *Classroom Manual* and the *Shop Manual.* Each photograph included on the *Instructor's Resource* CD cross-references the figure number in either the *Classroom Manual* or the *Shop Manual.* These high-resolution photographs are suitable for projection or reproduction.

Because of the comprehensive material, hundreds of high-quality illustrations, and inclusion of the latest technology, these books will keep their value over the years. In fact, *Fuel System and Emission Control* will form the core of the master technician's professional library.

How to Use This Book

WHY ARE THERE TWO MANUALS?

This two-volume text—*Fuel Systems and Emission Controls*—is not like most other textbooks. It is actually two books, a *Classroom Manual* and a *Shop Manual* that should be used together. The *Classroom Manual* teaches you what you need to know about fuel system and emission control theory, systems, and components. The *Shop Manual* will show you how to repair and adjust complete systems, as well as their individual components.

WHAT IS IN THESE MANUALS?

There are several aids in the *Classroom Manual* that will help you learn more.

- Each chapter is based on detailed learning objectives, which are listed in the beginning of each chapter.
- Each chapter is divided into self-contained sections for easier understanding and review. This organization clearly shows which parts make up which systems, and how various parts or systems that perform the same task differ or are the same.
- Most parts and processes are fully illustrated with drawings and photographs.
- A list of Key Terms is located at the beginning of each chapter. These are printed in **boldface type** in the text and are defined in a glossary at the end of the manual. Use these words to build the vocabulary needed to understand the text.
- Review Questions follow each chapter. Use them to test your knowledge of the material covered.
- A brief summary at the end of each chapter helps you review for exams.

The *Shop Manual* has detailed instructions on the test, service, and overhaul of automotive fuel and emission control systems and their components. These are easy to understand and often include step-by-step explanations of the procedure. Key features of the *Shop Manual* include:

- Each chapter is based upon ASE/NATEF tasks, which are listed in the beginning of each chapter.
- Helpful information on the use and maintenance of shop tools and test equipment.
- Detailed safety precautions.
- Clear illustrations and diagrams to help you locate trouble spots while learning to read the service literature.
- Test procedures and troubleshooting hints that help you work better and faster.
- Repair tips used by professionals, presented clearly and accurately.

WHERE SHOULD I BEGIN?

If you already know something about automotive fuel and emission control systems and how to repair them, you will find that this book is a helpful review. If you are just starting in automotive repair, then the book will give you a solid foundation on which to develop professional-level skills.

Your instructor will design a course to take advantage of what you already know, and what facilities and equipment are available to work with. You may be asked to read certain chapters of this manual out of order. That is fine; the important thing is to fully understand each subject before you move on to the next. Study the vocabulary words, and use the review questions to help you comprehend the material.

While reading the *Classroom Manual,* refer to your *Shop Manual* and relate the descriptive text to the service procedures. When working on actual automotive fuel and emission control systems, look back to the *Classroom Manual* to keep basic information fresh in your mind. Working on such complicated modern fuel and emission system isn't always easy. Take advantage of the information in the *Classroom Manual,* the procedures in the *Shop Manual,* and the knowledge of your instructor to help you.

Remember that the *Shop Manual* is a good book for work, not just a good workbook. Keep it on hand while you're working on a fuel or emission control system. For ease of use, the *Shop Manual* will fold flat on the workbench or under the car, and it can withstand quite a bit of rough handling.

When you perform actual test and repair procedures, you need a complete and accurate source of manufacturer specifications and procedures for the specific vehicle. As the source for these specifications, most automotive repair shops have the annual service information (on paper, CD, or Internet formats) from the vehicle manufacturer or an independent guide.

Acknowledgments

The publisher sincerely thanks the following vehicle manufacturers, industry suppliers, and individuals for supplying information and illustrations used in the Chek-Chart Series in Automotive Technology.

Allen Testproducts
American Isuzu Motors, Inc.
Automotive Electronics Services
Bear Manufacturing Company
Borg-Warner Corporation
Champion Spark Plug Company
DaimlerChrysler Corporation
DeAnza College, Cupertino, CA
Fluke Corporation
Ford Motor Company
Fram Corporation
General Motors Corporation
  Delco-Remy Division
  Rochester Products Division
  Saginaw Steering Gear Division
  Buick Motor Division
  Cadillac Motor Car Division
  Chevrolet Motor Division
  Oldsmobile Division
  Pontiac-GMC Division
Honda Motor Company, LTD
Jaguar Cars, Inc.
Marquette Manufacturing Company
Mazda Motor Corporation
Mercedes-Benz USA, Inc.
Mitsubishi Motor Sales of America, Inc.
Nissan North America, Inc.
The Prestolite Company
Robert Bosch Corporation
Saab Cars USA Inc.
Snap-on Tools Corporation
Toyota Motor Sales, U.S.A., Inc.
Vetronix Corporation
Volkswagen of America
Volvo Cars of North America

The publisher gratefully acknowledges the reviewers of this edition:

Kenneth Mays, Central Oregon Community College
Katherine Pfau, Montana State University, Billings

The publisher also thanks Series Advisor James D. Halderman.

Contents

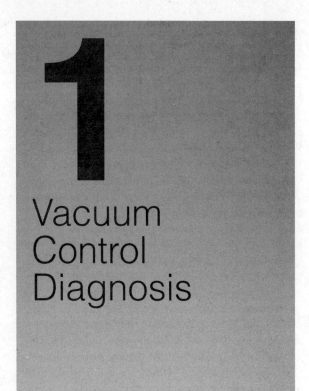

1 Vacuum Control Diagnosis

OBJECTIVES

Upon completion and review of this chapter, you will be able to:

- Test engine vacuum and determine causes of low or abnormal vacuum levels.
- Test various vacuum control components.
- Identify vacuum sources on an engine.
- Locate and use vacuum diagrams to determine if the engine vacuum system is functioning as designed.

ENGINE VACUUM

In a spark ignition automotive engine, the amount of air allowed to enter the engine controls the speed of the engine. One or more valves located in the air induction system control the airflow rate on such engines. These "throttle plates" are located in the base-plate of a carburetor or throttle-body assembly. When the driver depresses the throttle, these plates open and allow the amount of air entering the engine to increase. As the engine runs, each piston in succession moves down on its intake stroke. Air flows into the cylinder through that cylinder's open intake valve. Since there is nearly always a greater potential for air to flow into the cylinder than there is for air to flow across the mostly closed throttle plate, a pressure drop occurs in the manifold. For decades, the automotive industry has referred to this pressure drop as *vacuum.* This term is used in spite of the fact it is really only a pressure that is lower than the atmospheric pressure.

Today, a newer term is coming into our industry. Most modern literature refers to *manifold pressure* instead of manifold vacuum. Manifold vacuum and manifold pressure are two ways of referring to exactly the same thing. Manifold vacuum is like referring to a 16-ounce glass with only 8 ounces of water in it as being half-empty. Manifold pressure is like referring to that same glass of water as being half-full.

There are many things that affect the amount of vacuum that an engine is able to produce. All of the factors that affect engine vacuum fall into three major categories: atmospheric conditions, engine design, and engine condition.

Atmospheric Conditions

Altitude

Above the surface of the Earth, there is a 100-mile-thick blanket of air. The weight of this blanket of air pushes down on every square inch of the planet's oceans with 14.7 pounds of force. Since the land areas of the Earth are above sea level, the blanket of air is thinner and therefore exerts less force. The higher the altitude, the thinner the layer and the fewer pounds per square inch of force that is exerted.

Weather

There are enormous currents of air that move around the planet. As they move, they reduce or increase the mass of air over every spot on the planet. This increases and reduces the pressure applied to every point on the planet. However, changes in weather seldom have a large enough effect on atmospheric pressure to significantly affect the operation of an engine.

Engine Design

Everything about the design of the engine affects the vacuum created by the engine. The primary factor is the swept volume of the cylinders versus the amount of air that makes it past the throttle plates during the intake stroke of each cylinder. For instance, imagine it were possible to completely seal off the throttle, allow no air to bypass the piston rings, and make sure that all other valves are closed during the intake stroke. The pressure in the cylinder and the intake manifold would drop very low; the vacuum would be high. Even under these impossible conditions, there would not be a total vacuum. The molecules of air already in the cylinder and manifold would simply stretch apart, causing a pressure drop.

The number of cylinders, the swept volume of the cylinders, the compression ratio, camshaft design, intake manifold design, and exhaust manifold design all affect the engine vacuum. There are no hard-and-fast rules about how much vacuum to expect on a given engine or engine design. Generally, "high-performance" engines have lower vacuum than "low-performance" engines. For our purposes, keep in mind that a "high-performance" engine is one designed to excel at one or more aspects of performance. A low-emission engine has high performance in that area. A muscle-car engine has high performance in that area. A fuel-economy engine has high performance in that area.

Engine Condition

How much vacuum an engine produces depends on how efficiently it does its job. A badly worn engine or out-of-tune engine cannot produce as much vacuum as one in good condition and a good state of tune. Piston rings, valves, the condition of the fuel-injection system or carburetor, ignition timing and electronic timing control systems, and exhaust system all affect engine vacuum. Each has a predictable effect on engine vacuum. This predictability makes engine vacuum testing one of the essential starting points in diagnosing a driveability problem.

VACUUM SOURCES

Vacuum drawn directly from a tap on the intake manifold downstream of the throttle plates, figure 1-1, is called manifold vacuum. Manifold vacuum is most commonly used to diagnose engine condition. The amount of swept volume is greatest when the throttle is closed. Therefore, manifold vacuum is greatest at idle. Manifold vacuum decreases as the throttle is opened, so at wide-open throttle, there is very little vacuum in the intake manifold. On some high-performance fuel-injected applications the pressure in the manifold may even be slightly above atmospheric, even without a turbocharger or supercharger.

Vacuum drawn from an opening just above the throttle valve is called ported vacuum. When the throttle valve is closed during idle or deceleration, there is no significant vacuum at the port, figure 1-2. As the throttle opens, it uncovers the port. Intake velocity and pressure differentials caused by the intake stroke create a vacuum at the ported vacuum tap, figure 1-3. This vacuum is almost always less than manifold vacuum. On older applications, ported vacuum was used to modify carburetor operation and alter ignition timing. On modern electronically fuel-injected and timed engines, ported vacuum is rarely used for control.

TESTING ENGINE VACUUM

Most diagnostic vacuum readings are taken while the engine is idling. A vacuum gauge, figure 1-4, is used to measure the difference between atmospheric pressure and intake manifold pressure, or vacuum. The gauge

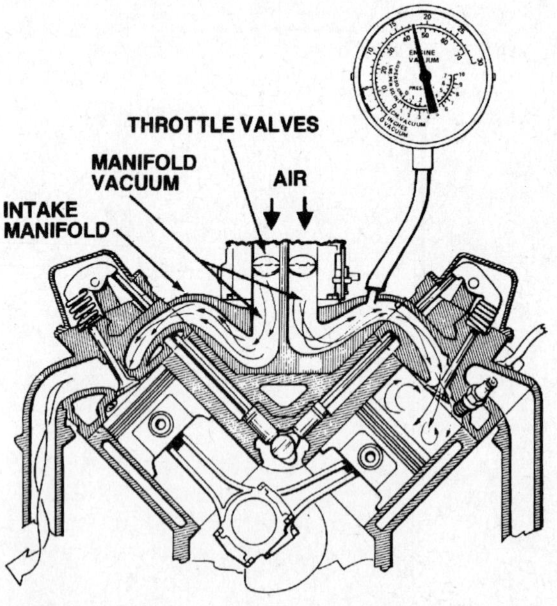

Figure 1-1. The vacuum gauge is connected to a tap on the intake manifold to measure manifold vacuum.

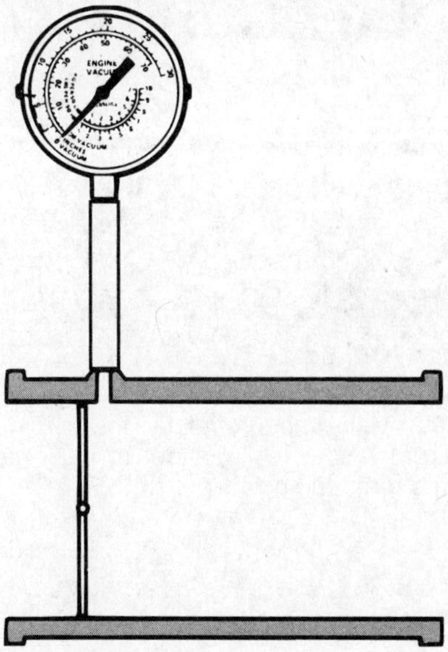

Figure 1-2. A ported vacuum tap is located above the throttle plates. When the throttle plates are closed the ported tap is not exposed to the vacuum of the manifold; therefore, ported vacuum is low.

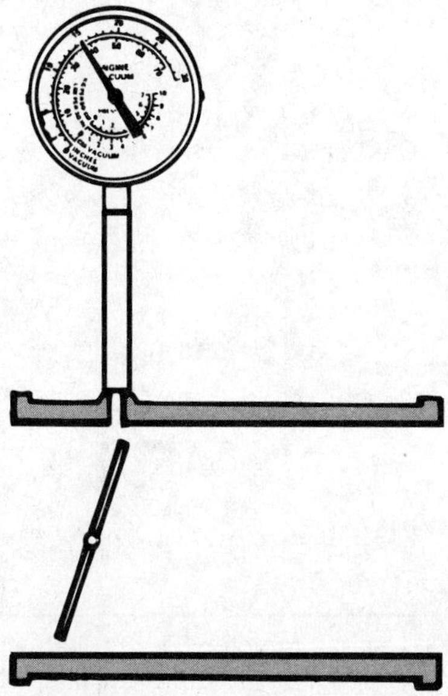

Figure 1-3. When the throttle plates are opened the ported vacuum tap is exposed to manifold vacuum and the vacuum reading on the gauge increases.

may be graduated in inches of mercury (in-Hg) or millimeters of mercury (mm-Hg). Of these two, the most common by far, even today, is inches of mercury.

The normal vacuum reading usually varies between 16 and 22 inches or 406 and 560 millimeters of mercury at sea level. Vacuum readings drop about one inch or 25 millimeters for every 1,000 feet (305 meters) increase in elevation above sea level, figure 1-5. It is important to know the basic specifications for the engine

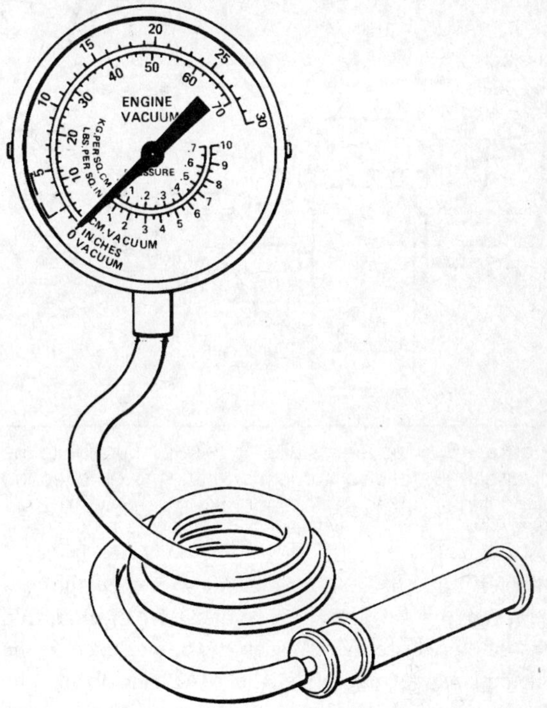

Figure 1-4. A typical vacuum gauge will read vacuum in inches of mercury and millimeters (or centimeters) of mercury. Most will also read pressures above atmospheric in pounds per square inch and kilopascals.

Altitude	Inches of Mercury
Sea Level to 1,000 Ft.	16 to 22
1,000 Ft. to 2,000 Ft.	15 to 21
2,000 Ft. to 3,000 Ft.	14 to 20
3,000 Ft. to 4,000 Ft.	13 to 19
4,000 Ft. to 5,000 Ft.	12 to 18
5,000 Ft. to 6,000 Ft.	11 to 17

Figure 1-5. A normal engine produces approximately the vacuum readings shown at a given altitude.

to be tested. These generally are provided in the factory shop manual or in an independent specification manual. You must know what a normal vacuum reading should be in order to determine abnormal conditions. This is especially important on late-model applications where there is a wide range of intake manifold configurations.

Using a Vacuum Gauge

To test engine vacuum, connect the vacuum gauge to a source of manifold vacuum. Some engines have a plug in the intake manifold; remove it and replace it with an

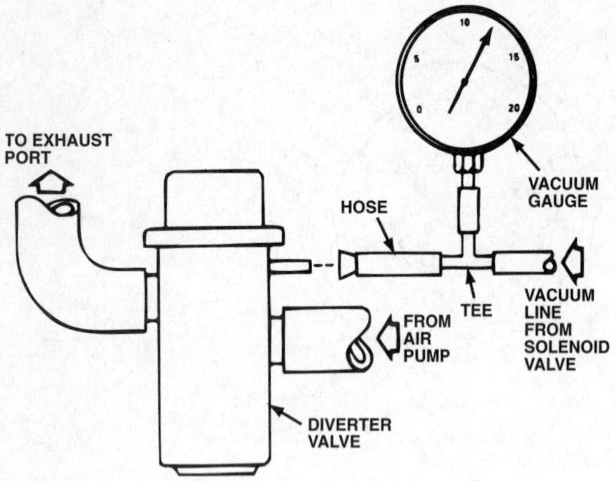

Figure 1-6. The tee installed in this vacuum line to the diverter valve allows vacuum operation to be checked without interrupting system operation. (Courtesy of Ford Motor Company)

adapter fitting. It is a good idea not to connect the vacuum gauge to the tap or hose connected to the manifold absolute pressure (MAP) sensor. The presence of the vacuum gauge may affect the MAP sensor and the MAP sensor may affect the reading on the vacuum gauge. Connect the vacuum gauge to the adapter fitting with a rubber hose. If the engine has no manifold plug, connect the vacuum gauge to the vacuum source of a system using manifold vacuum, such as the exhaust gas recirculation (EGR), evaporative emission (EVAP), or air-injection systems. Using a special adapter T-fitting, it is possible on many cars to connect a vacuum gauge into the vacuum line to the power brake booster. Connect the vacuum gauge to the EGR or EVAP system by installing a tee in the system line, figure 1-6. Connect the vacuum gauge line to the tee so the gauge monitors manifold vacuum without disturbing the system. All connections must be tight and free of leaks. Engine vacuum should be tested with the engine at normal operating temperature.

Some late-model applications have no provision for tapping manifold vacuum. Often the vacuum can still be tested indirectly by using a scan tool or voltmeter. This technique is covered in Chapter 2.

Other Special Equipment

You need hand tools and a vacuum gauge with a length of rubber tubing as well as the following supplies for some vacuum tests:

- A tachometer to measure engine speed when vacuum tests are specified at certain speeds, figure 1-7
- A remote starter switch to crank the engine during vacuum tests made at cranking speed, figure 1-8

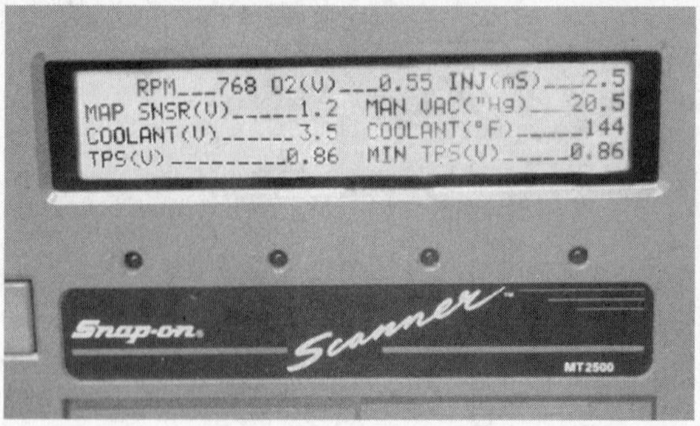

Figure 1-7. Many vacuum-related tests are done at a specified rpm. A scan tool provides an easy way to get an accurate rpm reading.

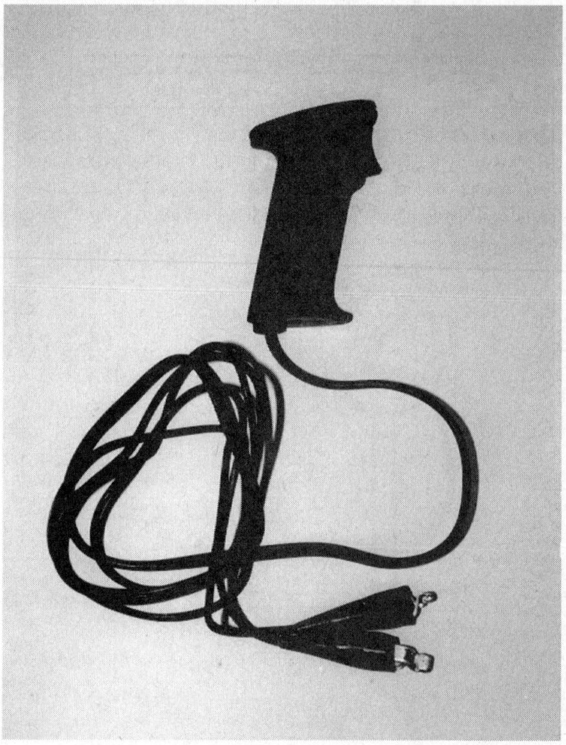

Figure 1-8. Using a remote starter switch allows the technician to crank or start the engine while still working under the hood.

- Jumper wires to disable some ignition systems, figure 1-9
- A propane bottle with appropriate fitting to check for vacuum leaks, figure 1-10
- Small plugs to close any vacuum hoses disconnected during testing; wooden or plastic golf tees work well for this, figure 1-11

You also need an ignition timing light to check and adjust ignition timing. Even on applications where timing

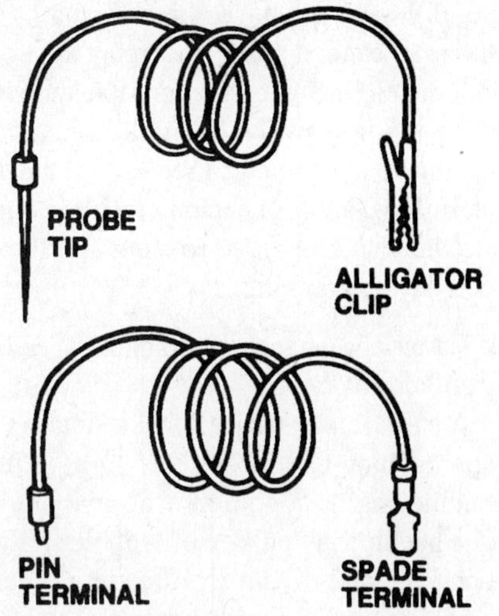

Figure 1-9. An assortment of jumper wires allows circuits to be disabled or bypassed while running engine-cranking tests.

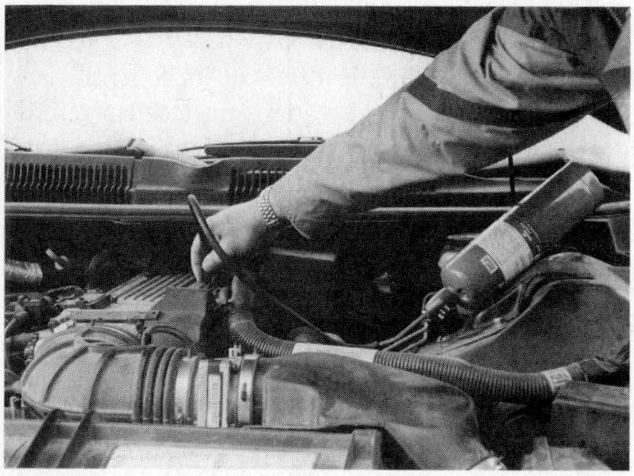

Figure 1-10. Propane metered though a special, though inexpensive valve is a safe fuel to use when checking for vacuum leaks.

is not adjustable, such as distributorless ignition systems, incorrect ignition timing can have a very serious effect on manifold vacuum.

Cranking Vacuum Test

This is a valuable tool to determine engine compression and intake system conditions. Cranking vacuum that is high and constant indicates the engine is mechanically sound and has a properly sealed intake manifold. Low vacuum at cranking speed indicates poor engine compression or vacuum leakage at manifold and carburetor gaskets.

If the vehicle is fuel injected, do not use this test. Simply skip this part of the diagnostic process. If the

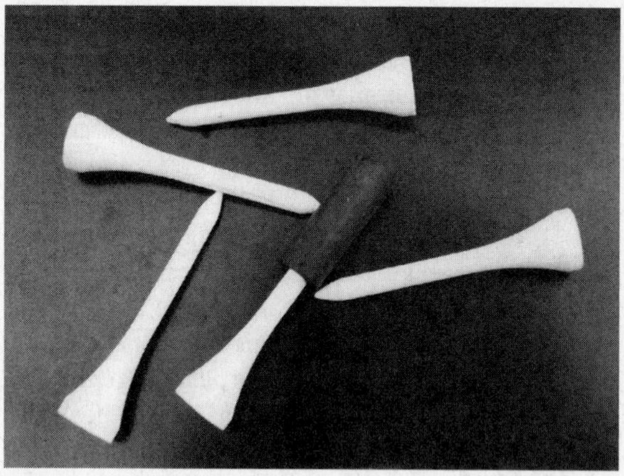

Figure 1-11. Vacuum line plugs can be purchased for only a few dollars. Many technicians use golf tees and they work quite well.

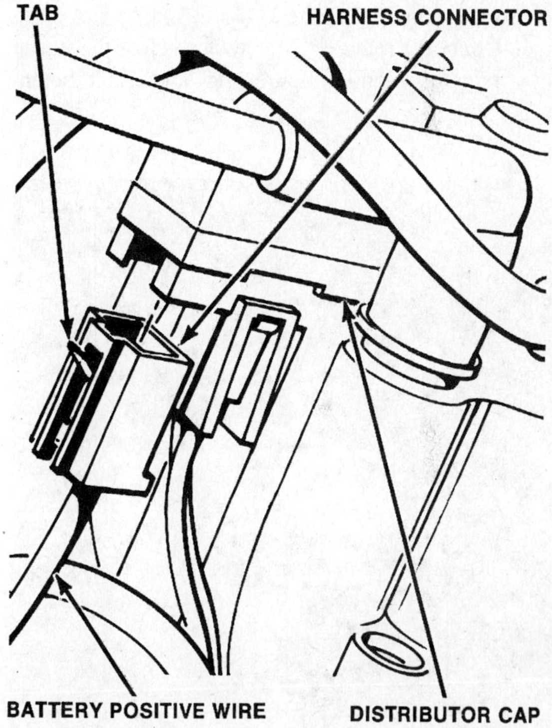

Figure 1-12. Disable the GM HEI ignition system, as well as other similar ignition systems, by disconnecting the battery feed wire from the ignition module.

vehicle is carbureted, this test may be performed. However, remember to back off the idle speed control.

1. Connect the vacuum gauge to a manifold vacuum source.
2. Disconnect the battery wire from the ignition system, figure 1-12. Many technicians will remove the coil wire and ground it. While this is often technically acceptable, the potential spark is

a fire hazard and could shock the technician. Disabling the ignition by disconnecting the battery usually also disables the injection system. If the injectors are heard to click during the cranking test, disable them by removing the injector fuse.

3. Ensure the throttle is as closed as possible. Back out the throttle stop screw (if adjustable) and release the automatic choke to seat the throttle valves tightly. Do not seat the throttle plates on post-1980 carburetors as this may cause damage to the throttle bore or throttle plates, figure 1-13.
4. If the carburetor is pre-1980 and has an idle speed air screw, turn the screw inward until it seats.
5. Connect a tachometer to the ignition to measure engine-cranking speed. For applications where the battery wire is removed instead of the secondary being grounded, the tachometer will not work. On these, measure cranking engine speed on a scan tool, figure 1-14.
6. Connect a remote starter switch across the starter relay or solenoid, figure 1-15, and crank the engine. If you do not have a remote starter switch, or if you cannot reach the relay or solenoid, crank the engine using the ignition switch. If using the ignition switch, hang the vacuum gauge from the hood so you can see it from the driver's seat, or have another person crank the engine.
7. Note the vacuum gauge reading and the engine speed during cranking.

Figure 1-13. Do not close the throttle plates all the way on post-1980 carburetor applications or any fuel-injected applications. The throttle bores on many of these applications have a coating that can be damaged when the throttle plates are closed all the way.

When the starter is engaged, the cranking speed should be steady and normal.

On most carbureted applications, a steady vacuum reading approaching five inches (125 mm) or more indicates that the engine condition is normal or even excellent. The needle may pulsate or wobble slightly, but it should be even and rhythmic. A reading of less than five inches (125 mm) indicates that there is a leak. Perform a cylinder leakage or compression test as explained in Chapter 2 to pinpoint the problem. If the reading is less than one inch (25 mm), back out the carburetor idle speed screw or disconnect the throttle stop solenoid to increase cranking vacuum. Repeat the test.

On fuel-injected applications, or 1980s/1990s carburetors where the throttle plates cannot be sealed, any relatively steady vacuum reading greater than zero is acceptable. If there is no reading and the engine does not start, refer to the cylinder leakage and compression test section of Chapter 2. If the engine does start and run, begin additional tests.

PCV System Test

These simple tests indicate whether the PCV system is working properly.

Test One

1. Connect the vacuum gauge and disable the ignition system, as in step 2 of the cranking vacuum test.

Figure 1-14. Disabling the ignition system to perform the engine-cranking test may keep a standard tachometer from working. For these applications, read engine rpm on a scan tool.

Figure 1-15. Using a remote starter switch to crank the engine during the cranking vacuum test allows the technician to take readings without needing someone else to crank the engine.

2. Crank the engine and pinch the PCV hose at the same time with a pair of pliers. This prevents air from leaking into the intake manifold.
3. Note the gauge reading. It should increase if the PCV system is open and working properly.

Test Two

Pull the valve and put your thumb over the PCV opening with the engine running at idle in order to feel the vacuum. If you feel vacuum, it is not working properly.

The PCV is buried and inaccessible on many late-model vehicles. In this case, test number one is impractical and the technician should use test number two instead.

On fuel-injected and post-1980 carbureted applications that display very little vacuum during cranking, this test should not be performed. Instead, if the PCV value is suspected to be defective, inspect the related hoses, gaskets, and grommets, then replace the PCV valve.

Vacuum Tests

Use this test to determine the air-induction efficiency of the engine. Problems with manifold vacuum may be caused by leaks in the intake system, valve train problems, and cylinder head or cylinder problems.

1. Re-enable the ignition system if you have not already done so.
2. Connect the vacuum gauge to a manifold vacuum source.
3. Connect a tachometer to the engine.
4. Start the engine and run until normal operating temperature (upper radiator hose hot and pressurized) is reached.
5. If the engine is carbureted, screw the idle speed adjusting screw until the engine is running at its lowest stable idle speed. If the engine is fuel injected, shut the engine off once it is warmed up and disable the idle speed control device. Generally, it is best to follow the manufacturer's guidelines when disabling the idle speed control. There are three common categories of idle air control devices.
   a. **Direct current motor control:** All three major U.S. manufacturers as well as many import manufacturers have used this method. A reversible dc motor is used to open and close the throttle plate by moving the throttle linkage. These motors should be fully retracted so that the throttle linkage is resting on the mechanical throttle stop screw.
   b. **Solenoid control:** When electrically disconnected, this device generally defaults to the closed position. These devices are installed to create an air passageway parallel to the main path of induction airflow. This creates a bypass around the throttle plates. Found primarily on Ford fuel-injection applications, when this device is disconnected, the engine does not idle or idles at a speed too low to be stable. After disconnecting the idle speed control on these applications, it is necessary to mechanically block the throttle open to obtain a stable idle speed.
   c. **Stepper motor control:** Although not technically accurate, the stepper motor behaves like a multiposition solenoid. Like the solenoid control, stepper motor controls are installed in an air passageway parallel to the primary airflow path across the throttle plates. These devices have up to 256 positions. Unlike the solenoid device, they remain in their last position when electrically disconnected. Follow manufacturers' recommendations for

the specific model you are working on to ensure it is fully closed. Several specialty tool companies make universal tools to drive the stepper motor to its fully closed position. After disconnecting the idle speed control on these applications, it may be necessary to mechanically block the throttle open to establish a stable idle. *Never disconnect an idle speed control motor with the engine running or with the ignition switch in the "on" position.*

6. With the engine running at its lowest stable idle speed, note the gauge reading and needle action.

Disabling the idle speed control is important when performing vacuum diagnosis because this system is designed to compensate for fluctuations in intake airflow that may result from the very conditions you are trying to diagnose.

The gauge needle should hold a steady, normal reading between 16 and 22 inches or 406 and 560 millimeters of mercury at sea level, figure 1-16. If the needle fluctuates one or more inches (25 mm or more) from the average reading, there is a high probability of a problem with proper seating of one or more engine valves, figure 1-17. Each time a valve fails to seat properly, the needle fluctuates or flutters. This condition may also indicate a leaking head gasket. Perform a cylinder leakage test to distinguish between these two possibilities.

A vacuum gauge reading a few inches or several centimeters below normal, along with a slight flutter of the needle, may indicate worn intake valve guides. On many older overhead valve engines, double-check valve guide conditions by removing the engine valve covers and squirting motor oil at the tops of the valve guides with the engine running. If a large cloud of blue oil smoke appears in the exhaust and the vacuum gauge reading increases, the valve guides are worn. On most modern, especially overhead cam, engines, this may be too time consuming. For these applications, an ignition analyzer oscilloscope is more practical. Pinpointing worn valve guides with a scope is discussed in Chapter 2.

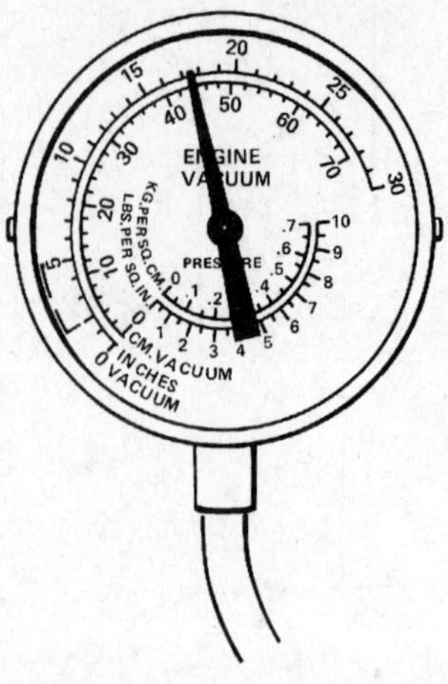

Figure 1-16. With the engine at an idle, the vacuum gauge should hold steady between 16 and 22 inches (406 and 560 millimeters) of mercury.

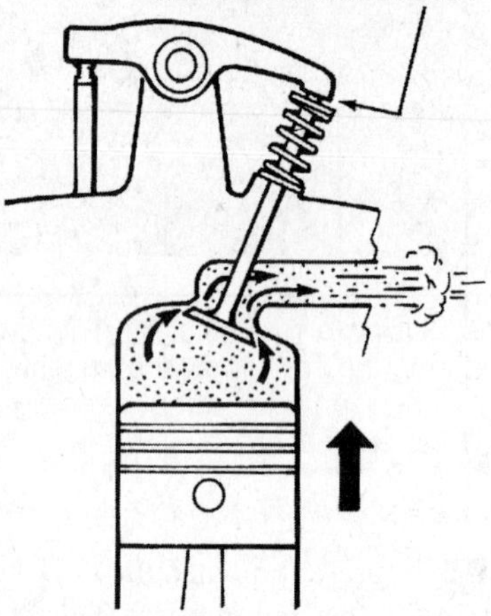

Figure 1-17. When a valve does not seat, power is lost and vacuum is reduced.

Finding Vacuum Leaks

Looking for vacuum leaks on modern engines is a difficult procedure. This is true primarily because modern engines have high-velocity air-induction systems. The speed of the air through the intake system minimizes any fluctuations or variations that might have been seen on older engines. Remember that a vacuum leak is defined as unwanted air entering the air-induction system between the throttle plates and the intake valve. Although traditionally done for decades, spray solvents and lubricants can be a very dangerous fire hazard when used to check for vacuum leaks.

The best technique for finding a vacuum leak is to connect a tachometer and a vacuum gauge to the engine. Start the engine and allow it to warm and stabilize at the lowest possible rpm. Introduce a small flow of propane into the throttle bore. If the vacuum increases and the rpm rises, it may indicate that one or more cylinders is running lean. The rise in rpm may also occur if ignition components are deteriorated. To distinguish between ignition problems and vacuum leaks, raise the rpm to a high idle, perhaps 1,200 to 1,400 rpm. Reintroduce the propane. If the resulting increase in rpm is about the same as at idle, the problem is ignition related; if the change is considerably less than at idle, there is a vacuum leak.

To pinpoint a vacuum leak, flow a small amount of propane at confined spots around the intake. Watch the vacuum gauge and the tachometer. Listen to the engine. If there is any change in vacuum, rpm, or the sound of the engine, the point where the propane is flowing is the point of the vacuum leak.

On some "V" engines, the intake manifold is often also the top oil sealing cover for the engine. To detect if the vacuum leak is under the intake, remove the PCV valve and allow propane to flow into the valve cover. Remember to disable the O_2S when using propane. If there is a change in vacuum, rpm, or the sound of the engine, there is a vacuum leak on the underside of the intake manifold.

Leaks under the intake on engines of this design often cause blue smoke at times of high manifold vacuum, such as deceleration. Sometimes this smoke is mistakenly diagnosed as an indication of valve guide wear.

There is also a new method for checking vacuum leaks. The Vacutec, from EMI-TECH, Inc., tests vehicle components for vacuum leaks while the engine is turned off by pumping nontoxic smoke through the various systems, figure 1-18.

The Vacutec works on many systems. A variety of adapters accompanies the device for connection to most models. It allows the technician to accurately vary the amount of pressure and the amount of time the Vacutec feeds smoke to the system. A remote trigger starts the device and, if a leak is present, the smoke begins to appear in about 20 seconds. The following is a list of the systems EMI-TECH reports the Vacutec can test:

- Exhaust leaks
- Oil leaks
- Cylinder sealing
- Intake manifold leaks
- Cooling systems
- Carburetor or throttle body
- Power brake booster
- EGR valve
- PCV valve
- Canister purge
- Evaporative system
- HVAC controls

For example, to inspect the EGR valve for leaks, apply an external vacuum source to the EGR valve by connecting the Vacutec to the exhaust system using 0.5 psi of smoke. If smoke is then present in the intake system, the EGR valve leaks.

In about 20 seconds' test time per cylinder, the Vacutec combines a cylinder compression and leak-down test in one operation. To begin this procedure, remove the spark plugs as you would during a conventional cylinder test. Once removed, insert the Vacutec nozzle from the flow meter into a cylinder using an adapter grommet provided. With both valves closed, watch the black ball inside the flow meter and press the remote trigger. The ball should rise, then lower to less than one liter per minute to indicate a good cylinder. However, if the ball remains high or marginal, the cylinder needs repair. Repeat the test to see if smoke leaves the intake manifold, throttle body, or carburetor, which would indicate a bad intake valve. A blown head gasket allows smoke to escape an adjacent cylinder through the spark plug opening. Bad pistons or rings allow smoke to escape into the crankcase and out the valve cover oil filler tube.

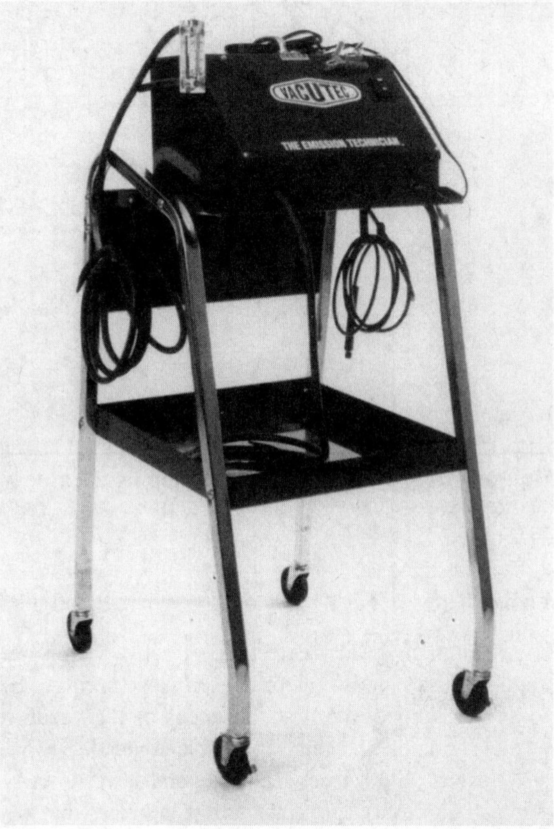

Figure 1-18. Vacutec uses nontoxic smoke to test for vacuum leaks in a variety of systems.

Carburetor or throttle body problems are just as easy to diagnose and troubleshoot. Test for vacuum pressure with the air cleaner connected. If, after hooking up the Vacutec, smoke is detected at the base of the air filter, remove the air filter to further inspect the vehicle. Locate leaks using plastic wrap around the carburetor to keep smoke from flooding under the hood. These leaks generally originate around the gasket.

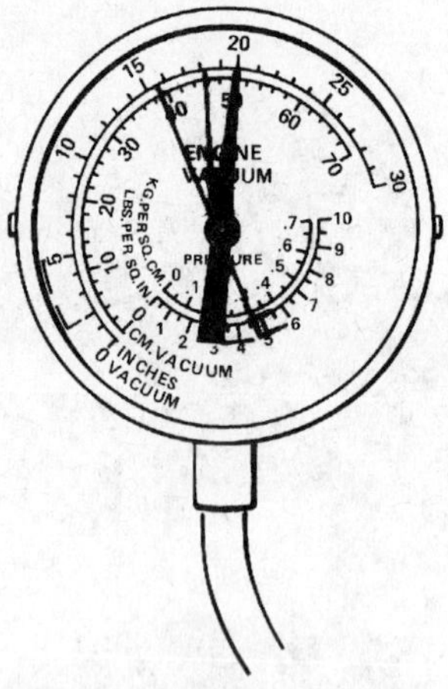

Figure 1-19. If the gauge needle drops back irregularly with the engine at an idle, one or more valves may be sticking.

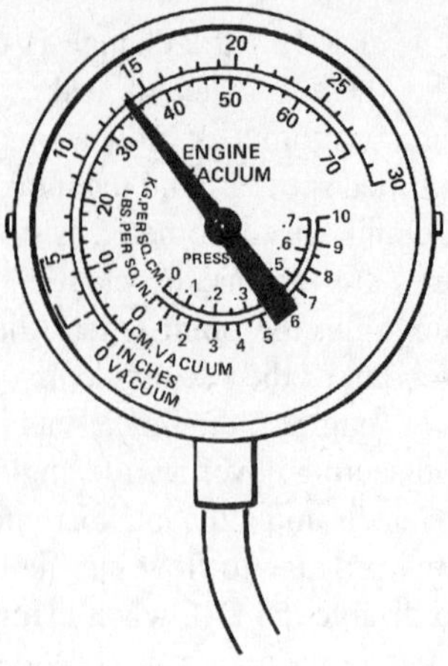

Figure 1-20. A low, steady vacuum reading can indicate problems with ignition timing.

Valve Action Test

The vacuum gauge also detects incorrect valve timing, sticking valves, and weak or broken valve springs. To do the test, repeat the first four steps of the vacuum leak test and note the gauge reading and needle action.

A reading that is between five and seven inches (125 and 180 mm) low on the gauge indicates that the valve timing may be late. This is caused by a timing belt or chain that slipped or jumped a tooth.

A consistent, but not rhythmic, drop from a normal reading indicates one or more valves are failing to close properly, figure 1-19. The action of the needle is similar to that of a leaking valve, but not as consistent. There are two primary reasons for the valve to fail to close properly: The valve is sticking in the guide or the valve spring is weak or broken.

To distinguish between weak and broken valve springs, increase the engine idle speed to about 2,000 rpm. If the needle fluctuates rapidly between 12 and 24 inches (305 and 610 mm), and the fluctuations increase in speed as rpm increases, the valve springs are weak. If the fluctuation is rapid but the regularity does not change with engine speed, a valve spring is probably broken.

Ignition Timing Test

If the gauge needle remains steady at idle, but at a point two or three inches (50 or 75 mm) below a normal reading, figure 1-20, ignition timing may be the problem. On modern engines, changes in ignition timing while the engine is running (advance and retard for various operating conditions) are controlled by the computer. However, on most distributor ignition applications, base timing is still adjustable by the technician. Time the ignition to the manufacturer's specifications with a timing light or a magnetic timing probe.

Exhaust Restriction Test

With the vacuum gauge and engine tachometer connected, slowly open the throttle until engine speed stabilizes at 2,000 rpm. The needle should rise to a little over a normal idle vacuum reading. Allow the engine to run at 2,000 rpm for several seconds. If the vacuum begins to drop, this is an indication of exhaust restriction.

A better and more conclusive test involves using the positive pressure range found on most vacuum gauges. Remove the oxygen sensor. If the engine has been running, use extreme caution as the temperature of the oxygen sensor (O_2S) may be several hundred degrees. Use adapters to connect the vacuum/pressure gauge to the O_2S port. Start the engine and run at 2,000 rpm for 30 seconds. The pressure shown on the gauge should not rise above 2 psi.

Piston Ring Test

Before testing for leaking piston rings, make sure the engine has shown normal readings on all other tests. The crankcase oil also must be in good condition. Diluted or worn-out oil causes an incorrect reading which indicates a loss of compression when there is no such loss.

Connect the gauge and engine tachometer just as you did for the vacuum leak test. Open the throttle quickly and hold until engine speed reaches 2,000 rpm. Quickly close the throttle. The needle should jump four or more inches (100 or more mm) over the normal reading if the piston rings are in good condition. A reading increase of less than four inches (less than 100 mm) indicates there may be low compression. Do a compression test to verify the vacuum gauge findings.

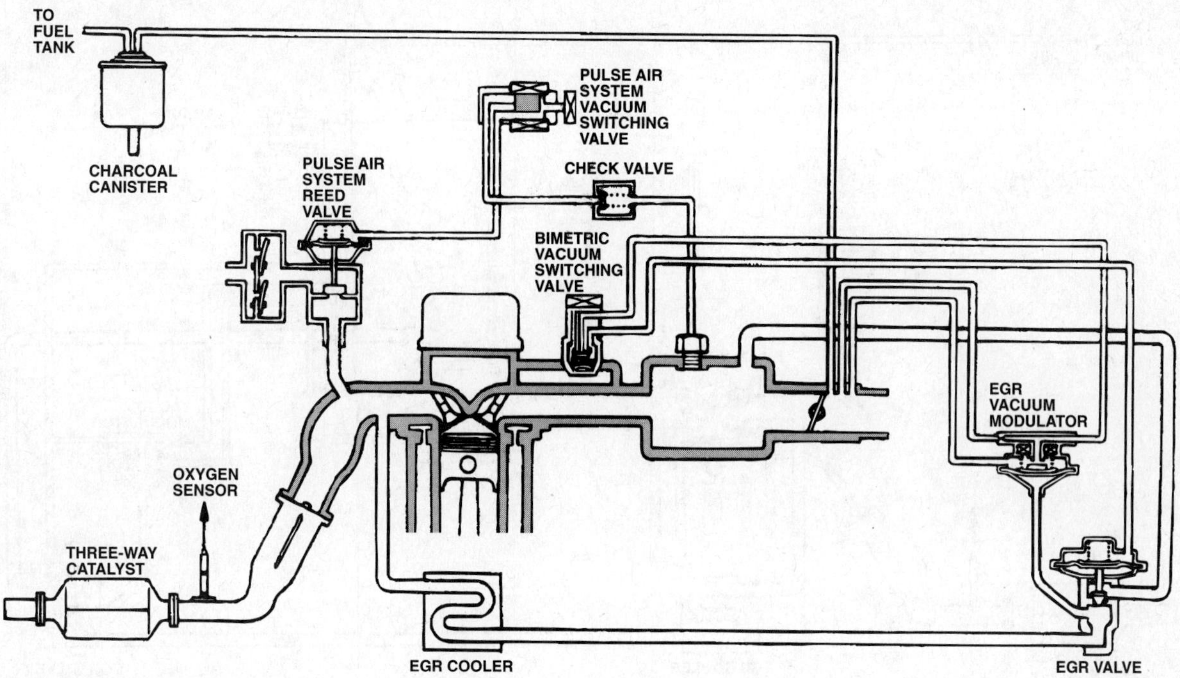

Figure 1-21. California vacuum system diagram. (Provided courtesy of Toyota Motor Sales U.S.A., Inc.)

VACUUM DIAGRAMS

On 1970s, 1980s, and some 1990s model vehicles, vacuum-controlled systems are used to regulate engine operation for better emission control. Vacuum systems may also be used to control air conditioners, headlamp doors, power brakes, and other devices. Several of these systems were often used on the same engine. Many were quite complex and often connected to one another.

During the 1990s, emission controls began to depend less on vacuum control and far more on electronic controls. Engine compartments were once crisscrossed with vacuum lines and hoses, each with a specific job. Now they are comparatively neat with most of these jobs being done by just a few wires.

When vacuum lines are disconnected to test or service engine parts, be sure they are reinstalled correctly. To help you properly route and connect vacuum lines, manufacturers are required to provide vacuum diagrams for each engine and vehicle combination.

Like electrical wires, vacuum hoses can be color-coded for easier identification. The vacuum diagram may be printed in color, or the color may be printed near the line in the drawing.

Although the use of electronic engine management systems and fuel-injection systems has allowed many manufacturers to considerably reduce the number of vacuum lines, the routing and connection pattern of vacuum lines can vary a great deal on a given engine during a single model year. One diagram may apply to an engine sold only in California, figure 1-21. A different diagram is used when the same engine is emission-certified for federal or 49-state use, figure 1-22. A third pattern may be used for engines operating in high-altitude areas.

The problem is compounded when manufacturers regulate changes in emission systems and calibrations during a single model year. This results in two or more identical engines in the same geographical region having slightly different devices and systems, figure 1-23. In this case, attempting to route and connect vacuum lines correctly without the help of a factory vacuum diagram is next to impossible, even for an experienced technician.

This diversity of routings is the reason why engine vacuum diagrams are generally included on vehicle emission control information (VECI) decals or provided on a separate underhood decal.

TESTING VACUUM COMPONENTS

Intake manifold vacuum operates a wide range of systems and devices on late-model vehicles. These vacuum systems can be as complex as the electrical system. A vacuum source for testing is needed to correctly diagnose and correct problems in vacuum control systems. In some cases, vacuum is taken directly from the intake manifold while the engine is running. This vacuum measurement is made with a simple vacuum gauge teed into

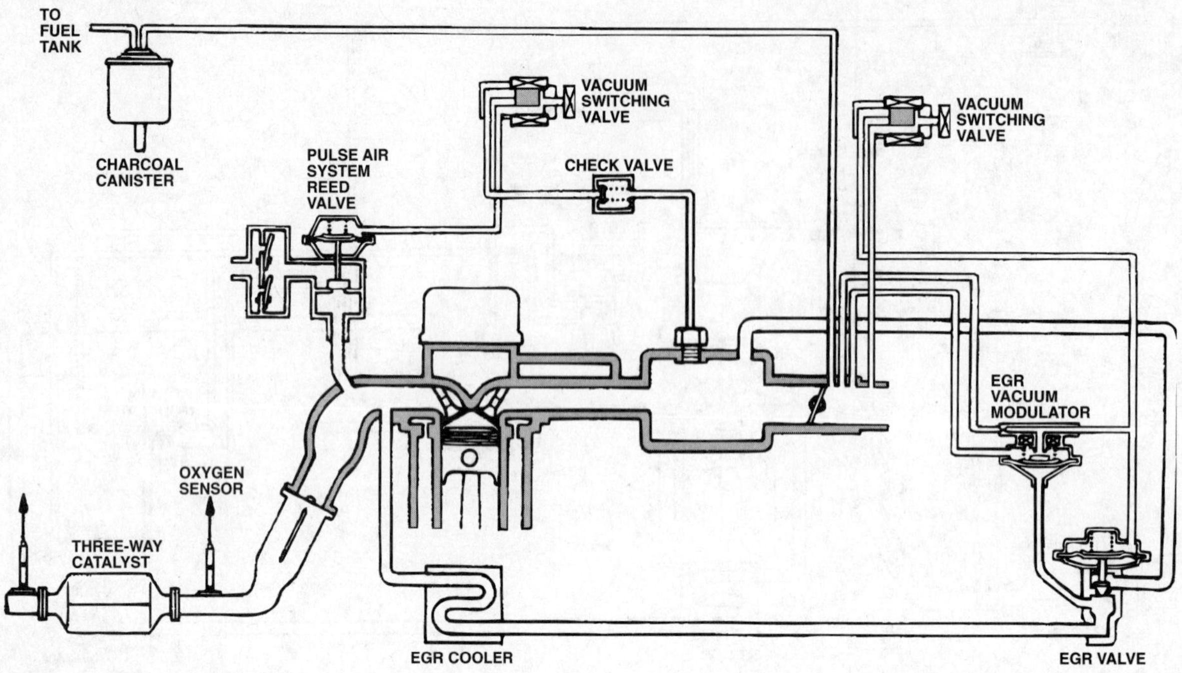

Figure 1-22. Federal vacuum system diagram. (Provided courtesy of Toyota Motor Sales U.S.A., Inc.)

the control system line. A hand vacuum pump or a console vacuum pump may also provide vacuum.

Hand Vacuum Pump

Small hand-operated vacuum pumps are used for vacuum testing. These have a built-in vacuum gauge. In many cases, the component is tested without removing it from the engine. The hand vacuum pump is connected to the component being tested by a hose, figure 1-24. Operate the pump lever or plunger to apply the correct amount of vacuum required for the test. The gauge indicates the amount of vacuum being applied by the pump. For some tests, you may need a separate vacuum gauge in addition to the one attached to the pump, figure 1-25.

Console Vacuum Pump

Professional quality engine analyzers often have a vacuum test device in the console. This includes a built-in vacuum pump, a vacuum gauge, and the necessary controls to operate the pump for testing vacuum components. These pumps are also available as stand-alone devices. The console vacuum pump serves the same function as a hand vacuum pump. Connect the part to be tested to the vacuum pump outlet on the console panel with a vacuum line, then turn on a switch on the console to apply vacuum. Most vacuum pumps have a shutoff valve to allow trapping vacuum in the component to check for leakage.

Vacuum Control Devices

Vacuum control devices can be divided into three general groups:

1. Vacuum diaphragm motors (often called actuators) are used to operate a mechanical component, such as an EVAP purge valve or an EGR valve, figure 1-26. Test these by applying vacuum to the inlet port and noting the diaphragm or actuator movement. On later-model applications, most of these functions have been replaced with motors, solenoids, and servos controlled by the onboard computer.
2. Vacuum valves and switches, figure 1-27, are used to control vacuum flow. Most are operated electrically or thermostatically. Test them by applying vacuum to the inlet or other specified port. Then operate the valve or switch either electrically or by temperature and measure vacuum flow through the device.
3. Vacuum delay valves contain a simple restrictor. These valves are in the vacuum line between the vacuum source and the device to be acted on, figure 1-28. They are used to delay vacuum flow. Test a vacuum delay valve by applying

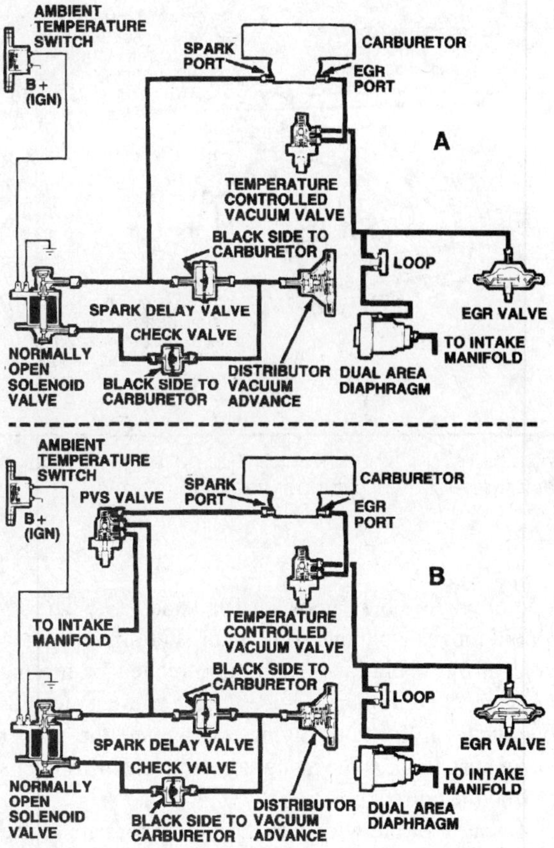

Figure 1-23. These two vacuum diagrams for a Ford V-8 show a ported vacuum switch (PVS in upper left of lower diagram) that was added during the model year on some vehicles. (Courtesy of Ford Motor Company)

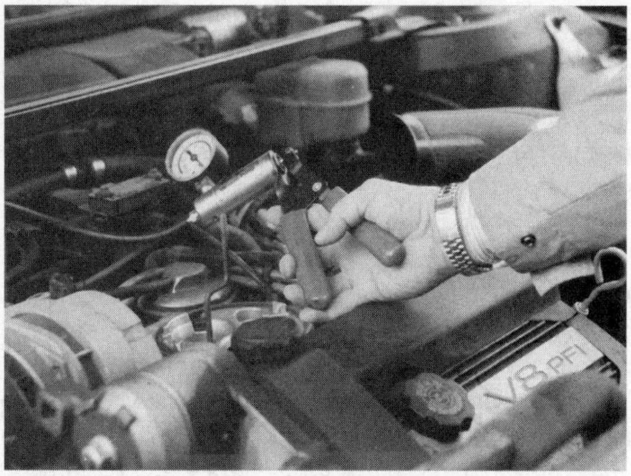

Figure 1-24. Here a hand-held vacuum pump is used to test the diaphragm of a standard vacuum-operated, electronically controlled EGR valve.

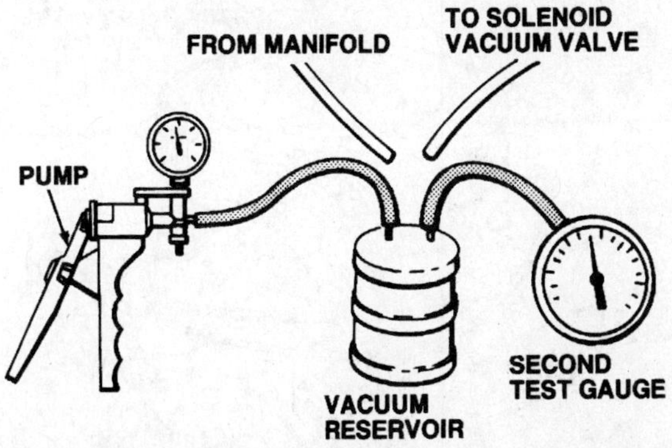

Figure 1-25. A second vacuum gauge may be required to test some components. (Courtesy of Ford Motor Company)

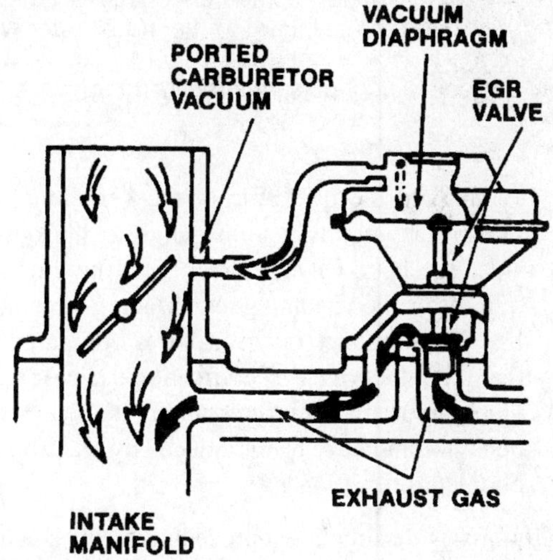

Figure 1-26. The vacuum diaphragm opens the EGR valve to recirculate exhaust gases into the intake mixture.

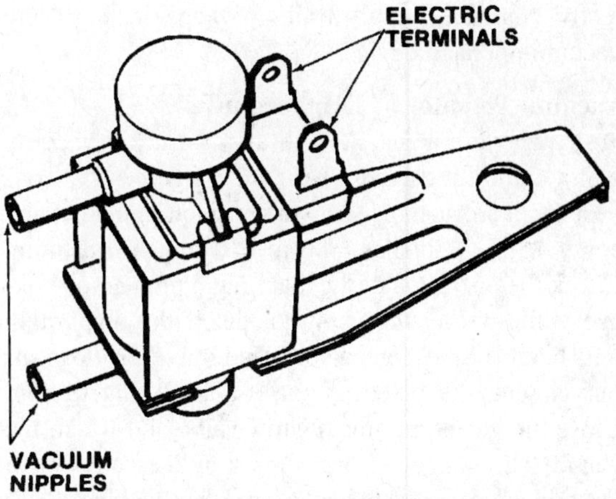

Figure 1-27. On many applications, vacuum is used to control most of the emission-related devices. A solenoid-operated valve, activated by the onboard computer, will often control the application of vacuum to these devices.

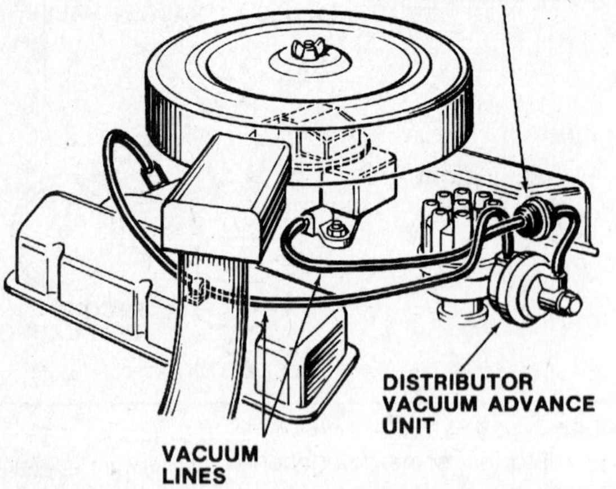

Figure 1-28. Although computer-controlled ignition timing has been the standard in the industry for well over a decade, some older applications use vacuum delay devices to control the rate of timing advance.

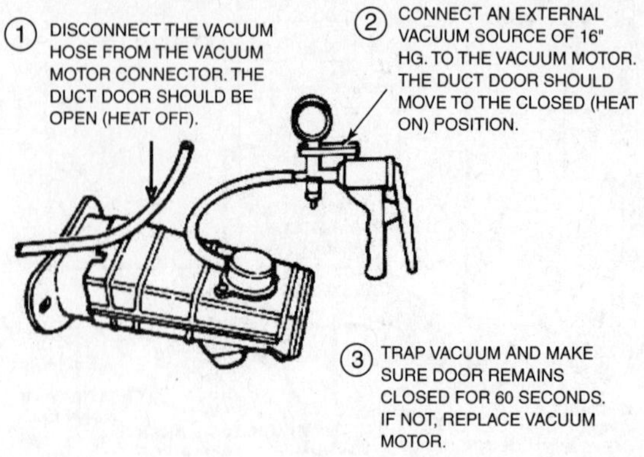

1 DISCONNECT THE VACUUM HOSE FROM THE VACUUM MOTOR CONNECTOR. THE DUCT DOOR SHOULD BE OPEN (HEAT OFF).
2 CONNECT AN EXTERNAL VACUUM SOURCE OF 16" HG. TO THE VACUUM MOTOR. THE DUCT DOOR SHOULD MOVE TO THE CLOSED (HEAT ON) POSITION.
3 TRAP VACUUM AND MAKE SURE DOOR REMAINS CLOSED FOR 60 SECONDS. IF NOT, REPLACE VACUUM MOTOR.

Figure 1-29. A hand vacuum pump is used to test an air cleaner, heated air door vacuum motor.

vacuum to the inlet side of the valve and attaching a second vacuum gauge to the outlet side. To check valve performance, the time required for the vacuum gauge readings to stabilize on both sides of the valve is compared to the manufacturer's specifications. These devices are often erratic and unpredictable in their behavior and have been replaced by electronics on later applications.

The following are simple test procedures using a hand-operated vacuum pump for each type of vacuum control device. These same principles are applied when testing vacuum controls for any vehicle. Use the manufacturer's specifications and vacuum diagrams to correctly interpret the results when testing any vacuum-operated device.

Vacuum Actuator or Diaphragm Test

Test a vacuum actuator, such as a warm-air vacuum motor on the air cleaner, as shown in figure 1-29. Connect the hand vacuum pump to the diaphragm inlet and apply 15 to 20 inches (380 to 510 mm) of vacuum. Check the operation of the diaphragm plunger or linkage while noting the pump gauge. If the diaphragm unit being tested contains a bleed valve or clean air purge, plug the bleed or purge hole during testing. Close the vacuum pump shutoff valve and watch the gauge. The gauge reading should hold steady for at least one minute if the diaphragm is not leaking.

Solenoid-operated Vacuum Valve Test

An electric solenoid is often used to control vacuum valves in emission-control systems. This is particularly true of engines built between 1980 and 1995. The solenoid moves a plunger to let in or shut off vacuum to control the operation of emission-related or heating and air conditioning devices. Test these solenoid-operated valves by applying vacuum to the proper valve port. Look in the manufacturer's test instructions to find the correct port.

Because of the wide variation in system and valve operation, follow the manufacturer's recommended test procedure exactly. For example, the valve may be closed with the solenoid de-energized. In this case, the hand pump gauge holds vacuum. Or, the valve may be open with the solenoid de-energized. If so, the pump gauge does not hold vacuum.

Apply battery voltage to the solenoid to energize it, figure 1-30. Do this with a jumper lead or by grounding the solenoid. Be sure to consult the manufacturer's specifications on these devices since some vacuum control devices are designed to receive a pulsed current rather than a continuous current. A continuous current may burn open the winding of the solenoid. When the solenoid is energized and vacuum is applied, the pump reading should change from that shown with a de-energized solenoid.

To check the switching action of a solenoid, first learn whether it blocks vacuum when energized (normally open) or de-energized (normally closed). The 2-port solenoid shown in figure 1-31 is tested either by applying vacuum with a hand vacuum pump or by blowing air into the outlet port. Either the vacuum inlet port or the vent should be open with the solenoid de-energized. Connect the solenoid to battery voltage and repeat the procedure. With the solenoid energized, the opposite port (vent or inlet) should now be open. Determine whether the solenoid must be energized or

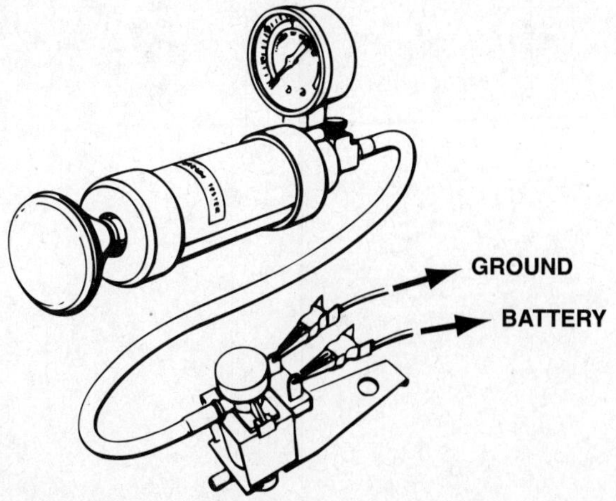

Figure 1-30. A hand-held vacuum pump being used to test a solenoid-operated vacuum valve. Keep in mind that some of these valves are open when power is applied and some are open when power is not applied. If connected in reverse, it will burn out a diode internally.

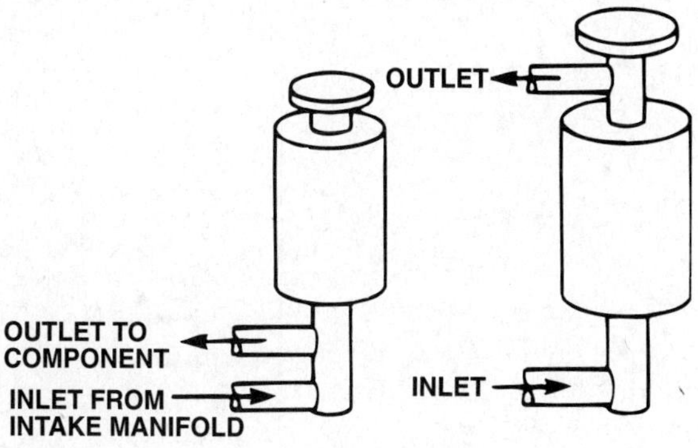

Figure 1-31. Typical 2-port solenoid.

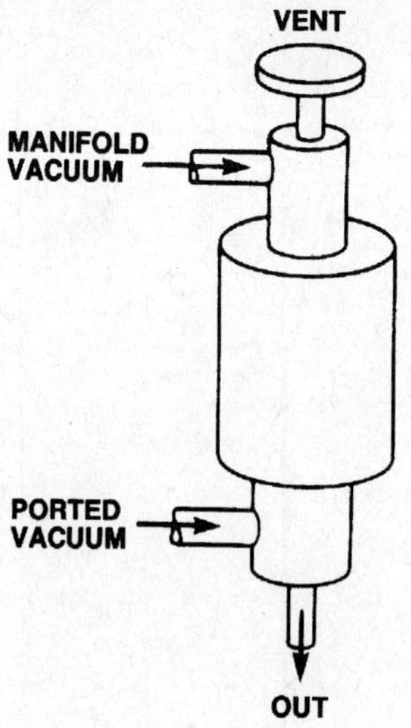

Figure 1-32. Typical 3-port solenoid.

de-energized to block the inlet port, then apply 15 to 20 inches (380 to 510 mm) of vacuum to the inlet port and trap the vacuum. If the vacuum leaks down, the solenoid is defective.

Some solenoids are designed to have a slight vacuum bleed even when the valve is closed. Be sure to check the manufacturer's specifications.

A 3-port solenoid, figure 1-32, generally switches between ported and manifold vacuum to a single outlet. Connect a vacuum gauge to the outlet port and apply vacuum to each inlet port as you energize and de-energize the solenoid with battery voltage. Vacuum, as well as leakage, is checked at the outlet just as with a 2-port solenoid.

2

Fuel and Emission System Test Equipment

OBJECTIVES

Upon completion and review of this chapter, you will be able to:

- Perform a compression test.
- Perform a cylinder balance test.
- Perform a cylinder leakage test.
- Perform a calibration and leak test on an emission analyzer.

INTRODUCTION

Driveability problems were the curse of the industry during the 1980s. As the industry tried to adapt to new and ever-changing rules governing emission controls, many vehicles suffered from virtually incurable problems. During this time, manufacturers of test instruments were challenged to deliver thorough but user-friendly test instruments to the auto repair industry. In some cases they succeeded, while in others they failed. This chapter explores a variety of test instruments and their use. Some of the tests described in this chapter are routine, and should be performed at regular intervals; others need be done only when driveability symptoms demand it.

Whether routine or troubleshooting, these tests may require various pieces of test equipment, ranging from a simple pressure gauge, figure 2-1, to a $60,000 engine analyzer, figure 2-2.

THE PRESSURE GAUGE AS TEST EQUIPMENT

Two of the tests used in troubleshooting driveability problems use pressure gauges. The first test involves checking the condition of the cylinders. No engine can properly operate without good cylinder integrity. The compression gauge and cylinder leakage gauges are used to test the condition of an engine's cylinders, figure 2-3. The compression gauge uses a check valve to trap pressure in the gauge between compression strokes during the test while the cylinder leakage gauge measures the pressure leaking from a cylinder. Cylinder leakage testers are generally marked in percentage of leakage, figure 2-4.

The second test is fuel pressure. In the days of carbureted engines, the fuel-pressure test was often overlooked. Today, with fuel-injected engines, testing fuel pressure is even more important. There are three common pressure standards for measuring fuel pressure: psi (pounds per square inch of pressure), kPa (kiloPascals of pressure), or bar (1,000 kPa), and Kg/cm2 (kilograms per square centimeter).

Many vacuum gauges are also marked with a positive pressure scale, figure 2-5. Since these compound

Figure 2-1. Testing the fuel-injection system pressure with a pressure gauge.

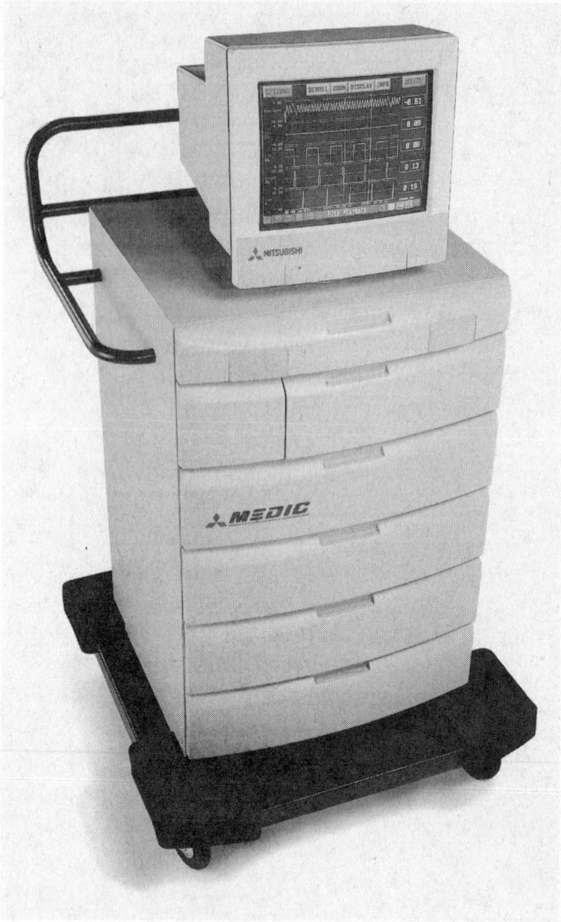

Figure 2-2. The classic engine analyzer has evolved from an enhanced high-voltage oscilloscope to a computerized diagnostic center. Some of these units can cost as much as $60,000.

gauges (vacuum and pressure) are calibrated to read low pressures (up to 10 psi or 69 kPa), they can test mechanical fuel pump pressure on carbureted engines. Testing electric fuel pump pressure on fuel-injection systems requires a pressure gauge that reads high pressure. This is because the electric fuel pumps used with late-model, fuel-injection systems are generally rated at a 39-psi (269-kPa) output. However, since some fuel-injection applications have operating pressures as high as 100 psi, any gauge purchased for testing injection fuel pressure should be able to read at least that high. Most pressure gauges are most accurate at their midrange; therefore, the most accurate high-pressure readings are obtained with a pressure gauge that reads from 0 to 100 or 0 to 150 psi (0 to 690 or 0 to 1,035 kPa).

Fuel Pump Testing

In a carbureted engine, the fuel pump must supply gasoline to the carburetor in the volume and at the pressure specified by the manufacturer. When the volume is too high, too low, or uneven, carburetion problems result. High pressure and volume force a carburetor float needle from its seat, which results in a

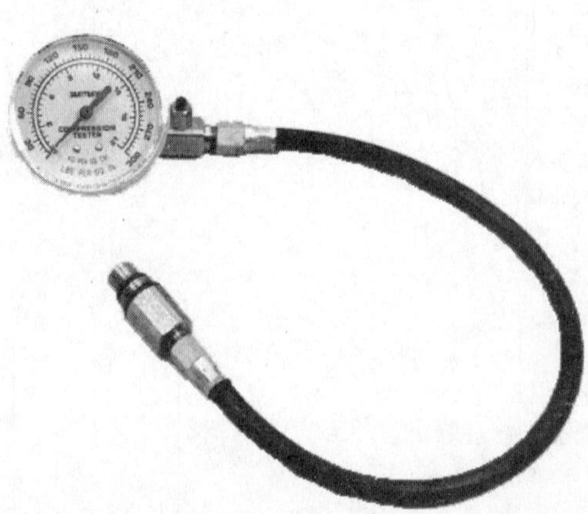

Figure 2-3. The compression gauge uses a check valve to trap pressure in the gauge between compression strokes. Cylinder pressure during a compression test is a good indicator of cylinder condition.

Figure 2-4. A cylinder leakage tester indicates how much a cylinder may be leaking.

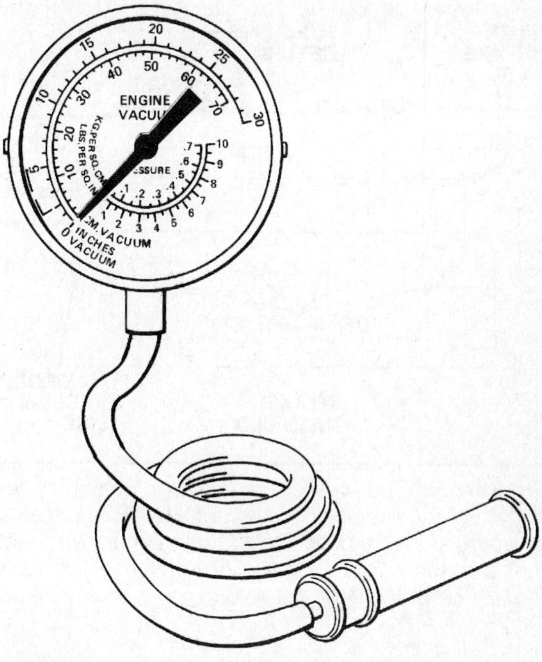

Figure 2-5. Most combination vacuum and pressure gauges can read pressures up to 10 psi.

high fuel level in the bowl. This causes the engine to run rich. Low pressure or volume starves the engine and results in a lean mixture and a misfire.

Any mechanical or electric fuel pump may be tested for volume and pressure with the proper pressure gauge. Do not remove the fuel pump from the engine to test it. Fuel pump pressure and volume test procedures are provided in the next chapter of this *Shop Manual.*

Fuel-Injection System Pressure Testing

Excessive or inadequate fuel pressures or volumes are one of the most commonly overlooked problems on fuel-injected engines. The fuel supply circuit of the fuel-injection system is equipped with a pressure regulator that controls pressure and directs excessive volumes of fuel back to the tank. On electronic fuel-injection systems, the oxygen sensor detects the effects of high fuel pressure or excessive volume and compensates. However, the ability of the system to compensate is limited. The most common causes of high fuel pressure are a defective fuel-pressure regulator or a restricted fuel return line to the fuel tank.

Low fuel pressure or volume cannot be compensated for. A bad fuel pump, restriction in a line, or a restricted fuel filter can cause low pressure.

When testing fuel-injection system pressure, connect the gauge to a line through which the fuel flows. The goal is to measure the pressure of the fuel as it flows through the system. On many fuel-injection systems, there is a Schrader valve provided. If no test fitting is provided, connect the gauge into the inlet line of the fuel rail or throttle body with a T-fitting. It is sometimes easier to place the T-fitting at the fuel filter.

Fuel-injection system pressure testing involves checking the operation of the pressure regulator, which keeps fuel pressure uniform at the injectors and returns any excess fuel to the tank. The regulator is located on the return side (downstream) of the injectors, figure 2-6. It is usually on the fuel rail of a port-injection system or built into the throttle body of a throttle-body injection (TBI) system. Most pressure regulators include a manifold vacuum diaphragm, figure 2-7, and alter fuel pressure to compensate for intake manifold pressure changes.

Most tool manufacturers sell fuel-injection pressure gauges. They are typically priced from $30 to over $500 depending on the capability and number of adapters. More complete information about the testing of fuel pressures and the diagnosis of fuel supply components in fuel-injection systems is in Chapter 3 of this *Shop Manual.*

Engine Compression Testing

The power output of any engine depends on the ability of its cylinder to create vacuum and seal combustion. One of the measurements of this ability is called compression. Compression is actually a measurement

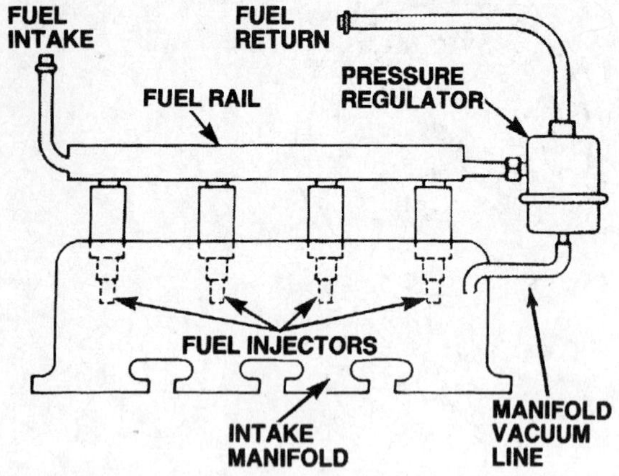

Figure 2-6. The pressure regulator is located downstream of the injectors in most fuel-injection systems. The regulator bleeds off excess fuel volume to regulate fuel pressure. The excess fuel returns to the tank. (Courtesy of DaimlerChrysler Corporation)

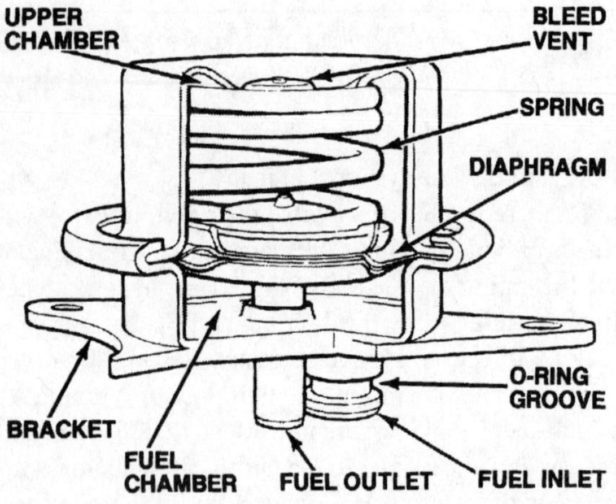

Figure 2-7. Injection system fuel-pressure regulators use a spring-loaded diaphragm. Fuel pressure on one side of the diaphragm acts against spring tension on the other side of the diaphragm. When the pressure of the fuel becomes greater than the tension of the spring, the diaphragm is deflected. The movement of the diaphragm opens a valve and excess fuel is routed back to the fuel tank.

of how well each cylinder is sealed by the piston rings, valves, cylinder head gasket, and the spark plug, figure 2-8. If any of these points are not properly sealed, compression is lost, causing a drop in power output for that cylinder.

The compression pressure in each cylinder and variations in pressure among the cylinders are measured by a pressure gauge equipped with a special Schrader valve designed to trap the maximum pressure measured in the gauge. Compression pressure is mea-

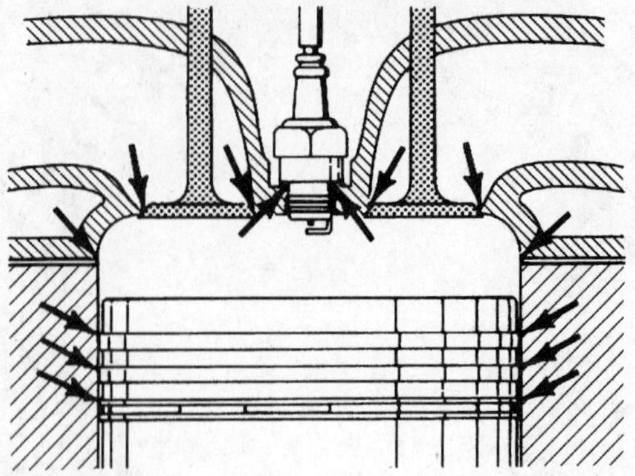

Figure 2-8. Good compression depends on proper cylinder sealing at the points shown.

sured in pounds per square inch (psi), kiloPascals (kPa), or bar.

Manufacturers offer compression specifications for their engines. These are found in the engine specification section of the factory shop manual and in many independent service manuals. Specifications are usually stated in one of two ways:

- Compression specifications may require the lowest-reading cylinder to be within a certain percentage (generally 75 percent) of the highest cylinder. If the highest cylinder reading is 160 psi (1,103 kPa), for example, the lowest cylinder reading must be at least 120 psi (827 kPa).
- Compression specifications are sometimes stated as a minimum figure, with a certain allowable variation between the cylinders. If the minimum is 130 psi (896 kPa), and the variation is 30 psi (207 kPa), the compression in each cylinder is satisfactory if it is a least 130 psi (896 kPa), but no more than 160 psi (1,103 kPa).

In both methods of specifying proper compression, the cranking speed of the engine is critical. Make sure the battery is fully charged and all possible loads are shut off so the engine cranks at a normal speed. Then use the following general procedure:

1. Place the transmission in Neutral or Park and set the parking brake. Start the engine and run at fast idle for several minutes to bring it to its normal operating temperature (upper radiator hose hot). Shut off the engine.
2. Disable the ignition to prevent sparks, fire, or electrical shocks when the engine is cranked. With breaker-point and most electronic ignitions, disconnect the coil high-tension wire from the distributor cap and ground it. Disable Delco-Remy HEI ignitions by disconnecting the battery

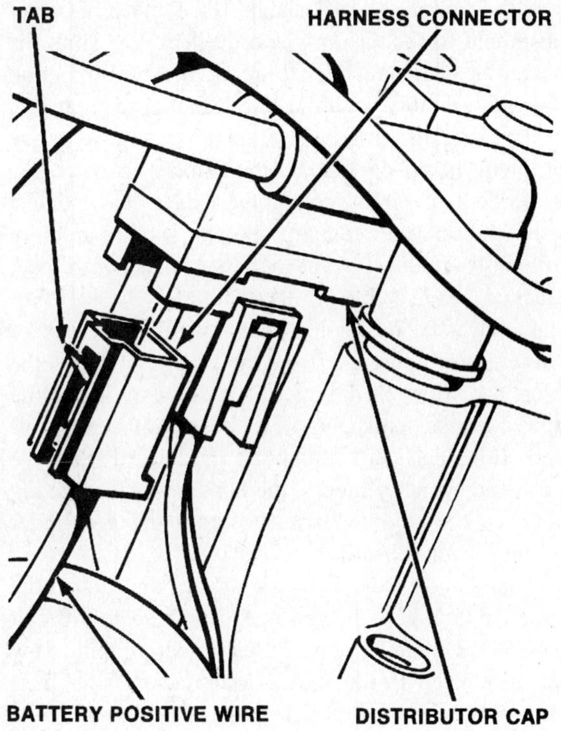

Figure 2-9. Disable an electronic ignition system for the compression test by disconnecting the battery power wire from the ignition module. For most applications, this not only disables the ignition system, but disables the injectors as well.

Figure 2-10. The throttle must be held open during the compression test.

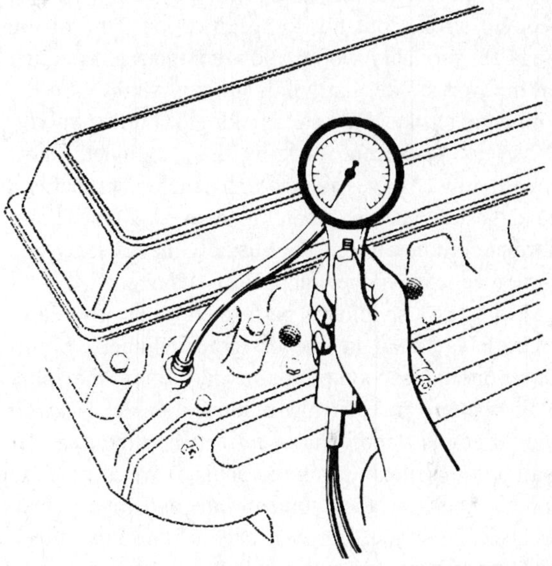

Figure 2-11. This compression tester has a built-in starter switch. Crank the engine at least four times when testing each cylinder.

wire from the distributor, figure 2-9. Disable the GM distributorless ignition systems as follows:

- *On distributorless engines*—Unplug the crankshaft position sensor on the engine or on the ignition module.
- *Distributorless ignition systems (DIS) on late 1986 and later V-6 and 4-cylinder engines*—Unplug the 2-wire connector from the electronic control module for the system (at the base of the coils).
- Keep in mind, on OBD II vehicles, disabling the ignition system to perform the compression test may set a trouble code and cause the malfunction indicator lamp (MIL) to be on the next time the engine is started. Be sure to use a scan tool to clear these codes and turn off the MIL when testing and repairs are completed.

3. Remove the air cleaner or air intake duct to the throttle body. Block the linkage to hold the throttle wide-open, figure 2-10.
4. Remove the cables from all spark plugs. Loosen each plug one turn. Blow the dirt from around the spark plugs with compressed air, then remove the spark plugs.
5. Thread the compression tester adapter into the number one spark plug opening finger-tight. Connect the tester to the adapter, figure 2-11.
6. Connect a remote starter switch according to the manufacturer's instructions for switch use.
7. Crank the engine for at least four complete compression strokes. Many technicians crank the engine up to 10 compression strokes. Whatever number you choose, it should be consistent for each cylinder. This gives the most comparable and highest possible reading. Note and record the gauge reading on the *first* and *last* strokes.
8. Release the compression gauge pressure to return the gauge to zero. Disconnect the tester

from the adapter and remove the adapter from the cylinder.

9. Repeat steps 5 through 8 for each remaining cylinder.
10. Compare the highest, or last stroke, reading for each cylinder with those specified by the manufacturer.

Compression Test Results

The compression pressure should be within the manufacturer's specifications for each cylinder. When limits are given for pressure differences between cylinders, the reading should fall within these limits.

Compare the readings taken on the first and last strokes for each cylinder. Compression is normal when the gauge shows a steady rise to the specified figure with each compression stroke. If the compression is low on the first stroke and builds up with each succeeding stroke, but not to specifications, the piston rings are probably worn. A low compression reading on the first stroke that builds up only slightly on the following strokes indicates sticking or burned valves. When two adjacent (side-by-side) cylinders have equally low compression readings, there is probably a head gasket leak between the two. A higher-than-normal compression reading usually means excessive carbon deposits in the combustion chamber.

More tests should be performed when low compression is found in one or more cylinders. Squirt about one tablespoon of engine oil through the spark plug opening in each low-reading cylinder. The oil should act as a temporary seal between the cylinder wall and the rings. Crank the engine several times to spread the oil on the cylinder walls and rings. Then, recheck the compression for each of these cylinders. This procedure, commonly referred to as a *wet compression test,* does not work with horizontal engines, such as those used in some Porsche, Volkswagen, or Subaru models.

If compression increases on the WCT test, worn rings or cylinder walls is the problem. Oil should increase compression about 5 percent. If compression does not increase, bad valves or a leaking head gasket is probably at fault. If adding oil does not improve the compression of two adjacent low-reading cylinders, there is probably a head gasket leak between them.

CYLINDER LEAKAGE TESTING

Many technicians feel the cylinder leakage test is superior to the compression test even though both tests have the same goal. With the piston at top dead center (TDC) on the compression stroke, connect the cylinder leakage tester to cylinder number one. Measure the percentage of leakage. Use a piece of heater hose held to the ear to locate the leakage. Stick the heater hose in the open bore of the fuel-injection throttle assembly or carburetor. If there is a great deal of air escaping through here, there is an intake valve problem. Insert the hose in the exhaust. If there is a great deal of air escaping, there is a bad exhaust valve. Observe the radiator coolant. If there are a lot of bubbles in the coolant, there is a head gasket leak, cracked head, or block. Next, remove the oil filler cap. Place the heater hose in the oil filler neck or valve cover opening. If excessive air is heard, the rings are allowing the air to bypass. Now place the hose over the spark plug holes of the adjacent cylinders. If a significant amount of air is heard from one of the adjacent cylinders, the head gasket is leaking between the cylinder being tested and the cylinder where the air is heard.

The sound of the air is significant only when the cylinder leakage indicated is excessive. The difficult part of the cylinder leakage test is determining how much leakage is too much. Manufacturers such as Honda say over 5 percent leakage is too much. Some technicians feel less than 15 percent is acceptable. The cylinder leakage test was performed because there is an obvious problem. Generally speaking, the types of problems that would be caused by 5–10 percent leakage are not noticed. The technician looks for a problem resulting from cylinder leakage which is probably greater than 15 percent.

ENGINE POWER BALANCE TESTING

The power balance tests are:

- Cylinder power balance test
- Injector/carburetor balance test

These tests indicate if an individual cylinder or a group of cylinders is not producing its share of power. The tests consist of shorting out the suspected spark plug or plugs so there are no power strokes from the cylinder or cylinders being tested. Various conditions, including engine rpm and manifold vacuum, change when the cylinders are not firing. The changes indicate whether that cylinder or group of cylinders has been doing its share of the engine's work.

If an engine is in good condition, all of its cylinders are doing the same amount of work. The changes in engine conditions caused by disabling a cylinder or group of cylinders should be about the same as the changes caused by shorting any other cylinder or group. The changes are measured in terms of engine rpm drop, manifold vacuum drop, or a combination of these factors.

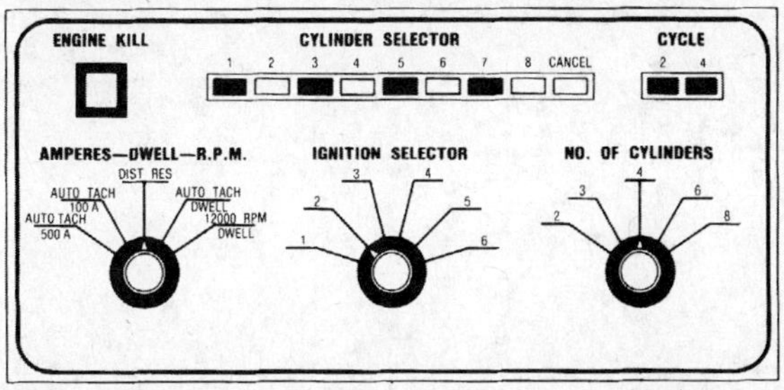

Figure 2-12. On this console control panel, the pushbuttons marked "Cylinder Selector" allow the technician to prevent certain cylinders from firing. The numbers refer to the firing order, not the cylinder number. On more modern units, this panel is a "virtual panel" on a computer display screen.

Power Balance Test Equipment

Power balance tests are usually made with an engine analyzer console which lets the technician control the firing of spark plugs with pushbuttons, figure 2-12. The numbers on the pushbuttons refer to cylinder *firing order*, not to the cylinder numbering. For example, the number five pushbutton controls the fifth cylinder in the engine's firing order, not cylinder number five. Some consoles have specially marked scales making it easier to measure the changes. In many newer engine analyzers, a computer touch-screen, a light pen, or a keyboard has replaced the pushbuttons.

Older-model console units may have a single knob controlling the individual spark plugs. This type of console may not be safe for use with an electronic ignition system. Be sure the equipment is compatible with the vehicle being tested.

Power Balance Test Precautions

If the vehicle tested has a valve-controlled EGR system, disconnect it during the power balance tests. Otherwise, the cycling of the EGR system affects the engine rpm and interferes with test results.

If the vehicle tested has a catalytic converter, try to limit the amount of unburned fuel reaching the converter. Various test equipment manufacturers have different methods of testing a converter-equipped car:

- One recommendation is to short each plug for no more than 15 seconds and then let the engine run normally for 30 seconds between tests. This keeps the unburned fuel from building up in the system.
- Another recommendation is to perform the test as quickly as possible without pauses. This is to keep the unburned fuel from building up.
- Engines with electronic controls, which include exhaust gas oxygen (O_2) sensors and injection systems or feedback carburetors, should be in open-loop mode for accurate power balance testing. When the cylinder is shut down, excess (unburned) oxygen enters the exhaust and creates a false signal. This drives the system fully rich and causes inaccurate power balance results. Follow the manufacturer's instructions when performing a balance test on an electronically controlled engine.
- If the engine has an electronic idle speed control system, it is not possible to accurately test at idle. The idle speed control system allows a momentary decrease in speed and then automatically returns it to normal. Some manufacturers recommend not performing a power balance test on such an engine or doing it at 1,500 to 2,000 rpm. Always check the vehicle manufacturer's directions and equipment manufacturer's recommendations.

On many OBD II applications, open loop on a warm engine occurs for only a few seconds after the engine is started. They set diagnostic codes and enter an alternative strategy for engine management when the misfire caused by the power balance tester is detected. On engines such as this, refer to the manufacturer's recommendations for alternative ways of testing power balance. Also refer to the diagnostic codes read through a scan tool. The codes identify which cylinders are providing less power. Scan Tools and OBD II are discussed more thoroughly in Chapter 11.

Cylinder Power Balance Test

In this test, individual cylinders are shorted out one by one while the rest of the cylinders fire. To make this test:

1. Connect the engine analyzer according to the equipment maker's instructions.
2. If the engine has an EGR system, disconnect and plug the vacuum line from the EGR valve.
3. If the engine has electronic controls, place it in open-loop operation according to the manufacturer's directions. On many systems, this is done by disconnecting a specified sensor.
4. Start the engine and bring it to normal operating temperature (upper radiator hose should be hot).
5. Run the engine at fast idle (about 1,000 to 1,500 rpm) or at other test speed as specified.
6. Press the button to kill one cylinder. Note and record the engine rpm drop and manifold vacuum drop. Release the button.
7. Repeat step 6 for each remaining cylinder.
8. After testing all cylinders, compare the results. Refer to the following paragraphs to interpret the readings.

If the changes in engine rpm and manifold vacuum are about the same for each cylinder, the engine is in good mechanical condition.

If the changes for one or more cylinders are noticeably different, the engine has a problem. The fault may be mechanical, or it may be in the ignition or fuel systems. You will have to make further tests to pinpoint the problem. Some of the remaining engine tests may help you to find the fault.

Special Procedures for Distributorless Engines

Most cylinder power analyzers are not compatible with distributorless ignition systems. Several engine analyzer manufacturers have attempted to integrate distributorless ignition testing into the cylinder power analyzer. Each of the manufacturers chose a slightly different path. Some opted for short-cuts which work only on certain applications, while others took no short-cuts but their equipment is incredibly expensive. Refer to the manufacturer's specifications.

There is, however, an alternative procedure which requires caution and skill. Once mastered, this test is very useful: Cut a piece of ⅛-inch vacuum hose into four, six, or eight sections. Each is about an inch long. Turn off the engine. One at a time, so as not to confuse the firing order, remove a plug wire from the coil pack. Insert a segment of the vacuum hose into the plug wire tower of the coil and set the plug wire back on top of the hose. Start the engine once all the segments are installed. Touching the vacuum hose "conductors" with a grounded test light kills the cylinder so the technician can note rpm drop. Again, the cylinder with the smallest drop in rpm is the weakest cylinder.

When power balance testing a vehicle equipped with OBD II, keep in mind that this test sets fault codes which must be cleared when the tests and repairs are completed. Also, many OBD II vehicles have engine control systems which cannot be disabled. The computer works so fast that it may be impossible to detect changes in the idle or vacuum when a cylinder is shut down. Many OBD II vehicles have built-in support for performing a power balance test utilizing a scan tool, figure 2-13. The PCM is programmed to maintain idle speed and timing advance as the power balance test is being performed. This method is also safer to perform because the scan tool will disable the injector for each cylinder; thus, no fuel will be dumped into the cylinder. Because of this no damage to the catalytic converter will occur due to overheating. Always check and see if this method is available on a vehicle before trying to perform a manual power balance test.

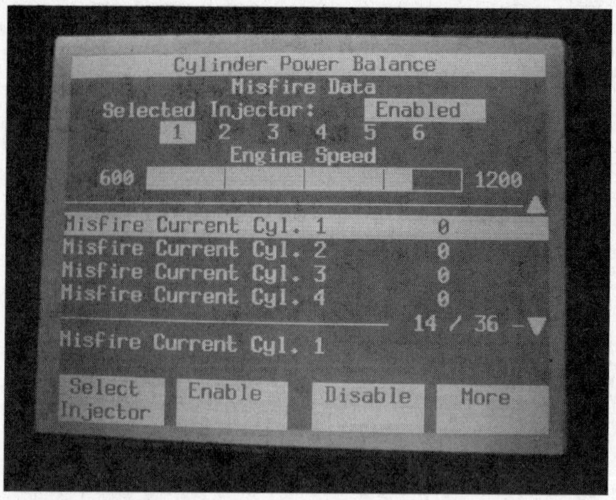

Figure 2-13. Many OBD II systems have scan tool support for performing a power balance test. This is the safest and most accurate method of performing a power balance test.

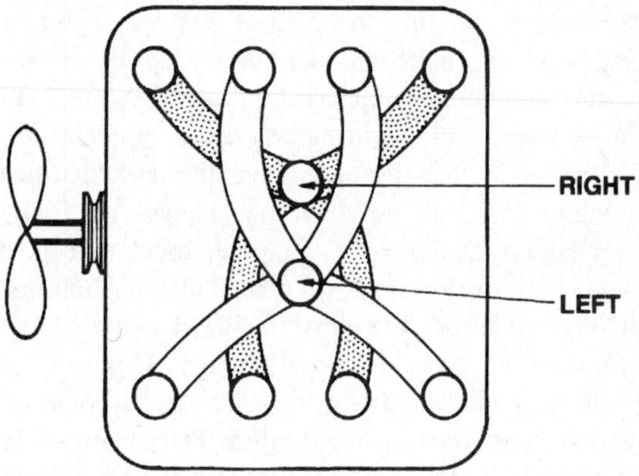

Figure 2-14. Most carbureted and many throttle-body injected V-6 and V-8 applications use a 2-plane manifold. Each of the two barrels feeds a separate group of cylinders.

Throttle-Body Injection Balance Test/ Carburetor Fuel Balance Test

This test applies to V-type engines with divided, or 2-plane, intake manifolds. The *Classroom Manual* explains that with a divided, or 2-plane, manifold, half of the cylinders are fed by half of the throttle-body injection unit or carburetor, figure 2-14.

During the test, the group of cylinders fed by one half of the carburetor is shorted and engine rpm noted. The second group of cylinders is then shorted and engine rpm noted. If the two rpm readings are about equal, the carburetor barrels are well balanced. If the rpm readings vary, the carburetor is out of balance.

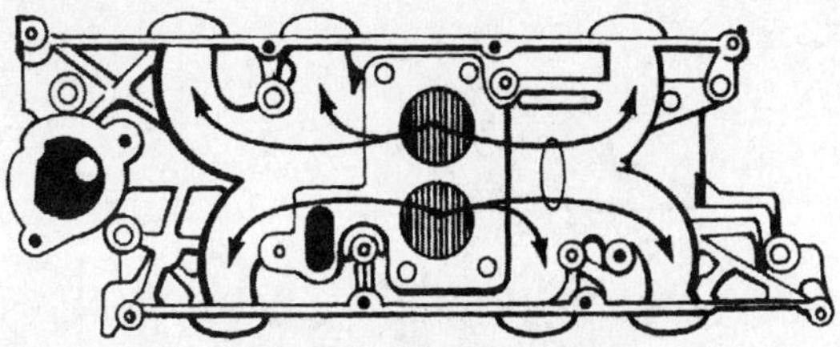

Figure 2-15. Some V-6 applications, such as the throttle-body injected 2.9-Liter Ford engines of the mid-1980s, use a divided single-plane manifold. On these engines, the left plane feeds the left side of the engine and the right plane feeds the right side.

To perform this test, the following knowledge is required:

- How the intake manifold is designed
- How the engine cylinders are numbered
- The cylinder firing order

There are two manifold designs tested with this procedure:

- V-6 engine manifolds are usually a split single-plane design, figure 2-15. Each manifold section feeds the cylinders on that side of the engine.
- V-8 engines with 2-plane manifolds use the design shown in figure 2-14.

Using the correct cylinder numbering sequence and firing order, determine which console buttons control the cylinders fed by one half of the carburetor, figure 2-16. In almost all cases, depressing every other console button shorts the cylinders fed by one half of the carburetor. To perform the carburetor balance test:

1. Bring the engine to normal operating temperature (upper radiator hose should be hot), then run it at fast idle.
2. Again, follow the manufacturer's directions for testing an engine with a catalytic converter or electronic controls.
3. Short out the cylinders fed by one half of the throttle body injection unit or carburetor and note the engine rpm. Let the engine run normally for about 60 seconds.
4. Short out the cylinders fed by the other half of the throttle-body injection unit or carburetor and note the engine rpm.
5. Compare the two engine speeds. If the difference is greater than 10 rpm, the carburetor is probably out of balance, assuming no other mechanical or ignition problems are present.

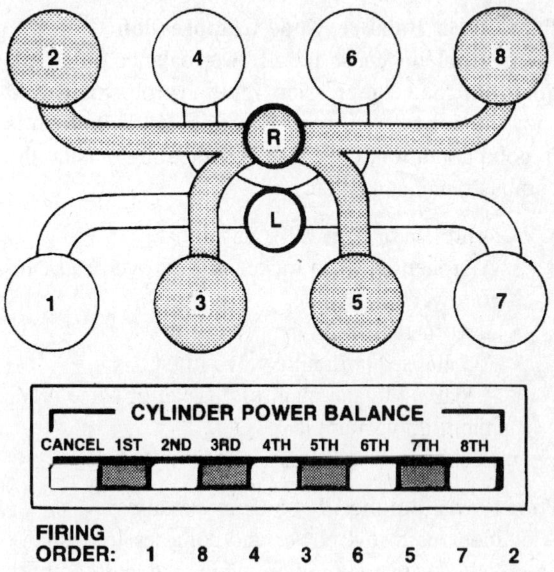

Figure 2-16. In a 2-plane manifold using either a carburetor or a throttle-body injection system, depressing the odd-numbered buttons on the engine analyzer, the cylinders being fed by the left barrel on the carburetor or throttle-body unit are shorted.

Comparing Power Balance and Compression Test Results

Comparing the results of a compression test and a power balance test often helps find a problem a single test cannot pinpoint. Also, the emission devices and engine accessories, in addition to transverse engine layout on many late-model engines, make access to the spark plugs difficult. This adds to the time needed for a compression test. A power balance test can be performed much faster. If power balance is good and the engine has no other apparent problems, omit the compression test. If power balance is bad, perform a compression test (sometimes on only one or two cylinders) to help identify the problem. Keep in mind, in the unlikely event that all cylinders have equally poor compression, the power balance test may lead the technician to believe all is well when in reality all the cylinders have low compression. Compression is an actual and independent measurement of the condition of each cylinder, while power balance is a comparison.

Good Power Balance, Low Compression

If an engine has good balance but low compression, it is probably an evenly worn, high-mileage engine. Symptoms may also include excessive crankcase blow-by, lack of power, and poor economy. A worn, slipped timing chain or slipped timing belt may also reduce compression.

Poor Power Balance, Good Compression

Many problems cause poor power balance on an engine with good compression. Ignition problems should show up during oscilloscope testing. If the ignition is in good condition, the cause is something outside the combustion chamber, such as:

- A broken or bent valve pushrod
- A broken or worn rocker arm or overhead cam follower
- A worn camshaft lobe
- A collapsed hydraulic valve lifter
- A leaking intake manifold gasket or some other manifold vacuum leak
- Leaking valve guides

Poor Power Balance, Poor Compression

A cylinder with low power and compression usually has a valve or piston problem which reduces combustion pressure and temperature. This could be a burned valve or piston, broken or leaking piston rings, a leaking head gasket, or leaking valve guides. Two adjacent cylinders with these symptoms usually have head gasket leakage between them.

INFRARED EXHAUST ANALYZERS

Infrared exhaust gas analyzers, as their name implies, use infrared light to examine a vehicle's exhaust for the presence and amounts of various gases. There are four types of infrared exhaust gas analyzers in use: 2-gas, 3-gas, 4-gas, and 5-gas. Two- and 3-gas analyzers were the mainstays of the industry during the 1970s and 1980s. Today they are used primarily as an alternate, and only on older models. Four-gas analyzers, figure 2-17, are the most common at present, but in some locations 5-gas analyzers are becoming popular.

Figure 2-17. A BAR97 5-gas emissions analyzer. (Courtesy of SPX OTC)

Figure 2-18. The probe sucks the exhaust into the machine for analysis.

2-Gas, 3-Gas, 4-Gas, and 5-Gas Analyzers

A 2-gas infrared exhaust gas analyzer measures the carbon monoxide (CO) and hydrocarbons (HC) in the engine exhaust. Carbon monoxide readings are given in percentages, with typical readings in the range of 1 to 6 percent. Hydrocarbon readings are measured in parts per million (ppm) because it is found in far smaller amounts in an exhaust sample. A typical reading for HC is in the range of 0 to 500 ppm. One ppm equals 0.0001 percent.

All models of infrared analyzers work in essentially the same manner. The analyzer measures CO and HC in the engine exhaust by comparing the air from the tailpipe with the surrounding clean air. It does this by drawing the two air samples into separate glass tubes. An infrared light beam shines through both tubes, but the beam is refracted (bent) by the impurities in the exhaust sample. The amount of refraction is measured by the analyzer and converted to electrical signals to drive the analog meters or digital display.

The analyzer gathers its exhaust sample from a probe inserted in the tailpipe, figure 2-18. On vehicles

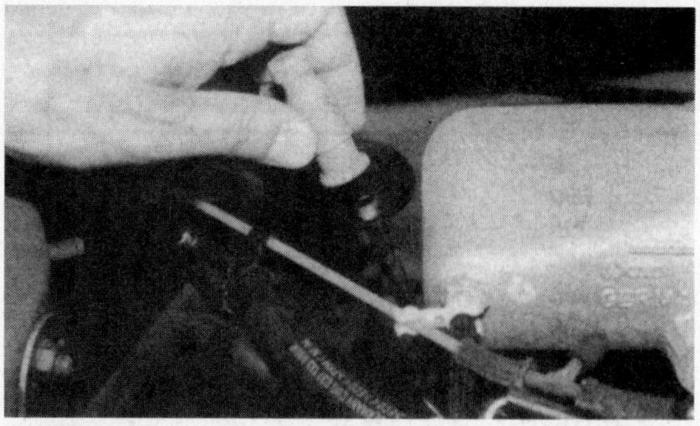

Figure 2-19. A few applications, such as this Volkswagen, have an exhaust gas diagnostic test port to sample exhaust gases upstream of the catalytic converter.

with catalytic converters, most of the CO and HC produced by the engine is removed before it reaches the tailpipe. Some vehicles, primarily European applications from the 1970s and early 1980s, had an exhaust sample port at the exhaust manifold, figure 2-19. This allowed sampling of the "unconverted" exhaust. Government regulations concerning making emission control and fuel systems tamper-resistant eliminated the need for these upstream sampling taps.

In recent years, infrared analyzers also measuring carbon dioxide (CO_2) and oxygen (O_2) have come into use. Like their 2-gas counterparts, these analyzers measure CO and HC as well. Carbon dioxide is measured by infrared comparison, which is the same way CO and HC are measured. Oxygen is measured by an oxygen sensor, similar to the O_2S used on engines with feedback fuel controls.

Carbon dioxide and oxygen are measured as a percentage of total exhaust volume, and although neither is toxic or a pollutant, their presence in the exhaust provides important information about what is happening in the combustion chamber. The typical 4-gas analyzer measures CO_2 from 0 to 20 percent and O_2 from 0 to 25 percent.

The latest gas to be added to the exhaust analyzers is oxides of nitrogen (NO_x). This is not a single gas, but rather a blend of gases. Oxides of nitrogen occur in the exhaust when the combustion temperature rises above 2500°F. These are insidious gases frequently linked to photochemical smog and acid rain. Internal combustion gasoline engines are capable of producing very high levels of these gases. The output of this gas tends to increase as the engine efficiency increases.

Operating Precautions

All infrared analyzers are used according to the general guidelines provided here, but always follow the equipment manufacturer's specific instructions.

- Test in a well-ventilated area for safety reasons as well as to prevent impurities, such as exhaust in the air, from polluting the reference air sample.
- Keep the analyzer out of drafts and in an environment with a low humidity. Some instruments give inaccurate results if air is allowed to flow across the circuit boards.
- If the vehicle has a manifold heat control valve, make sure the valve moves without binding.
- Make sure there are no leaks in the exhaust system. A leak dilutes the sample, resulting in an inaccurate test result.
- Do not connect a high-velocity exhaust ventilation system to one exhaust pipe on a dual-pipe system while taking a sample from the other pipe. This causes air to be drawn in through the tailpipe and makes the sample inaccurate.
- Disconnect the air pump if the vehicle has air injection. On converter-equipped vehicles, follow the manufacturer's instructions.
- Let the analyzer warm up for the specified time after it is turned on before taking a measurement.

Filter Changes

Infrared exhaust analyzers pull the exhaust test sample through a filter system, figure 2-20. This removes any extra moisture and particles from the sample. Routine filter service is required to ensure the exhaust sample has no leftover particles from previous tests. Different analyzers use different types of filters. Check the manufacturer's instructions for filter cleaning instructions.

Most analyzers have an exhaust flow indication of some type to signal when filter service is required. This signal may be a flashing warning light or a simple visual gauge, figure 2-21, containing a small floating ball.

Zero and Span Adjustments

Infrared analyzers are affected by altitude. The higher the altitude, the lower the density of the exhaust and reference samples. Most analyzers have some form of adjustment, such as ZERO and SPAN points on the displays or dials. The span-set lines on many analyzer meters are calibrated for use at an altitude of about 600 feet (183 meters). Figure 2-22 shows the difference in span settings required to calibrate typical meters. Each test equipment manufacturer has a procedure and an altitude correction for span adjustment that must be followed.

The following general procedure shows how the span adjustment is made on many units:

1. Warm up the analyzer and let the unit stabilize for 15 minutes. Some later-model gas analyzers

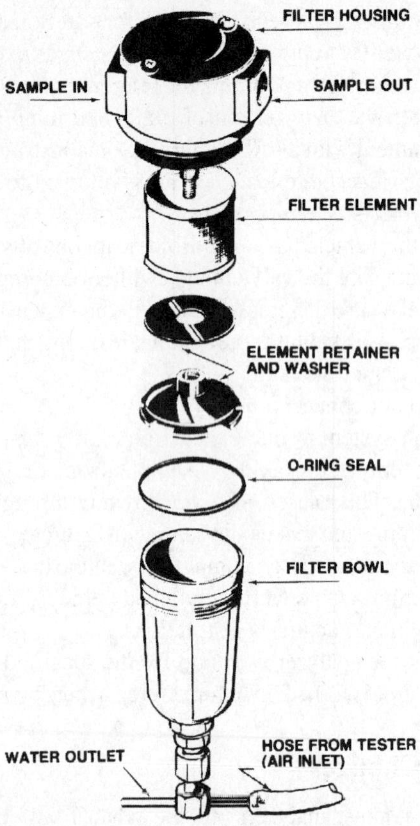

Figure 2-20. Typical filter arrangement on exhaust gas analyzers.

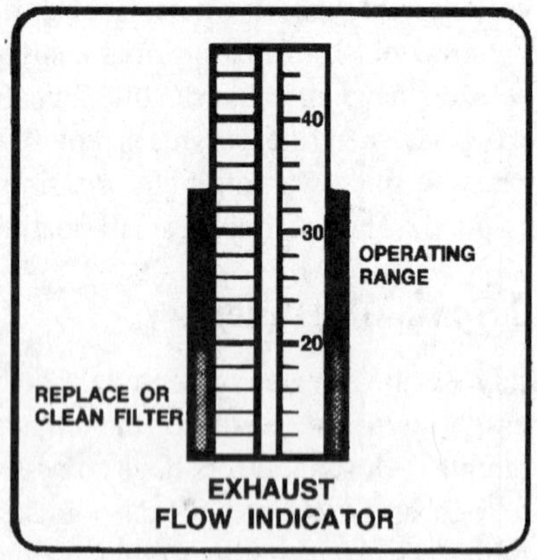

Figure 2-21. Most exhaust gas analyzers use a light to notify the technician when the flow of exhaust gases is reduced due to a restricted emission analyzer filter. Some older models use a gauge instead.

may require less warm-up time. Be sure to check the operating manual.

2. Zero the scales with the exhaust probe sampling fresh air. Some analyzers can be adjusted while in use.

ELEVATION IN FEET	HC SPAN SETTING PPM	CO SPAN SETTING %
0	880	2.4
600	900	2.5
1200	920	2.6
1900	940	2.7
2400	960	2.8
3200	980	2.9
4000	1000	3.0
4400	1030	3.1
4800	1040	3.2
5300	1060	3.3
6000	1090	3.4

Figure 2-22. This is a typical altitude correction chart for exhaust gas analyzer span adjustment. Correction factors vary for different analyzers. Most modern, microprocessor based, exhaust gas analyzers automatically perform this calibration.

3. Turn the test-or-calibrate knob on the console panel to the calibrate position and hold. Some panels have switches that do not require holding.
4. Adjust the HC meter needle to the set line.
5. Depress and hold the span button. Use the spanset knob to move the meter scale needle to the span-set position on the meter dial.
6. Repeat the procedure for each additional gas to be measured, then return the test-or-calibrate knob to the test position.

Recheck the setting after the analyzer has been in use for an hour. Many technicians feel it is a good idea to readjust the unit each time it is used.

Periodic Calibration

Most infrared analyzers can be electromechanically calibrated. They are also designed for certified gas calibration. Some states require this procedure, particularly when the machine is used for mandatory emission testing, as in Texas and California. Special calibration equipment is built into, or must be attached to, the analyzer unit, figure 2-23. The equipment allows a premeasured percentage of propane or hexane gas to enter the tester's infrared system. The gas is certified to contain known and stated amounts of HC and CO. Calibration gas may also include a reference percentage of CO_2 and oxides of NO_x. Oxygen is calibrated using the atmosphere as the sample.

The analyzer compares the gas content in the gas against its reference sample and gives a reading on the display meter. Compare this reading to the stated value of the calibrated gas. If the analyzer reading is within the specified limits of the gas sample concentration, the analyzer is correctly calibrated. If the reading is

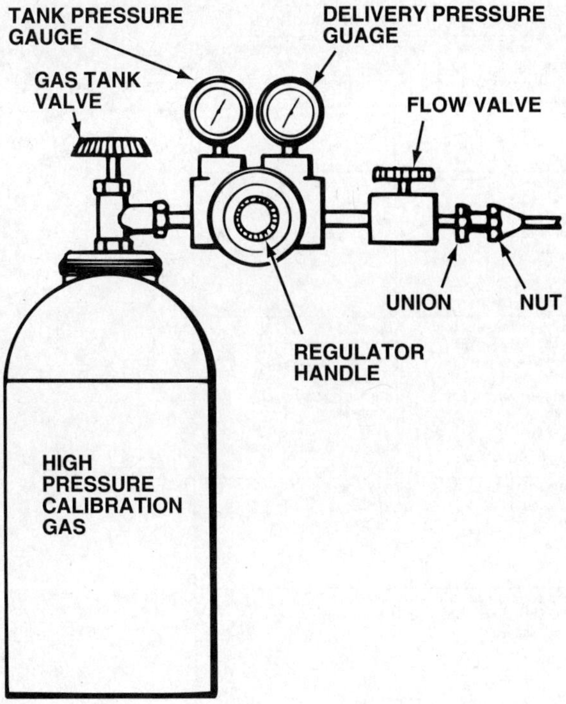

Figure 2-23. Special gas calibration equipment is required for maintaining the accuracy of gas analyzers which do not have automatic calibration capability.

outside the specified limits, the analyzer must be calibrated. This procedure should be left to a trained calibration technician or performed strictly in accordance with the manufacturer's instructions.

Onboard Calibration

Late-model analyzers may have a replaceable cylinder of pressurized calibration gas. The gas in the cylinder contains precise, known quantities of the measured gases. The analyzer draws a sample from the cylinder. Compare this measurement with the known concentration of gas. If the test sample is not within specified limits, the unit must be calibrated.

BASIC EXHAUST ANALYZERS

Federal, state, and local laws govern automobile emission levels. These standards contain HC and CO specifications for various vehicles. They may differ according to geographic location and the type of vehicle tested. The emission limits are often different for brand new vehicles, tune-up and adjustment of the same vehicles in use, and simple vehicle inspection.

In many states, CO_2 is also sampled. When there are low levels of CO_2 in the exhaust, it is assumed the sample is diluted by an exhaust leak and the test is invalid.

Tune-up service books are another place to find specifications. Analyzer manufacturers also include general specifications along with specific test procedures in the operating instructions accompanying their equipment.

For an accurate analysis of fuel combustion, sample exhaust both at idle and at 2,500 rpm. The engine must be at normal operating temperature when the test is made. If a dynomometer is used, test emissions at 2,500 rpm under simulated highway load conditions.

Normal HC and CO Readings

Normal emissions for late-model vehicles should be no more than the maximum level allowed by the emission laws which apply to the engine being tested. Expect the emission levels to be lower on older vehicles after a tune-up.

The following table, figure 2-24, shows idle HC and CO emission levels for various model years. For engines with air injection, these figures represent readings taken with the air pump disconnected. These readings do not represent state maximum levels; they are provided only as a reference for what typical readings might be on a well-tuned engine.

Model Year	% CO	ppm HC
65	4.0	700
70	2.0	150
75	0.60	60
80	0.50	45
85 and up	0.30	30

Abnormal HC and CO Readings

Carbon monoxide is an indicator of rich-lean and HC is an indicator of combustion quality. The level of CO in the exhaust gas of a running engine is related to the air-fuel mixture. Too much CO emission is a direct result of an overly rich air-fuel mixture. There is either too much gasoline or not enough air in the mixture. The problem is in the fuel system and is usually caused by one of the following conditions:

- A dirty air filter
- Incorrect idle speed
- A leaking fuel injector
- High system pressure in a fuel-injection system
- Clogged or dirty air mixture passages in the carburetor
- Rich fuel-injection adjustments
- Rich carburetor mixture adjustment
- A sticking or improperly adjusted choke
- A high float level in the carburetor
- A leaking power valve in the carburetor

	EMISSION	IDLE	IDLE OFF	CRUISE 1800-2000	AIR/FUEL RATIO	POSSIBLE CAUSES	RELATED SYMPTOMS
1	CO	3%	3%	3%	Rich AFR Below 10:1	Vacuum leak to map sensor Fuel injectors leaking Excessive fuel rail pressure Vacuum diaphragm bad Faulty sensor Fuel level too high	Black smoke or sulphur odor Poor fuel economy Surge/hesitation Stalling Engine not preconditioned Engine in open loop
	HC	250 ppm	280 ppm	300 ppm			
	CO_2	7-9%	7-9%	7-9%			
	O_2	.2%	.2%	.2%			
2	CO	1.5%	1.5%	1.0%	Rich AFR at low speed 10-12:1	Engine oil diluted with fuel Cold engine PCV valve defective Fuel injectors leaking Faulty ECT sensor	Poor fuel economy Sooty spark/black smoke Rough idle/surge hesitation Vapor canister purge valve bad Vapor canister saturated
	HC	150 ppm	150 ppm	100 ppm			
	CO_2	7-9%	7-9%	11-13%			
	O_2	0.5%	0.5%	1.0%			
3	CO	0.5%	0.5%	0.1%	Lean AFR Over 16:1	Check ignition primary/secondary Vacuum leak (intake runner) EGR valve, vacuum lines Poor cylinder sealing (rings/head) Fuel injectors restricted Improper timing Exhaust valve leak	Rough idle Misfire - high speed Detonation (cruise 2000 rpm) Surging Idle hunting (computer) Overheating NOTE: Excessive misfire and overheating can damage converter.
	HC	200 ppm	200 ppm	250 ppm			
	CO_2	7-9%	7-9%	7-9%			
	O_2	4-5%	4-5%	4-5%			
4	CO	2.5%	1.0%	0.8%	Lean AFR at High Speed Over 16:1	Air cleaner heater door closed Fuel injectors restricted Fuel pump pressure low	Rough idle Misfire Surging/hesitation
	HC	100 ppm	80 ppm	50 ppm			
	CO_2	7-9%	7-9%	7-9%			
	O_2	2-3%	2-3%	2-3%			
5	CO	0.3%	0.3%	0.3%	AFR 13:1 - 15:1	Engine not preconditioned Air management system not disabled Converter not warmed	None No driveability symptoms
	HC	100 ppm	80 ppm	50 ppm			
	CO_2	10-12%	10-12%	10-12%			
	O_2	2.5%	2.5%	2.5%			

Figure 2-24. These 4-gas readings can help with emission failure diagnosis.

High HC levels indicate unburned fuel in the exhaust. This is usually caused by a lack of ignition, or by incomplete combustion. The cause of high HC emissions is often traced to the ignition system. However, engine mechanical problems or carburetion problems also increase HC emissions. Some common causes of high HC levels are:

- Ignition system faults which cause a misfire. This includes defective spark plug wires and fouled spark plugs.
- Leaking vacuum hoses, vacuum controls, or gaskets.
- An excessively rich *or* lean air-fuel mixture.
- Clogged fuel injectors that cause a lean misfire.
- Incorrect ignition timing. This can even happen on applications which do not have adjustable ignition timing.
- Defective valves, valve lash, valve springs, lifters, guides, or camshaft.
- Defective rings, pistons, or cylinders.
- Low engine compression.

In addition to these problems, high HC and CO readings at the same time may be caused by one of the following conditions:

- An excessively rich air-fuel mixture
- Defective PCV system or catalytic converter
- Defective air pump

Complexities Added by the Oxygen Sensor

The oxygen sensor (O2S) causes a lot of distortion in the CO and HC readings when it is working properly. The job of the oxygen sensor is to signal the engine computer when the engine is running lean. The computer compensates by enriching the mixture. Anything adding additional air to the combustion process on a non-oxygen sensor engine causes the HC to increase, and generally causes the CO to decrease. On an oxygen sensor-equipped engine, the additional oxygen is detected by the oxygen sensor, which reports its presence to the computer. The computer thinks the engine is running lean and there is a likelihood that HC emissions increase. To bring the HC emissions down and reduce the potential for lean misfire, the computer enriches the mixture.

Extra oxygen is also present in the exhaust gases when combustion quality deteriorates. Low compression results in low heat generated during combustion. The low heat may cause large quantities of the fuel to not be burned. If all the fuel is not burned, oxygen is left over. The computer assumes the oxygen is from a lean air-fuel mixture and responds by enriching the mixture.

Exhaust leaks also affect oxygen sensor readings. Air leaks into the exhaust system just as the exhaust gases leak out. This distorts the oxygen sensor readings and causes the engine to run rich just because there is an exhaust leak.

When high CO readings are found on an oxygen sensor-equipped vehicle, look for sources of extra air getting into the air induction system, extra air left over by the combustion process, or extra air leaking into the exhaust. When troubleshooting high CO on an oxygen sensor-equipped engine, use not only the list for high CO, but also the list for high HC and the list for both high HC and high CO.

Carbon Dioxide Analysis

An engine running in perfect condition with an intact exhaust system gives a CO_2 reading of about 15 percent. As the quality of combustion decreases, the amount of CO_2 in the exhaust gases also decreases. Anything over 10 percent should be considered acceptable; however, the closer to 15 percent the better. When the reading is less than 10 percent, look at the items on the high HC list. If the reading is below 7 percent, yet the engine seems to be running fairly well, there is most likely an exhaust leak.

Oxygen Analysis

On a catalyst-equipped engine, high CO readings are often masked by the converter. Since CO is the primary reference for rich/lean running engines, this analysis is often inconclusive. Oxygen readings are not as complete as CO readings. A typical O_2 reading is 0.5 percent to about 1 percent. If the reading is higher than 1.0 percent and CO_2 is less than 10 percent, suspect the engine is running lean. This corresponds to a low CO reading. If the reading is less than 0.5 percent and the CO_2 reading is low, suspect the engine is running rich. If the CO_2 reading is normal, suspect nothing related to air-fuel ratio.

Oxides of Nitrogen Analysis

Accurate NO_x readings require the use of a dynamometer. At this point, very little is known about NO_x as a diagnostic tool; however, evolving emission laws may soon require that NO_x levels stay below a regulated point at the tailpipe, figure 2-25.

Basic Test Procedure

1. Connect the exhaust gas analyzer according to the manufacturer's instructions.
2. Check the analyzer manufacturer's manual for guidelines, or state or local inspection regulations, if applicable.
3. Disconnect and plug the air-injection outlet line, if applicable.
4. Calibrate the analyzer if necessary. Insert the sampling probe into the tailpipe.
5. Take all readings with the air cleaner installed.
6. Take all readings with the oxygen sensor connected.
7. Watch the meter and record the HC, CO, CO_2, and O_2 readings. All should be within manufacturer's specification or the legal limits.
8. Increase engine speed to 2,500 rpm and hold. Record the readings again. CO and HC should be as low as, or lower than, the readings taken at idle. The O_2 reading should not change significantly at the higher speed, while CO_2 readings typically climb closer to 15 percent.
9. Return the engine to idle speed. Record all four readings again. After a momentary increase on deceleration, CO and HC readings should return to the levels recorded in step 7. The CO_2 reading should slowly decrease while the O_2 reading increases.

Fuel Injection Idle Mixture Adjustment

Since 1980, no domestic fuel-injection system is adjustable. Since 1984, no fuel-injection system on a passenger car delivered in North America is adjustable. Older import fuel-injection systems have adjustable air-fuel ratios.

Bosch D-Jetronic applications which have an adjustable air-fuel ratio have a potentiometer on the side of the control unit. This potentiometer fine-tunes the air-fuel ratio during a tune-up.

On Bosch L-Jetronic systems, turning a large screw adjustment on the airflow meter performs the CO adjustment. The screw alters the amount of air allowed to bypass the airflow meter flap, changing the air-fuel ratio.

The CO adjustment on the Bosch CIS system is located in the airflow sensor near the fuel distributor. A long, three-millimeter Allen wrench is required. After completing the minimum idle and curb idle adjustments and after again ensuring the engine is thoroughly warmed up, insert a CO exhaust gas analyzer into either the tailpipe or the sampling port located ahead of the catalytic converter as indicated in the EPA sticker under the hood.

Be sure to consult the vehicle's service manual for specific procedures for each application where mixture is adjustable.

HC and CO Readings at Idle

Listed below are combinations of common HC and CO readings and their possible causes. Test results with 3- and 4-gas analyzers are explained later in this chapter. These readings and causes assume the engine does not have an oxygen sensor or the computer is forced into open loop, a mode of operation where the oxygen sensor is ignored.

1. High HC, low CO:
   a. Very lean air-fuel ratio, causing misfire
   b. Manifold vacuum leak
   c. Open, or high resistance, spark plug cable or fouled plug, causing misfire

EXHAUST EMISSIONS

ENGINE CONDITIONS	HC	CO	CO_2	O_2	NOx
Ignition Misfire	Large Increase	Some Decrease	Some Decrease	Some-Large Increase	Some-Large Decrease
Compression Loss	Some-Large Increase	Some Decrease	Some Decrease	Some Increase	Some-Large Decrease
Rich Mixture	Some-Large Increase	Large Increase	Some Decrease	Some Decrease	Some-Large Decrease
Lean Mixture	Some Increase	Large Decrease	Some Decrease	Some Increase	Some-Large Increase
Very Lean Mixture	Large Increase	Large Decrease	Some Decrease	Large Increase	Some-Large Decrease
Slightly Retarded Timing	Some Decrease	No Change or Some Increase	No Change	No Change	Large Decrease
Very Retarded Timing	Some Increase	No Change	Some-Large Decrease	No Change	Large Decrease
Advanced Timing	Some Increase	No Change or Some Decrease	No Change	No Change	Large Increase
EGR Operating	No Change	No Change	Some Decrease	No Change	Large Decrease
EGR Leaking	Some Increase	No Change	No Change or Some Decrease	No Change	Some Decrease or No Change
Air Injection Operating	Large Decrease	Large Decrease	Some-Large Decrease	Large Increase	No Change
Catalytic Converter Functional	Some Decrease	Some Decrease	Some Increase	Some Decrease	Some Decrease 3-way cut
Exhaust Leak	Some Decrease	Some Decrease	Some Decrease	Some Increase	No Change
Worn Engine	Some Increase	Some Increase	Some Decrease	Some Decrease	No Change or Slight Decrease
Flat Camshaft	No Change or Some Decrease	Some Decrease	Some Decrease	No Change or Some Decrease	No Change or Some Decrease

ENGINE CONDITION'S EFFECT ON EXHAUST EMISSIONS

Figure 2-25. Changes in these 5-gas analyzer measurements can help with emission failure diagnosis.

2. High HC, normal CO:
   a. Low compression
   b. Vacuum leak
   c. Arcing ignition points
3. High HC, high CO:
   a. Extremely rich air-fuel ratio
   b. Leaking fuel-injector or high fuel pressure
   c. Choke stuck partly closed
   d. Misadjusted carburetor or incorrect injection metering
4. Normal HC, high CO:
   a. Restricted PCV valve
   b. Restricted air filter

HC and CO Readings at 2,500 rpm

1. High HC, low CO:
   a. Open spark plug cable or fouled plug causing misfire
   b. High-speed misfire
   c. Floating exhaust valve

2. High HC, normal CO:
   a. Vacuum leak
   b. Arcing or misaligned ignition points
3. Normal HC, high CO:
   a. Restricted air filter
   b. Choke stuck partly closed
   c. Defective high-speed or power circuits in the carburetor
   d. Leaking injector or high fuel pressure

When using an infrared analyzer for idle adjustment, remember that each adjustment affects the air-fuel mixture and the exhaust analyzer readings. If the engine runs richer, CO should increase. If the engine runs leaner, CO should decrease. If the mixture is so lean it results in a misfire, HC increases.

Other Tests

More special tests can be made with an infrared exhaust analyzer. The following paragraphs outline the principles for some of the most common ones.

Throttle-position Sensor/Accelerator Pump Test
Quickly opening the throttle causes the throttle-position (TP) sensor to tell the fuel-injection computer that extra fuel is required for acceleration. On a carbureted application, quick opening of the throttle causes the accelerator pump to spray fuel into the intake manifold. Both of these actions overcome the flat spot in acceleration that results from the momentary drop in manifold vacuum. Test the operation of the accelerator pump or TP sensor as follows:

1. Bring the engine to normal operating temperature (upper radiator hose should be hot) and run at slow idle speed.
2. Watch the CO meter. Quickly open and release the throttle.
3. The engine speed should increase without hesitation. The CO meter reading should also increase.
4. If the CO reading does not increase, or if it decreases before increasing, test the TP sensor with a scan tool or voltmeter. If the engine is carbureted, the accelerator pump linkage may need adjustment, or the pump needs repair or replacement.

Power Valve Test
Test the operation of the manifold absolute pressure (MAP) sensor, mass air flow (MAF) sensor, or carburetor power valve or power system as follows:

1. Bring the engine to normal operating temperature (upper radiator hose hot) and run at approximately 2,000 rpm.
2. Watch the CO meter. Quickly open the throttle completely.
3. The CO meter reading should show a slight but quick increase. This indicates the power valve has opened.
4. If the CO reading does not increase, the power valve is defective.

Since the power valve test is made at a throttle opening above about 2,000 rpm, the engine computer has already responded to the TP sensor. The accelerator pump on a carbureted application operates at this higher engine speed.

PCV System Test
Test the PCV system as follows:

1. Bring the engine to normal operating temperature (upper radiator hose hot) and run at specified idle speed.
2. Remove the PCV valve from the engine. Do not disconnect it from the hose to the intake manifold, the throttle body, or the carburetor.
3. To check for a strong vacuum, the technician's thumb is placed over the open end of the valve. Also notice the rpm drop.

Air Pump Test
Test the air-injection system as follows:

1. Bring the engine to normal operating temperature (upper radiator hose hot) and run at approximately 1,000 rpm. Record the HC and CO readings.
2. Hold the engine speed constant while disconnecting the air supply hose between the pump and manifold.
3. Compare the two HC and CO readings obtained. If the air supply system is working properly, there should be an increase in both HC and CO with the air injection disconnected. If there is little or no increase in the readings, inspect the air pump and its connecting lines.

2-Gas Analysis

The HC and CO emissions from vehicles built with catalytic convertors from the late 1970s to the present have been reduced to very low levels. Tailpipe readings taken with an infrared analyzer are usually less than 1 percent CO and 0 to about 40 or 50 ppm HC if the engine and emission controls are in good shape. Unless the engine has a definite performance problem, the traditional methods of using HC and CO diagnosis are of little value. However, 4-gas analysis reveals more about combustion efficiency and engine performance.

4-Gas Analysis

It is difficult to measure HC and CO emissions from late-model engines at the tailpipe. A 4-gas analyzer, however, measures O_2 and CO_2 in the exhaust, as well as HC and CO. Both O_2 and CO_2 are measured by percentage.

A certain amount of O_2 is normal and necessary in the exhaust. Oxygen is required for combustion, and the catalytic converter uses O_2 to oxidize HC and CO into harmless CO_2 and H_2O. But, too much O_2 at the tailpipe may indicate combustion problems.

High HC from a noncatalyst engine indicates unburned fuel in the exhaust, which often means a lean-misfire problem. With a converter, high O_2 indicates a lean condition. An O_2 reading above 2 percent may indicate a lean condition; if O_2 is above 4 percent, a lean-mixture condition is certain.

When comparing CO and O_2 readings, low O_2 with high CO indicates a rich mixture. High O_2 with low CO indicates a lean mixture. If O_2 is greater than CO, and CO is 0.5 percent or more, the converter may be malfunctioning. However, if CO is greater than 0.5 percent and also greater than O_2, the air-fuel mixture is rich, but the converter may be properly functioning.

At the stoichiometric point of 14.7 to 1 air-fuel ratio, O_2 should be low, along with HC and CO. CO_2, however, should be high: at least 8 to 10 percent, ideally above 13 percent. A lean air-fuel ratio causes CO_2 to decrease as O_2 increases. At a 14.7 to 1 ratio, CO and CO_2 percentages should equal about 14.7 percent, total.

Good combustion produces low HC and CO readings within specifications. Additionally, CO_2 readings of 8 to 10 percent (ideally above 13 percent) and O_2 of 1 to 2 percent indicate the combustion process is healthy and the burn quality is good.

PROGRAMMED SPECIAL TEST ANALYZERS

Single-purpose analyzers are used by many states which have mandated periodic vehicle emission testing programs. The analyzer, figure 2-26, contains a program which performs all of the test functions required by the particular state program.

After inspecting the vehicle to determine that all required emission control components are installed and properly connected, connect the analyzer to the engine. After the analyzer warms up, its screen displays a request for data concerning the make, model, year, engine, and other factors. From this point, the display screen directs the technician in completing the programmed test sequence. The microprocessor evaluates the data received from the engine accoring to its programmed memory tables, displays its findings, and produces a printout.

If problems arise during the test sequence, the microprocessor either aborts the test or asks the operator to verify certain steps. For example, when the test sample is out of limits, the display may request the operator to check the position of the tailpipe probe to determine if the sample is diluted.

Figure 2-26. State emissions-testing programs use analyzers that are designed to run a given test sequence to pass or fail a vehicle.

SOURCES OF SPECIFICATIONS AND SERVICE PROCEDURES

Before successfully troubleshooting an automotive fuel or emission system, access to the necessary specifications and service procedures is required. Complete and accurate specifications are the only tool the technician has for determining how the system *should* operate. Without specifications, all the sophisticated test equipment in a shop is useless. Test results mean nothing until they are compared to specifications indicating what the test results should be. Service procedures go hand in hand with specifications. The technician needs to know the steps peculiar to a particular system in order to obtain the correct result.

Specifications

The main sources of tune-up and emission service specifications are:

- Manufacturers' service manuals
- Independent manuals

Manufacturers usually publish a new service or factory shop manual for each model year, figure 2-27. However, the information is often divided and scattered throughout the manual, separated by the repair and overhaul information that is also included.

Figure 2-27. Factory service manuals provide specifications for one year only. There are usually separate manuals for each model produced.

Independent manufacturers of replacement components test equipment and service manuals often print books of specifications, such as the *Car Care Guide* from Chek-Chart Publications. The specifications may apply to a single model year or they may cover a span of years. Figure 2-28 shows part of a page from a typical specifications book.

Procedures

Instructions for the repair and test procedures performed are also found in manufacturers' shop manuals, as well as manuals offered by independent publishers. This information is usually presented in a step-by-step sequence similar to this manual.

Reference manuals containing the procedures often use troubleshooting charts and diagrams to help eliminate potential causes of the problem. Figures 2-29 and 2-30 illustrate two of the many different approaches to providing the necessary information.

Engine Decals or Labels

According to federal law, all automotive manufacturers must provide a Vehicle Emissions Control Information (VECI) decal or label in the engine compartment of every vehicle produced. This decal takes different forms, depending upon the manufacturer, but they all must list basic engine and emission control specifications and procedures peculiar to the vehicle, figure 2-31. Because of running changes made in a model year, changes in engine calibration to correct a driveability problem, and other factors, the information on the decal may not agree with printed specifications for the vehicle. Remember, the VECI decal is supposed to be the final word and should always be followed if it conflicts with printed data given in a manual.

Changes made according to factory-approved procedures may affect emission control system operation. If so, the law requires a supplemental decal, figure 2-32, be affixed beside the VECI decal. This indicates a change has been made affecting basic specifications or emission-control system vacuum line routing and connections. A clear plastic shield can be placed over the decal for protection.

Reference Manual Sources

Factory shop manuals are available directly from the vehicle manufacturers or from authorized dealers. Most import dealers and many domestic dealers carry factory manuals in their parts departments. Some manufacturers include their entire vehicle line in one set of manuals; other publish separate manuals for each model.

Most domestic vehicle shop manuals may be ordered from the manufacturer by mail. The Vehicle Manufacturer Service Information matrix is provided by the National Service Task Force (NASTF). This matrix provides information on obtaining tools, training, and service information from the various vehicle manufacturers. Most manufacturers have a website where this information can be obtained. Refer to Appendix D of the *Classroom Manual* for a list of manufacturers' websites.

ACURA

1987-98 All Models
SLX—See Isuzu

CHEK-CHART

UNDERHOOD SERVICE SPECIFICATIONS

AAITU2

AAITU2

ELECTRICAL AND IGNITION SYSTEMS

BATTERY

BCI equivalent shown, size may vary from original equipment. Check clearance before replacing, holddown may need to be modified.

		STANDARD		OPTIONAL	
Model	Year	BCI Group No.	Crank. Perf.	BCI Group No.	Crank. Perf.
Integra	1987-89	45	410	—	—
	1990-93	25	435	—	—
	1994-98	51R	405		
Legend	1987-89	24F	580	—	—
	1990	24F	585	—	—
	1991-95	24	585	—	—
NSX	1991-94	24F	550	—	—
w/MT	1995-97	35	435	—	—
w/AT	1995-97	24F	585	—	—
Vigor	1992-94	24F	550	—	—
2.2CL	1997	24F	550	—	—
2.5TL	1995-98	35	440	—	—
3.2TL	1996-98	25	440	—	—
3.5RL	1996-98	24	550	—	—

GENERATOR

Application	Year	Rated Output	Test Output (amps & eng. rpm)
1590cc	1987	55	58-68 @ 2000
		65	58-68 @ 2000
1590cc	1988-89	65	40-50 @ 2000
1678cc	1991-93	80	40 min. @ 2000
1797cc, 1834cc	1990-91	70	74-85 @ 2000
	1992-93	80	40 min. @ 2000
	1994-97	90	60 min. @ 2000
2165cc	1997	90	75 @ 2000
2451cc	1992	100	78 min. @ 2000
	1993-94	100	55 min. @ 2000
	1995-97	100	75 @ 2000
2494cc, 2675cc	1987-90	70	74-85 @ 2000
2977cc, 3206cc	1991-92	110	102 @ 2000
2977cc	1993-94	110	40 min. @ 2000
	1995-97	120	85 @ 2000
3206cc, 3474cc	1993-97	110	60 min. @ 2000

REGULATOR

Application	Year	Test Temp. (deg. F/C)	Voltage Setting
All	1987-98	—	13.5-15.1

STARTER

Engine	Year	Cranking Voltage (min. volts)	Max. Ampere Draw @ Cranking Speed
All	1987-90	8.0	350
1678cc, 1834cc	1991-93	8.0	350
1797cc, 1834cc	1994-95	8.0	360
1797cc, 1834cc: 1.2 kw	1996	8.0	270 max.
1.4 kw	1996	8.0	360 max.
2451cc: 1.6 kw	1992-94	8.0	350
2.0 kw	1992-94	8.0	400
2451cc	1995-96	8.5	380
2977cc, 3206cc	1991-95	8.5	350

SPARK PLUGS

Engine	Year	Gap (inches)	Gap (mm)	Torque (ft-lb)
1590cc, 1834cc	1987-98	.039-.043	1.0-1.1	13
1797cc	1994-98	.047-.051	1.2-1.3	13
1678cc	1991-94	.051	1.3	13
2165cc	1997	.039-.043	1.0-1.1	13
2451cc	1992-94	.043	1.1	13
	1995-98	.039-.043	1.0-1.1	13

SPARK PLUGS Continued

Engine	Year	Gap (inches)	Gap (mm)	Torque (ft-lb)
2494cc	1987	.039-.043	1.0-1.1	13
2675cc	1987-90	.039-.043	1.0-1.1	16
2977cc, 3206cc, 3474cc	1991-98	.039-.043	1.0-1.1	13

IGNITION COIL

Resistance (ohms @ 70°F or 21°C)

Engine	Year	Windings	Resistance (ohms)
1590cc	1987	Primary	1.215-1.485
		Secondary	9040-13,560
1590cc	1988-89	Primary	.3-.5
		Secondary	9440-14,160
1678cc	1991-93	Primary	0.6-0.8
		Secondary	9760-14,640
1797cc, 1834cc	1994-98	Primary	0.6-0.8
		Secondary	12,800-19,200
1834cc	1990-93	Primary	.6-.8
		Secondary	9760-14,640
2165cc	1997	Primary	.45-.55
		Secondary	16,800-25,000
2451cc	1992-98	Primary	0.3-0.5[3]
		Secondary	10,800-16,200[3]
2494cc, 2675cc	1987-89	Primary	.3-.4
		Secondary	9040-13,560[1]
2675cc	1990	Primary	.35-.42
		Secondary	16,000-24,000[1]
2977cc, 3206cc, 3474cc	1991-98	Primary	0.9-1.1[2]

1 Measured between upper right side cavity terminal and secondary terminal.
2 Measured between two electrical leads of each coil.
3 Primary resistance measured between upper right and lower left terminal (under lock) of ignition coil. Secondary resistance measured between upper right and high voltage terminal.

BASE TIMING

At slow idle and Before Top Dead Center, unless otherwise specified.

1987 models, disconnect and plug distributor advance hoses.

1988-89 1590cc, remove the main fuse box cover and jumper terminals Br and Br/Bl.

1988-89 V6, 1990-98 All: Connect a jumper between the terminals of the 2-pin light grey or blue timing connector under dash.

1988-94 2451cc, 2675cc, 2977cc, 3206cc, access the ignition timing adjuster in the control box under hood. Turn the screw on the unit to adjust timing. Rivets must be drilled from timing adjuster mounting to gain access to screw.

1995-98 2451cc, 2977cc, 3206cc, connect inductive clamp to loop wire by ignitor.

1995-98 2451cc, 2977cc, 3206cc, 3474cc, if timing is not correct, replace ECM.

Engine	Year	Man. Trans. (degrees)	Auto. Trans. (degrees)
1590cc	1987	0± 2	0± 2
1590cc	1988-89	12± 2	12± 2
1678cc	1991-93	16± 2	16± 2
1797cc	1994-98	16± 2	16± 2
1834cc	1990-98	16± 2	16± 2
2165cc	1997	15± 2	15± 2
2451cc	1992-97	15± 2	15± 2
2494cc	1987	3	3
w/vacuum hoses connected	1987	23± 2	18± 2
2675cc	1987-90	15± 2	15± 2
2977cc, 3206cc, 3474cc	1991-98	15± 2	15± 2

DISTRIBUTOR PICKUP

Engine	Year	Resistance (ohms)
1590cc: Pickup coil	1987	650-850
Camshaft position sensor	1987	650-850
1590cc: Pickup coil	1988-89	350-550
Crankshaft position sensor	1988-89	700-1000

Figure 2-28. A portion of a page from a typical tune-up specifications manual.

Building a library of factory shop manuals is highly desirable, but it also gets very expensive. Independent publishers, such as Chek-Chart, compile automotive specifications and procedures for all domestic passenger cars in single service and repair manuals. Import cars and light-duty trucks are generally covered in separate manuals, but some specification books often cover all three categories.

Aftermarket manufacturers of tune-up components such as spark plugs, carburetor kits, and other fuel- and emission-control system parts also publish books containing specifications and procedures. However, information is usually limited to the particular part supplied, so additional information is needed for comprehensive testing.

Noise

SYMPTOM: Noise

INSPECTION FLOW

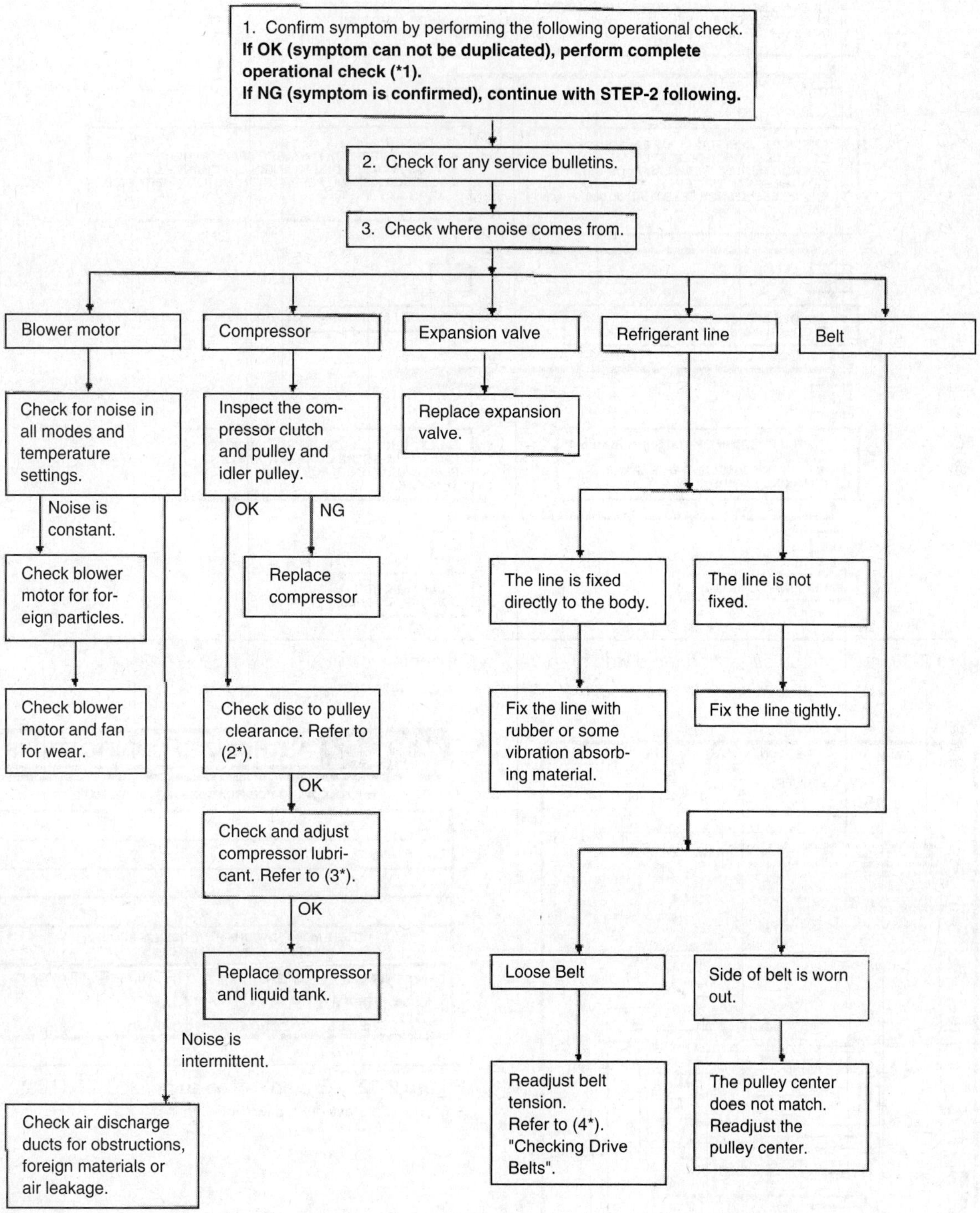

Figure 2-29. Troubleshooting flow charts are used to help the technician diagnose problems. (Courtesy of Nissan North America, Inc.)

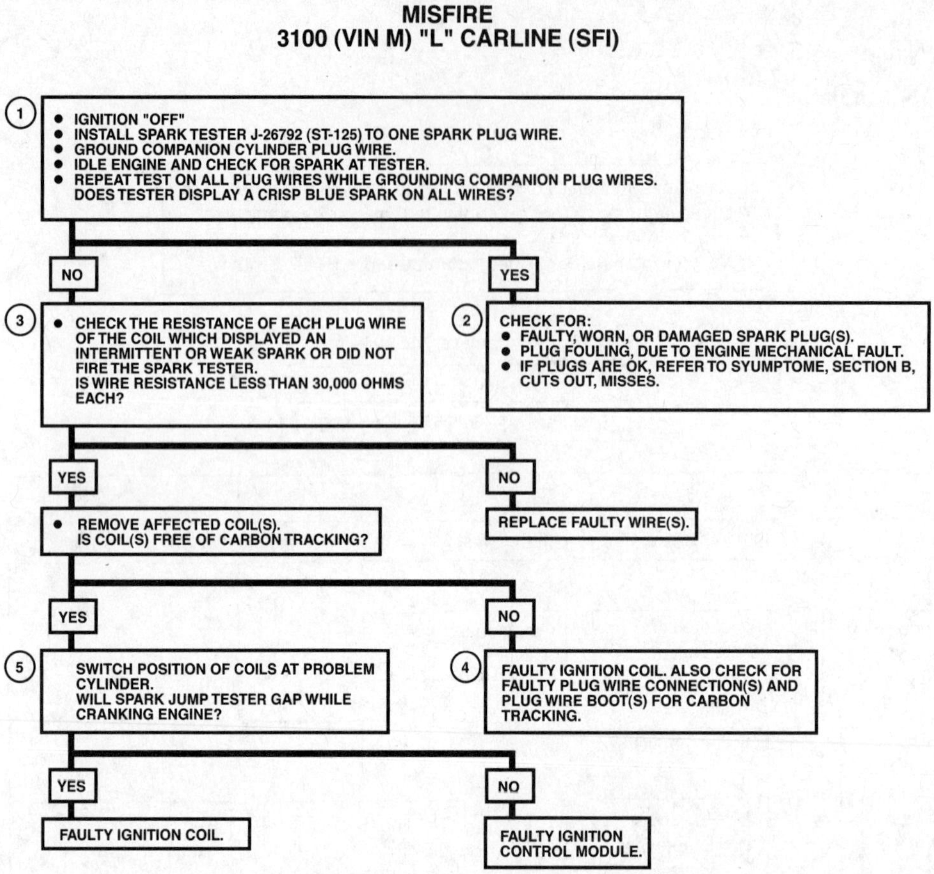

Figure 2-30. This troubleshooting table was prepared by Chevrolet. (Courtesy of General Motors Corporation)

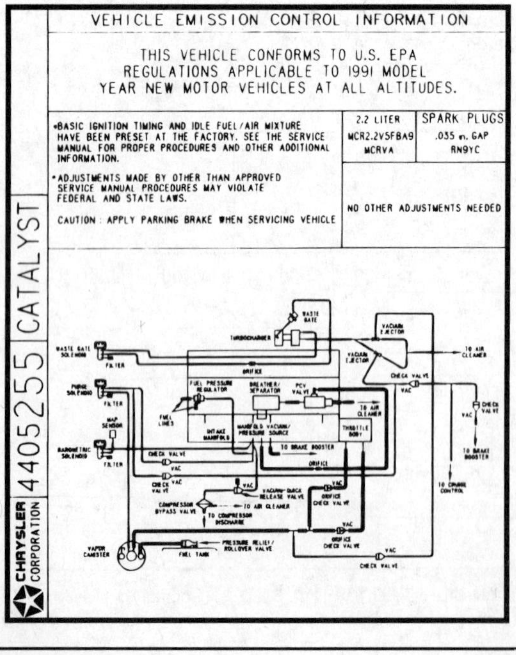

Figure 2-31. A typical VECI decal or sticker. (Courtesy of DaimlerChrysler Corporation)

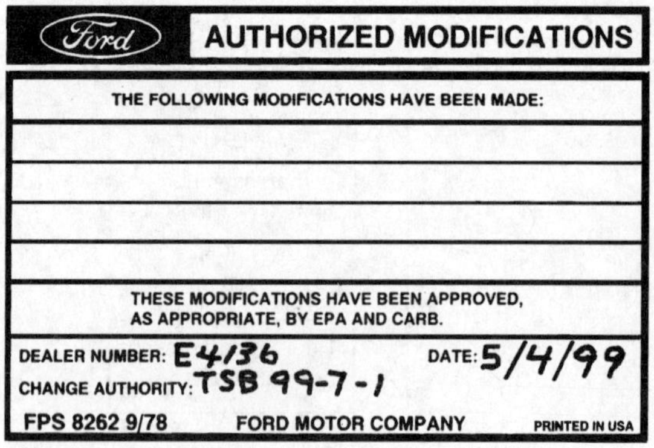

Figure 2-32. Example of an authorized modifications sticker. (Courtesy of Ford Motor Company)

3

Fuel System Inspection and Testing

OBJECTIVES

Upon completion and review of this chapter, you will be able to:

- Inspect fuel tanks, lines, and systems and determine if there are any leaks present.
- Service fuel filters.
- Perform an electric fuel pump test.

FUEL SYSTEM INSPECTION

The fuel system has many parts and connecting lines, figure 3-1. Any of these parts may develop leaks or restrictions. Fuel system parts should be checked for damage and safety defects before and during service operations. Inspect the vehicle by following the general procedures provided in the following paragraphs.

Fuel Tank, Filler Tube, and Filler Cap

The evolution of fuel evaporative controls has gone from venting the fuel tank into the engine crankcase or vapor storage canister to a complex system which is monitored and controlled by the Powertrain Control Module (PCM). The inspection procedure will vary according to the components and system used. Many of the basic procedures will remain the same, figure 3-1.

1. Inspect the tank for dents, cracks, or leaks.
2. Check the fuel tank vent system, figure 3-2, to ensure that the vents are not plugged. Plugged vents can cause the tank to collapse as the fuel is pumped out or as the temperature inside the tank decreases.
3. Inspect all vent tubes and their connecting lines for damage and worn or cracked hoses.
4. Check the condition of the filler tube and filler cap, figure 3-3. Ensure a good seal between the cap and the filler tube.
5. Check the operation of the filler cap release valve if test equipment is available. Testers are available to pressure-test the cap. In areas using the I/M 240 smog test, the tester will interface with the test computer, figure 3-4.
6. Look for fuel leaks at the filler tube, at vent line connections, and around the fuel tank gauge sending unit.
7. Check the tank holddown straps to ensure that they are secure. If loose, tighten to specifications. Overtightening may break the straps or crush the tank.
8. If necessary, inspect the tank for dirt and sludge by removing a plug or by disconnecting a line at the tank bottom, if there is one. Late-model

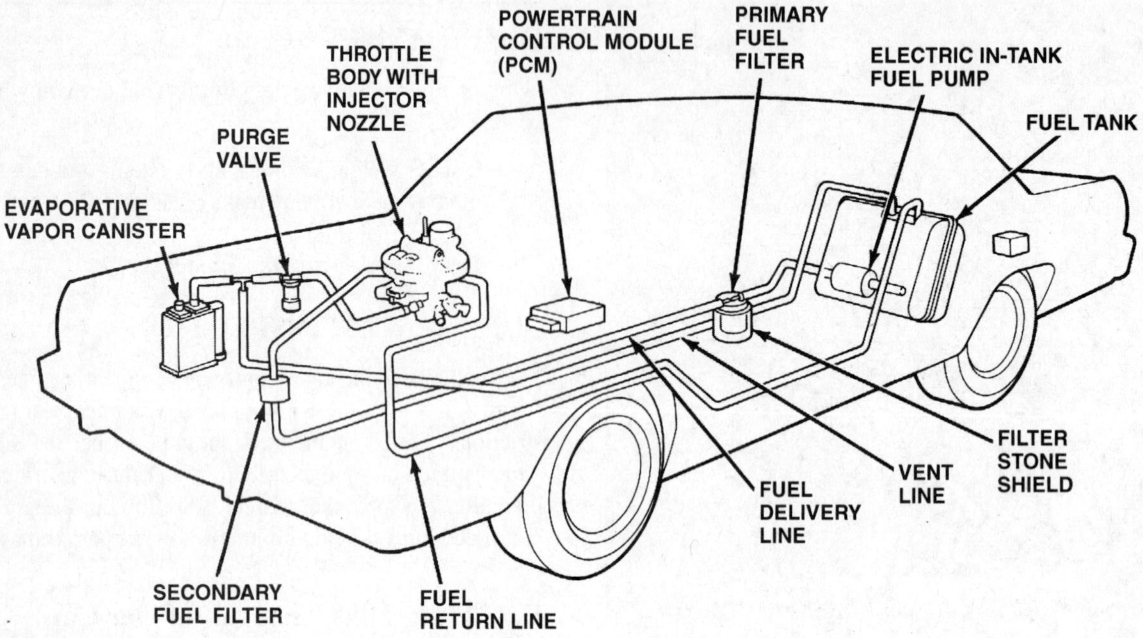

Figure 3-1. The fuel system of a throttle-body injected automobile. (Courtesy of Ford Motor Company)

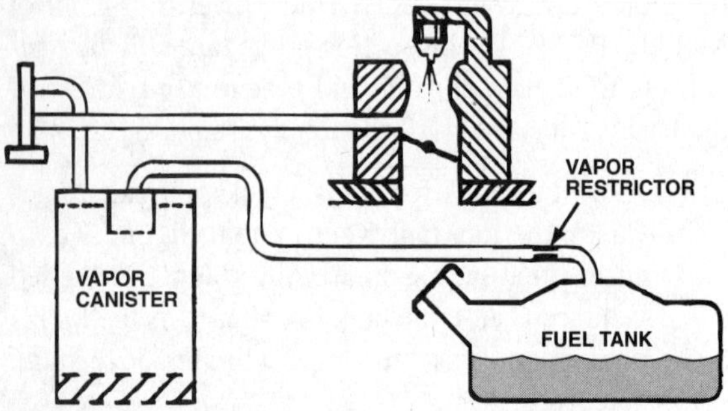

Figure 3-2. System using a thermal vacuum valve to control canister purge.

fuel tanks generally have no drain and the fuel will have to be siphoned out. If cleaning is required, remove the tank following the manufacturer's instructions. Do not steam clean tanks made of polyethylene plastic or expose them to extreme heat. If the tank connot be cleaned, it must be replaced.

Fuel Lines

1. Find the fuel supply and vapor return lines. Check for loose connections and pinched or crushed lines.
2. Inspect all flexible lines for weathering, oil damage, or kinks. Remove any soft hose and inspect its inside for deterioration.

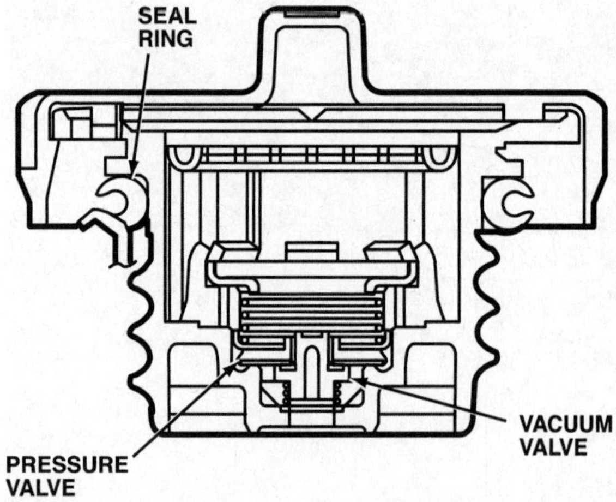

Figure 3-3. Check the seal ring on the fuel cap.

Mechanical Fuel Pump

1. Check the pump for oil leaks. If an oil leak is found, check the pump-mounting gasket for breaks.
2. Check the mounting bolts for looseness.
3. Check the diaphragm breather hole in the pump casting. It should be dry. If oil is present, the seal around the diaphragm is probably leaking. If gasoline is present, the diaphragm is broken and fuel is passing through it. If so, check the engine oil be-

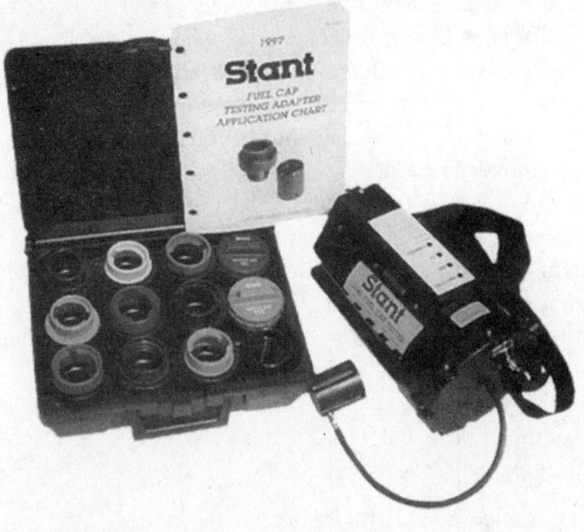

Figure 3-4. Filler cap tester.

cause gasoline may have entered the crankcase. If it has, change the oil and the oil filter.

4. Inspect all fuel line and hose connections at the pump for damage or loose connections. Repair as needed. Replace damaged hoses.

Electric Fuel Pump

1. If the pump is outside the tank, make sure it is securely mounted. Tighten the pump or bracket fasteners if they are loose. If the pump has an external shield, make sure it is in good condition and securely fastened.
2. Check all electrical connections and the condition of the wiring to the pump. Repair as required.
3. Inspect all fuel lines and hose connections at the pump for damage and loose connections. Repair as required. Replace damaged hoses.

Fuel Filter

1. Inspect the hose connection on inline filters for leaks. Ensure that the filter is installed in the correct direction, figure 3-5. If installed backwards, it restricts fuel flow.
2. If a carburetor inlet filter is used, disconnect the line at the carburetor and take off the filter, figure 3-6. Be sure it is not plugged. Inspect inlet gaskets and replace if damaged or leaking. Replace a broken filter spring.

Throttle-Body Injection (TBI) Unit

1. Check the fuel inlet fitting for leaks; tighten if loose. If leaks continue after tightening, install new inlet fitting gaskets or O-rings.

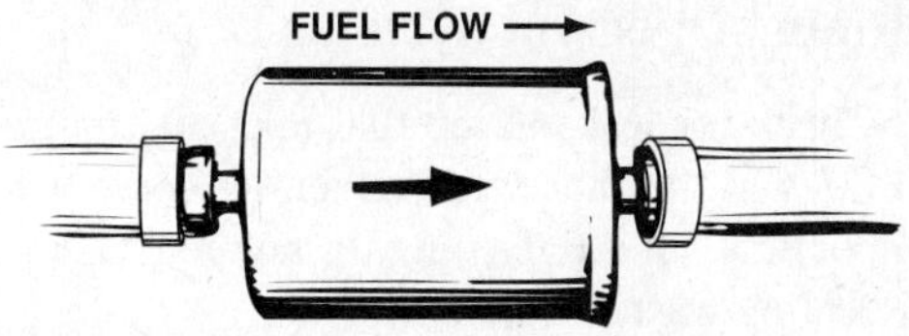

Figure 3-5. Inline fuel filters must be properly installed or they restrict fuel flow.

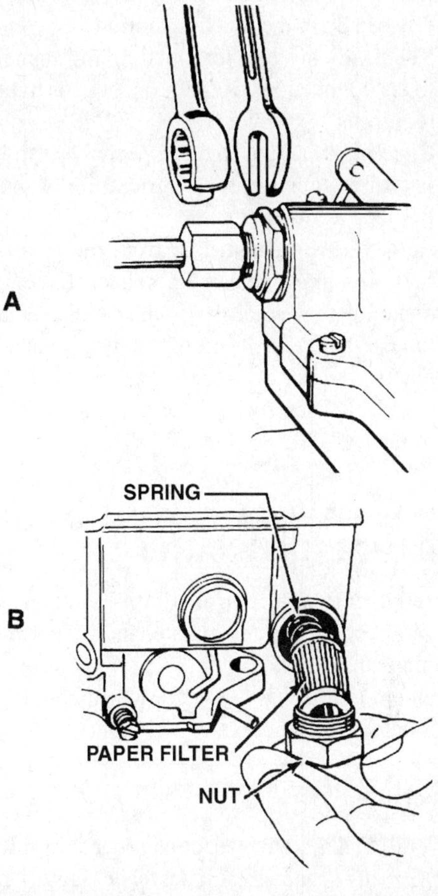

Figure 3-6. Disconnect the fuel line from the carburetor with two wrenches (A). Remove the fuel filter (B).

2. Inspect all vacuum lines for proper connections. Use a vacuum diagram to check line and hose routing, if necessary.
3. Check all vacuum lines for deterioration, oil damage, or hardening. Replace as required.
4. Check the TBI unit mounting fasteners for proper torque. This is a common location for vacuum leaks to develop. Tighten to specifications if loose. Do not overtighten, or you may warp or crack the base of the casting.
5. Check fuel return lines for the same conditions as the fuel delivery lines.

Multiport Injection Systems

1. Check for leaks at the fuel inlet and return fittings at the fuel rail. Tighten if loose. If leaks continue after tightening, install new O-rings or sealing washers in the fittings.
2. Check all fuel-injector electrical connections. Make sure they are properly routed and the connections are tight and sealed. Leaking injector seals create a vacuum leak, which results in a lean condition. Make sure the injectors are properly seated in the intake manifold.
3. Check all air cleaner ducting for damage or loose conncetions. Repair, replace, or tighten as required.
4. Check all vacuum lines, especially at the fuel-pressure regulator, for deterioration or damage. Replace as required.
5. Remove the vacuum line from the port on the fuel-pressure regulator and check for evidence of liquid fuel. Fuel in this line indicates a ruptured diaphragm in the pressure regulator. Replace regulator as needed.
6. Check fuel return lines for the same conditions as the fuel delivery lines.

EVAPORATIVE EMISSION CONTROL SYSTEM MAINTENANCE

Evaporative emission control (EVAP) systems generally should not require any maintenance in normal service other than periodic inspection, canister filter replacement (if there is one), or replacement of damaged or deteriorated hoses.

Inspection

The overall fuel system inspection procedure at the beginning of this chapter includes checking all EVAP hoses for damage or deterioration. Replace all defective hoses with the proper hose and tighten all connections. If necessary, install a new spring or worm-type clamp to hold the hose securely to its fitting.

If liquid gasoline is present in the canister, inspect and check the separator according to the manufacturer's specified procedure.

Purge Tests

The inspection of the system varies depending on the type of component used to control the canister purge.

If the system uses a thermal vacuum valve, test for vacuum to the canister when engine temperature is warm, and blocked when cold.

If a canister purge solenoid is present, test for vacuum to the canister when engine temperature is warm, and blocked when cold. This solenoid may be a normally open type that is energized when cold, or a normally closed type that is energized when warm. It may also be an on/off or pulse width modulated control.

With OBD II, the testing becomes more complicated because the system is more complicated, figure 3-7. The PCM is not only controlling the operation, but is continually testing and monitoring the system. If it is not responding correctly, it will set a diagnostic trouble code (DTC). Because of the complexity of the OBD II systems, it is recommended that the manufacturer's test procedures be followed.

Regardless of whether the system is controlled by a computer or by engine vacuum, a canister not properly purging causes:

- Incorrect air-fuel ratio
- An increase in exhaust emissions
- Loss of fuel economy
- Driveability complaints

Whenever these problems arise, test the EVAP purge valve using the manufacturer's specified procedure.

Canister Filter Replacement

Canister maintenance requires only that the oiled foam or fiberglass filter at its bottom, figure 3-8, be replaced periodically. Recommended replacement intervals range from 15,000 to 100,000 miles (24,000 to 160,000 kilometers) or 12 to 60 months depending on the model. Since outside air is drawn through this filter, replace it more frequently when the vehicle is driven under dusty or dirty conditions. To change a canister filter:

1. Label and disconnect all hoses from the canister.
2. Loosen the holddown clamps and remove the canister.
3. Pull the old filter from the bottom of the canister and discard it. Some filters may be held by a retaining bar or ring.
4. Install the new filter pad under the retainer.
5. Reinstall the canister, tighten the clamps, and reconnect the hoses to the fittings from which they were disconnected.

ELECTRIC FUEL PUMP TESTS

The pressure and volume tests for electric fuel pumps used with carbureted engines are identical to those for mechanical fuel pumps. If the pumps is tested with the engine not running, bypass the pump control wiring and supply battery voltage directly to the pump.

Tests for electric fuel pumps used with fuel-injection systems are more involved. All late-model vehicles with these pumps have computerized engine control systems, and fuel pump testing normally oc-

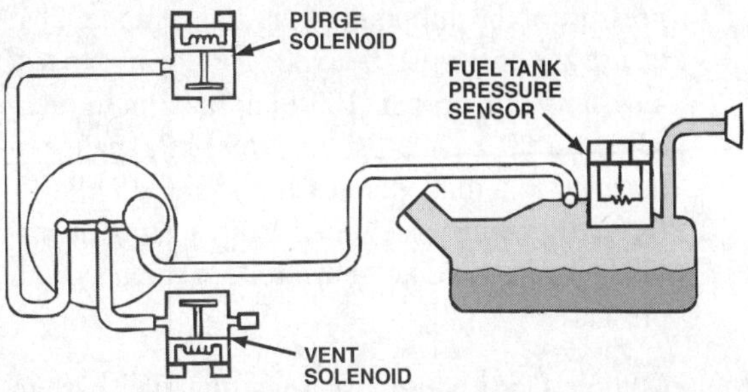

Figure 3-7. OBD II enhanced EVAP system.

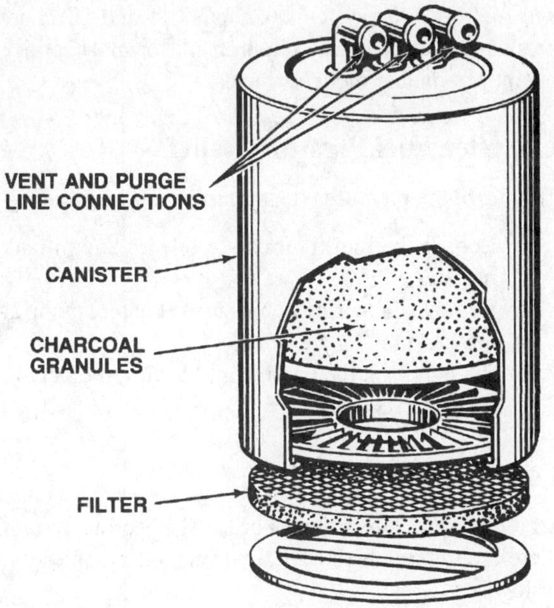

Figure 3-8. Some carbon canisters have replaceable filters.

curs only in the large context of diagnosing a no-start or engine driveability complaint. Consult the shop manual for complete instructions on testing a particular pump. The following section gives general test information.

Injection System Fuel Pressure and Volume

The impeller-type electric fuel pumps used with fuel-injection systems can produce fuel pressures and delivery volumes far higher than required for engine operation. For this reason, the pressure and volume tests for these pumps measure values other than actual pump output. The pressure reading is the system-regulated pressure as supplied to the injectors. The volume measure, when this test is specified, is taken at the fuel return line rather than the delivery line.

The fuel pressure gauge can be connected to a Schrader valve type port if equipped. This port is located on the fuel rail, fuel line, or throttle body of some Ford TBI systems.

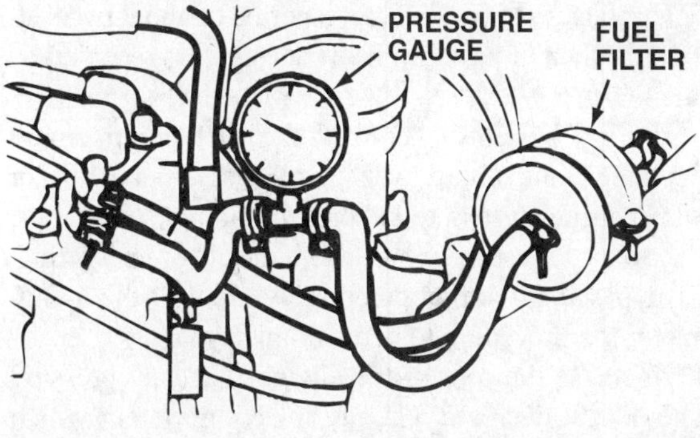

Figure 3-9. Checking fuel pressure with an inline adapter.

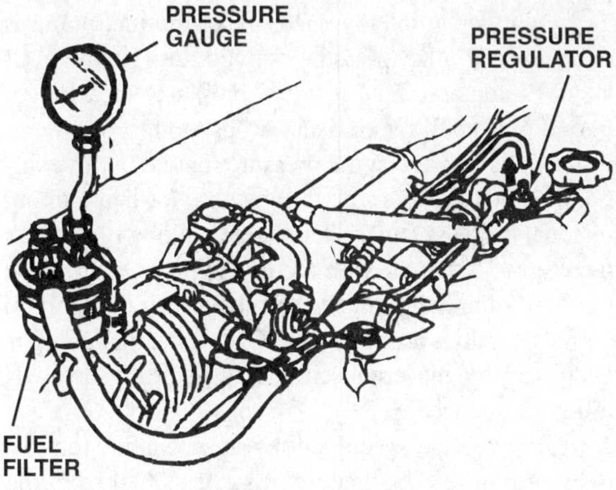

Figure 3-10. Remove the inner bolt from the outlet line on top of the fuel filter. Thread an adapter in its place and connect the pressure gauge.

Pressure must be relieved before the gauge is connected if the gauge is placed in-line. Some systems do not provide a service port. Adapters are available to be placed in the fuel line, which have a service port provided, figure 3-9.

Honda provides a 6-mm bolt in the center of the bolt connecting the outlet fuel line to the fuel filter. After the pressure is relieved, this bolt can be removed and an adapter installed to connect the gauge, figure 3-10.

After the fuel-pressure gauge is connected, energize the fuel pump, or if possible simply start the engine. Allow the engine speed to stabilize. Compare the gauge readings to factory specifications. The specifications may be for pressure with or without the vacuum line connected to the fuel-pressure regulator. Confirm the test requirements.

If the engine will not start, the fuel pump can be energized following manufacturer's specifications. Most late-model systems energize the fuel pump when the ignition is first turned on, for a period of about two seconds. It then turns off until an rpm signal is received by the PCM.

Earlier systems energized the fuel pump circuit with the crank circuit by using oil pressure, airflow, or the charging system to indicate a running engine.

Both of these systems have a means of energizing, or bypassing, the fuel pump relay. Consult the manufacturer's procedure to perform this bypass.

Increase the rpm of the engine. The fuel pressure should not decrease. On any injection system which has a vacuum line connected to a fuel-pressure regulator, the pressure should rise by 5 to 10 psi as manifold vacuum drops. If the fuel pressure does not rise or the pressure drops, perform a fuel volume test.

When the engine is turned off, or the fuel pump is de-energized, the pressure should drop slowly. It should maintain some residual pressure for a longer period of time. If the pressure drops too fast or too far, it indicates leakage at the pressure regulator valve, in-pump check valve, or at an injector. Not maintaining residual pressure can cause longer cranking time when the engine is started with a cold engine. This may be masked if the fuel pump is energized with the ignition on for the initial prime by the PCM. If the fuel pump is energized by the crank circuit, the starting delay is more pronounced.

If fuel volume specifications are available, remove the return hose. This line returns excess fuel from the engine to the tank. Place an approved, graduated fuel receptacle to catch the fuel and energize the fuel pump. If specifications are not available, or the return line is not accessible, connect a fuel-pressure gauge to test for pressure. Operate the vehicle under the conditions when the symptoms occur. If the fuel pressure drops when the symptoms occur, a volume problem exists. Check the fuel supply system, including filters, for restrictions. Confirm the fuel pump electrical system is staying energized. If there are no restrictions or electrical problems, replace the pump.

Fuel Pressure Relief

Low-pressure, throttle-body injection systems, such as those used on many GM, Ford, and 1986 and later Chrysler vehicles, operate at 10 to 15 psi (69 to 103 kPa). Other injection systems operate at pressures of 50 psi (345 kPa) or more. All systems retain some pressure when the engine is off. These facts have two important meanings for service work:

1. Use a test gauge indicating the proper pressure range. As discussed in Chapter 2 of this manual, use a gauge indicating the maximum system pressure at the midpoint of the gauge range. That is, use a 0- to 30-psi (0- to 207-kPa) gauge to test a system with 15-psi (103-kPa) maximum pressure; use a 0- to 100-psi (0- to 690-kPa) gauge to test a system that operates at 50 psi (345 kPa).
2. Before opening a fuel line for any testing or service, the technician must relieve fuel system pressure.

Manufacturer's procedures for releasing fuel system pressure are summarized in the following sections. Use these procedures again when replacing fuel system parts (Chapter 4) and service electronic engine control and fuel-injection systems (Chapters 11 and 12). Fuel pressure relief is a basic requirement for most service operations on late-model vehicles.

Chrysler Fuel Pressure Relief

To relieve the pressure on Chrysler injection systems:

1. Loosen the fuel filler cap to release any in-tank pressure.
2. Remove the wiring connector to the fuel pump at the fuel tank.
3. Start and run the engine until it dies, then crank for three seconds.

Ford Fuel Pressure Relief

To relieve pressure from a Ford fuel-injection system there are two methods. The first method involves the following:

1. Loosen the fuel filler cap to release any in-tank pressure.
2. Attach Ford pressure gauge, Part No T80L-9974-A or equivalent, figure 3-11, to the Schrader valve on the throttle valve or fuel rail.
3. Place the fuel drain tube in a suitable container and press the pressure release button.

The second method involves tapping or disconnecting the inertia switch to disable the fuel pump and start the engine until it dies, then cranking it for three seconds. Remember to reset or reconnect the inertia switch when finished.

Gauges similar to the Ford part previously mentioned are available from the aftermarket and may be used on any injection system which has a Schrader valve in the fuel system.

GM Fuel Pressure Relief

To relieve the pressure from a GM fuel-injection system:

1. Loosen the fuel filler cap to release any in-tank pressure.

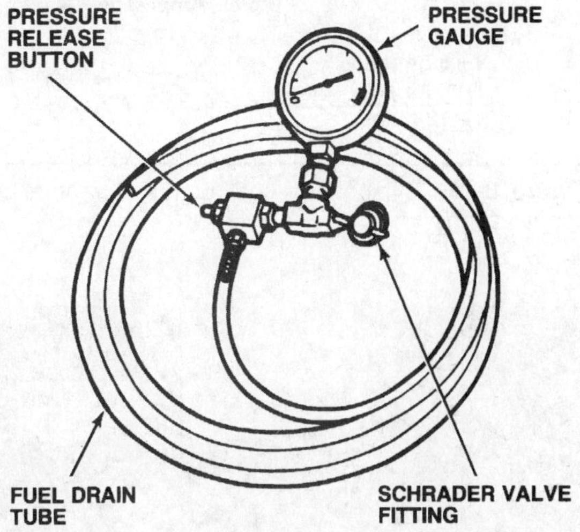

Figure 3-11. A combination pressure gauge and pressure relief valve for servicing Schrader valve-equipped fuel-injection systems.

2. Attach a pressure gauge with a pressure release, figure 3-11, to the Schrader valve on the fuel rail or line.
3. Place the fuel drain tube in a suitable container and press the pressure release button.
4. Remove the fuel pump fuse or disconnect the fuel pump circuit at the connector by the fuel tank.
5. Start and run the engine until it dies, then crank for three seconds.

Import Vehicle Fuel Pressure Relief

To relieve the pressure from most import fuel-injection systems:

1. Loosen the fuel filler cap to release any in-tank pressure.
2. Attach a pressure gauge with a pressure release, figure 3-11, to the Schrader valve on the fuel rail or line.
3. Place the fuel drain tube in a suitable container and press the pressure release button.
4. Remove the fuel pump relay.
5. Start and run the engine until it dies, then crank for three seconds.

Some imports complete the fuel pump energizing circuit through a set of contacts in the vane airflow meter. To do this:

1. Disconnect the VAF electrical connector.
2. Start and run the engine until it dies. Do not crank the engine as this pressurizes the system again.

Gauges similar to the Ford part previously mentioned are available from the aftermarket and may be used on any injection system that has a Schrader valve in the fuel system.

To release the pressure from a Ford low-pressure injection system without a Schrader valve:

1. Open the trunk and unplug the electrical connector at the inertia switch.
2. Crank the starter for about 15 seconds to dissipate any residual system pressure.

GM Fuel Pressure Relief

To relieve the pressure from injection systems used on GM vehicles:

1. Loosen the fuel filler cap to release any in-tank pressure.
2. Remove the fuel pump fuse from the vehicle's fuse panel. Start the engine and let it run until it dies.
3. Crank the starter for 3 seconds to ensure complete pressure relief.

General Test Procedures

To test system pressure on throttle-body injection systems, tee the pressure gauge into the throttle-body fuel delivery line, or attach the gauge to the Schrader valve supplied for that purpose. This applies to domestic vehicles. Many imports require disconnecting a fuel line to connect the gauge. Multiport, injection systems usually have a Schrader valve on the fuel distribution rail. Start the engine and make sure the pressure is within specifications at idle, and that it remains there as the engine speed is varied.

When fuel pump volume is specified, it should be measured at the return line from the throttle body, or the return line from the fuel pressure regulator on multiport injection systems. Disconnect the return line, attach a piece of hose to the return line fitting, and place the other end of the hose in a suitable container. Start the engine and confirm that the specified amount of fuel is delivered within the time allowed.

Electrical Circuit Tests

The first part of an electrical test, of course, is to get the wiring diagram. The technician must understand the circuit before performing any troubleshooting. After obtaining this information, test for voltage at the fuel pump under the conditions when it should have voltage. Keep in mind, fuses blow as a result of a malfunction. If a fuse blows, check the current draw of the pump.

If there is adequate voltage to the pump, check the ground circuit for excess voltage when there is voltage

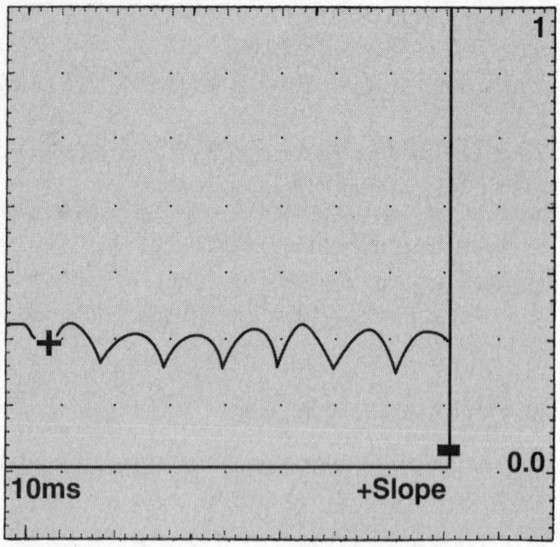

Figure 3-12. Fuel pump amperage. The humps are normal, caused by segments in the armature.

Typical Amperage Readings

TBI	**2 – 5 Amps**
PFI	**4 – 8 Amps**
CPI	**8 – 12 Amps**

Figure 3-13. Typical GM fuel pump current draw specifications.

Figure 3-14. Locate the fuel pump relay. Most late-model vehicles have relay centers that contain the relays.

on the power side. If there is less than 0.5 volt on the ground side, the problem is in the pump.

If there is low or no voltage to the pump, understand how the current travels, and test the circuit for possible open or resistance components.

Another approach to diagnosing a faulty fuel pump is to use an amp probe and lab scope. This allows the amount of current, and also the wave form of the current, to be measured, figure 3-12.

Testing a Fuel Pump Using Amperage Draw

One method of testing fuel pumps is by measuring current draw of the fuel pump using a DVOM or a DVOM with an inductive amperage probe. This method allows the technician to monitor current draw under operating conditions and even during a road test in many cases. Defective fuel pumps do not present problems until they have run for a period of time. By monitoring the fuel pump current draw a defective fuel pump can often be detected even when there are no driveability symptoms present at that time.

The difficulty of using this method is that most manufacturers do not list specifications for current draw. Therefore, the technician often has to rely on experience or by measuring the draw on a known good vehicle to compare with the subject vehicle. Some manufacturers may list some general specifications or current ranges for various fuel-injection systems, such as the ones shown in figure 3-13 for General Motors Systems.

If the amperage readings are too high, check for the following:

- Restricted fuel filter
- Other fuel line restrictions such as restricted return line or restricted inlet line
- If no restrictions are found, replace the fuel pump

If the amperage readings are too low, check for the following:

- High resistance connections
- Poor ground
- If there is not excessive resistance in the circuit then replace the fuel pump

Performing the Current Draw Test

The current draw test may be performed using different methods depending on the vehicle. In most cases the test can be performed at the fuel pump relay using a DVOM that can measure up to 10 amps. The following steps represent a typical procedure for measuring fuel pump draw.

1. Locate the fuel pump relay on the vehicle, figure 3-14.
2. Obtain the schematic for the fuel pump circuit or use the schematic shown on the side of the relay, figure 3-15.

Figure 3-15. For ISO-labeled relays terminal 86 is the power side of the coil, 85 is the ground side of the coil, 30 is common power for relay contacts, and 87 is the normally open contact.

Figure 3-16. Insert jumper wires in terminal 30 and terminal 87.

3. Remove the relay and install jumper wires in the relay terminals. For ISO-labeled relays these would be terminals 30 and 87, figure 3-16. If the relay is labeled differently be sure to consult a wiring diagram.
4. Install the DVOM in series between the jumper wires, making sure to select the amperage scale on the DVOM, figure 3-17.

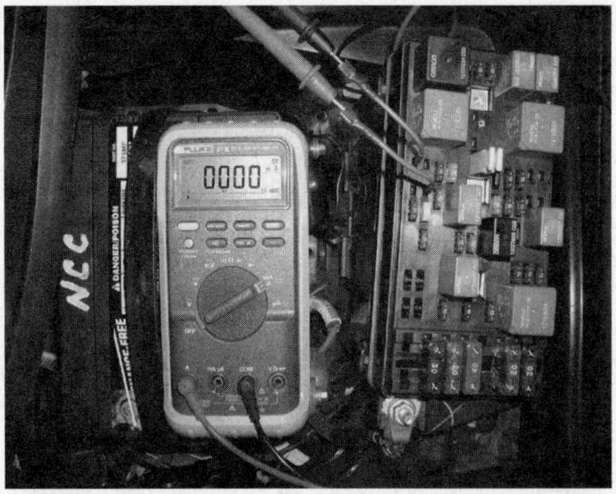

Figure 3-17. Install the DVOM in series between the jumper wires and set the DVOM to measure amperage.

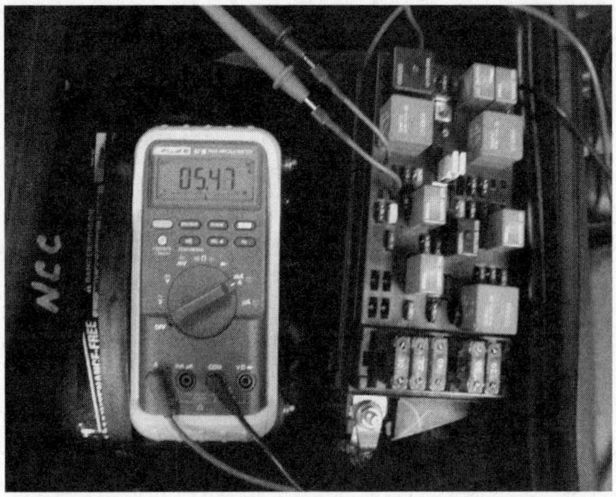

Figure 3-18. Turn the ignition on and read the amperage draw on the DVOM. Compare this to a known good vehicle or manufacturer's specifications.

5. Turn the ignition on. You should hear the fuel pump run and the DVOM should display the amperage draw of the fuel pump, figure 3-18. Allow the fuel pump to run for 15 seconds.
6. Compare these readings to those provided by the manufacturer or to a known good vehicle.

THERMOSTATICALLY CONTROLLED AIR CLEANER SERVICE

Some older Ford air cleaners use a thermostatic bulb and mechanical linkage, figure 3-19. This system is also used on many European imports and many late model domestic systems with TBI. All other air cleaners use a temperature sensor and vacuum motor, figure 3-20.

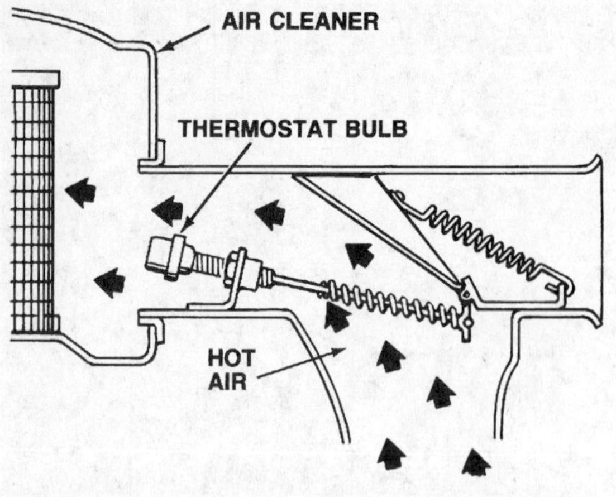

Figure 3-19. An air cleaner with a thermostatic bulb and mechanical linkage. Ford stopped using this system in 1977.

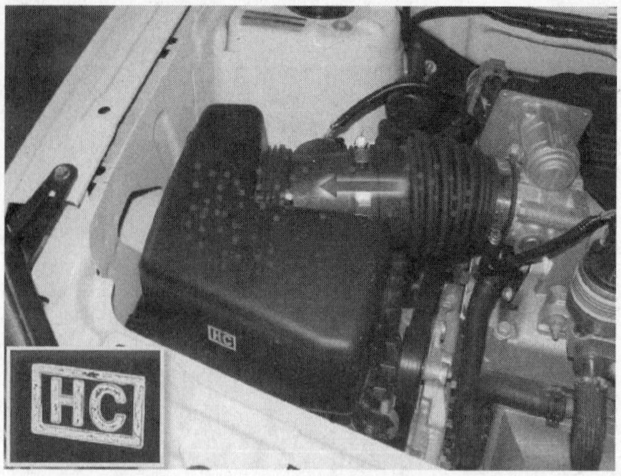

Figure 3-21. The hydrocarbon absorber is incorporated into the air cleaner housing and absorbs vapors from the throttle plates. (Courtesy of General Motors Corporation)

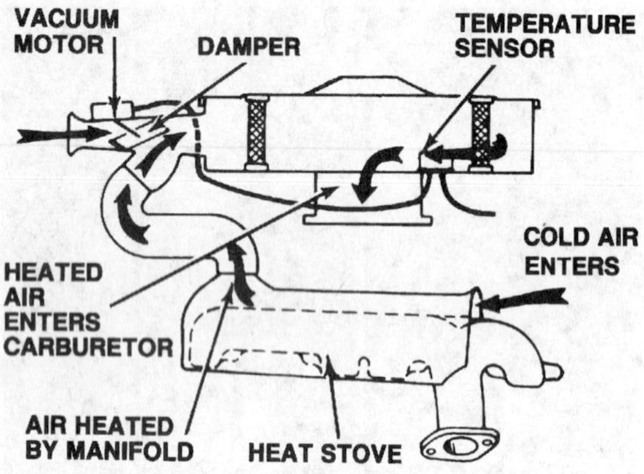

Figure 3-20. An air cleaner with a temperature sensor and vacuum motor.

HYDROCARBON ABSORBER

The hydrocarbon absorber is used on some ultra-low-emission vehicles (ULEV). This unit is designed to capture hydrocarbon emissions from the intake when the vehicle is shut off. This prevents these hydrocarbons from reaching the atmosphere.

Any hydrocarbon emissions that exist in the intake manifold are routed to the air cleaner housing, figure 3-21. These vapors are then captured by an activated charcoal layer that is part of the hydrocarbon absorber. The hydrocarbon absorber is part of the air filter housing. After the vehicle is started the hydrocarbon vapors are purged from the hydrocarbon absorber back into the engine.

The hydrocarbon absorber is nonserviceable. It is designed to last 150,000 miles or 15 years. To replace the unit the air filter cover must be replaced.

4

Fuel System Component Replacement

OBJECTIVES

Upon completion and review of this chapter, you will be able to:

- Remove and install a fuel tank following all safety requirements.
- Replace a fuel pump.
- Replace/repair fuel lines as necessary meeting manufacturer's specifications.
- Inspect and replace air cleaner elements.
- Clean and adjust (if necessary) throttle bodies.

FUEL SYSTEM SAFETY

Whenever a fuel system component must be replaced, safety is paramount. Gasoline vapor is highly explosive and can be ignited by a single spark. Observe the following safety practices at all times when working on or near any part of a fuel system:

1. Always work in a well-ventilated area.
2. Keep a fully charged fire extinguisher close at hand. Make sure it is the type that can be used on gasoline fires.
3. Never smoke, light matches, light a cigarette lighter, or use a cutting or welding torch around liquid gasoline or gasoline vapors.
4. Plug or cap all fuel lines. Use small plastic plugs or caps made especially for this purpose whenever possible. Rags stuffed into hoses or screws into lines are not adequate substitutes for a properly installed plug or cap.
5. Do not apply open flame or heat to, or near, the fuel tank, filler neck, or fuel lines.
6. Always disconnect the *negative* (ground) battery cable from the battery before starting any fuel system repairs.
7. Immediately wipe up any gasoline spills with old cloths or paper towels. Keep a pail of sand or absorbent compound nearby and lightly sprinkle over areas where spills were cleaned up. Dispose of the cloths or towels properly.
8. Use battery-operated lights where possible when looking for fuel leaks or working around open fuel lines. If using an electric lamp in a socket or drop cord, it should have a reflector guard to prevent accidental breakage.

Fuel-Injection System Pressure Relief

All fuel-injection systems used on late-model vehicles operate under high pressure. Pressure must be maintained in the injection system lines when the engine is off to ensure easy, quick restarts. This means residual pressure from 10 to 50 psi (70 to 345 kPa) is always present in the system. Opening any fuel-injection system connection without first relieving system pressure, as described in Chapter 3 of this *Shop Manual,* is a safety hazard. This is because gasoline under pressure spurts out all over the technician and any nearby components.

Chapter 3 contains manufacturer specific procedures for fuel pressure relief. Most methods are based on one of the following principles:

1. If the system has a Schrader valve on the fuel rail or throttle body, connect a hose with a valve depressor and release fuel into a suitable container.
2. Disable the electric fuel pump electrically and start the engine. Run the engine until it runs out of fuel and stalls. Then crank the engine for 3 to 5 seconds to ensure that all pressure is released.

Review the manufacturer procedures in Chapter 3 of this *Shop Manual* before relieving fuel pressure on any vehicle.

FUEL TANK REPLACEMENT

With increased use of in-tank electric fuel pumps, tank removal and reinstallation has become a common service job. Vehicle designs vary greatly between models and years, so the correct way to repair or replace the fuel tank also varies considerably. There are a few general steps covering *all* vehicles. These steps should be used if the manufacturer's instructions are not available.

In some cases, the sequence of steps may need to be changed. Whenever this procedure is used, think through the next step before actually doing it. This way, it is possible to determine whether or not the steps are in the correct order for the vehicle in question. Confirm that the tank must be removed. Some vehicles have an access panel over the top of the mounted fuel pump. Above all, work carefully and remember the safety tips already mentioned.

1. Remove the filler cap to release any pressure from the tank before disconnecting any fuel lines. If working on a fuel-injected vehicle, relieve the pressure in the system. Refer to the preceding section and to Chapter 3 of this *Shop Manual* for the procedure.
2. Disconnect the negative battery cable from the battery post.
3. Raise the vehicle on a hoist or on jackstands to gain enough room to work under it with ease. Always position jackstands properly; do not rely on a hydraulic jack or even a hoist to hold the vehicle safely.

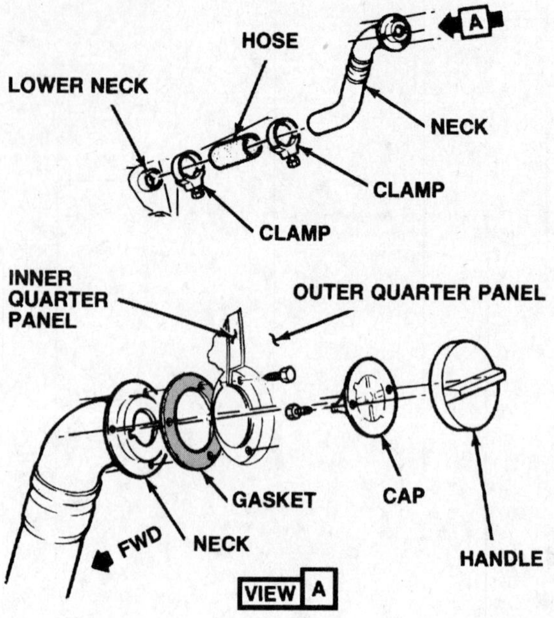

Figure 4-1. Some filler necks are connected to the tank by a hose. Once disconnected from the tank, the neck may not have to be removed.

4. Drain the fuel tank into a closed metal or plastic container. Some tanks have a drain plug for this purpose. Most tanks do not, however, and the fuel will have to be siphoned or pumped out. To do this, disconnect the fuel line at the inlet side of the fuel pump or at a convenient point on the chassis and connect a siphon or pump hose to it. Local ordinances may apply to draining the fuel tank; be sure to observe them.
5. Disconnect any electrical harnesses or wiring connectors from the fuel gauge sending unit or in-tank pump, if there is enough room to do so at this time. It may be necessary to postpone this step until the tank can be partially lowered from the underbody.
6. If the vapor separator is accessible without difficulty, disconnect the lines or remove the separator unit.
7. Inspect the filler tube connection to the fuel tank. Some are connected with a length of hose and clamps, figure 4-1. Others are soldered to the tank, or installed in the tank with a large rubber grommet or seal, figure 4-2. Disconnect the filler tube if possible.
8. Disconnect the upper end of the filler tube from the body panel or mounting bracket, figure 4-1. Remove the tube and gasket, if possible, by twisting them out of the rubber retainer or grommet. On some vehicles, it is not necessary to re-

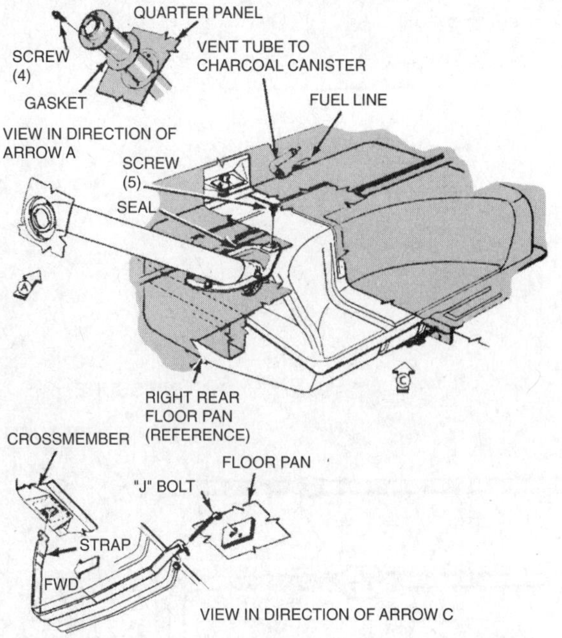

Figure 4-2. Filler necks may be installed with a rubber grommet or seal.

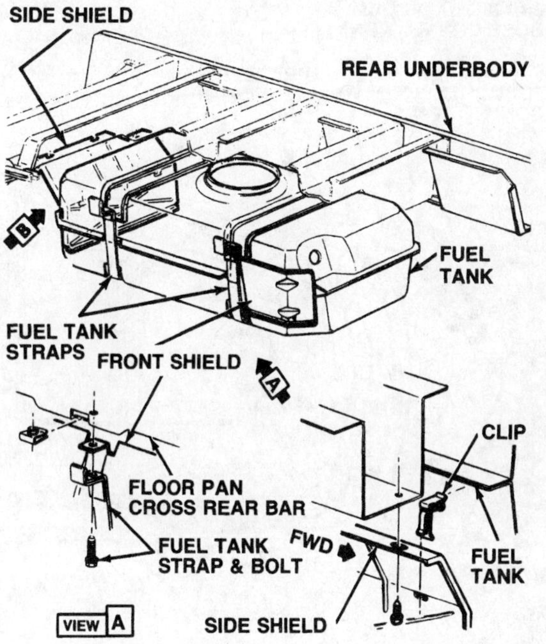

Figure 4-3. Some fuel tanks may use shields to prevent damage from rocks.

move the filler tube to replace the fuel tank once the two have been disconnected.

9. Unbolt and remove any shields possibly interfering with tank removal, figure 4-3. With some vehicles, particularly station wagons, it may be necessary to remove a rear wheel and brake drum, or unbolt a shock absorber or strut and let the rear axle drop slightly. This might be necessary to remove a side-mounted tank.
10. Put a transmission jack or other type of jack under the fuel tank to support its weight. Loosen and remove the tank retaining strap and fasteners, figures 4-2 and 4-3.
11. Lower the tank enough to disconnect any other wires, straps, or devices connected to it. Once the tank is loose, carefully lower it to the ground.
12. Use a spanner wrench to remove the retaining cam ring from the gauge sending unit, the in-tank pump assembly, or the vapor separator on the old tank, figure 4-4. Inspect all of these parts for leaks. When reinstalling the unit in the tank, be sure to use all new gaskets.

To install a new tank, first remove any usable parts from the old tank and install them on the new one, using new gaskets. It is also a good idea at this time to replace any rubber hoses attached to the tank. This avoids future problems requiring the removal of the tank a second time.

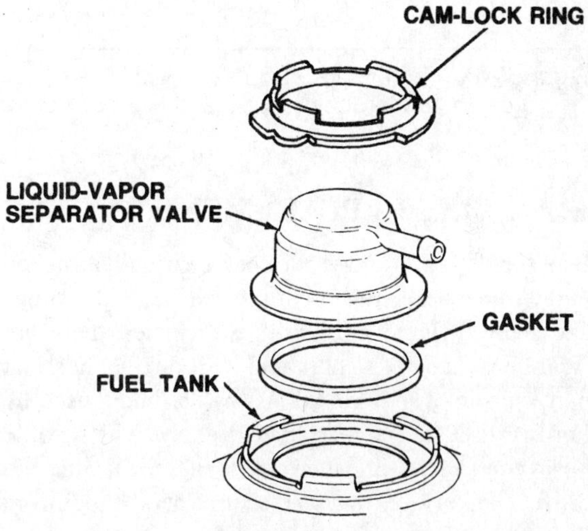

Figure 4-4. The Ford vapor separator is attached to the fuel tank. It has a cam ring and must be removed with a spanner wrench.

If necessary, attach the filler tube to the tank before reinstalling the tank in the vehicle. Using a jack, raise the tank into position under the vehicle. Install the tank straps and fasteners, drawing them up alternately while working the tank into its final position. Be sure the straps and fasteners are snug, but do not overtighten them. Reconnect the fuel lines and secure the filler tube to the vehicle body.

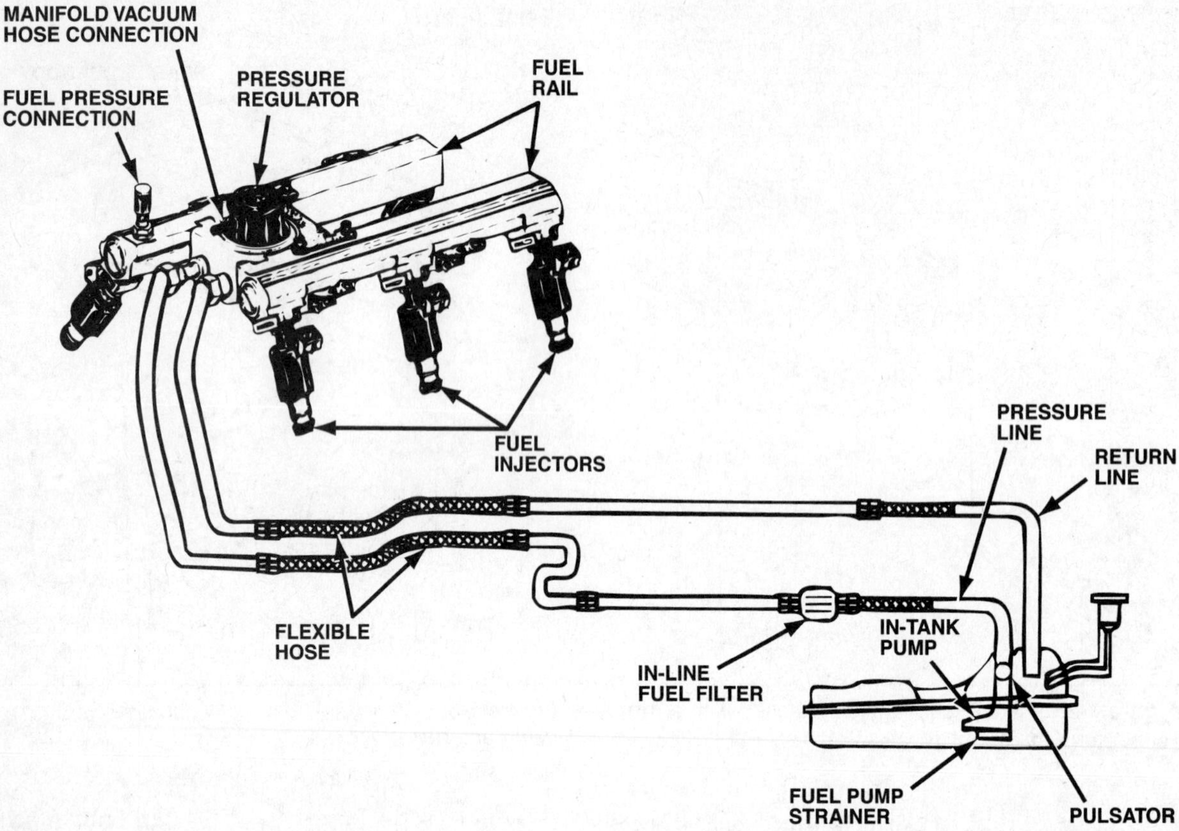

Figure 4-5. Flexible hose connects rigid lines to points of vibration. Clamps and screws hold these rigid lines to the frame and underbody. (Courtesy of General Motors Corporation)

FUEL LINE REPLACEMENT

Fuel supply return and vapor lines are usually fastened to the frame or the bottom of the vehicle, with clamps and screws, figure 4-5. On most vehicles, these are rigid lines made of steel tubing with flexible hoses at both ends to absorb vibrations (nylon tubing used by Ford is discussed in a separate section). The flexible hoses connect the rigid tubes to the engine fuel line and the fuel tank. Hoses may also be used at other points in the line where flexibility is needed.

Damaged or leaking sections of fuel or vapor lines must be replaced. This may require an entire new line or just the replacement of part of the old line. Fuel and vapor lines connecting the fuel pump to the throttle-body injection (TBI) unit are often replaced as preformed assemblies, figure 4-6. Any other repair must be made by cutting, squaring, and forming the replacement lines or sections from rolled tubing.

CAUTION: *Always use double-wrap brazed steel or other tubing made especially for use as a fuel line.*

Copper and aluminum tubing should not be used because they do not withstand normal vehicle vibration. When replacing hoses, only hose made specially for fuel systems should be used. The rubber used in vacuum or water hose does not resist chemical attack from gasoline. Most fuel-injected engines run fuel pressure over 40 psi. For this reason, do not use a hose designed for carburetor applications on a fuel-injection system.

Damaged hoses should be completely replaced with a new hose. Do not attempt to cut and splice hoses. Fuel line hose tends to wear out gradually. Each section installed in the fuel or vapor lines may eventually need replacement. Since the hose is flexible, a bad section usually can be replaced without removing the entire fuel line from the vehicle. Be sure to use the right type of hose.

Two types of brass fittings are commonly used when connecting two rigid fuel lines. The flared type, figure 4-7, may use either an inverted flare or an SAE 45-degree flare. A double flare should be used on the tubing to ensure a good seal and to prevent the fitting from cracking. Compression fittings, figure 4-8, use separate sleeves, tapered sleeves, or half-sleeve nuts. These are also good fuel line fittings.

Figure 4-6. The fuel pipe assembly is often available as a preformed replacement.

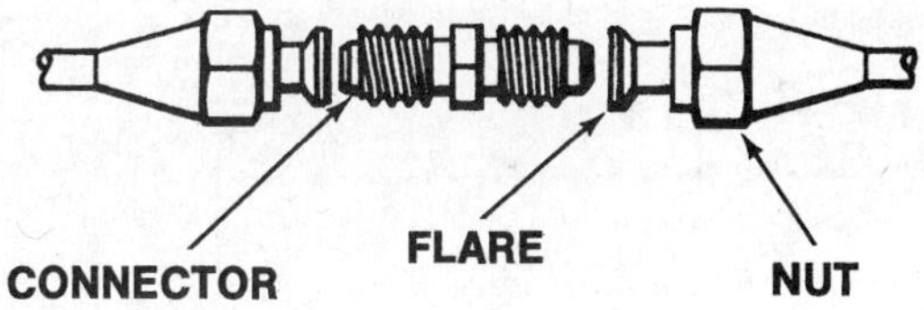

Figure 4-7. A flared fuel line fitting.

A third type of fuel line fitting, usually called a "banjo" fitting, is used on many European and Asian applications. The hose or line has a circular fitting, or collar, on the end. Washers are placed on either side of the fitting and a hollow bolt passes through the collar and into a threaded fixture in the other component.

In the past, it was acceptable to splice metal lines with rubber hose on low-pressure carburetor applications. On fuel-injection systems, damaged lines should be replaced with an original equipment design and quality part.

Repairing Nylon Fuel Tubing

Nylon fuel tubing is used on most late-model Ford vehicles. The tubing and special connectors are described in Chapter 4 of the *Classroom Manual.* Nylon fuel tubing and push-connect fittings cannot be repaired with hoses and clamps to replace a damaged or leaking section. Ford supplies approved service parts to safely correct any problem with nylon fuel lines. Replacement tubing and the required barbed connectors are available from Ford dealers. Specify the correct inner diameter of the tubing and obtain two connectors of the same size for each splice to be made. The following procedure is also used to replace damaged push-connectors. The tubing must be heated in boiling water; therefore, wear gloves when repairing nylon fuel lines.

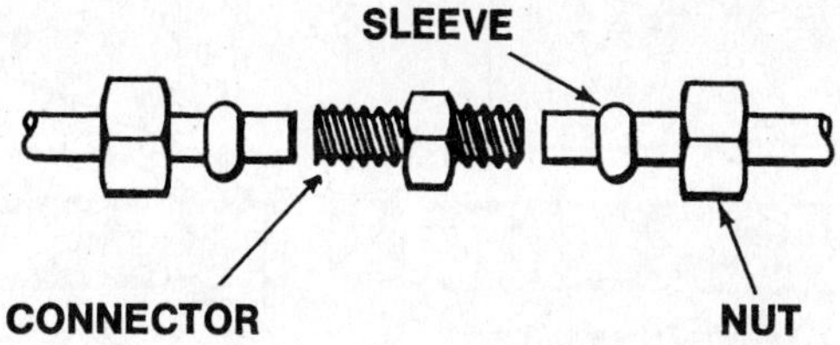

Figure 4-8. A compression fuel line fitting.

1. Relieve the system pressure as described in Chapter 3 of this *Shop Manual.*
2. Raise the vehicle on a hoist or jackstands to gain enough room to work under it with ease.
3. Remove the fuel line retaining clips on both sides of the damaged section of tubing to provide enough freedom to work easily with the line.
4. Cut out the damaged section of tubing. Use it as a guide and cut a length of replacement tubing to match.
5. Insert both ends of the replacement tubing into a container of boiling water for 1 to 2 minutes.
6. Remove the replacement section from the boiling water and immediately insert a barbed connector into each end of the tubing as far as it will go.
7. Insert one end of the original tubing to be spliced into the container of boiling water for 1 to 2 minutes.
8. Remove the tubing end from the water and immediately insert one barbed connector into one end of the tubing splice. Insert the barbed connector into the original tubing as far as it will go, figure 4-9.
9. Repeat steps 7 and 8 to complete the other end of the splice.
10. Reinstall any removed fuel line retaining clips.
11. Start the engine and check for leakage at the repaired area.

Electric Fuel Pumps

There are many variations in electric fuel pump placement and installation. For this reason, always refer to the service manual instructions before working on an electric fuel pump. The following paragraphs provide some general guidelines.

Electric fuel pumps are usually installed close to the tank because they push fuel better than they pull it. Be sure to relieve fuel-injection system pressure and then disconnect the negative battery cable before raising the vehicle on a hoist or with a jack. Always position jackstands properly; do not rely on a hydraulic jack or even a hoist to hold the vehicle safely.

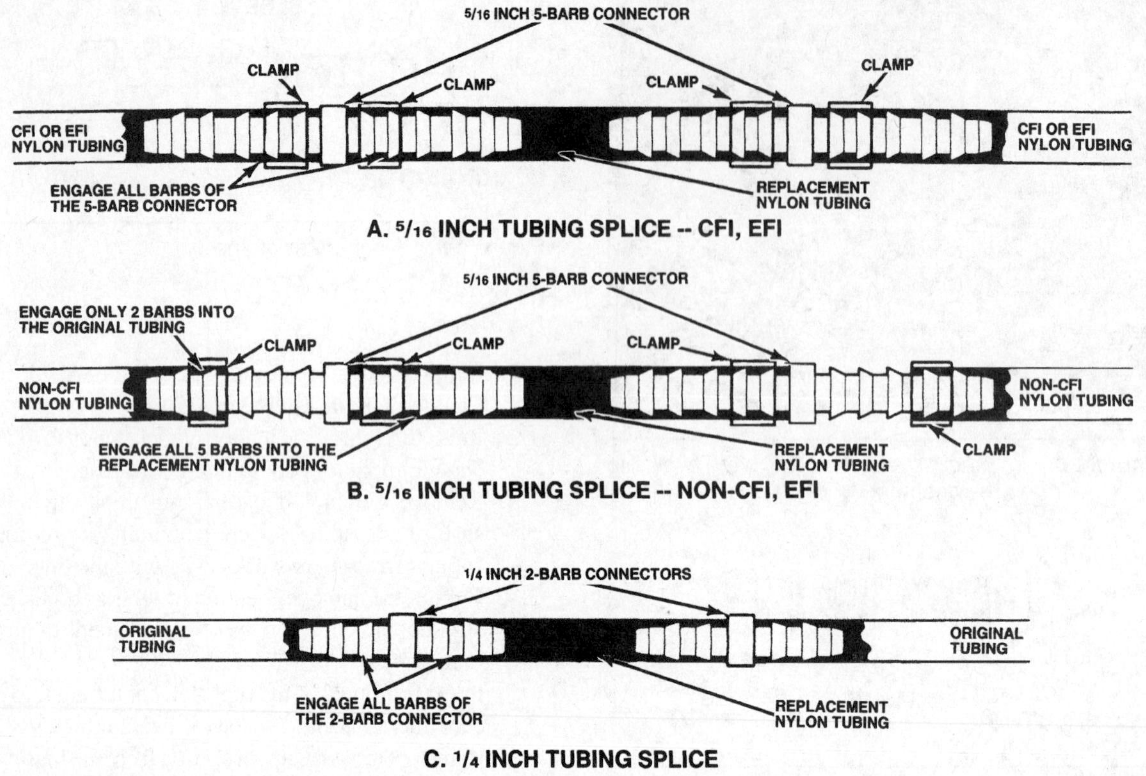

Figure 4-9. Splicing nylon tubing with service connectors. (Courtesy of Ford Motor Company)

Replacing In-Tank Electric Pumps

Access to an in-tank fuel pump usually requires removal of the fuel tank, as described earlier in this chapter. Once the tank is removed from the vehicle, most fuel pumps are removed by turning the retaining cam ring from the gauge-sending unit with a spanner wrench, figure 4-10. In a few instances, the pump may be a separate unit.

Before removing the retaining cam ring, clean the surrounding area to remove any accumulated dirt that might otherwise fall into the tank. Remove the pump assembly from the fuel tank. In most cases, the entire pump, filter, and gauge-sending unit is replaced with a new assembly. When the pump is installed as a separate unit, it may be possible to correct a fuel-supply problem simply by replacing the hose connecting the pump body to the fuel pressure fitting, figure 4-11. These hoses are a relatively common cause of inadequate fuel delivery.

Ford and some other manufacturers recommend the replacement of rubber hoses and clamps whenever the pump unit is removed from the tank. The fuel-soaked rubber hoses react chemically with air outside the tank and turn brittle when returned to the tank. This deterioration results in an inability to hold the required pressure and leads to premature failure. Always install new gaskets, seals, or O-rings and fuel pump filter, if applicable, when replacing an in-tank pump.

Replacing External Electric Pumps

External electric fuel pumps are mounted on bushings to reduce noise and vibration. They are usually located underneath the vehicle on a frame rail near the fuel tank, figure 4-12. Check bushings and replace as needed. Some vehicles have a protective cover over the pump housing that must be removed for access to the pump fasteners, figure 4-13. When installing a pump, always replace any seals, O-rings, or gaskets with new ones. After installation, run the engine and inspect for leaks.

FUEL FILTER REPLACEMENT

Most modern fuel-injected applications recommend fuel filter replacement at service intervals up to 100,000 miles. These recommendations always assume the vehicle is not operated in an adverse environment and is never serviced with contaminated fuel. Earlier applications recommended service intervals of 15,000 miles or 24,000 kilometers. A clogged filter de-

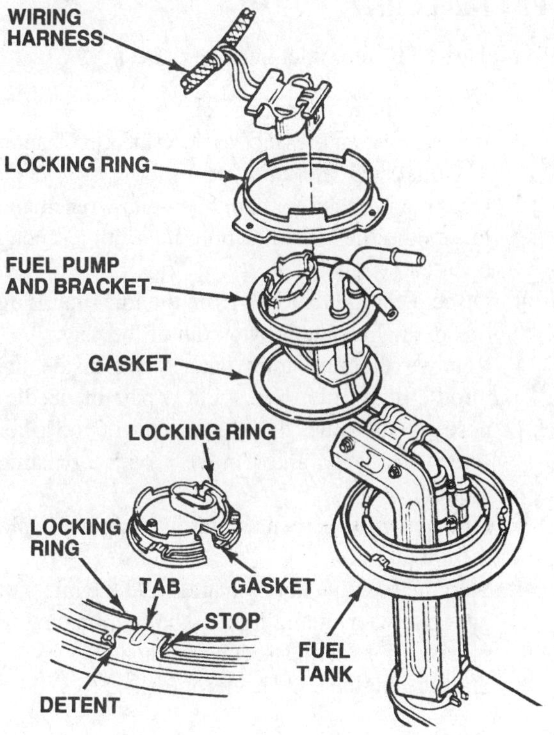

Figure 4-10. In-tank electric fuel pumps are held in the tank with a retaining cam or lock ring. (Courtesy of Ford Motor Company)

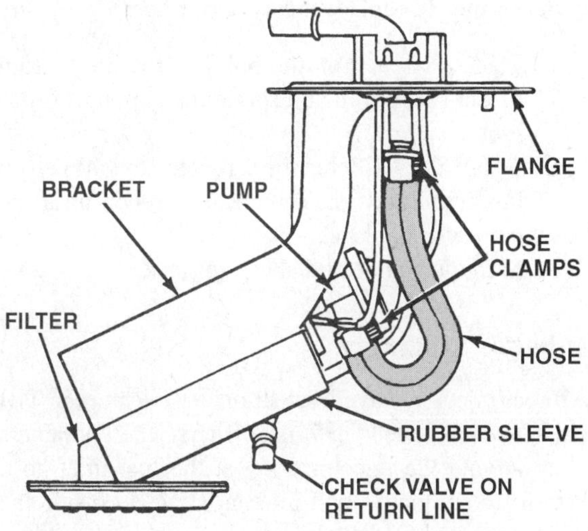

Figure 4-11. Electric in-tank fuel pump failure can often be traced to the hoses connecting the pump unit to the tank outlet. (Courtesy of Ford Motor Company)

livers adequate pressure, but with low volume. Fuel filters may be one of five types:

- In-tank
- Inline
- Carburetor or throttle-body inlet
- Fuel pump outlet
- Canister

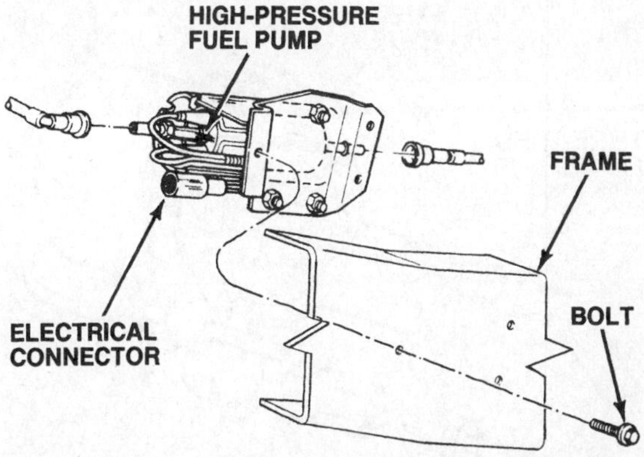

Figure 4-12. When an external electric fuel pump is used, it is generally located near the fuel tank on a frame rail under the vehicle.

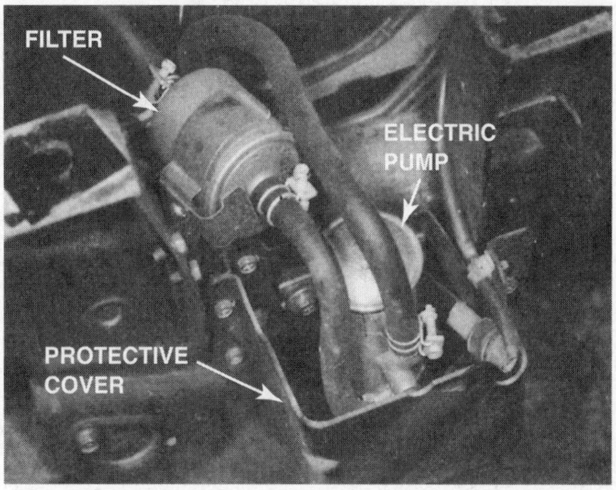

Figure 4-13. Before an electric fuel pump can be removed on some installations, a protective cover (arrow) over the pump housing must be removed.

Replacing In-Tank Filters

The in-tank filter is a frequent source of fuel delivery problems. To replace the filter, follow the procedure outlined earlier in this chapter for in-tank fuel pump replacement.

Replacing Inline Filters

The inline filters on carbureted models are generally located by the fuel tank under the vehicle or in the engine compartment. Fuel-injected models are located in the fuel line by the pump on externally mounted pumps, figure 4-13, on the frame, or in the engine compartment, figure 4-14. Some filters are difficult to access. Late-model Chryslers with the returnless fuel system mount the filter above the fuel tank.

To prevent leakage, always install new hose clamps or sealing rings along with the new filter. Also, replace

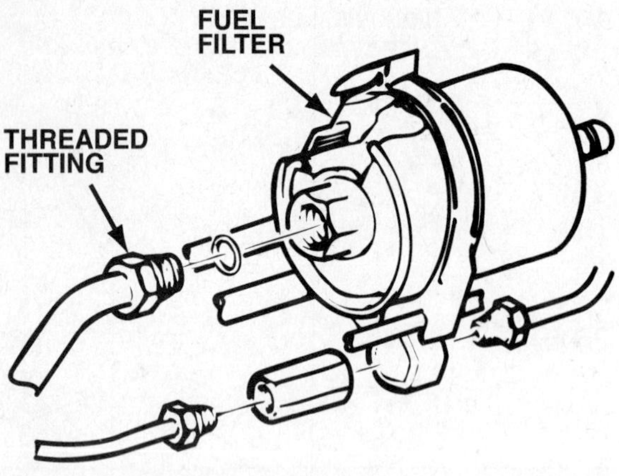

Figure 4-14. Removing an inline fuel filter installed between the fuel pump and the carburetor.

spring-type hose clamps with new ones when replacing the fuel filter. Do not push too hard on the filter case when installing the hoses because the thin metal or plastic case can buckle. If this happens, the filter element may become partially restricted, or the filter may leak internally.

To change a disposable inline fuel filter on fuel-injected engines:

1. Relieve system pressure as described in Chapter 3 of this *Shop Manual.*
2. If the filter is located on a frame rail, raise the vehicle on a hoist or jackstands to gain enough room to work under it with ease.
3. Place a drain pan under the filter, then disconnect the fuel line at both ends of the filter canister with a flare nut wrench. Some filters use threaded flex hoses and require two wrenches, one to hold the filter hex and the other to loosen the fuel line fitting nut.
4. Loosen the filter canister mounting bracket fasteners, if used. Remove the filter from the bracket.
5. Replace the O-rings in both fuel line fittings, if used.
6. Start each fuel line by hand to prevent crossthreading of the nut. Be sure the flow arrow points toward the engine.
7. Install the filter canister in its mounting bracket. Tighten the bracket fasteners, then securely tighten each fuel line nut with the proper flarenut wrench.
8. If the vehicle is raised on a hoist or jackstands, lower it to the ground.
9. Start the engine and check for leaks.

TBI Inlet Filter

To replace a TBI unit inlet nut filter screen:

1. Remove the air cleaner assembly
2. Relieve system pressure as described in Chapter 3 of this *Shop Manual.*
3. Hold the inlet nut with an open-end wrench and loosen the fuel line attaching nut with a second open-end wrench.
4. Disconnect the fuel line from the inlet nut. Plug or cap the line and move it out of the way.
5. Remove the fuel inlet nut. Reach inside the throttle-body inlet bore with a pair of needle-nose pliers, grasp the slotted end of the filter screen assembly, and remove it with a twisting motion.
6. Fit a new filter screen assembly completely into the inlet bore.
7. Start the fuel line nut by hand. Hold the inlet nut with one wrench and tighten the fuel line nut to specifications with a second wrench.
8. Start the engine and check for leaks.

Replacing Fuel Pump Outlet Filters

Some carbureted imported engines with electric fuel pumps use a filter mounted inside the fuel pump at the outlet fitting. To replace this type of filter:

1. Place a wrench on the bolt head at the bottom of the fuel pump and rotate the cap until it unlocks.
2. Install a new filter in the cap. Reinstall the cap on the pump and rotate with the wrench until the cap locks in place.
3. Start the engine and check for leaks.

Injector Strainers

All fuel injectors have inlet strainers. Injectors in TBI units have so-called “external” filters. These strainers wrap around the injector body at the fuel inlet port. Port injectors have small internal strainers inside the fuel inlet of the injector body.

The inlet strainers for most injectors cannot be removed and cleaned by themselves, nor are replacements available separately from the complete injector assembly. A clogged or damaged strainer usually means that the whole injector must be replaced. However, cleaning procedures for injectors often will cure a problem caused by a partially clogged strainer. Chapter 8 of this *Shop Manual* explains injector cleaning in more detail.

Figure 4-15. Paper air cleaner filters should be replaced at the intervals specified by the manufacturer, or sooner if the vehicle is used under dusty conditions.

Figure 4-16. Air cleaners may use one or more wing nuts or self-locking nuts to keep the cover in place.

AIR CLEANER SERVICE

Paper air cleaner filters, figure 4-15, should be inspected as part of an engine lubrication service. They should be changed at the interval recommended by the manufacturer, or as needed if clogged and dirty upon inspection. Dirty or clogged filters restrict the airflow to the cylinders. This may result in an overly rich air-fuel mixture and cause poor engine performance. A restricted air filter may also set fault codes on a computer-controlled engine.

Filter Replacement

Multiport fuel-injection (MFI) systems use a plastic air filter housing. In most cases, clips hold the housing together. Clean the exterior of the housing. Open the clips holding the air cleaner together. Remove the old air filter and wipe out the interior of the housing with a clean, damp cloth. Install the new filter. Fasten the housing clips.

To change a paper air filter element on older TBI or carbureted engines, remove the housing cover. This is held in place by one or more wing nuts or self-locking nuts. Unscrew the nuts, figure 4-16, and lift off the cover. In many newer applications, a plastic housing held in place by clips has replaced the metal air cleaner. Remove the old filter and wipe the inside of the housing with a clean, damp cloth. Remove all dirt without allowing any to enter the carburetor or TBI unit. Do not try to clean a paper air filter element; replace it with a new one.

Some air cleaner covers are marked with an arrow pointing to the front of the engine or to the air cleaner snorkel. Install the cover and nuts. Do not overtighten as this can warp a carburetor or throttle body.

To change the paper air filter on a remote air cleaner, figure 4-17, disconnect any ducting which can interfere with cover removal. Unsnap the cover clips or loosen the cover attaching screws. Separate the cover from the housing and remove the filter element. Wipe the inside of the housing with a clean damp cloth and install a new filter element. Reposition the cover and tighten the screws or snap the clips in place. Then reconnect any ducting removed and make sure that it is properly seated and sealed.

PCV Filter Service

Crankcase ventilation air in a closed PCV system is drawn through a hose from the air cleaner. When this air is taken from the dirty side of the air cleaner, a separate PCV or crankcase ventilation filter is installed in the air cleaner housing, figure 4-18. Remove and replace it with a new filter whenever the air cleaner filter is changed.

When the PCV inlet air is taken from the clean side of the air cleaner, a separate PCV filter is not used. A flame arrester may be installed at the air line inlet for safety. Clean or replace the flame arrester as specified by the manufacturer.

Air Cleaner Removal and Replacement

The air cleaner housing assembly must be taken off to service the carburetor or the TBI unit. Modern air cleaners have several vacuum lines, PCV hoses, and fresh air ducts attached. Disconnect these lines and

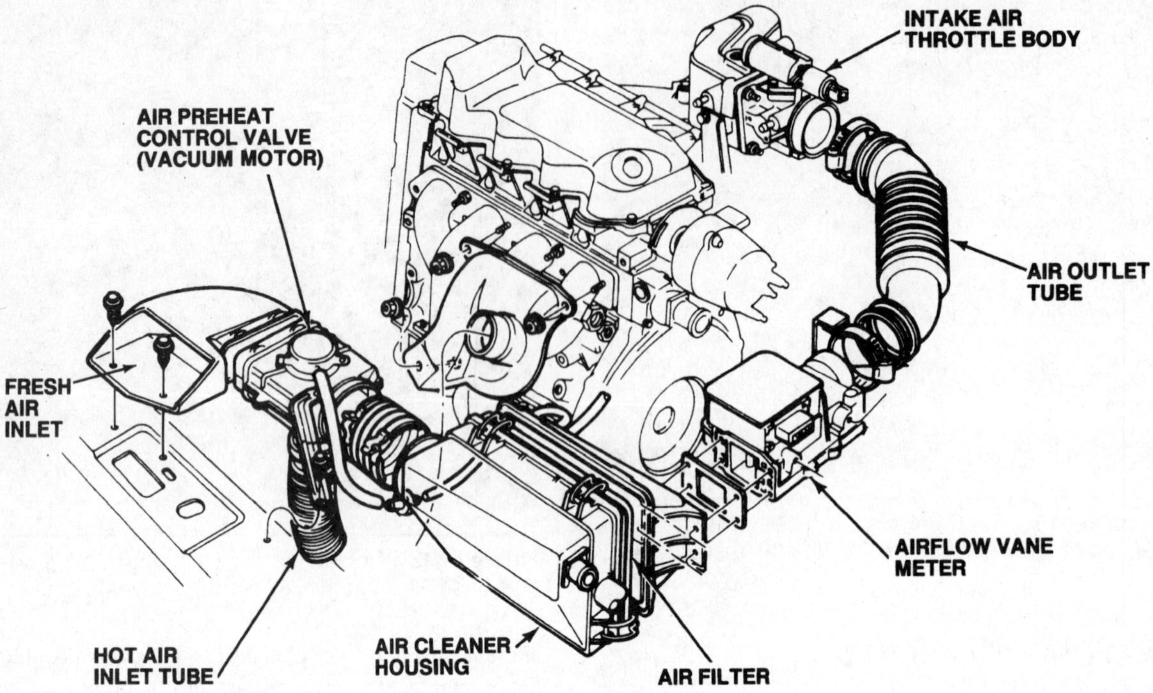

Figure 4-17. The air intake duct must be removed from a remote air cleaner before the cover can be unsnapped or removed. (Courtesy of Ford Motor Company)

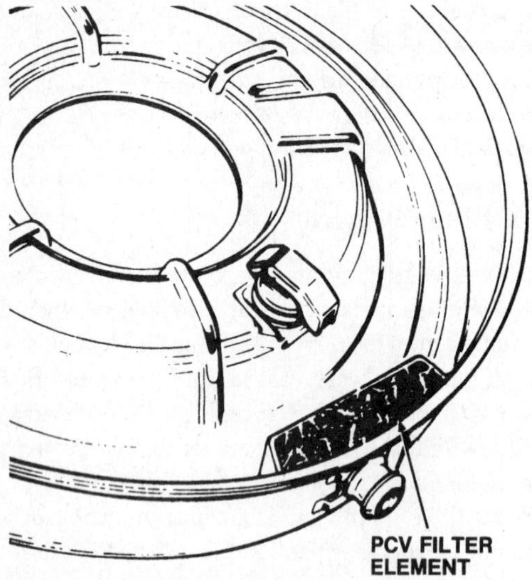

Figure 4-18. The PCV filter in an air cleaner housing.

Figure 4-19. Various hoses and connections must be disconnected before the air cleaner housing can be removed.

hoses at the air cleaner housing, figure 4-19. Mark them as needed for correct assembly.

Unclamp and disconnect the air inlet duct at the end of the snorkel tube, if so equipped. Unscrew the wing nuts or self-locking nuts and lift the entire air cleaner housing from the carburetor or TBI unit. Some vehicles may have a support bracket between the air cleaner housing and some point on the valve cover. Some GM air cleaners on 4-cylinder engines have a bracket underneath the snorkel tube attached with a screw. Remove any bracket attaching nuts or screws as required before removing the air cleaner housing.

A gasket is used between the air cleaner and the carburetor airhorn or TBI unit to prevent air leaks.

This gasket may remain on the airhorn or TBI unit when the cleaner housing is removed. If not, check if it is attached to the bottom of the air cleaner housing. Replace the gasket if it is cracked or deteriorated. If necessary, install a new gasket when reinstalling the air cleaner housing.

Remote air cleaner housings such as those found on most multiport fuel-injection systems seldom require removal, unless the engine is to be removed, there is a component located under the remote housing requiring service, or the ductwork is damaged and needs replacing. In most cases, removal of this type of housing is simply a matter of disconnecting the attached ducting and vacuum lines (if any), then removing the mounting screws.

Figure 4-20. The fuel inlet and return lines must be disconnected before a TBI unit can be removed.

Carburetor and TBI Removal and Installation

The carburetor or TBI unit must be removed from the intake manifold for service. The following procedures include all necessary steps for carburetor or TBI unit removal and replacement. The exact sequence in which these steps are done may vary according to the carburetor or TBI unit.

TBI Removal

To remove a carburetor or TBI unit:

1. Disconnect the air cleaner hoses and vacuum lines. Remove the air cleaner assembly.
2. Disconnect the throttle linkage. Three different types of connections may be used: a ball and socket, a rod and clip, or a cable linkage. Remove the necessary clips or connectors and move the linkage to one side.
3. On TBI units, relieve fuel system pressure as described in Chapter 3 of this *Shop Manual.*
4. Disconnect the fuel inlet line from the TBI unit. Also, disconnect the fuel return line from a TBI unit, figure 4-20. Plug or cap the lines to prevent leakage.
5. Label and disconnect all vacuum lines attached to the carburetor or TBI unit. Wrap a piece of masking tape on each line and identify the connection before removing the line. Color-coded golf tees or clothespins also may be attached to the ends of the lines. This makes correct vacuum line replacement easier.
6. Unplug all electrical connectors attached to the carburetor or TBI unit and label as needed.
7. Remove the TBI unit attaching fasteners. Lift the assembly straight up and off the manifold.
8. If the TBI unit is stuck to the manifold flange gasket, tap the throttle flange from side to side. Use a small soft-faced hammer and tap gently to break the seal. *NEVER* pry on the TBI unit body with a screwdriver.
9. Cover the manifold flange openings with a clean shop cloth to prevent dirt or dust from entering the engine. Remove the flange gasket or rubber insulator.
10. Scrape the base of the TBI unit to remove any gasket residue. Clean with low-pressure, compressed air.

TBI Installation

To install a TBI unit:

1. Remove the cloth from the manifold flange openings. Install a new gasket or the rubber insulator on the flange.
2. Lower the TBI unit onto the flange and install the mounting fasteners hand-tight. Tighten the fasteners alternately and evenly to specified torque to prevent warping the TBI unit base.
3. Reconnect the throttle linkage and check its operation.
4. Reconnect all electrical connectors.
5. Attach the vacuum lines to their proper fitting. Use a vacuum diagram if necessary to make sure the lines are properly routed and connected.
6. Connect the fuel inlet and return lines.
7. Reinstall the air cleaner and connect all vacuum lines and hoses to the housing. Tighten the wing or self-locking nuts on the air cleaner studs.
8. Connect any required air cleaner ducts and be sure all brackets are secure.
9. Start and run, checking for fuel leaks.

AIR INTAKE THROTTLE-BODY SERVICE

Although manufacturers do not specify preventive maintenance for the intake air throttle body on MFI systems, throttle-body coking is a common problem. This takes the form of a thin sludge or varnish deposit accumulating on, or around, the throttle plate, in the throttle-body bore, or in the idle air bypass valve, figure 4-21. These deposits take the form of a thin oil film and are caused by pulsation of induction air combined with engine oil and combustion by-products entering the throttle body through the PCV system. Sludge deposits gradually reduce airflow and may cause hard cold starting, stalling at idle or during acceleration, a rough or rolling idle, or a stumble under light acceleration conditions.

Figure 4-21. As can be seen in this photo, some throttle bodies can become severely coked.

Air Intake Throttle Body Cleaning

To prevent driveability problems resulting from deposit buildup, the air intake throttle body should be cleaned at a minimum of 25,000 mile intervals. However, some throttle bodies use a special "sludge-tolerant" coating on those areas inside the unit most likely to accumulate deposit buildup. Such throttle-body designs are identified with a decal and should not be cleaned, since any damage to the coating affects the idle airflow calibration.

Some manufacturers suggest leaving the throttle body on the intake air plenum during cleaning, but a thorough cleaning generally requires throttle-body removal to assure both sides of the throttle plates are clean. Some late-model GM and Ford throttle bodies are an integral part of the air intake manifold. Since manifold removal does not guarantee any better access to the throttle plates, it is faster and easier to clean this type of throttle body on the engine, provided there is sufficient access for service once the air intake duct is disconnected from the throttle body.

A good-quality carburetor spray cleaner removes the bulk of the deposits, but make sure that it does not contain methylethylketone; this agent is stronger than required for general cleaning. Also avoid carburetor choke cleaners, as they cause deterioration of internal O-rings. To prevent damage to the throttle-position sensor (TPS) and idle air control (IAC) valve, either cover them with plastic bags or remove them from the throttle body. If these components are removed, many throttle bodies can be cleaned with a cold immersion-type cleaner, just like a carburetor. Before using an immersion cleaner, however, make sure the throttle body does not contain any internal rubber parts or throttle shaft sealed ball bearings. An immersion cleaner chemically reacts with rubber, causing it to swell, harden, or distort. It also washes away bearing grease or dilutes its lubrication properties. Manufacturers recommend specific cleaners that are safe to use.

A variety of cleaning tools, such as small bottle brushes, toothbrushes, cleaning pads, lint-free cloths, and cotton swabs, are used to remove the deposits once the cleaner has been applied and allowed to soak. When cleaning a throttle body, always wear protective gloves to prevent skin contact with the cleaner, and also wear eye protection.

Before cleaning a throttle body, perform a minimum idle speed or minimum air rate test following the manufacturer's procedure. If the test results are out of specification, the throttle body probably is dirty and needs cleaning. Do not make any adjustments at this time or they will be incorrect. Wait until the unit has been cleaned and reinstalled to make required adjustments.

Throttle-body cleaning procedures differ by manufacturer, but most include the following steps:

1. Start the engine and warm to normal operating temperature.
2. Disconnect the air intake duct and remove the throttle body, figure 4-22.
3. Spray sufficient cleaner into the throttle body to thoroughly wet both the bore and throttle plates. Do not spray any plastic or electrical parts. Let the cleaner soak for at least 15 minutes, figure 4-23.
4. Soak a brush or pad with cleaner; hold the throttle plates wide open and scrub the bore and plates with moderate pressure to remove the sludge buildup, figure 4-24. Pay particular attention to those areas where the throttle shaft contacts the bore wall, as well as the edges of the throttle plates.
5. If necessary, repeat steps 3 and 4 until all throttle-plate sludge has been removed.

Figure 4-22. Disconnect the air intake duct. The throttle body may have to be removed to facilitate cleaning.

Figure 4-23. Spray the appropriate cleaner on the throttle plates and bore and allow to soak.

6. Rinse the throttle body with another thorough spraying of cleaner and dry the unit with low-pressure compressed air, figure 4-25.
7. With the throttle wide open, let it snap shut. Lightly open the throttle to check for any sticking of the throttle plates, figure 4-26. If the throttle does not operate smoothly, repeat the procedure starting at step 3.

Figure 4-24. Use a brass brush to aid in removing the deposits.

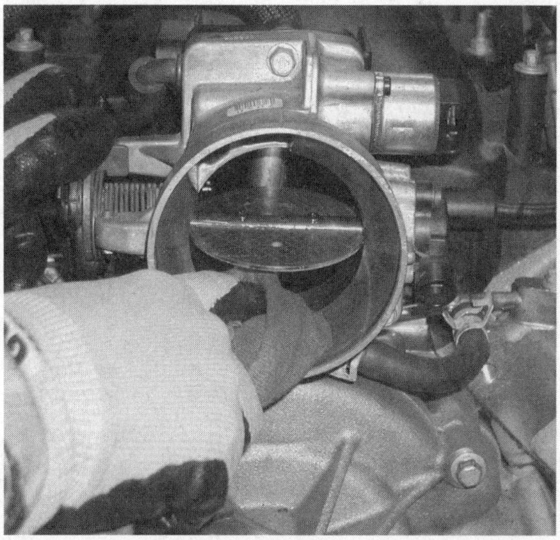

Figure 4-25. After scrubbing the deposits, use a clean dry rag to remove the cleaner.

8. Once the throttle operates smoothly, reinstall the throttle body and connect the intake air duct.
9. Following the manufacturer's procedure, check and adjust minimum idle rpm or airflow.

AIR INTAKE THROTTLE-BODY REMOVAL AND INSTALLATION

The air intake throttle body must be removed from the intake air plenum for service. The following procedures include all steps necessary for air intake throttle-

Figure 4-26. After cleaning the throttle plates and bore, operate the throttle by hand to check for any binding or sticking.

body removal and replacement. The exact sequence in which these steps are done may vary according to the injection system.

Air Intake Throttle-Body Removal

1. Drain the radiator below the level of the intake manifold. Disconnect any coolant hoses at the air intake throttle valve.
2. Disconnect or remove the following items:
   - Negative battery cable
   - Air inlet duct
   - All electrical connectors and vacuum lines at the throttle body
   - The accelerator linkage, throttle valve, and speed control cables from the accelerator bracket and throttle lever
3. Loosen and remove the bolts holding the air intake throttle body to the intake air plenum.
4. Remove the throttle body from the plenum and discard the flange gasket. Break the seal with a soft-faced hammer if necessary.

Air Intake Throttle-Body Installation

1. Carefully clean all gasket residues from the throttle body and intake air plenum mating surfaces. If scraping is required, use a dull tool to avoid damage to the aluminum sealing surfaces. Do not allow any gasket residue to enter the intake air plenum.
2. Fit a new gasket and position the throttle body on the air plenum. Install and tighten the throttle body mounting bolts to specifications.
3. Reinstall and reconnect all items listed in step 2 of the removal procedure.
4. Reconnect any coolant hoses disconnected from the throttle body. Refill the cooling system.
5. Start the engine and run at idle. Readjust the accelerator, throttle valve, and speed control cable, if required.

INTAKE MANIFOLD REMOVAL AND INSTALLATION

Removing an intake manifold is more complicated with V-type engines than with inline design. It is also more complicated with port fuel-injected engines than with carbureted or TBI engines. Basically, the manifold must be stripped of all wires, lines, and accessories before it can be removed. The following procedures include the steps necessary for intake manifold removal and replacement. The exact sequence in which the steps are done may vary from one engine to another.

Disconnect or remove the following:

- Negative battery cable
- Air cleaner assembly
- Carburetor or TBI unit fuel lines, vacuum hoses, and accelerator linkage
- All electrical connectors on the carburetor or TBI unit or intake manifold
- Crankcase vent hose at the valve cover
- Canister vapor hose
- Distributor vacuum lines
- Power brake line
- EGR valve vacuum line
- Exhaust pipe at the manifold flange
- Oxygen sensor
- Turbocharger and its inlet and outlet pipes and brackets, if equipped
- Any other lines, hoses, or electrical connections interfering with manifold removal

Removing the Intake Manifold—TBI or Carbureted Inline Engine

The carburetor or TBI unit can be removed before or after the manifold is removed from the engine. This procedure covers intake and exhaust manifold removal as an assembly. If the intake and exhaust manifolds are on opposite sides of the cylinder head, disconnecting or removing some of the items specified in the following may not be required.

To remove the intake manifold:

1. Disconnect the negative battery cable.
2. Disconnect or remove parts, lines, hoses, and wires that attach to the manifold.

3. Remove fuel lines, vacuum hoses, and accelerator linkage.
4. Remove all electrical connectors on the manifold and the carburetor or TBI unit.
5. Remove all vacuum hoses connected to the manifold. Mark their location.
6. Remove the turbocharger and its inlet and outlet pipes and brackets, if equipped. Also remove any other lines, hoses, electrical connections, or brackets that interfere with manifold removal.
7. Remove the manifold attaching fasteners, and clamps, if used. Remove the intake and exhaust manifold as one unit if they are bolted together.
8. Pull the manifold away from the engine block and discard the gaskets. Break the seal by gently tapping with a soft-faced hammer if necessary.
9. Inspect the manifold. When the intake and exhaust manifolds are removed as an assembly, bolts at the center of the assembly usually hold the two units together. Remove the bolts and separate the manifolds, if necessary.
10. Inspect or clean any EGR passages in the manifold.

Installing the Intake Manifold—Carbureted or TBI-Equipped Inline Engine

1. If a new manifold is to be installed, transfer all necessary parts and fittings from the old manifold to the new one, using new gaskets as required. Reassemble the intake and exhaust manifolds to each other, if both were removed. Clean gasket surfaces when reusing.
2. Scrape and clean the gasket surface on the cylinder head to remove all gasket residues. If the old manifold is to be reinstalled, scrape and clean its gasket surface also.
3. Install a new gasket on the manifold side of the cylinder head. Be sure that the holes align; then install the manifold assembly.
4. Clean, oil, and torque all fasteners to specifications. If the manufacturer does not provide a specific tightening sequence, start at the center of the manifold and work toward the ends.
5. Reinstall and reconnect all items previously removed.
6. Reinstall and reconnect the carburetor or the TBI unit.
7. Reconnect any cooling system hoses disconnected from the intake manifold. Refill the cooling system.
8. Start the engine and let it run at idle. Check for leaks. Readjust the idle speed to specifications after installing a new manifold.

Removing the Intake Manifold—TBI-Equipped or Carbureted V-Type Engine

The carburetor or TBI unit can be removed before or after the manifold is removed from the engine.

1. Disconnect the negative battery cable.
2. Drain the radiator and remove the air cleaner.
3. Disconnect or remove parts, lines, hoses, and wires attached to the manifold.
4. Remove any belt-driven accessories, brackets, and air conditioning hoses which interfere with manifold removal.
5. Remove the manifold attaching fasteners and lift the manifold from the engine. Use a hoist or have someone help with this step as a manifold is awkward to handle and heavy.
6. Remove and discard the intake manifold gasket and seals.
7. Inspect and clean any EGR passages in the manifold, figure 4-27.

Installing the Intake Manifold—Carbureted V-Type or TBI-Equipped Engine

1. Cover the intake ports with clean, damp shop cloths and clean all manifold, cylinder block, and cylinder head mating surfaces to remove all gasket and sealant residue.
2. When installing a new manifold, transfer all necessary parts and fittings from the old manifold to the new one, using new gaskets as required.
3. Install a new manifold gasket and end seals. Use sealant only as specified by the manufacturer.
4. Reinstall the manifold with a hoist or the help of an assistant. Some manifolds are attached with stretch-type bolts that must not be reused; use new

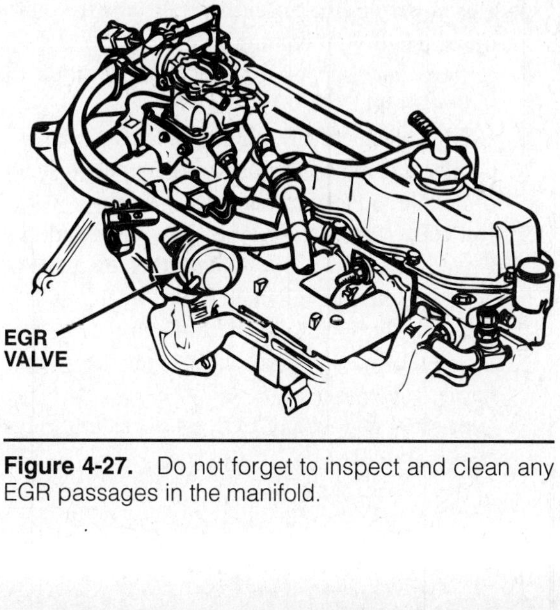

Figure 4-27. Do not forget to inspect and clean any EGR passages in the manifold.

bolts as required. Install and torque the fasteners to specifications following the manufacturer's tightening sequence. Tighten the bolts in several stages; do not apply full torque in one step.
5. Reinstall and reconnect all accessory equipment removed during teardown.
6. Reinstall the carburetor or TBI unit.
7. Refill the cooling system. Start the engine and let it run at idle. Check for leaks.
8. Torque the manifold bolts again when the engine is hot, if so specified by the manufacturer.
9. Readjust the idle speed and mixture to specifications as required when installing a new manifold.

Multiport Fuel-Injection Manifold Replacement

Manifold designs used with MFI systems vary considerably and removal procedures differ by design. There are some typical designs. The 2-piece manifold shown in figure 4-28 is typical of those used on inline engines.

Inline Manifold Removal

To remove a typical inline MFI manifold:

1. Disconnect negative battery cable.
2. Drain the engine coolant below the level of the manifold.
3. Unclamp and disconnect the air cleaner outlet tube at the air throttle body.
4. Disconnect the accelerator and speed control cables (if so equipped) from the accelerator bracket and the throttle lever.
5. Disconnect all PCV, EGR, and vacuum lines at the manifold fittings.
6. Unscrew and disconnect the EGR tube from the upper manifold if required.
7. Remove any support bracket bolts attached to the manifold.
8. Unplug all electrical connectors. If possible, disconnect the main injector wiring harness instead of the individual connectors.
9. Relieve system pressure as described in Chapter 3 of this *Shop Manual;* then disconnect and plug the fuel inlet and return lines.
10. Remove the manifold attaching fasteners. Remove the manifold assembly with the wiring harness and gasket.
11. Clean and inspect any EGR passages in the intake manifold.

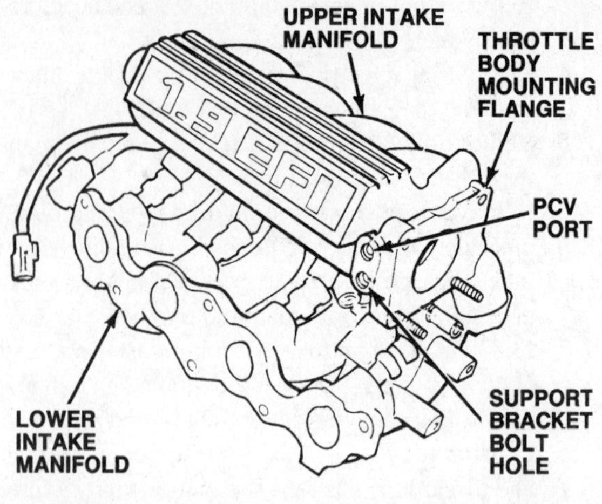

Figure 4-28. Typical 2-piece fuel-injection manifold for inline engines. (Courtesy of Ford Motor Company)

Inline Manifold Installation

To install an inline MFI intake manifold:

1. Clean all gasket residue from the mating surfaces with a wire brush, including the manifold attaching studs. Lubricate fasteners with clean engine oil.
2. Fit the manifold onto the cylinder head. Install the fasteners finger-tight.
3. Reconnect the fuel inlet and return lines. Install the EGR tube and tighten the compression nut finger-tight. Tighten the manifold and EGR tube fasteners to specifications following the sequence specified by the manufacturer.
4. Reconnect all electrical connectors and vacuum lines at the manifold or throttle body.
5. If a support bracket is used, reattach the bracket to the manifold.
6. Connect the accelerator and speed control cables to the accelerator bracket and the throttle lever.
7. Reinstall the air throttle body.
8. Connect the air cleaner outlet tube to the air throttle body.
9. Refill the cooling system.
10. Start the engine and run at idle until engine temperature stabilizes. Check for coolant leaks.

Ford MFI Manifold Removal

The 2-piece manifold assembly shown in figure 4-29 is typical of those used by Ford. To remove and install this type of manifold:

1. Drain the cooling system below the level of the manifold.

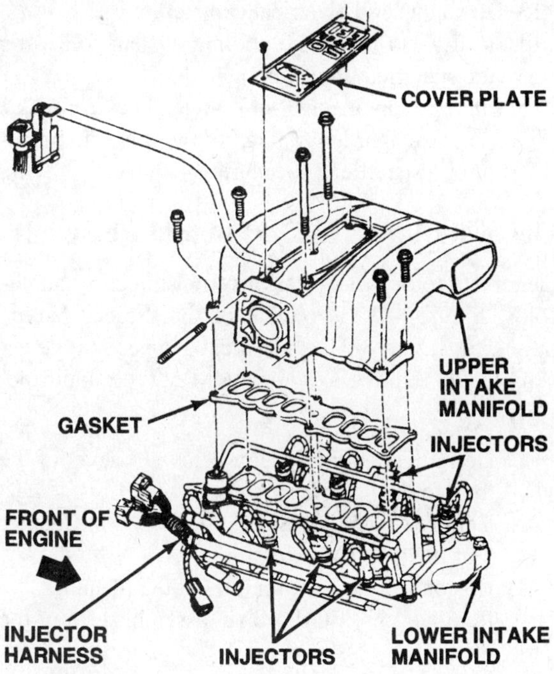

Figure 4-29. Typical 2-piece Ford fuel-injection manifold for V-type engines. (Courtesy of Ford Motor Company)

2. Relieve system pressure as described in Chapter 3 of this *Shop Manual.*
3. Unplug all electrical connectors at the air intake throttle body. Unclamp and disconnect the air intake duct from the air intake throttle body.
4. Label and disconnect all upper intake manifold vacuum lines at their connection points to the vacuum tree, the EGR valve, and the fuel pressure regulator.
5. Disconnect the canister purge fitting at the air intake throttle body.
6. Disconnect the PCV hose from the rear of the upper intake manifold.
7. Disconnect the accelerator and speed control cables (if so equipped) at the air intake throttle body. Cable linkage replacement is necessary when broken, or the cable is frayed or binding.
8. Disconnect the automatic transmission throttle valve (TV) cable, if there is one, and remove the accelerator cable bracket.
9. Disconnect all ignition cables at the spark plugs.
10. Disconnect the upper radiator hose at the water outlet housing. Disconnect all cooling system hoses from the manifold and throttle body.
11. Unplug all sensor wiring connectors.
12. Disconnect the injector wiring harness.

Figure 4-30. This is a 2-piece intake manifold. The upper intake plenum has been removed to gain access to the fuel injectors.

13. Remove the upper intake manifold retaining bolts. Remove the upper intake manifold and air intake throttle-body assembly.
14. When servicing fuel injectors, it is possible to remove the fuel rail and injectors at this point, without removing the lower manifold. To remove the injectors, disconnect the crossover fuel hose at the fuel rail and then remove the fuel rail attaching fasteners. Carefully pull the fuel rail, with the injectors attached, from the lower intake manifold.
15. To remove the lower intake manifold fasteners, remove and lift the manifold from the engine, figure 4-30. A hoist or an assistant should be used for this step.
16. Insert clean shop cloths into the intake ports and clean all gasket and sealant residue from the cylinder block, cylinder head, and lower intake manifold mating surfaces.

Ford MFI Manifold Installation

When installing a new lower intake manifold, transfer all necessary parts and fittings from the old manifold to the new one, using new gaskets as required.

1. Install new manifold gaskets and end seals. Use sealant only as specified by Ford.
2. Reinstall the lower intake manifold.
3. Install and torque the fasteners to specifications following the recommended tightening sequence.

Tighten the bolts in several stages; do not apply full torque in one step.

4. If the fuel rail and injectors were removed from the lower intake manifold, install a new O-ring on the top and bottom of each injector. Lightly lubricate the O-rings with fresh oil to ease assembly. Install each injector in its port with a light downward twisting motion; then position the fuel rail over the injectors and carefully install the assembly onto the injectors.
5. Reinstall the upper intake manifold and air intake throttle body on the lower intake manifold. Install and torque the bolts to specifications, following the recommended tightening sequence.
6. Reconnect the injector wiring harness and all sensor wiring connectors.
7. Connect all cooling system hoses to the manifold, the air intake throttle body, and the water outlet housing.
8. Install the distributor and connect the ignition cables to the spark plugs.
9. Install the accelerator cable bracket and connect the accelerator and speed control cables to the air intake throttle body.
10. Connect the automatic transmission TV cable, if used.
11. Reconnect the PCV hose and canister purge hose fittings.
12. Reconnect all vacuum lines.
13. Reconnect all electrical connectors.
14. Refill and bleed the cooling system following the specified procedure.
15. Start the engine and let it idle. Check for leaks. Set basic ignition timing following the manufacturer's specified procedure.

General Motors MFI Manifold Removal

General Motors uses various fuel-injection manifolds. Figure 4-30 is typical of the 2-piece V-type manifold; figure 4-31 is a typical 1-piece V-type assembly. To remove a 2-piece GM V-type manifold, figure 4-30:

1. Drain the cooling system below the level of the manifold.
2. Relieve system pressure as described in Chapter 3 of this *Shop Manual.*
3. Disconnect all vacuum lines at the plenum.
4. Remove the EGR valve assembly from the plenum.
5. Remove the accelerator and transmission TV cable bracket from the plenum.
6. Unbolt the air intake throttle body at the plenum.
7. Unbolt and remove the throttle cable bracket and the ignition wire plastic shields.
8. Unbolt and remove the plenum and gaskets.
9. Disconnect the fuel inlet and return lines at the fuel rail. Remove the fuel rail bracket bolt.

Figure 4-31. This is a 1-piece intake manifold. Many newer engines use plastic intakes instead of cast iron or aluminum.

10. Disconnect the pressure regulator vacuum line and unplug the electrical connectors at the injectors.
11. Remove any belt-driven accessory units that may interfere with manifold removal.
12. Unbolt the fuel rail and carefully remove it as an assembly from the lower intake manifold.
13. Remove the valve covers from the cylinder heads.
14. Disconnect all cooling system hoses.
15. Unplug all electrical connectors at the lower intake manifold.
16. Remove the lower intake manifold fasteners. With a hoist or assistant, remove the manifold assembly from the block.
17. Insert clean shop cloths into the intake ports and clean all manifold, cylinder block, and cylinder head mating surfaces to remove all gasket or sealant residue.
18. Clean and inspect any EGR passages in the manifold.

General Motors MFI Manifold Installation

When installing a new lower intake manifold, transfer all necessary parts and fittings from the old manifold to the new one, using new gaskets as required.

1. Install new lower intake manifold gaskets and end seals. Use sealant only as specified by the manufacturer.
2. Reinstall the manifold using a hoist or the help of an assistant.
3. Install and tighten the fasteners to specifications following the recommended sequence. Tighten the bolts in several stages; do not apply full torque in one step.
4. Install a new O-ring on each injector tip. Position the fuel rail and injector assembly over the lower intake manifold and carefully insert each injector into its port. Seat the fuel rail assembly and install attaching fasteners.
5. Reconnect all cooling system hoses and all electrical connectors to the lower intake manifold.
6. Install the valve covers to the cylinder heads.
7. Reinstall any belt-driven accessories removed.
8. Connect the injector electrical connectors. Reconnect the vacuum line to the pressure regulator.
9. Reconnect the fuel inlet and return lines at the fuel rail and install the fuel rail bracket bolt.
10. Install the plenum with new gaskets. Tighten fasteners to specifications following the recommended sequence.
11. Install the ignition wire plastic shields and throttle cable bracket.
12. Attach the throttle body to the plenum.
13. Install the accelerator and TV cable bracket to the plenum.
14. Reconnect all vacuum lines at the plenum.
15. Install the EGR valve assembly.
16. Refill and bleed the cooling system.
17. Start the engine and run at idle. Check for coolant leaks.
18. Check ignition timing and adjust as needed.

Removing and Replacing a Turbocharger Manifold

The turbocharger is usually mounted on the exhaust side of the engine and the impeller inlet is connected to the exhaust manifold by a cast elbow or by steel tubing. Air intake ducting is connected to the turbine inlet, and steel tubing connects the turbine outlet to the intake manifold or to the intercooler.

Turbocharger manifold installations vary greatly to fit the engine and chassis layout and to accommodate the fuel-injection system or carburetor. Even the early Buick carbureted installations varied considerably, depending on whether the engine was used in a front-wheel-drive or rear-wheel-drive chassis.

There is not just one general procedure for removing and replacing manifolds used with turbocharger installations. Before servicing a turbo, study the manufacturer's procedures.

Typically, the following items are disconnected or removed to service the turbocharger.

- Exhaust tubing or inlet elbow at the turbo inlet
- Turbo exhaust outlet
- Turbo wastegate, wastegate linkage, and vacuum lines
- Air intake ducts, carburetor or TBI plenum and air inlet tubing at the turbine inlet
- Turbine outlet tubing to the manifold or to the intercooler
- Intercooler tubing to the intake manifold

Other than the additional parts used in a turbo installation, manifold removal and replacement are similar to the same service tasks on a normally aspirated engine. Chapter 9 of this *Shop Manual* explains turbocharger service in more detail. Follow the procedures in that chapter when working on a turbocharger installation. Also, observe the following special precautions when servicing a turbocharger:

- Always clean the area around a turbocharger with a noncaustic solution before removing it.
- Work carefully when removing the turbocharger to prevent bending, nicking, or otherwise damaging the compressor wheel blades.

- Cover all turbocharger openings with duct tape to prevent dirt and contamination from entering the assembly while it is off the engine.
- When installing an inlet or outlet elbow, make sure that the turbocharger rotating assembly does not bind once the elbow is fitted.
- Use new gaskets and *faithfully* observe the recommended torque specifications.

THROTTLE LINKAGE

Many vehicles use a cable-type linkage that cannot be adjusted. One end of the cable passes through the cowl and connects to the accelerator pedal rod by a retainer, figure 4-32. The other end connects to the carburetor or throttle-body throttle lever. Brackets on the engine help to position the cable properly.

Cable linkage replacement is necessary when broken, or when the cable is frayed or binding. When necessary, remove the cable end from the accelerator pedal arm. Disconnect the cable at the carburetor or throttle body and remove it from the cowl. Insert the new cable through the cowl. Connect the carburetor end first; then depress the accelerator pedal and install that end.

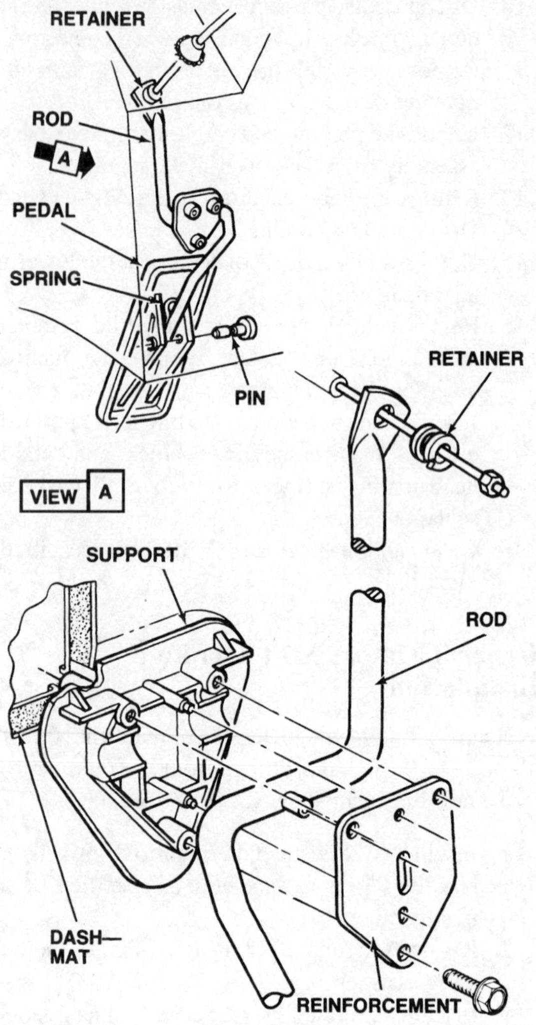

Figure 4-32. Typical cable linkage installation at accelerator pedal end.

5

Electronic Engine Control System Testing and Service

OBJECTIVES

Upon completion and review of this chapter, you will be able to:

- Retrieve codes on a non-OBD II equipped vehicle.
- Obtain codes on an OBD II equipped vehicle.
- Identify current and history codes.
- Identify generic versus manufacturer-specific codes.
- Diagnose and repair faults that cause stored trouble codes.

INTRODUCTION

Electronic engine control testing and service require a thorough understanding of the:

- Functional and operational relationships between control systems and their components
- Equipment and techniques used to test these components

While this chapter provides an overview of some functional and operational relationships, it primarily addresses the various types of tests, equipment, and procedures associated with electronic engine control diagnosis. Therefore, the *Fuel Systems and Emission Controls Classroom Manual* is highly recommended both as a prerequisite study and as a reference with respect to the functions and operations of systems and components.

TROUBLESHOOTING PRINCIPLES

Early engine control systems managed one or two functions, such as ignition timing and fuel metering. Most of today's systems are multifunctional and fully integrated. This means they monitor and manage a variety of engine functions, such as ignition, fuel metering, EGR, air injection, and evaporative emissions. Many electronic systems also integrate transmission, antilock brakes, traction control, and other electronic systems into a single processor. The operation, testing, and service principles of all engine control systems are basically the same, no matter how many vehicle functions the system manages.

It is important not to overlook basic troubleshooting procedures when diagnosing an engine control system. Adding electronic controls to an engine does NOT change the basic structure or operation of the engine. Always begin by performing a thorough visual inspection of all components, wiring, connectors, and fluid levels before checking the control system. Use Chapter 14 of the *Classroom Manual* as a reference regarding system functions while studying this chapter.

AREA TESTS AND PINPOINT TESTS

Chapter 1 discusses the method of testing from the general to the specific. This method is very important when it comes to diagnosing engine control systems. Always start by performing general system tests, or *area tests.* Area test results identify which *pinpoint tests* should be performed. Pinpoint tests isolate the root cause within a particular subsystem, circuit, or component. All manufacturers have organized their system troubleshooting procedures in this manner. This method of diagnosis makes it possible to break down the testing of any engine control system into three overall steps:

1. Perform an area test (overall system check) based on the driver's complaint.
2. Use the results of the area test to do one or more specific pinpoint tests as required.
3. Perform the necessary service based on the results of the pinpoint tests, followed by an operational check to verify that the system is working properly.

ELECTRONIC SYSTEM SERVICE PRECAUTIONS

In addition to the general shop and vehicle system safety precautions discussed in the safety summary at the front of this volume, servicing an electronic control system requires following the precautions listed below. Following these precautions helps prevent damage to vehicle systems and test equipment.

Electrical Connections

1. Check for and retrieve any stored trouble codes from the computer before disconnecting the battery, since that will erase the stored codes from the computer memory. All programmed electronic devices such as radios, clocks, and air conditioning controls will have to be reset once the battery is reconnected.
2. Make sure the ignition switch is OFF before disconnecting the battery. If the switch is ON, the resulting high-voltage surge may damage electronic components. For the same reason, do not disconnect or reconnect any electrical connectors with the ignition ON unless the manufacturer's procedure specifies doing so.
3. Make sure all connectors are clean and properly connected. If they are not, high resistance can affect the operation of electronic systems that function with low voltage and very low current levels. Paying attention to connectors and wiring may solve many problems.

Oxygen Sensors (O_2S)

1. Do not short across the terminals of a 2-wire or 3-wire O_2 sensor, or connect any test equipment to an O_2 sensor unless the equipment is approved for this use. A high-impedance graphing multimeter (GMM) or lab scope with an input impedance of at least 10 megaohms per volt can be used with most O_2 sensors. The current draw of a low-impedance meter will lower the signal voltage.
2. Avoid using rubber lubricants, belt dressings, or other sprays containing silicone near an O_2 sensor. The silicone compounds tend to collect on the ambient air side of the sensor. This causes an incorrect voltage signal that the computer generally interprets as a lean-mixture signal. As a result, the computer will overenrich the mixture.
3. Leaded gasoline has the opposite effect on O_2 sensors as silicone spray. The lead collects on the exhaust side of the sensor, causing an incorrect voltage signal that the computer will interpret as a rich-mixture signal. As a result, the computer excessively leans out the mixture.

Exhaust System

1. Do not modify the exhaust system of any vehicle with electronic engine controls. Removing the muffler or catalytic converter, or installing headers, changes the exhaust backpressure. This may affect the operation of the exhaust gas recirculation (EGR) system, which operates with a specific amount of backpressure.
2. Catalytic converters operate at very high temperatures and remain hot long after the engine is shut off. Working on a vehicle with a converter requires special precautions:
   a. Avoid contact with the converter to avoid serious personal injury.
   b. Correct misfiring conditions at once to avoid excessive unburned fuel from entering into the converter.
   c. Do not crank the engine for more than 15 seconds without starting it, because this will allow fuel to enter the converter.
   d. After performing a power balance test, run the engine for at least 30 seconds at fast idle to clear the converter.

Cooling System

Many engine control functions are activated by coolant temperature. Do NOT remove the thermostat or install one that does not meet the manufacturer's temperature specifications. If the engine does not warm up properly, the vehicle may not cycle out of open loop and may record trouble codes.

Electrostatic Discharge

1. Solid-state components are sensitive to electrostatic discharge (ESD). A special wrist strap that plugs into a grounded mat helps prevent electrical damage to ESD-sensitive components during service. Removing the wrist strap, or stepping off the grounded mat, breaks the continuity of the path of equalization for a possible static charge. Because a static charge starts to build at this time, you must remain grounded when using this equipment.
2. Electronic components are wrapped in staticfree packing materials. Do not unwrap these components prior to installation. To prevent ESD damage, connect the wrapping or package to a known good ground on the vehicle with an alligator clip lead before removing the component from its package.
3. Avoid wearing clothes made from synthetic materials. Such clothes are especially prone to the buildup of an electrostatic charge on the body, as well as inadvertent discharge of the buildup to an electronic component.
4. When not wearing a grounded wrist strap, always touch a known good ground before handling the component. This should be repeated often while handling the component, and especially after sliding across a seat, sitting from a standing position, or walking a distance.
5. Do not touch the electrical terminals of a solid state component unless instructed to do so by a specific diagnostic procedure.
6. Always use a high-impedance graphing multimeter (GMM) when performing electrical checks. Make sure to connect the GMM ground lead first.

SELF-DIAGNOSTIC SYSTEMS

All automotive manufacturers produce electronic engine control systems with self-diagnostic capabilities. The systems of GM, Ford, and Chrysler are used as examples throughout this chapter.

The computers used in late-model engine control systems are referred to as self-diagnostic because they are programmed to check their own operation, as well as the operation of each sensor or actuator circuit. The extent to which a system can do this depends upon the system design. Generally, the computer does the following:

- Recognizes when a particular signal is not being furnished
- Recognizes when a signal is improbable for given conditions, such as input from a barometric pressure sensor that indicates the vehicle is being driven at an altitude of 25,000 feet when it is at sea level
- Recognizes when a signal is out of limits for too long, such as too rich or too lean of an O_2S signal
- Tests sensor and actuator circuit continuity by sending a test voltage signal or monitoring a return voltage signal
- Turns on the malfunction indicator lamp (MIL) on the instrument panel to alert the driver when the system needs service.

DIAGNOSTIC TROUBLE CODES

Following basic preliminary visual inspections, the first step in troubleshooting an engine control system is retrieving diagnostic trouble codes (DTCs). Retrieving DTCs is the *area test* for the electronic control system.

When properly used, DTCs efficiently organize the diagnostic process. While DTCs are valuable diagnostic aids, they do NOT provide all the answers. DTCs provide clues to where the problem lies and indicate that a particular circuit or component should be checked. This means that DTCs are general indicators that narrow down the possibilities; they do not point directly to the precise cause of a problem.

Kinds of Trouble Codes

There are two major diagnostic trouble code categories, Hard Faults and Memory Faults. Accurate diagnosis depends on understanding the differences between these two types of DTCs. Note that the manner in which hard fault and memory fault codes are retrieved and output varies with the manufacturer. It is very important to reference the appropriate procedures when retrieving DTCs.

Hard Faults

A DTC that is set during an *on-demand* self-test of the computer with either the key on engine off or with the engine running is a fault that is present at the time of the test. A DTC categorized as a hard fault is typically addressed immediately by performing a pinpoint test.

Memory Faults

Faults that occur intermittently or only during certain vehicle operations are stored in the computer's *keep-alive memory* (KAM) for a set number of warm-up and/or drive cycles; they are retrieved in the same manner as hard fault codes. A memory code may be an indication that a subsystem or component is functioning erratically. Pinpoint tests are used to isolate memory code concerns, but only after all hard faults have been repaired. It is common to have memory and hard fault codes that are related. Often when a hard fault is repaired, the memory concern is repaired as well. It is important to remember that pinpoint tests are designed to resolve hard faults. Therefore, pinpoint tests do not always identify a concern when performed in conjunction with a memory code.

Manufacturers use various terms to describe their trouble codes, which can be confusing. For example, Ford calls memory DTCs *continuous codes.* Other manufacturers use the term *stored codes* to describe memory DTCs.

Diagnostic Data Link Connectors

All vehicles are equipped with a diagnostic data link connector (DLC), which is an interface between the vehicle's computer and the test device used to retrieve codes, figure 5-1. In compliance with the second generation of onboard diagnostic protocols known as OBD II, all vehicles have standard DLC configurations and locations (SAE standard J1962). The DLC on vehicles equipped with OBD II control systems is found in the passenger compartment beneath the instrument panel on the driver's side. In addition to a standardized DLC location and configuration, OBD II protocols also provide standards for:

- Diagnostic repair tools (SAE standard J1962)
- Diagnostic terms, abbreviations, and acronyms (SAE standard J1930)
- Diagnostic test modes (SAE standard J1979, J2190)
- Diagnostic trouble codes and definitions (SAE standard J2012)

DLC locations and configurations vary on vehicles produced before the incorporation of OBD II control systems. Some pre-OBD II vehicles have the DLC located in the engine compartment, while others locate the DLC in the passenger compartment. Be sure to refer to appropriate service resources for DLC locations and configurations when preparing to diagnose a pre-OBD II vehicle.

Retrieving Trouble Codes

As previously mentioned, the DLC provides an interface between the vehicle's computer and a code retrieval device. Each manufacturer has equipment that it produces and/or endorses as follows:

- GM: Tech 2 diagnostic scan tool, figure 5-2. The Tech 2 succeeds the Tech 1 scan tool, and is the latest diagnostic tool for GM vehicles. The Tech 2 is designed specifically for newer, faster OBD II control systems using Class 2 data.
- Ford: New Generation STAR (NGS) and the Service Bay Diagnostic System (SBDS).
  - The NGS is a hand-held scanner similar to GM's Tech 2. As with the Tech 2, it retrieves DTCs, per-

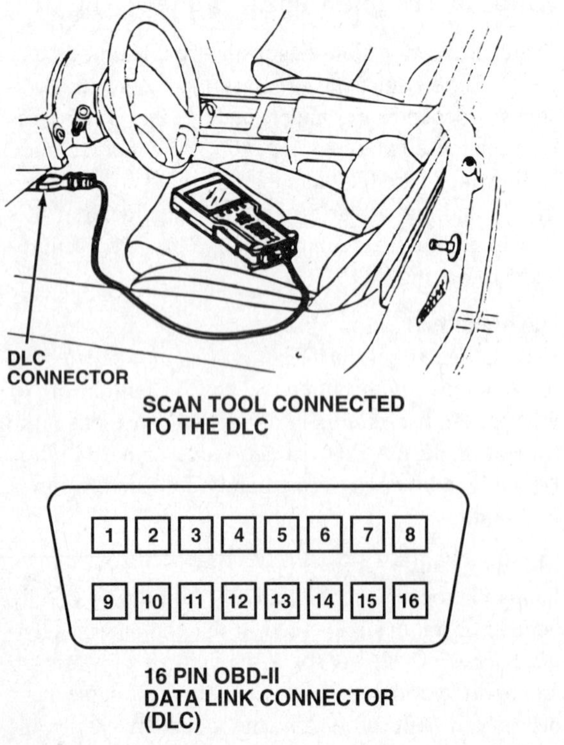

Figure 5-1. OBD II data link connector.

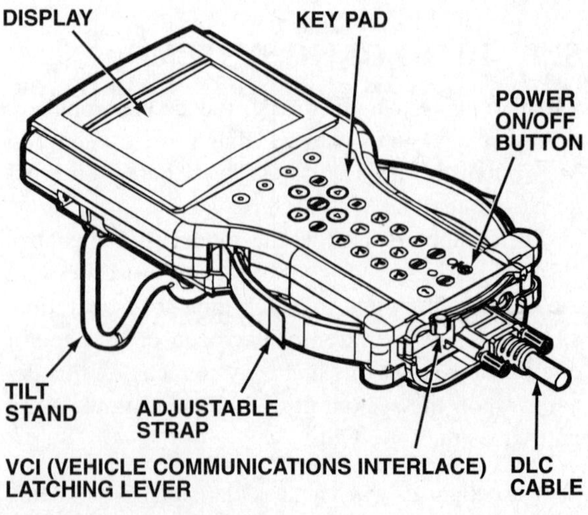

Figure 5-2. GM's Tech 2 scan tool.

forms various function tests, and can be used as a scanner to observe operational data, figure 5-3.

- SBDS is a larger multipurpose diagnostic device. Like NGS, it retrieves DTCs, performs function tests, and displays operational data, but it also contains guided diagnostic software, procedural information, electrical schematics and service information, figure 5-4.
- Chrysler: Diagnostic readout box (DRB) III, figure 5-5.

As with GM's Tech 2, Chrysler's DRB III is an enhanced version of the original DRB, and a successor to the DRB II. The DRB II tester can be used on all systems from 1984 to 1993. The DRB III was developed for OBD II vehicles, but can be used on 1984 to 1993 vehicles with additional software. The DRB III also can be programmed for use on other manufacturers' OBD II vehicles.

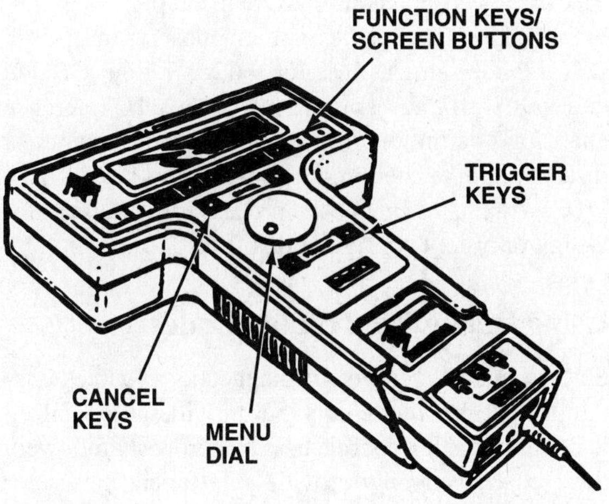

Figure 5-3. Ford's NGS.

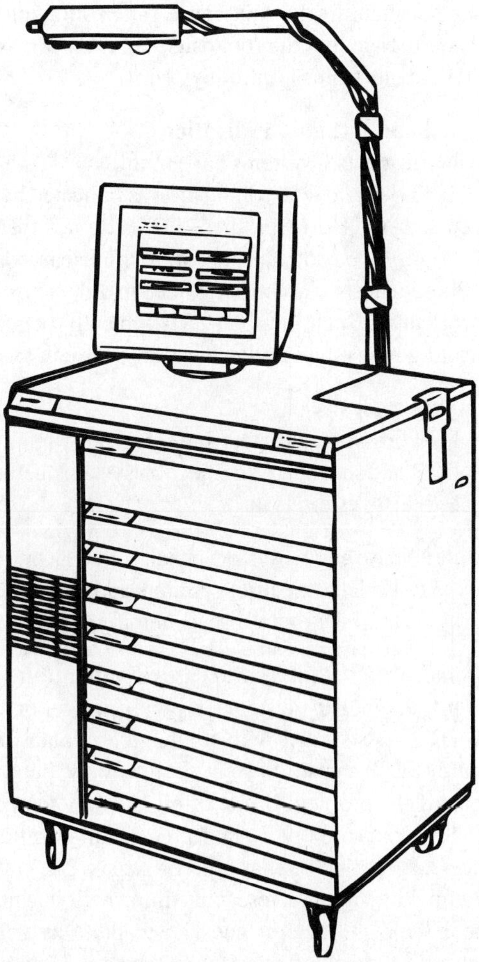

Figure 5-4. Ford's SBDS.

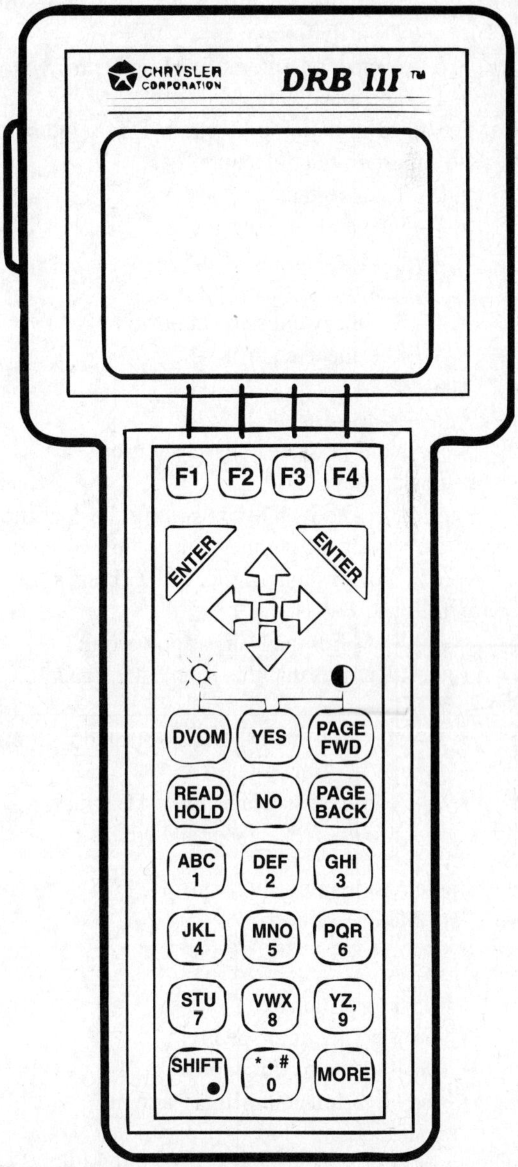

Figure 5-5. Chrysler's DRB III.

Keep in mind, scan tool devices are available from independent aftermarket manufacturers. These devices perform many of the same functions and operate similarly to the devices just discussed.

Displaying and Interpreting Trouble Codes

DTCs are displayed and read on the screen of a scan tool device. OBD II-compliant vehicles output standardized 5-character alphanumerical DTCs (in accordance with SAE standard J2012). The alphanumeric characters are defined as follows:

- The prefix letter indicates the DTC function:
  - P—Powertrain
  - B—Body
  - C—Chassis
- The first number identifies who is responsible for the DTC definition:
  - 0—Society of Automotive Engineers (SAE)
  - 1—The manufacturer
- The third digit for powertrain DTCs indicates the subgroup relationship:
  - 0—Total system
  - 1—Fuel and air metering
  - 2—Fuel and air metering
  - 3—Ignition misfire
  - 4—Auxiliary emission controls
  - 5—Idle/speed control
  - 6—PCM
  - 7—Transmission
  - 8—Non-electronic engine control powertrain
- The fourth and fifth digits specify the condition or area (such as engine temperature or engine speed). For example, DTC P0193 breaks down as follows:
  - P—Identifies the DTC as powertrain
  - 0—Identifies that this is an SAE standard DTC
  - 1—Identifies the DTC is associated with the fuel/air control system
  - 93—Further associates the DTC with the fuel rail pressure sensor and circuit

An example of a manufacturer-specific DTC is as follows (Ford Motor Company):

- P1364:
  - –P–Powertrain
  - –1–Manufacturer specific
  - –3–Ignition misfire
  - –64–Ignition coil primary circuit malfunction

Each manufacturer publishes a list of trouble codes used with each vehicle make, model, and system. It is important to refer to the appropriate DTC definitions and directives when diagnosing a DTC concern.

Pre-OBD II vehicles do not have such rigid DTC formats. Prior to the adoption of OBD II protocols, each manufacturer developed its own DTC format. The first electronic control systems output 2-digit numeric codes. The process of diagnosis with 2-digit DTCs is the same as it is for today's 5-digit OBD II codes. It is still necessary to refer to a DTC reference chart for a definition of the code and for pinpoint test direction. Some later pre-OBD II systems have 3-digit DTCs. Again, the diagnostic process is the same as for 2-digit or 5-digit configurations.

Using Diagnostic Trouble Codes

Service manuals and other diagnostic resources contain troubleshooting charts that provide an organized and logical testing approach. When properly followed, these charts generally provide an efficient means for identifying the root cause for a DTC. However, it is important to remember, as was pointed out earlier in this chapter, that a DTC identifies a concern area, but it does not identify the root cause. The following examples illustrate that the root cause for a DTC may be outside the electronic control system.

Rich or Lean Exhaust Indication

Most engine control systems have multiple DTCs associated with rich and lean combustion as indicated by the oxygen sensors. However, these DTCs do not specifically identify the root cause of the rich or lean condition. Pinpoint tests must be performed in order to isolate the root cause. Such pinpoint tests typically begin by confirming base engine-related conditions such as:

- Vacuum leaks
- Fuel pressure and metering integrity
- EGR and secondary air injection system operation
- Exhaust system leaks.

If the entire base engine system checks out OK, the next step is to test related electrical circuits and components one-by-one until the root cause is identified, figure 5-6.

High or Low Coolant Temperature Indication

Onboard diagnostic systems typically have one or more DTCs associated with the engine coolant temperature (ECT) sensor. System designs vary, but most indicate if the problem is one of high or low temperature. However, to properly interpret the information it is necessary to know what type of sensor the system uses and how it operates. This information can be found in the manufacturer's service manuals as well as a number of other aftermarket information manuals and resources.

	Test Step	Result ▶	Action to Take
H40	DTCs P1131, P1132, P1151 AND P1152: UPSTREAM HO2S NOT SWITCHING. DTC P1130: FUEL SYSTEM NOT SWITCHING AT FUEL TRIM (RICH OR LEAN)		
	Diagnostic Trouble codes (DTCs) P1131 bank 1 (Cylinder 1) and P1151 bank 2 indicate the fuel / air ratio is correcting rich for an overly lean condition. The HO2S voltage is less than 0.45 volt. DTCs P1132 bank 1 (Cylinder 1) and P1152 bank 2 indicate the fuel / air ratio is correcting lean for an overly rich condition. The HO2S voltage is greater than 0.45 volt. DTC P1130 and P1150 indicate the fuel control system has reached maximum compensation for a lean or rich condition and the HO2S is not switching. DTC / HO2S Reference List — HO2S-11 = DTCs P1131, P1132 and P1130 — HO2S-21 = DTCs P1151, P1152 and P1150 Possible causes: Fuel system — Excessive fuel pressure. — Leaking fuel injector(s). — Leaking fuel pressure regulator. — Low fuel pressure or running out of fuel. — Contaminated fuel injector(s). — Vapor recovery system. Induction system — Air leaks after the MAF. — Vacuum leaks (vacuum lines and gaskets). — Restricted air inlet. — PCV system. — Improperly seated engine oil dipstick. EGR system — Leaking gasket. — Stuck open EGR valve. — Leaking diaphragm. Base engine — Oil overfill. — Cam timing. — Cylinder compression. — Exhaust leaks before or near HO2Ss. • Check intake air system for leaks, obstructions and damage. • Check air cleaner element, air cleaner housing for blockage. • Verify integrity of the PCV system. • Check for vacuum leaks. • **Are any of the above concerns present?**	Yes ▶ No ▶	REPAIR as necessary. COMPLETE PCM Reset to clear DTCs. RERUN Quick Test. GO to H42.

Figure 5-6. DTC troubleshooting chart. (Courtesy of Ford Motor Company)

As with the previous example of a rich/lean exhaust indication, pinpoint tests must be performed to identify the root cause. The DTC itself does not identify the fault. For example, a DTC indicating a high-temperature condition can result from:

- A thermostat that is stuck closed or opening at too high of a temperature
- Low coolant level, blocked radiator airflow, clogged hoses, or restrictions in the water jackets or passages—all of which could cause the engine to overheat
- A defective coolant temperature sensor, its wiring, or its connector
- A defective computer

A DTC indicating that engine temperature is too low could result from:

- A thermostat that has been removed, or one that opens at too low a temperature
- Defective ECT wiring, connector, or sensor
- A defective computer

Open-loop and Closed-loop Faults

Some conditions indicated by a DTC are of a nature that prevents the engine control system from switching into closed-loop operation. Other problems may exist in the closed-loop mode, but do not produce a DTC until the problem that keeps the system in open-loop operation is corrected. With this understanding, it is vital to check system performance after making repairs. Failure to check system performance could result in a comeback.

TROUBLESHOOTING AND TESTING PROCEDURES

This section covers the troubleshooting and testing techniques associated with the three major domestic manufacturers.

General Motors

Onboard Diagnostic System (OBD) Check

As previously discussed, the first step in the diagnostic process is a thorough visual inspection of the powertrain control module (PCM), wiring, grounds and related components. The idea of the inspection is NOT to overlook the obvious by making sure connections are secure and that there is no damage to wiring and components before proceeding with testing.

Following the visual inspection, GM recommends performing the "OBD system check"—an organized approach to diagnosing concerns associated with the electronic control system, figure 5-7. The GM OBD system check begins by observing the operation of the malfunction indicator lamp (MIL). The MIL should illuminate when the ignition key is turned ON. An improperly operating MIL needs to be addressed before proceeding with other system diagnostics. If the MIL is operating properly, the next step is to install the scan tool and retrieve DTCs. If DTCs are displayed, the technician is directed to the applicable pinpoint test for further testing. If no DTCs are displayed, the next step is to observe various engine functions and operating parameters during operation, or resort to methods of diagnosis by symptom (more about these methods of diagnosis later in this chapter).

Figure 5-8 shows some typical trouble codes. With the increasing sophistication of GM engine control systems during the past decade, their self-diagnostic capabilities have expanded greatly. They now are able to display more than 100 different trouble codes. Models equipped with a body computer module (BCM) can display more than 50 additional codes dealing with body electrical problems. These facts point out that it is impossible to remember all the possible codes and what they mean. It is necessary to work with the specifications and test procedures for a given vehicle to diagnose problems with accuracy. Make sure to write down the codes in order to remember them and because the computer memory will have to be cleared to determine whether the code is continuous or intermittent.

As previously mentioned, the OBD system check begins by observing the operation of the malfunction indicator lamp (MIL). The MIL should illuminate when the ignition key is turned ON. The following are variations of how an MIL can operate improperly:

- The lamp does not light during the bulb check (key on, engine off). If this happens, a special diagnostic chart is provided to troubleshoot the lamp before moving on to other tests, see figure 5-9.

A POWERTRAIN ONBOARD DIAGNOSTIC (OBD) SYSTEM CHECK

STEP	ACTION	VALUE(S)	YES	NO
1	1. Turn on the ignition switch, and leave the engine off. 2. Observe the malfunction indicator lamp (MIL). Is the MIL on?	—	Go to Step 2	Go to no malfunction indicator lamp
2	1. Turn off the ignition switch. 2. Install a scan tool. 3. Turn on the ignition switch. 4. Attempt to display PCM data with the scan tool. Does the scan tool display PCM data?	—	Go to Step 3	Go to data link connector diagnosis
3	Attempt to start the engine. Did the engine start and continue to run?	—	Go to Step 4	Go to engine cranks butdoes not run
4	Display diagnostic trouble codes (DTCs). If any DTCs are stored, save the freeze frame and fail record information using the Capture into feature on the scan tool. Are any DTCs stored?	—	Go to applicable DTCs	Go to Step 5
5	Display DTC fail records with the scan tool. If any DTC fail records are stored, save the freeze frame and fail record information using the capture info feature on the scan tool. Are any fail records stored?	—	Go to applicable DTCs	Go to Step 6
6	Compare PCM data value displayed on the scan tool to the Engine Scan Tool Data List. Are the displayed values normal or close to the typical values?	—	System OK	Go to indicated component system checks

Figure 5-7. Powertrain onboard diagnostic (OBD) system check. (Courtesy of General Motors Corporation)

- The lamp glows dimly without displaying any codes. There is either a bad ground at the computer or the voltage supplied to the computer is less than 9 volts. Check the ground and charge the battery.
- The lamp flashes erratically, displaying unlisted trouble codes. There is probably voltage interference. It is most likely coming from the ignition system, a CB radio, an open diode in the A/C compressor clutch circuit, or a misrouted CCC wire harness that is too close to other engine wiring.
- The lamp does not light but the owner has a driveability complaint. In this case, a special "driver comment" diagnostic chart is provided to troubleshoot the most likely causes of specific symptoms.
- The lamp does not light on carbureted engines. The serial data line from the ECM and the MIL are connected to the same terminal in the DLC. Since both receive voltage from a driver circuit in the ECM, an open circuit in the lamp driver circuit will prevent the lamp from lighting or codes from being displayed. A diagnostic chart is available for troubleshooting this.

DTC	Description	Domestic RPO L82	Export Unleaded RPO L82	Export Leaded
P0336	24X Reference Signal	B	B	—
P0341	CMP Sensor Circuit Performance	B	B	—
P0401	EGR System Flow Insufficient	A	D	—
P0403	EGR Solenoid Circuit	B	D	—
P0404	EGR System Performance	B	D	—
P0405	EGR Pintle Position Circuit Low Voltage	B	D	—
P0420	TWC System Low Efficiency	A	D	—
P0440	EVAP System	A	A	—
P0442	EVAP Control System Small Leak Detected	A	A	—
P0446	EVAP Canister Vent Blocked	A	A	—
P0452	Fuel Tank Pressure Sensor Low Voltage	B	B	—
P0453	Fuel Tank Pressure Sensor High Voltage	B	B	—
P0502	VSS Circuit Low Input	A	A	—
P0503	VSS Circuit – Intermittent Input	A	A	—
P0506	Idle Control System Low RPM	B	B	—
P0507	Idle Control System High RPM	B	B	—
P0530	A/C Refrigerant Pressure Sensor Circuit	D	D	—
P0560	System Voltage	D	D	—
P0601	PCM Memory	B	B	—
P0602	PCM Not Programmed	B	B	—
P0654	Engine Speed Output Circuit	D	D	—
P0705	Trans Range Switch Circuit	D	D	—
P0706	Trans Range Switch Performance	D	D	—
P0711	Trans Fluid Temperature (TFT) Sensor Circuit Performance. Refer to Automatic Transaxle Diagnosis.	D	D	—
P0712	Trans Fluid Temperature (TFT) Sensor Circuit Low Voltage. Refer to 4T65-E Automatic Transaxle Diagnosis.	D	D	—
P0713	Trans Fluid Temperature (TFT) Sensor Circuit Low Voltage. Refer to Automatic Transaxle Diagnosis.	D	D	—
P0719	Brake Switch Circuit Low. Refer to 4T65-E Automatic Transaxle Diagnosis.	D	D	—
P0724	Brake Switch Circuit High. Refer to Automatic Transaxle Diagnosis.	D	D	—
P0740	Torque Converter Clutch Enable Solenoid Control Circuit. Refer to Automatic Transaxle Diagnosis.	A	A	—
P0742	Torque Converter Clutch System Stuck On. Refer to Automatic Transaxle Diagnosis.	B	B	—
P0751	Shift Solenoid 1 (1–2 Shift Solenoid) Performance/Stuck Off. Refer to Automatic Transaxle Diagnosis.	A	A	—

Figure 5-8. Basic GM diagnostic trouble codes. (Courtesy of General Motors Corporation)

NO MALFUNCTION INDICATOR LAMP

STEP	ACTION	VALUE(S)	YES	NO
1	Was the powertrain onboard diagnostic system check performed?	—	Go to Step 2	Go to a powertrain onboard diagnostic (OBD) system check
2	Attempt to start the engine. Does the engine start?	—	Go to Step 3	Go to Step 6
3	Check the fuse for the instrument cluster ignition feed circuit. Is the fuse OK?	—	Go to Step 4	Go to Step 15
4	1.Turn the ignition on. 2. Probe the ignition feed circuit at the cluster connector with a J34142-B test light to ground. Is the J34142-B test light on?	—	Go to Step 5	Go to Step 12
5	1.Turn the ignition off. 2. Disconnect the PCM. 3. Turn the ignition on. 4. Jumper the MIL control circuit at the PCM connector to ground and observe the MIL. Is the MIL on?	—	Go to Step 10	Go to Step 11
6	Check the PCM ignition feed and battery feed fuses. Are both of these fuses OK?	—	Go to Step 7	Go to Step 14
7	1. Turn the ignition off. 2. Disconnect the PCM. 3. Turn the ignition on. 4. Probe the ignition feed circuit at the PCM harness connector with a J34142-B test light to ground. Is the J43142-B test light on?	—	Go to Step 8	Go to Step 13

Figure 5-9. Use this chart to diagnose problems when the MIL does not illuminate. (Courtesy of General Motors Corporation)

Pinpoint Tests

Once the DTCs are recorded and it is determined whether or not they are continuous or intermittent, proceed with the diagnostic chart provided for each code. Some circuits have more than one code depending on the nature of problem. Each DTC has its own separate chart to help locate the problem. If there is no code for a particular circuit or subsystem, refer to the appropriate diagnosis by symptom chart.

Figure 5-10 shows the basic test equipment needed for performing pinpoint tests. While the GM service special tool numbers are identified for many of the items, some can be easily constructed with a few basic materials, and others can be purchased from aftermarket suppliers.

Before proceeding with a pinpoint test, make sure all vacuum lines are in good condition, are properly routed, and are not leaking. Also make sure all electrical connectors are secure and clean.

It is important to note that a diagnostic chart is useless if the fault does not exist at the time of testing. Intermittent faults that were recorded earlier, but are not present during testing, cannot be solved with the charts. Some codes require specific engine-operating conditions for a given length of time before they are recorded in memory. An intermittent code may be recorded in memory, but it must occur during testing if you are to find the cause.

The following is an example of how a DTC chart is used to perform a pinpoint test. DTC P0102 (Mass Air Flow sensor circuit low frequency) is used for this example. Also for the purpose of this example, assume all preliminary checks have been performed with no problems found, and the DTC is a hard-fault. Refer to the schematic in figure 5-11 as you review the following test steps.

1. The first step is to check the following non-electronic areas that can cause this DTC to be set:
   - Objects blocking the MAF sensor inlet
   - Vacuum leaks at the throttle body
   - Vacuum leaks at EGR flange and pipes
   - Crankcase ventilation valve faulty, missing, or incorrectly installed
2. If no problem is found with step number 1, step number 2 is to:
   - Turn the ignition OFF.
   - Disconnect the MAF sensor connector.
   - Turn the ignition ON (engine OFF).
   - Use a DMM to measure the voltage between the MAF signal circuit and battery ground.
3. If this voltage is near the specified value (approximately 5.0 volts), the third step is to connect a test light between the MAF sensor ignition feed and ground circuits at the MAF sensor harness connector.
4. If the test light illuminates, the next step is to check the connection at the MAF sensor. If the connection is OK, the direction is to replace the MAF sensor and recheck system operation.

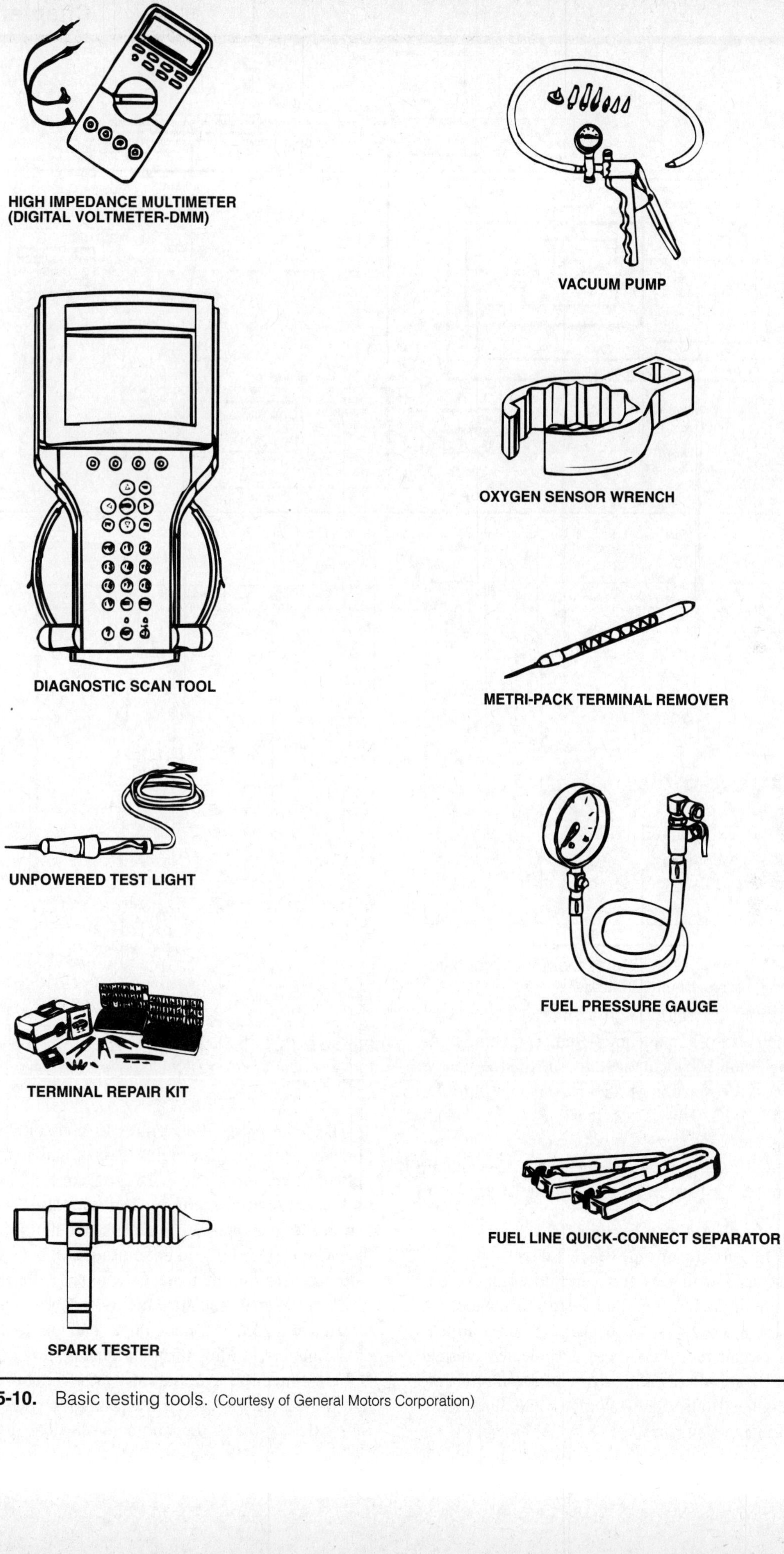

Figure 5-10. Basic testing tools. (Courtesy of General Motors Corporation)

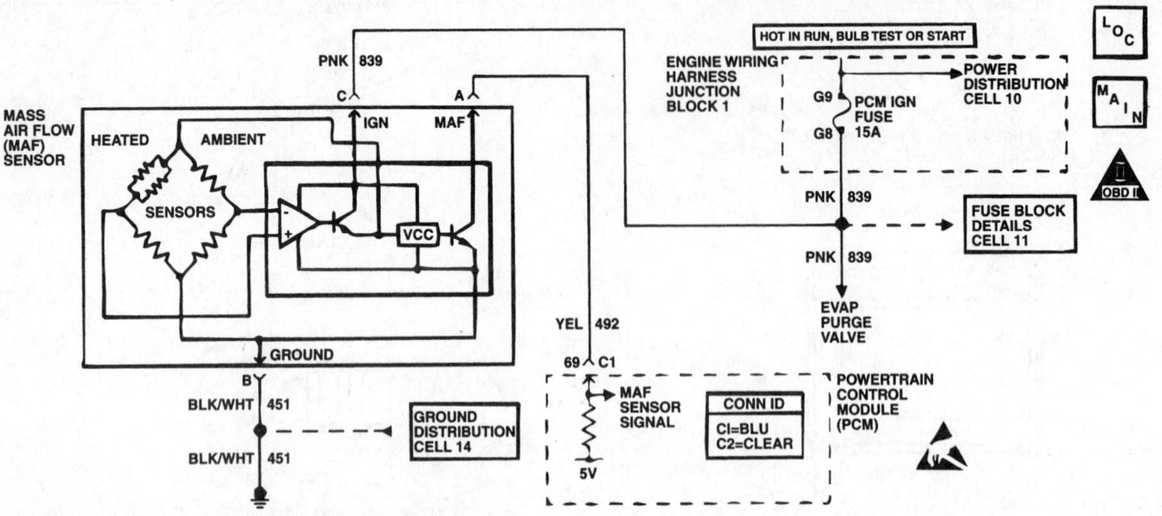

Figure 5-11. Mass air flow sensor circuit. (Courtesy of General Motors Corporation)

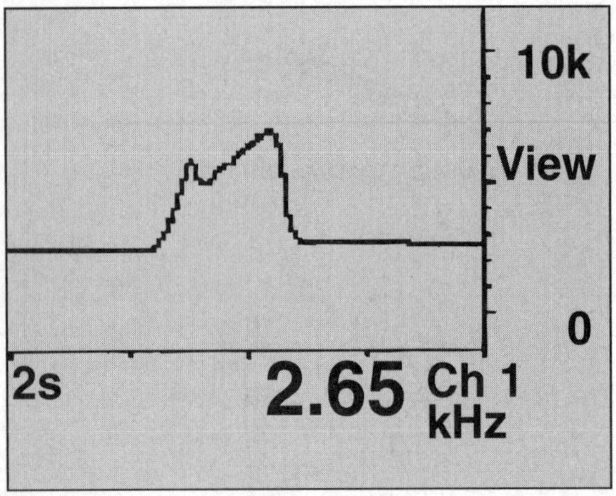

Figure 5-12. Normal waveforms for MAF on snap acceleration. (Courtesy of Snap-on Vantage®)

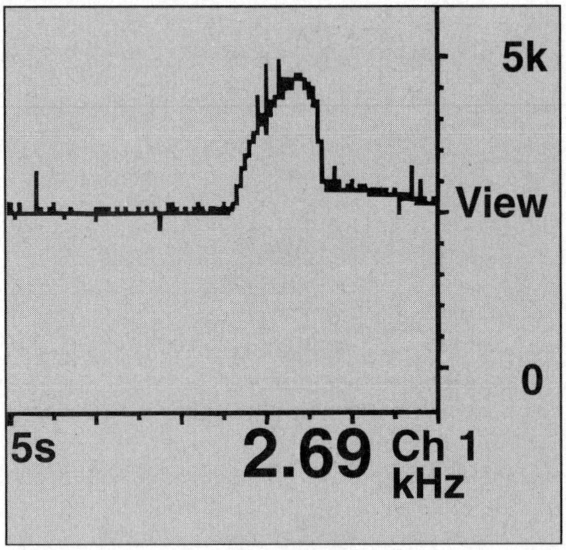

Figure 5-13. This MAF signal with rpm increase shows faulty sensor or connectors. (Courtesy of Snap-on Vantage®)

5. When testing a frequency signal, a graphing multimeter allows you to actually see the frequency. To test the reaction of airflow, snap the throttle, figure 5-12. This is a typical waveform of all mass airflow sensors, both voltage and frequency. A MAF that does not have a clean signal, figure 5-13, is an indication of a faulty sensor or connectors.

Testing EFI Integrator and Block Learn

The PCM used with GM fuel-injected engines contains a pair of functions called integrator and block learn. These are responsible for making minor adjustments to the air-fuel ratio of a fuel-injected engine, similar to the mixture control solenoid dwell on a carbureted engine, figure 5-14. Integrator and block learn represent injector on-time.

The term *integrator* applies to a method of temporarily changing the fuel delivery, and it functions only in closed loop. The PCM program contains a base fuel calculation in memory. The integrator and block learn functions interface with this calculation, causing the ECM or PCM to add or subtract fuel from the base calculation according to the O_2 sensor feedback signal.

This information can be monitored as a number between 0 and 255 by connecting a scan tool to the serial data transmission line through the data link connector (DLC). Since the average integrator function is one-half the maximum, or 128, the scan tool reads the base fuel calculation as the number 128. The integrator

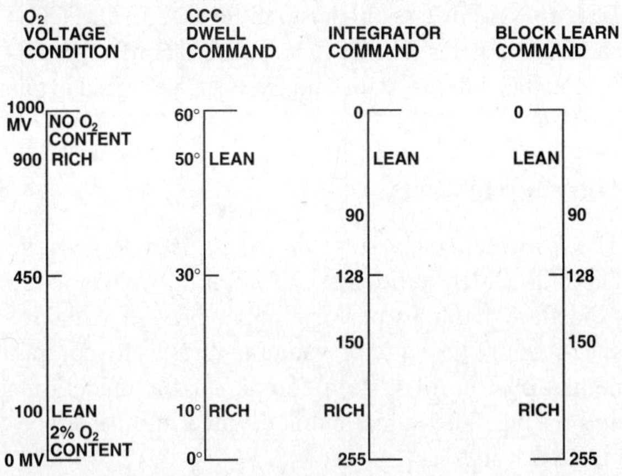

Figure 5-14. Integrator and block learn functions. (Courtesy of General Motors Corporation)

function reads the O_2 sensor output voltage, adding or subtracting fuel as required to maintain the 14.7 to 1 ratio. An integrator reading of 128 is neutral, which means that the O_2 sensor is telling the PCM that a 14.7 to 1 air-fuel ratio has been burned in the cylinders. If the scan tool reads a higher number, the PCM is adding fuel to the mixture. If the number is less than 128, fuel is being subtracted.

This corrective action, or integrator function, is effective only on a short-term basis. The integrator allows the computer to make short-term—minute-by-minute—corrections in fuel metering. Such corrections may be necessary, for example, when a car is driven from low altitude across a high mountain pass and back to low altitude in an hour or two.

Block learn represents the long-term effects of integrator corrections, although the corrections it makes are not as great as those made by the integrator. The function takes its name from division of the operating range of the engine for any given combination of rpm and load into 16 cells or "blocks," figure 5-15. Fuel delivery is based on the value stored in memory in the block corresponding to a given operating range. As with the integrator, the number 128 is the base fuel calculation and no correction is made to the value stored in that cell or block. Every time the engine is started, block learn starts at 128 in every cell and corrects as required to produce a situation in which the O_2 sensor reads the results as a 14.7 to 1 air-fuel mixture burning in the cylinders. Block learn also watches the integrator. If the integrator increases or decreases, block learn makes a correction in the same direction. This causes the integrator correction to be reduced gradually until it returns to 128.

Both integrator and block learn are a form of adaptive memory designed to allow the computer to fine-tune injection. Both also have predetermined limits

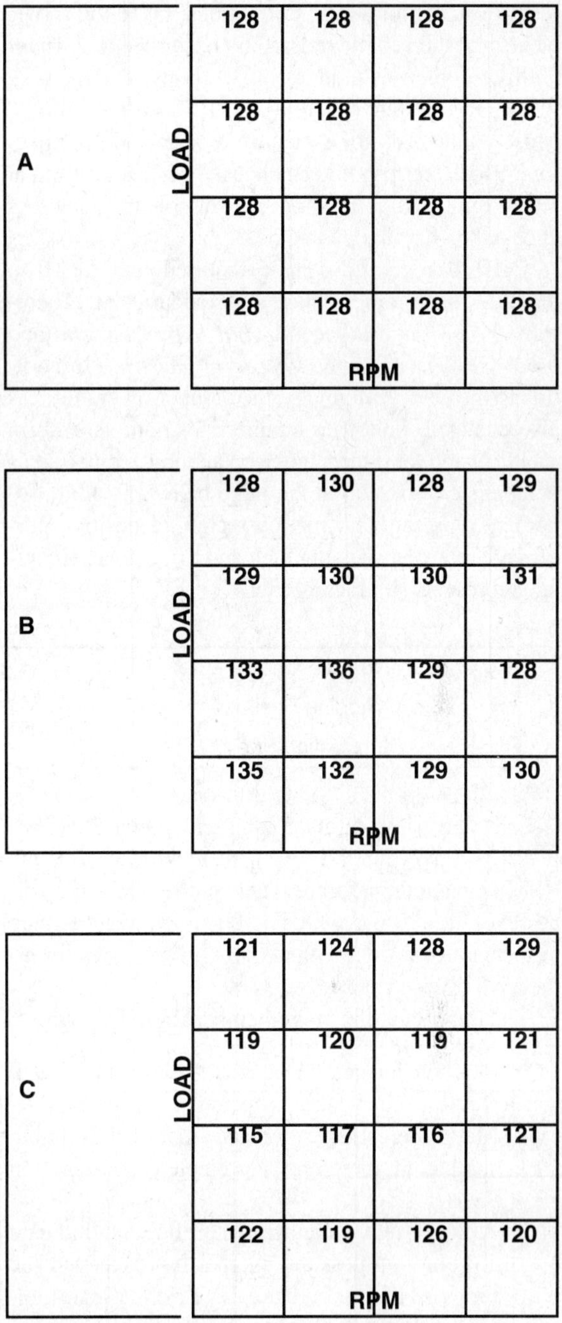

Figure 5-15. Fuel delivery blocks. (Courtesy of General Motors Corporation)

called "clamping" that vary according to engine design. If block learn reaches the limits of its control without correcting the condition, the integrator also goes to its limits. At this point, the engine starts to run poorly. Clamping is necessary, however, to prevent driveability problems that might result under certain conditions. For example, stop-and-go traffic on a crowded roadway could last long enough for block learn to make long-term corrections. Once the traffic clears and the vehicle returns to cruise speeds, a driveability problem would result. A high integrator or block learn number reading on the scan tool should cause you to look for a vacuum leak or sticking injectors. Block learn should be checked both at idle and at 3,000 rpm, just as mixture control solenoid dwell is checked on carbureted engines.

OBD II regulations have required that fuel-trim numbers and nomenclature be standardized. Therefore, *integrator* has become *short-term fuel trim* and *block learn* has become *long-term fuel trim.* In addition to the new terminology, the numbering system has also changed. Fuel-trim numbers for both short-term fuel trim and long-term fuel trim are now expressed as percentages. As seen in the table below, a value of 0 percent represents no fuel correction, a positive percentage represents adding fuel, and a negative percentage represents subtracting fuel.

0%	No Adjustment
+%	Adding Fuel
−%	Subtracting Fuel

Usually any short-term or long-term percentage between −10 percent and +10 percent indicates that the PCM is maintaining proper fuel control. Once the values exceed −10 percent or +10 percent and continue to increase or decrease, the PCM will set codes for either rich or lean exhaust.

The important thing to remember about fuel trim is:

- Plus percentages indicate that fuel is being added; therefore, the vehicle is operating on the lean side.
- Minus percentages indicate that fuel is being subtracted; therefore, the vehicle is operating on the rich side.
- A plus or minus fuel-trim value does not indicate a problem unless the number has exceeded the program limits. In that case, a rich or lean fuel-trim code should be set.

FORD

Ford has used various engine control systems since 1978. These include feedback carburetor control (FBC), microprocessor control unit (MCU), and five generations of electronic engine controls—EEC-I, EEC-II, EEC-III, EEC-IV, and most recently EEC-V. The FBC and MCU systems are primarily for fuel-metering control; the EEC systems are more comprehensive and represent increasing degrees of system integration. The self-diagnostic capabilities built into EEC-III and EEC-IV systems are similar to those used in GM's C-4 and CCC systems. The EEC-V system is Ford's OBD II-compliant control system, and the system covered in this segment.

Diagnostic Tests

The Ford control system *area test* used to retrieve DTCs is called a *quick test.* As with any diagnostic procedure, begin by performing a thorough visual inspection of the battery, vacuum lines, and components and control system components, connections, and wiring. The quick test is divided into three specialized tests:

- Continuous memory self-test
- Key-on engine-off (KOEO) on-demand self-test
- Key-on engine-running (KOER) on-demand self-test

As previously mentioned in this chapter, Ford endorses the New Generation START (NGS) tester and the Service Bay Diagnostic System (SBDS) as the tools of choice for retrieving DTCs and performing data link diagnostics. However, effective aftermarket scan tools are available (scan tools are covered in more detail later in this chapter).

Continuous Memory Self-Test

The continuous memory self-test, which is a function of the powertrain control module (PCM), is always active whenever the key is ON. This differs from the KOEO and KOER self-tests that are activated on demand only. A fault does not need to be present at the time of testing for continuous DTCs to be output. This makes continuous memory codes especially valuable for diagnosing intermittent concerns. Continuous memory codes are retrieved through the data link connector (DLC) via a scan tool. Only a system pass, continuous code, or an incomplete OBD II drive cycle (P1000) DTC will display. Continuous memory codes are retrieved first, before retrieving KOEO and KOER codes. The reason for retrieving memory codes first is to ensure that they will not be unintentionally cleared (erased) during the service of KOEO or KOER DTC concerns. As with the previous GM examples, hard fault concerns (KOEO and KOER) are addressed before memory concerns. Figure 5-16 shows the diagnostic subroutine for retrieving memory codes.

KOEO On-Demand Self-Test

The KOEO test is a functional test of the PCM performed on demand with the key on and the engine off. The test performs checks on various sensor and actua-

	Test Step	Result	Action to Take
DSR1	QUICK TEST: VEHICLE PREPARATION		
	• Apply parking brake. • Place shift lever firmly in PARK (NEUTRAL for manual transmission). • Block all drive wheels. • Turn off all electrical loads. • Is vehicle prepared to run Quick Test?	Yes ▶ No ▶	GO to DSR2. REPEAT this test step.
DSR2	RETRIEVE AND RECORD ANY CONTINUOUS MEMORY DTCS		
	NOTE: The purpose of recording any continuous memory diagnostic trouble codes (DTCs) at this time is to ensure that they will not be unintentionally cleared during any key on engine off (KOEO) or key on engine running (KOER) DTC repair (by disconnecting the PCM, vehicle battery, etc.). As always, KOEO and KOER DTCs will be addressed before continuous memory DTCs. • Key off. • Connect scan tool to data link connector (DLC). • Key on, engine off. • Retrieve and record any continuous memory DTCs (MIL and non-MIL). NOTE: If unable to access continuous memory DTCs or any scan tool communication concern exists, go to Section 5, Pinpoint Test Step QA1. For more information on retrieving MIL and non-MIL DTCs, refer to Section 2, Diagnostic Methods (Continuous Memory Self-Test). • Key off. • **Were any continuous memory DTCs present?**	Yes ▶ No ▶	RECORD on a piece of paper, etc.) the continuous memory DTC(s). These DTC(s) will be addressed after KOEO and KOER self-tests. GO to DSR3. GO to DSR3.

Figure 5-16. Continuous memory code diagnostic subroutine. (Courtesy of Ford Motor Company)

tor circuits. A fault must be present at the time of testing for a DTC to be set.

KOER On-Demand Self-Test

The KOER test is also an on-demand functional test of the PCM, but it is performed with the engine running and the vehicle stopped. This test checks various inputs and outputs during operating conditions at normal temperature. The brake pedal position, transmission control, power steering, and 4×4 low switch tests are all part of the KOER self-test and MUST be performed during the test operation if applicable. As with the KOEO test, a fault must be present at the time of testing for a DTC to be set. Figure 5-17 shows the diagnostic subroutine for both the KOEO and KOER self-tests.

Pinpoint Tests

Ford's pinpoint test procedures are similar to those used by GM, but they are provided in a column format, figure 5-18 instead of the linear chart form. The first column describes the test step, the second column gives possible results, and the third column tells you what action to take based on results obtained. When each pinpoint test and service sequence is completed, repeat the self-tests to ensure that the problem is resolved and that no DTCs remain in the system.

CHRYSLER

The Chrysler engine-control system was introduced on late-1983 front-wheel-drive Chrysler vehicles with fuel injection or a turbocharger. This modular control system (MCS) used two separate modules with a variety of sensors and actuators, figure 5-19 to control fuel metering, ignition timing, EGR flow, and canister purge. The test capabilities of the original onboard diagnostic (OBD) system were expanded on 1985 engines to include additional fault codes and memory. All MCS systems were tested with a diagnostic readout box (DRB) connected to a test connector in the engine compartment.

This first-generation engine control and diagnostic system was replaced by the single-module engine control (SMEC) computer on some 1987 4-cylinder and V6 engines. The second-generation SMEC brought the two circuit boards used in MCS logic and power modules into a single housing. Its advanced microprocessor is smaller, faster, and more powerful than the MCS microprocessor, with electrically erasable memory (EEPROM) that can be programmed in the assembly plant. The SMEC processes instructions twice as fast as the previous MCS system.

Further refinement of Chrysler's engine computer came in 1989 with the introduction of the single-board

Test Step		Result ▶	Action to Take
DSR3	PERFORM KEY-ON ENGINE-OFF (KOEO) SELF-TEST		
	• Scan tool connected. • Start engine, if possible, and idle until it is at normal operating temperature. • Key off. • Key on, engine off. • Activate KOEO self-test.	Yes ▶	Go to the powertrain DTC charts (located after the diagnostic subroutines) for pinpoint test direction (begin diagnosis with first DTC output).
	NOTE: If unable to activate KOEO self-test or any scan tool communication concern exists, go to Section 5, Pinpoint Test Step [QA1]. If additional information is required for KOEO Self-Test, refer to Section 2, Diagnostic Methods. • After the completion of KOEO Self-Test, retrieve and record any KOEO DTCs. • If DTC P1000 is present, ignore it at this time. • Key off. • **Were any KOEO DTCs (except P1000) present?**	No ▶	GO to [DSR4].
DSR4	PERFORM KEY-ON ENGINE-RUNNING (KOER) SELF-TEST		
	• Scan tool connected. • Shift lever in PARK (A/T) or NEUTRAL (M/T). • Start engine. • Warm engine to normal operating temperature. After warming engine, leave engine running. • Activate KOER self-test.	Yes ▶	Go to the powertrain DTC charts (located after the diagnostic subroutines) for Pinpoint Test direction (begin diagnosis with first DTC output).
	NOTE: If engine stalls during KOER self-test, discontinue test and go to [DSR5]. If unable to activate or complete KOER self-test, or any scan tool communication concern exists, go to Section 5, Pinpoint Test Step [QA1]. If additional information is required for KOER self-test, refer to Section 2, Diagnostic Methods. • After the completion of KOER self-test, retrieve and record any KOER DTCs. • If DTC P1000 is present, ignore it at this time. • Key off. • **Were any KOER DTCs (except P1000) present?**	No ▶	GO to [DSR5].
DSR5	WERE ANY CONTINUOUS MEMORY DTCS RECEIVED IN [DSR2]?		
	• Previous diagnostic subroutines performed as directed.	Yes ▶	GO to [DSR5].
	• **Were any continuous memory DTCs received in [DSR2]?**	No ▶	Quick test complete. RETURN to symptom Chart in Section 3.

Figure 5-17. KOEO and KOER diagnostic subroutine. (Courtesy of Ford Motor Company)

TEST STEP		RESULT ▶	ACTION TO TAKE
DA1	SERVICE CODE 24: CHECK AMBIENT TEMPERATURE		
	• AMBIENT TEMPERATURE MUST BE GREATER THAN 50°F FOR THIS TEST.	YES ▶	GO TO [DA2].
		NO ▶	RERUN QUICK TEST.
DA2	CHECK FOR V REF AT THROTTLE POSITION SENSOR		
	• REFER TO ILLUSTRATION Q. • KEY OFF, WAIT 10 SECONDS. • DVOM ON 20V SCALE.	LESS THAN 4.0V OR GREATER THAN 6.0V ▶	GO TO PINPOINT TEST STEP [C1].
	• DISCONNECT TP SENSOR. • KEY ON, ENGINE OFF. • MEASURE VOLTAGE AT THE TP HARNESS CONNECTOR BETWEEN VREF AND SIGNAL RETURN.	4.0V TO 6.0V ▶	RECONNECT TP SENSOR, GO TO [DA3].

Figure 5-18. Ford pinpoint test. (Courtesy of Ford Motor Company)

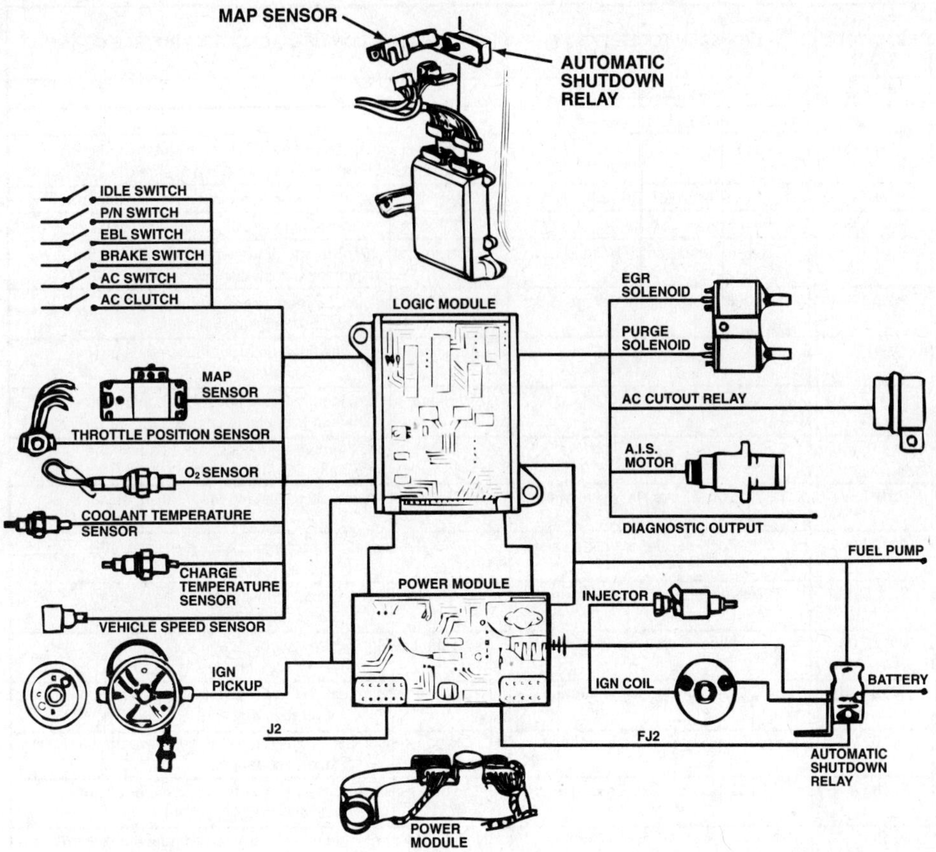

Figure 5-19. Chrysler modular engine control system. (Courtesy of DaimlerChrysler Corporation)

engine controller (SBEC) and the SBEC II in 1992. Through the use of application-specific integrated circuits (ASICs), the SBEC and SBEC II circuitry has been greatly simplified. The SBEC engine controller accepts a larger number of input signals and produces more output signals than earlier controllers. In addition to the increased number of functions, its enlarged memory capacity greatly expanded its OBD capabilities.

The SMEC and SBEC onboard diagnostics are accessed through a second-generation diagnostic readout box (DRB II, which replaced the original DRB tester). The DRB II can be used to test all systems from 1984 to 1993. The DRB III tester is used on 1994 and later Chrysler vehicles, but can be programmed for 1984–93 Chrysler vehicles and for other manufacturers' OBD II cars.

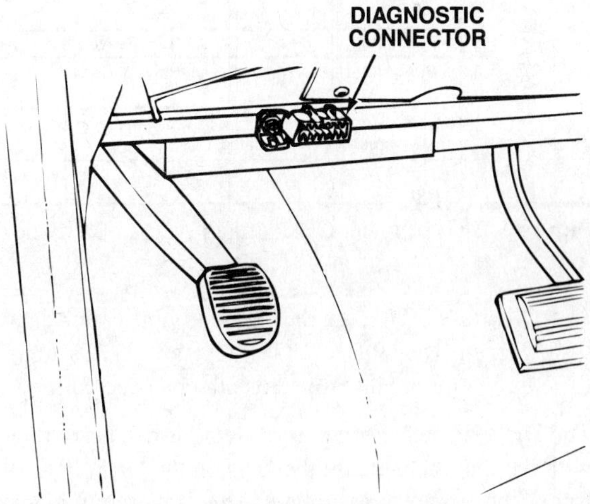

Figure 5-20. Chrysler OBD II data link connector. (Courtesy of DaimlerChrysler Corporation)

Diagnostic Trouble Codes

The DRB II and DRB III are the Chrysler-endorsed tools of choice for retrieving DTCs. The following steps outline the procedure used to retrieve DTCs using the DRB II tester:

1. Connect the DRB II to the data link connector (DLC) located in the passenger compartment at the lower edge of the instrument panel near the steering column, figure 5-20.
2. Turn the ignition ON and access the "Read Fault" screen.
3. Record all DTCs and "freeze frame" information shown on the DRB II.

GENERIC SCAN TOOL CODE	DRB SCAN TOOL DISPLAY	DESCRIPTION OF DIAGNOSTIC TROUBLE CODE
P0030	1/1 O2 Sensor Heater Relay Circuit	
P0036	1/2 O2 Sensor Heater Relay Circuit	
P0106 (M)	Barometric Pressure Out of Range	MAP sensor input voltage out of an acceptable range detected during reading of barometric pressure at key-on.
P0107 (M)	Map Sensor Voltage Too Low	MAP sensor input below minimum acceptable voltage.
P0108 (M)	Map Sensor Voltage Too High	MAP sensor input above maximum acceptable voltage.
P0112 (M)	Intake Air Temp Sensor Voltage Low	Intake air (charge) temperature sensor input below the minimum acceptable voltage.
P0113 (M)	Intake Air Temp Sensor Voltage High	Intake air (charge) temperature sensor input above the maximum acceptable voltage.
P0116		A rationality error has been detected in the coolant temp sensor.
P0117 (M)	ECT Sensor Voltage Too Low	Engine coolant temperature sensor input below the minimum acceptable voltage.
P0118 (M)	ECT Sensor Voltage Too High	Engine coolant temperature sensor input above the maximum acceptable voltage.
P0121 (M)	TPS Voltage Does Not Agree with MAP	TPS signal does not correlate to MAP sensor signal.
P0122 (M)	Throttle Position Sensor Voltage Low	Throttle position sensor input below the acceptable voltage range.
P0123 (M)	Throttle Position Sensor Voltage High	Throttle position sensor input above the maximum acceptable voltage.
P0125 (M)	Closed Loop Temp Not Reached	Time to enter closed loop operation (fuel control) is excessive.
P0131 (M)	1/1 O2 Sensor Shorted to Ground	Oxygen sensor input voltage maintained below normal operating range.
P0132 (M)	1/1 O2 Sensor Shorted to Voltage	Oxygen sensor input voltage maintained above normal operating range.
P0133 (M)	1/1 O2 Sensor Slow Response	Oxygen sensor response slower than minimum required switching frequency.
P0134 (M)	1/1 O2 Sensor Stays at Center	Neither rich nor lean condition is detected from the oxygen sensor input.
P0135 (M)	1/1 O2 Sensor Heater Failure	Oxygen sensor heater element malfunction.
P0137 (M)	1/2 O2 Sensor Shorted to Ground	Oxygen sensor input voltage maintained below normal operating range.
P0138 (M)	1/2 O2 Sensor Shorted to Voltage	Oxygen sensor input voltage maintained above normal operating range.
P0139 (M)	1/2 O2 Sensor Slow Response	Oxygen sensor response not as expected.
P0140 (M)	1/2 O2 Sensor Stays at Center	Neither rich nor lean condition is detected from the oxygen sensor input.
(M) Check engine lamp (MIL) will illuminate during engine operation if this diagnostic trouble code was recorded. (G) Generator lamp illuminated.		

Figure 5-21. Chrysler OBD II diagnostic trouble codes. (Courtesy of DaimlerChrysler Corporation)

4. To erase DTCs, use the "erase trouble code" data screen. NOTE: Do NOT erase any DTCs until problems are investigated and repairs performed.

The DRB III uses a complex system of menus starting with the main menu. Use the keys on the DRB III and follow the prompts to negotiate to the areas required for testing an OBD II vehicle.

Figure 5-21 is a sample of Chrysler OBD II fault codes.

Pinpoint Tests

Chrysler's pinpoint tests are compiled and published in a "driveability test procedure" or "powertrain control diagnosis procedure" booklet. The format, figure 5-22, is similar to the one used by Ford. The first column describes the test step and shows required equipment connections, the second column gives test indications, and the third column identifies what action to take. When each pinpoint test and service sequence is completed, repeat the diagnostic test to ensure that the problem is resolved and that no fault codes remain in the system.

SCAN TOOLS

To this point, this chapter has limited its discussion of scan tool uses to that of retrieving DTCs. However, scan tools perform other valuable diagnostic functions. The name *scan tool* is derived from the fact that these devices, in addition to retrieving DTCs, watch or scan control system input and output signals through the

Step	Action	Value(s)	Yes	No
1	Was the powertrain onboard diagnostic (OBD) system check performed?	—	*Go to Step 2*	Go to the *powertrain OBD system check*
2	Was the instrument cluster system check in electrical diagnosis performed?	—	*Go to Step 3*	Go to the instrument cluster system check in electrical diagnosis
3	1. Turn off the ignition switch. 2. Disconnect the PCM. 3. Turn on the ignition switch. 4. Using a *J39200* digital multimeter, measure voltage between the affected PCM output circuit and the PCM harness connector and ground. Is the voltage near the specified value?	B+	*Go to Step 4*	*Go to Step 7*
4	1. Set the digital multimeter to the 10 amp scale and install the digital multimeter to measure current between the affected PCM output circuit and ground. 2. Monitor the current reading on the digital multimeter for at least 2 minutes. Does the current reading remain between the specified values?	0.05 amp-1.5 amps	*Go to Step 12*	*Go to Step 5*

Figure 5-22. Chrysler pinpoint test procedure. (Courtesy of DaimlerChrysler Corporation)

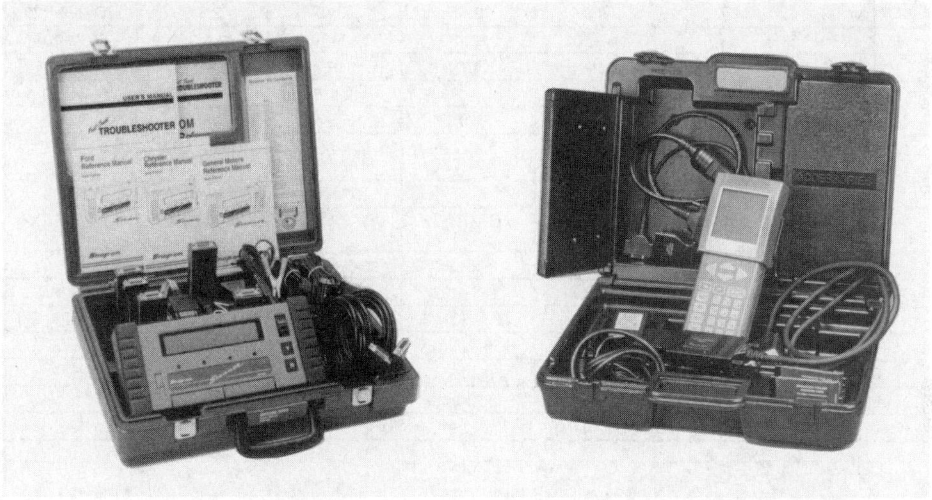

Figure 5-23. Scan tools interface directly with the PCM.

computer's data stream. Data stream information (or serial data as it is sometimes called) is a valuable diagnostic tool for the technician. Serial data is like eavesdropping on the microprocessor as it communicates with sensors and actuators, or even with other microprocessors.

Additionally, the discussion of scan tool devices has been limited to those produced and/or endorsed by the three major domestic manufacturers. There is, however, a variety of scan tool devices available from independent test equipment manufacturers. The scan tools developed by independent manufacturers can be used with different vehicles. This is made possible through the use of interchangeable cartridges and cable adapters. For example, inserting a Chrysler test cartridge into the scan tool and connecting the Chrysler connector adapter to its test lead allows any Chrysler vehicle with an engine controller to be checked. The test cartridge is programmed with Chrysler test procedures. The scan tool asks for vehicle identification, usually the VIN number. Once the required information is entered through the keypad, the scan tool is ready to perform specific test procedures or read individual sensor and actuator signals from the serial data stream. To use the same scan tool with a GM or Ford vehicle, simply change the cartridge and adapter, then enter the correct vehicle identification. Figure 5-23 shows one such aftermarket scan tool device.

Before continuing, it is important to note that while scan tools help make diagnosis faster and easier, they are NOT a substitute for a knowledgeable, thinking technician. The scan tool provides raw information

TROD/TR Sensors/Inputs	PCM/Breakout Box Pin/PID only	Measured/PID Values KOEO	Hot Idle	30 MPH	55 MPH	Units Measured/PID
TROD/TR	8	VBAT/OD (N)	VBAT/OD (N)	VBAT/OD	VBAT/OD	DCV/MODE
TRL/TR	7	VBAT/MAN1 (N)	VBAT/MAN1 (N)	.1/OD	.1/OD	DCV/MODE
IMRCM	8	5	5	5	5	DCV
TRD/TR	9	VBAT/DRIVE (N)	VBAT/DRIVE (N)	.1/OD	.1/OD	DCV/MODE
FLIV/FLI	12	1.7/50	1.7/50	1.7/50	1.7/50	DCV/%
FEPS	13	.5-.6	.5-.6	.5-.6	.5-.6	DCV
CKP (+)	21	0	400-425	770-90	1200-1400	Hz
PSP	31	.5	2.5(1)	.5	.5	DCV
TRR/TR	32	VBAT/REV(N)	VBAT/REV(N)	.1/OD	.1/OD	DCV/MODE
O2S12	35	0.1	(D)	(D)	(D)	DCV
TFT	37	.5-2/100-210	.5-2/100-210	.5-2/110-210	.6-2/110-210	DCV/DEG
ECT	38	.4-1/160-200	.4-1/160-200	.4-1/160-200	.4-1/160-200	DCV/DEG
IAT	39	1.7-3.5/50-120 (K)	1.7-3.5/50-120 (K)	1.7-3.5/50-120 (K)	1.7-3.5/50-120 (K)	DCV/DEG
FPM	40	3.5/50	3.5/50	3.5/50	3.5/50	DCV%
ACCS	41	.1/OFF	VBAT/ON(A)	.1/OFF	.1/OFF	DCV/OFF-ON
VSS (+)	58	0	0	65/30	125/55	Hz/MPH
O2S11	60	0	switching(C)	switching(C)	.4-1/160-200	DCV
FTP V/FTP	62	2.6/0	2.6/0	2.6/0	2.6/0	DCV/IN-H20
FRP V/FRP	63	3.35/50	2.8/39	2.8/39	2.8/39	DCV/PSI
CPP/PNP	64	.1/ON	.1/ON	5/OFF	5/OFF	DCV/OFF-ON
DPFEGR	65	.2-1.3	.2-1.3	.2-4.5	.2-4.5	DCV
KS	57	0	0	0	0	DCV
TSS	84	0	340-380/ 680-720	620-880/ 1160-1180	1090-1150/ 2150-2220	Hz/MPH
CID	85	0	5-7	11-15	17-21	Hz
ACP	86	VBAT/OPEN	VBAT/OPEN	VBAT/OPEN	VBAT/OPEN	DCV/OPEN-CLOSED
MAFV	88	0	.6-.9	1-1.6	1.3-2.3	DCV
TPV	89	.53-1.27	.53-1.27	1-1.3	1.1-1.9	DCV
BPP	92	.1/OFF	VBAT/ON(E)	.1/OFF	.1/OFF	DCV/OFF-ON
EFTA	P/D	50-120(K)	50-120(K)	50-120(K)	50-120(K)	DEG
GEAR	P/D	1	1	3	4	GEAR
LOAD	P/D	(L)	10-20	20-31	25-52	%

Figure 5-24. Typical reference values. (Courtesy of Ford Motor Company)

about sensors, actuators and their circuits. It is the technician's responsibility to accurately interpret the data. Understanding the function and operation of the system and its components is critical. Additionally, the proper interpretation of scan tool data only narrows down the concern to a particular circuit. It is still necessary to perform traditional electromechanical troubleshooting procedures to isolate the root cause of a concern.

When a scan tool is used as a data recorder, it captures a "snapshot" in its memory. This can be especially useful in pinpointing intermittent concerns. To capture/record data, operate the vehicle with the scan tool connected until the symptom occurs. When the symptom is present, capture the data by pressing the appropriate button on the keypad. The scan tool will record a specified number of data frames. A frame is an electronic picture from the computer that contains a sample of sensor and actuator signals. The scan tool continuously records this data and once triggered retains a specified number of frames prior to the problem while continuing to record frames until its memory is full. After making the recording, return to the service bay and play back the data. The analysis of the sensor and actuator data can be very helpful in pinpointing the intermittent concern. Most scanners can also be connected to a printer to produce a paper copy of the data for analysis. Refer to the appropriate service manual publication for the vehicle being serviced for the input and output values for various operating conditions. figure 5-24 shows a typical reference value chart for Ford.

The scan tool not only communicates with the computer, but it can override the computer and take control of the system, providing special test capabilities. This means that the technician can input certain conditions to the system being tested and then observe the reactions. For example, it is possible to check automatic transmission shift solenoid functions with the vehicle on a hoist.

Most scan tools perform the following:

- Retrieve and clear DTCs
- Read bidirectional data streams
- Record serial data
- Perform functional tests, such as air injection and EGR tests, timing tests, and back-up mode tests
- ABS system tests
- Tests on other electronic control systems such as antilock brake systems, supplemental inflatable restraint systems, and other systems that have data link communication capabilities

ELECTRONIC CONTROL SYSTEM SERVICE

The basic principles of testing, adjusting, and repairing components apply equally to all electronic engine controls systems. However, there are additional system requirements in terms of service. The following paragraphs discuss the components involved.

O_2 Sensor Testing and Service

Early engine control systems require periodic replacement of the O_2 sensor, either on a specified-mileage or time interval. Late-model systems have O_2 sensors warranted for 50,000 miles, and their replacement is not a scheduled maintenance service. However, if a sensor fails, it will need to be replaced.

O_2 sensor test procedures vary somewhat according to the system, as well as the type of sensor (1-wire, 2-wire, 3-wire, or 4-wire). While the methods recommended to test the sensor output voltage signal and the computer's response to the sensor signals may differ slightly, there are common precautions to observe whenever working with any O_2 sensor. For example:

- Do not short circuit or ground an O_2 sensor output lead. This can permanently damage the sensor.
- Do not check the sensor output with an analog voltmeter. Such testers have low impedance and draw enough current to lower the signal voltage.
- Do not check the sensor with an ohmmeter. This is not a valid test.
- Do not use silicone sprays near an O_2 sensor. The silicone compounds will collect and block off ambient air to the sensor, resulting in an incorrect signal.

Figure 5-25. O_2 sensor location. (4-Cylinder GM J-Car)

- Do not use leaded gas in the vehicle. The lead will collect on the exhaust side of the sensor and cause an incorrect signal.
- Do not incorrectly position the silicone boot, if one is used. The boot position can affect sensor operation and may melt if it contacts the hot manifold surface.

O_2 Sensor Removal and Replacement

O_2 sensors are similar in appearance to an 18-mm spark plug, but have one, two, three, or four wires protruding from the top. They generally are installed in the exhaust manifold, figure 5-25, or exhaust pipe. OBD II vehicles also have an O_2S immediately after the catalytic converter. Access to the sensor varies depending upon the particular engine and control system. Some are easily reached, while access to others is possible only from underneath the vehicle. If possible, use an O_2 sensor wrench for removing and installing a sensor. This specially designed slip-on socket accommodates the sensor leads and has an offset drive lug to accept a ½-inch ratchet or breaker bar. When replacing an O_2 sensor, it is a good idea to clean the manifold or exhaust pipe threads with an 18-mm thread chaser to make sure they are well formed and free of dirt and burrs.

All O_2 sensors require the use of a special antiseize compound when installed. The compound consists of liquid graphite and glass beads. The graphite burns away, leaving the beads in place on the threads to ease sensor removal. New sensors have this compound already applied but the threads of a used sensor should be cleaned before applying the compound. Sensor installation without the compound will make it extremely difficult, or even impossible, to remove in the future.

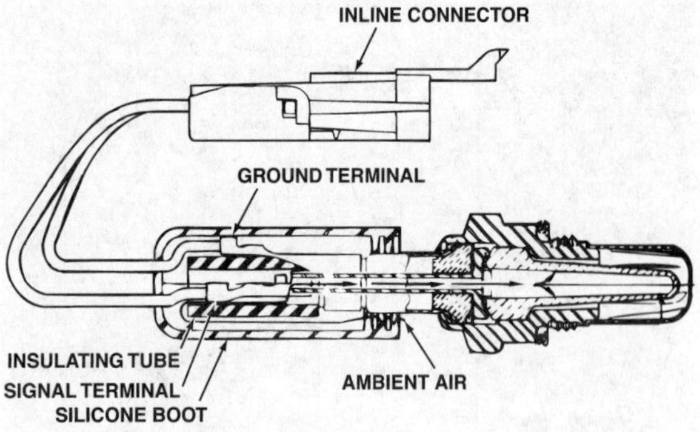

Figure 5-26. O_2S boot positioning. (Courtesy of General Motors Corporation)

O_2 sensors should be tightened to the exact torque specifications, which are generally in the 20 to 35 foot-pound (15 to 25 Nm) range. Exposure to the hot exhaust gases will make the sensor more difficult to remove. In some cases, it will require up to two or three times more torque for removal than is required for installation. Be especially careful in positioning any rubber boot used with a sensor. It should be located to protect the sensor without blocking the air vent, figure 5-26.

Fuel-Injector Solenoid Service

There are four methods for testing injectors:

1. Use a scan tool to observe fuel-trim data (*integrator* and *block learn data* in GM terminology).
2. Use a DMM to check injector resistance and operating voltages.
3. Measure current flow with an amp probe and GMM or lab scope.
4. Perform a flow-test.

The injectors must be pulled from the fuel rail or throttle body to check the flow pattern while cranking. If the flow pattern is erratic, or otherwise suspect, a variety of injector cleaning equipment and cleaners is available. However, be sure to check the manufacturer's recommendations before cleaning injectors. Ford maintains that its deposit-resistant injectors do not require cleaning; GM takes the position that cleaning its Multec injectors can damage them.

Injectors that use a ball and seat design with a director plate probably should not require cleaning. The fuel-metering parts of such injectors are shielded from the intake manifold gases that cause deposit buildup. Some cleaners contain a high percentage of methanol or other solvents that can damage the injector coil wire insulation, causing low resistance or shorted injectors. Such cleaners also may increase friction at critical bearing surfaces, resulting in premature wear. Although some suppliers of cleaners claim that their use will remove intake valve deposits, extensive studies performed by manufacturers show that cleaners have little or no effect on such deposits.

Throttle-Position Sensor Adjustment

Every electronic fuel-injection system uses a throttle-position sensor (TPS) to supply the engine control microprocessor with a signal proportionate to the opening angle of the throttle plate(s). The TPS generally is a variable potentiometer. A TPS that is out of specification can cause incorrect idle speeds or fuel metering. On some early-model EFI control systems, the TPS is adjustable; on most late-model systems, it is not. If the sensor is adjustable, it will have slotted mounting screw holes. These allow the sensor to be respositioned until a specified voltage reading (typically one volt or less) is obtained with the throttle in its closed position. After repositioning the sensor, move the throttle to its wide-open position and check the voltage reading. It should be slightly less than the reference voltage. If the sensor can be adjusted within specifications at both its closed and wide-open positions, it is good. If the voltage reading is satisfactory at the closed position, but not at the wide-open position, it is defective and cannot be properly adjusted.

Most throttle-position sensors are externally mounted. Adjustment (if possible) is made by loosening the sensor mounting screws and rotating the sensor housing until a specified voltage reading is obtained. The following procedure, used to check and adjust Ford level C sensors, is typical:

1. Connect the EEC-IV breakout box according to manufacturer's instructions.
2. Connect the (+) lead of a digital volt-ohmmeter (DVOM) to pin 47 and the—lead to pin 46.
3. Turn the ignition switch on, but do not start the engine.
4. Loosen the sensor retaining screws and rotate the assembly until the DVOM reading is 0.9 to 1.1 volt, then tighten the retaining screws.
5. Watch the DVOM while opening the throttle to its wide-open position and then back to idle. The DVOM should read 1.0 volt to 4.0 volts and back to 1.0 volt.

Solenoids and Relays

The resistance of an engine control system actuator must match the voltage and current capacity of the system in which it is used. For example, internal electronic switches in the computer called "drivers" activate all solenoids and relays used in a GM CCC system. Each driver is part of a group of four called "quad-drivers"; failure of one driver can damage the remaining drivers in the group. The solenoids and relays must have a minimum of 20 ohms resistance. The microprocessor drivers work with a maximum 0.8-ampere current at a maximum 16 volts system voltage. Ohm's law states that 16 volts ÷ 0.8 ampere = 20 ohms. Thus, if the solenoid or relay resistance is less than 20 ohms, the high current will destroy the microprocessor drivers.

A solenoid or relay that is shorted or grounded can also damage the microprocessor. They should be checked with an analog ohmmeter on its low-resistance scale. Check the solenoid resistance in both directions, because some use diodes to protect the microprocessor from reverse current. Use the higher resistance reading to verify solenoid condition.

Figure 5-27. Programmable read-only memory. (PROM) (Courtesy of General Motors Corporation)

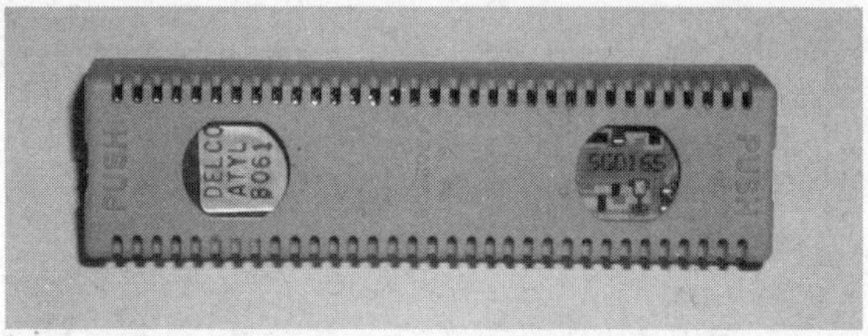

Figure 5-28. The MEMCAL combines the functions of the CALPAC and the PROM into a single unit.

PROM, MEMCAL, and CALPAC Replacement

The controller contains an interchangeable programmable read-only memory (PROM) chip, figure 5-27. The PROM is programmed with the necessary calibration data or program for a specific vehicle, engine, transmission, and accessory combination. By using interchangeable PROMs, the controller can be customized for use with a particular vehicle. Some controllers contain more than one PROM. This allows one basic controller to be used in a wide variety of models, reducing the manufacturer's costs and inventories. By using PROMs, the manufacturer can correct certain driveability problems more easily. A revised PROM can be manufactured and installed much less expensively than replacing the entire controller.

Controllers and PROMs are available separately, since a system problem can occur in either one. If a controller requires replacement, you probably will have to remove the PROM from the defective unit and install it in the replacement controller. Because the PROM is designed for a given vehicle, the replacement PROM to be installed must always carry the same part number as the one it replaces. Part numbers are stamped on the PROM and can be found in manufacturers' parts catalogs and service bulletins. With GM systems, a scan tool can read and display a PROM ID number to verify that the correct one is installed. Use of the incorrect PROM in a vehicle will result in false codes and driveability problems at the very least, and possible system damage at the other extreme.

Ford calls its PROMs "calibration assemblies." In addition to the basic PROM, GM also uses two other types: a CALPAC and a MEMCAL. The CALPAC (calibration package) is a smaller chip that is used in addition to the PROM with TBI fuel-injected engines. It contains backup logic that will control fuel delivery if the electronic control module (ECM) or powertrain control module (PCM) malfunctions. The CALPAC allows the engine to run using only throttle position and ignition reference pulses to calculate fuel and ignition changes. A missing or defective CALPAC, however, can cause a no-start or no-run condition and cause a trouble code to set.

A MEMCAL combines the functions of a PROM and CALPAC in a single unit, figure 5-28. If faulty, damaged, or incorrectly installed, the MEMCAL will set a trouble code. Like a PROM, a CALPAC or MEMCAL must be transferred when installing a replacement controller.

GM PROMs are housed in small dual-inline-package (DIP) devices that have two rows of pins. With early controllers, the PROM plugs directly into mating socket openings in the computer. With late-model controllers, however, the PROM is held in a small plastic carrier, figure 5-29. The carrier plugs into the controller and is removed with a special tool, figure 5-30. Use of the carrier and removal tool helps to avoid damaging the PROM pins during removal and installation. Some 1986 and later controllers have a single carrier to which the PROM and CALPAC are both permanently attached, and thus must be serviced as a unit. MEMCAL

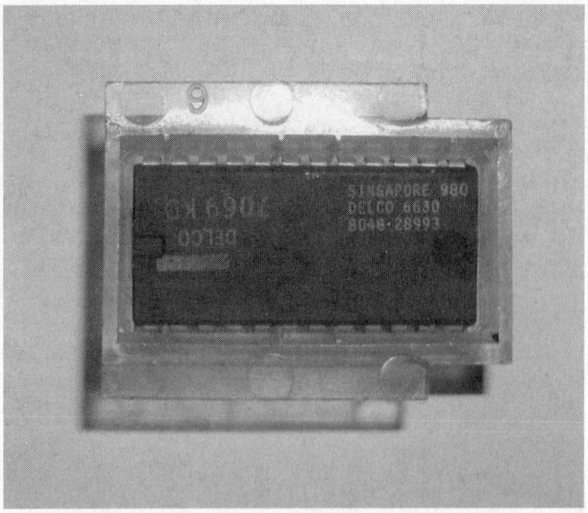

Figure 5-29. Some PROMS are inserted into a plastic carrier to ease installation and prevent damage.

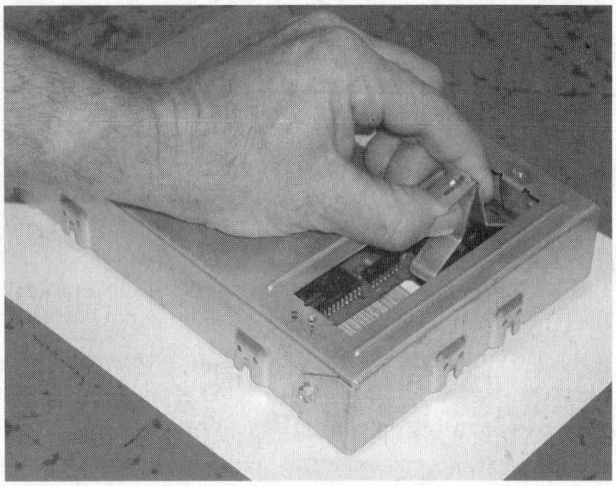

Figure 5-30. A special tool is provided to aid in removing the PROM and carrier from the PCM.

replacement involves unsnapping latches on either side of the unit for removal. When the new unit is installed, gently press downward on the MEMCAL while snapping the latches inward until they click into place.

When replacement of a GM PROM, CALPAC, or MEMCAL is required, always make sure the ignition is off before disconnecting the battery ground cable, and then remove the ECM or PCM from the vehicle's passenger compartment. If the ignition is on when the battery cables are disconnected or reconnected, a voltage surge may destroy the controller. The exact location of the ECM or PCM varies from vehicle to vehicle, but most are located behind the right-side kick panel or beneath the instrument panel. Trying to install a PROM with the ECM or PCM in the vehicle is a tempting time-saver, *but it can be fatal to the PROM.* Static electricity can irreparably damage memory chips. Simply sliding across the seat could create enough static electricity to wipe out the PROM when it makes contact with the controller pin sockets. (The same precautions should be taken with any engine control computer when servicing the PROM or calibration assembly.)

PROM package and controller socket designs differ from year to year or even from vehicle to vehicle. A PROM also is installed in different directions, depending upon whether the engine is carbureted or fuel injected. For these reasons, always refer to the manufacturer's instructions for a given vehicle before replacing a PROM. A CALPAC is replaced with a different removal tool, but the procedure and precautions are essentially the same. PROMs and CALPACs that are permanently attached to the carrier must be serviced as a unit.

When installing a PROM, CALPAC, or MEMCAL on late-model controllers, be especially careful about applying too much vertical force. The ECM or PCM circuit board is thinner than those used in earlier controllers and excessive force can cause the circuit board to flex, which may damage other components or even the circuit board itself.

Adaptive Memory and Long-Term Memory Recalibration

Various factors (such as production variations, engine wear, and variable fuel quality) make an engine's real needs different from the design values used to establish the initial control program stored in the microprocessor and PROM.

To compensate for these variables, an "adaptive memory" or "learning ability" feature in late-model engine control systems monitors operation of the engine and its related systems. Whenever a controlled value regularly falls outside design limits, the adaptive memory makes a small modification to the control program, reestablishing proper operation.

Adaptive memory program modifications are stored in keep-alive memory (KAM), a short form of RAM, and lost when power to the computer is disconnected. When this occurs on an older or high-mileage vehicle, the owner may complain of driveability problems until the car has been driven long enough to allow the adaptive memory to again fine-tune the engine operation. To "teach" the memory, the vehicle should be at operating temperature and driven at part throttle with moderate acceleration and idle conditions until normal performance returns. This usually requires about 20 miles of driving under varied operating conditions.

Some systems use a special electronically erasable programmable read-only memory, or EEPROM. The programming of the normal PROM is usually permanent. If defective, it is discarded. An EEPROM, however, can be erased with a special procedure and reprogrammed to correct any defects in the original programming or to improve driveability.

OBD II SUMMARY

This chapter has discussed the many commonalities associated with OBD II systems. Governmental agencies like the Environmental Protection Agency (EPA) and the California Air Resources Board (CARB) enlisted the help of various industry groups in an effort to mandate a uniform set of test capabilities and procedures to which all manufacturers must adhere, starting with 1994 models sold in California.

OBD II requires a standard 16-pin test connector for the engine control computer, with standard pin assignments. All domestic, European, and Asian vehicles sold in the United States must transmit a minimum list of standard trouble codes and computer system data to a scan tool. After meeting the minimum standard requirements, manufacturers can include other codes, data, and even particular tests relating only to their systems. The first OBD II systems installed on 1994 models were designed primarily to monitor the three-way catalyst's deterioration rate and spark plug misfires that lead to catalyst failures.

6

Testing Engine Management Input Devices

OBJECTIVES

Upon completion and review of this chapter, you will be able to:

- Identify and test switched inputs.
- Test temperature sensors and circuits.
- Test the throttle position sensor(s).
- Test Mass Air Flow sensors.
- Test oxygen sensors and oxygen sensor heater circuits.

USING TROUBLE CODE CHARTS

All OBD II vehicles (1996 and newer) and many pre-OBD II vehicles have trouble codes associated with many input sensors. The trouble code indicates that a problem has developed in the circuit or sensor that provides an input to the engine management computer. Faults in these circuits can be caused by faulty sensors, open circuits, shorted circuits, defective PCM, or excessive resistance in the circuit. It is important to remember that a trouble code does not necessarily mean that a particular sensor is bad. It is important that the technician always look at and test the entire circuit as well as the sensor that the trouble code relates to.

When a trouble code has set, the sensor often gets replaced, only to have the vehicle return for the same problem at a later time. This is especially prevalent on vehicles that have intermittent problems. Remember that in most cases a troubleshooting chart is designed to diagnose a fault or condition that is present at the time of testing. If the problem is intermittent and the fault is not present at the time of testing, the troubleshooting procedure will usually not lead you to a successful repair.

Another important part of code diagnostics charts is that they usually list the conditions (parameters) for setting the code. Knowing the conditions that need to be met in order for the code to set is very important to the technician. This is especially useful when diagnosing an intermittent code. Knowing these parameters is necessary so the technician can operate the vehicle and meet the criteria needed in order for the code to be run. Once a repair has been made the only way to validate the repair is to ensure that the vehicle is operated under the correct conditions so that the PCM will run the DTC. If the repair was successful, then the DTC will pass; if not, the DTC will fail.

The conditions for running the diagnostic test for an oxygen sensor heater follow. As can be seen, many conditions have to be met before the diagnostics will run. If the technician has made a repair or is trying to duplicate the code in the case of an intermittent problem, care will have to be exercised to ensure that the correct conditions are met in order for the diagnostics to run. For example, if the coolant temperature is above 95°F at startup, the test will not run. Therefore, the technician will need to let the vehicle cool down before the repair can be verified. Always study the code setting conditions before attempting to repair a code.

Conditions for Running the DTC

- No TP sensor, EVAP system, misfire, IAT sensor, MAP sensor, fuel trim, fuel-injector circuit, EGR pintle position, ECT sensor, CKP sensor, or MAF sensor DTCs are present.
- Engine run time is more than 3 seconds.
- System voltage is between 9 and 16 volts.
- Intake air temperature (IAT) is less than 35°C (95°F) at startup.
- Engine coolant temperature (ECT) is less than 35°C (95°F) at startup.
- IAT and ECT are within 6°C (11°F) of each other at startup.
- Average mass airflow for the sample period is less than 18 g/s.

Conditions for Setting the DTC

HO_2S 1 voltage remains within 150 mV of the bias voltage (about 450 mV) for a longer amount of time than it should. The amount of time ranges between 50 and 80 seconds and depends on engine coolant temperature at startup and average mass airflow since startup.

CURRENT CODES VERSUS HISTORY CODES

Current (hard) codes and history (soft) codes differ on OBD II and non-OBD II vehicles. It is important for the technician to understand the differences between these two systems. The MIL light is used to indicate that a problem has occurred within the engine management system and a code or codes has been set in the PCM's memory. The operation of this light and how codes are set differ between OBD II and non-OBD II equipped vehicles.

NON-OBD II VEHICLES

Generally most non-OBD II systems utilize hard (current) and soft (history) codes. In general, anytime a hard code sets the MIL will illuminate. That means that if the vehicle comes in with the MIL illuminated then in most cases the problem is present at that time. There are some exceptions to this rule in that some codes are "latched" codes, meaning that when the code sets the light will illuminate and stay illuminated for that ignition cycle even if the problem goes away. For most vehicles and codes, if the MIL is illuminated you can assume that the condition that caused the code to set is present at that time.

Soft (history) codes are set when a problem existed that caused the code to set; however, since the code set the problem has gone away. Therefore, in most cases, the presence of a soft (history) code usually indicates that the problem is of an intermittent nature. Whenever the technician finds soft (history) codes and no MIL illuminated, he or she should approach the problem as an intermittent condition.

OBD II VEHICLES

On OBD II vehicles the illumination of the MIL lamp does not always mean that the condition that caused the code to set is present at that time. The reason for this is because once an OBD II code is set the MIL will remain illuminated until the code successfully passes three consecutive trips. At that point, the MIL will turn off and the code will become a history code. This means that even though the MIL lamp is on at the time of repair the actual cause for the fault may not be present at that time. Therefore, care must be taken when performing code diagnostics using trouble charts. In order to validate the repair the technician will need to use a scan tool to ensure that the code has passed the current trip during a road test, if necessary.

History (soft) codes on an OBD II vehicle mean that while a current code was set at one time the problem has gone away and the code has successfully passed three consecutive trips. This indicates that the code in question is most definitely intermittent in nature. Troubleshooting these codes is much more difficult because of the intermittent nature. Care must be taken by the technician to validate the repair to ensure that the problem has been resolved.

TESTING SWITCHED INPUTS

Switched inputs are inputs to computers that are either on or off depending on the position of a switch controlling the circuit. Switch inputs may be pull-up inputs, figure 6-1, which means that a voltage is supplied through a switch to the PCM. Pull-down switched inputs, figure 6-2, are inputs that have a voltage supplied from the PCM. This voltage is pulled down to ground when the switch is closed.

Testing a Switched Input Using a Scan Tool

1. Connect the scan tool and select engine data. Scroll through the menu until the parameter of the switched input is shown, figure 6-3.

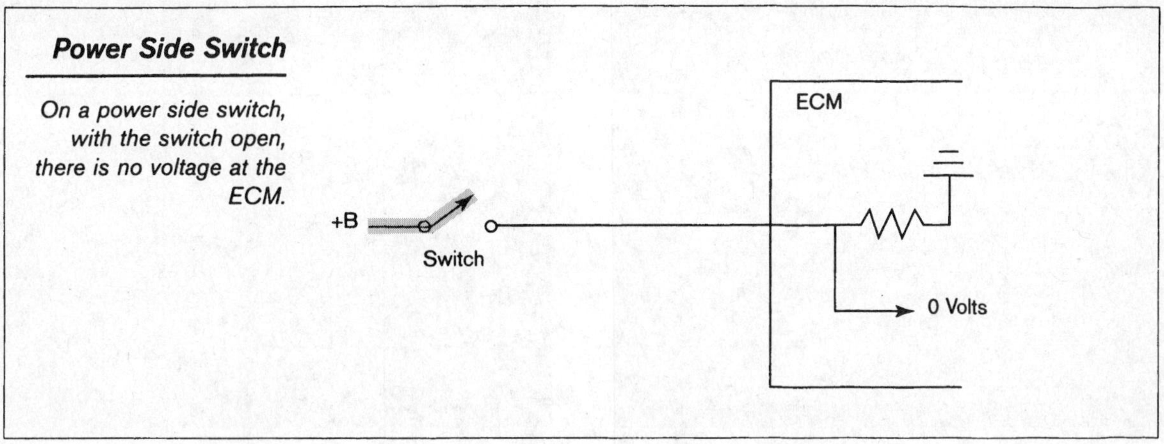

Figure 6-1. A pull-up switched input. (Provided courtesy of Toyota Motor Sales U.S.A., Inc.)

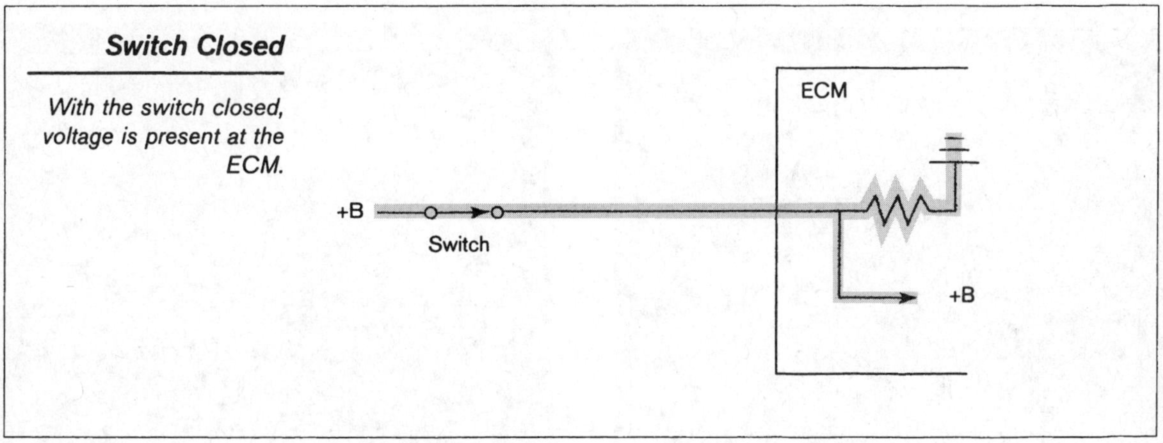

Figure 6-2. A pull-down switched input. (Provided courtesy of Toyota Motor Sales U.S.A., Inc.)

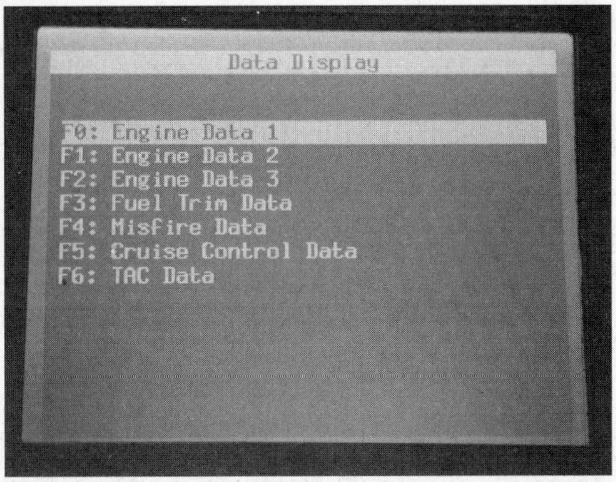

Figure 6-3. Connect the scan tool and choose the menu that shows the switched input that you are testing.

2. Observe the parameter's current state on the scan tool. Depending on the vehicle the scan tool may indicate "Yes or No" or "On or Off," figure 6-4.
3. Activate the switch while observing the scan tool to determine if the switch changes state as the switch is activated, figure 6-5.
4. If no change of state occurs, possible causes are: an open in the circuit, a short to ground in the circuit, a defective switch, or a defective PCM.
5. After repairing the circuit, verify that the switched input changes state with the scan tool.

Testing a Switched Circuit Using a DVOM

1. Using a schematic, back probe the circuit from the switched input at the PCM using a DVOM, figure 6-6.

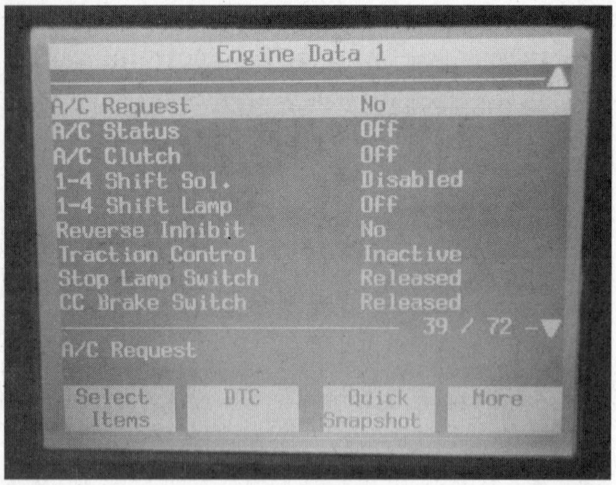

Figure 6-4. Observe the switch status on the scan tool. In this example, the AC request input from the AC control head is being tested.

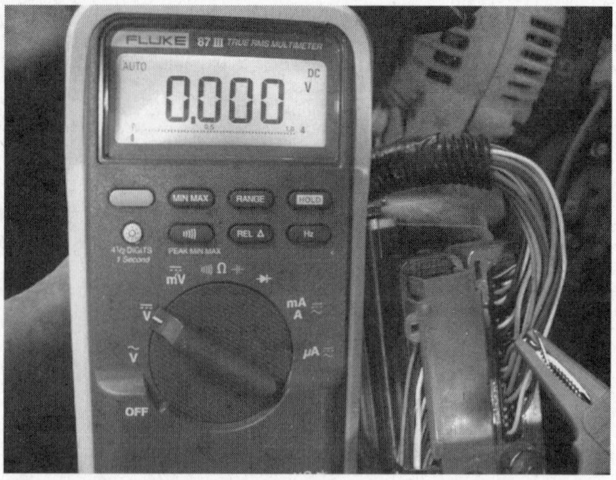

Figure 6-6. Using a schematic, identify the PCM terminal for the switched input. Carefully back probe the terminal and connect a DVOM on DC volts.

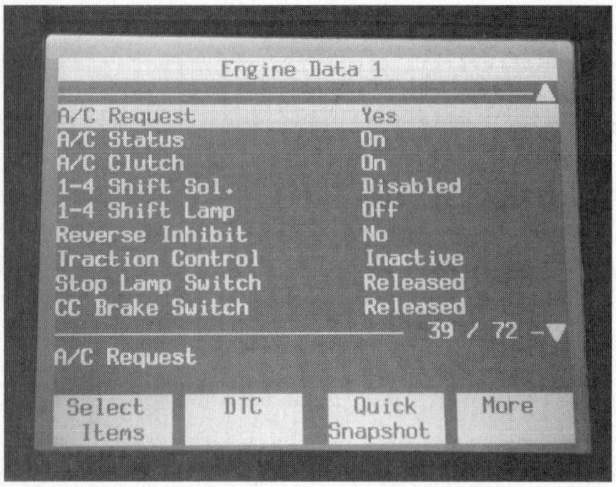

Figure 6-5. After requesting AC operation on the control head, check the AC request on the scan tool to determine if it has changed state.

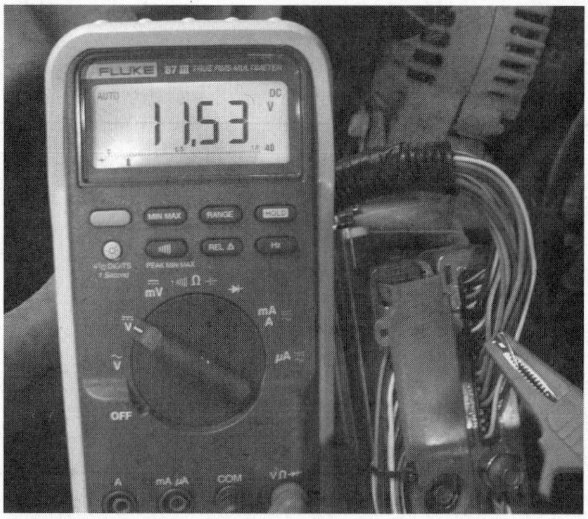

Figure 6-7. Since this is a pull-down circuit, voltage should be present without the switch activated.

2. Depending on whether it is a pull-up or pull-down circuit, the DVOM will measure close to zero volts or 12 volts, figure 6-7.
3. Activate the switch and observe the DVOM reading. The voltage should change from zero volts and go to approximately 12 volts if the circuit is a pull-up circuit. If the circuit is a pull-down circuit, the DVOM should change from 12 volts to zero volts, figure 6-8.
4. If no change of voltage occurs, the possible causes are: an open circuit, a circuit shorted to ground, a defective switch, or a defective PCM.
5. After repairs have been made, verify that the DVOM shows a change in voltage when the switch is activated.

ANALOG INPUTS

Analog inputs are inputs whose voltage signal is constantly variable. Analog inputs can be both AC-producing devices as well as DC devices. Analog input type sensors may be two-wire or three-wire sensors. Some examples of input devices that produce an analog signal are:

- Throttle-position sensor (TPS)
- Engine coolant temperature sensor (ECT)
- Intake air temperature sensor (IAT)
- Magnetic crankshaft position sensor (CKP)
- Magnetic vehicle speed sensor (VSS)
- Manifold absolute pressure sensor (MAP)
- Analog type mass air flow sensor (MAF)

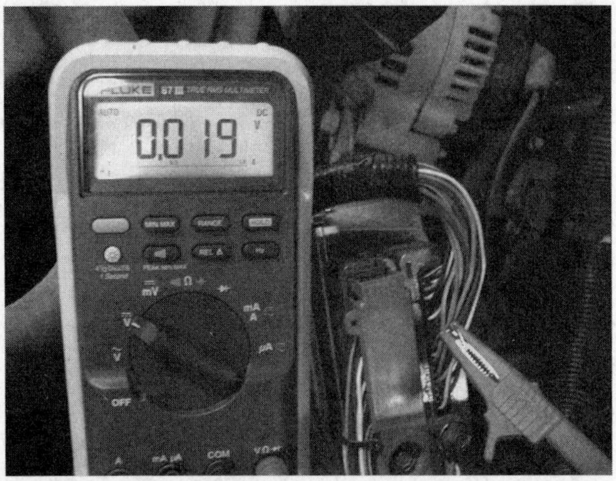

Figure 6-8. When the switch is activated, the voltage should drop to close to zero volts.

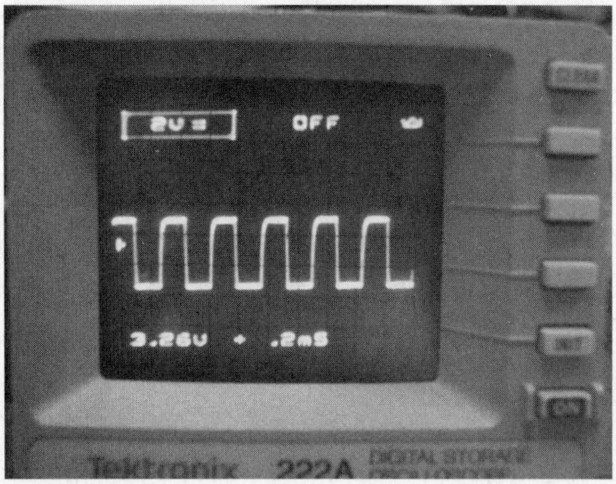

Figure 6-9. An oscilloscope is the best tool to test a digital input signal.

While this list does not include all the possible types of analog sensors, it encompasses the most commonly used sensors. With the exception of the magnetic type sensors, these sensors are provided a 5-volt reference from the PCM and the sensor modifies the 5-volt signal. Therefore, most of these sensors have a variable voltage output of between zero volts and 5 volts.

In order for the technician to successfully test these input devices, the technician needs to know what type of input device is being used and what the operating range of the sensor is. Most manufacturers give a written description of the input devices used on their systems. By reading the description of the sensor and studying a schematic for that particular sensor, a technician should be able to test analog inputs fairly easily. Testing analog inputs can be accomplished using a DVOM, scan tool, or an oscilloscope. Testing these individual input devices is covered in more detail later in this chapter.

DIGITAL INPUTS

Digital inputs are the third type of input sensors. These inputs produce an on-off voltage or digital voltage. They may operate at different voltage levels and produce signals that vary in frequency. Computers can easily read digital signals. Of the three types of input signals, digital inputs are the most difficult to test. In some instances a scan tool may be helpful. A DVOM is not always a useful tool because it will simply average the on and off values of a digital signal which in many cases produces a steady DC voltage reading on the meter. A DVOM that can measure frequency is often a better tool to try to measure a digital signal. The most effective tool in testing digital signals, an oscilloscope, will display the digital signal very clearly on the screen, figure 6-9. Technicians do not always have access to an oscilloscope. Therefore, when testing a digital signal other methods must be used such as measuring frequency or using a MIN-MAX record feature that some DVOMs have. Various testing methods will be covered later in this chapter.

TESTING TEMPERATURE SENSORS

Temperature sensors are thermistor devices that change resistance with temperature. In most automotive applications, negative coefficient thermistors are used. This means that as the temperature increases the resistance of the sensor decreases. Testing of temperature sensors is the same regardless of whether the technician is testing an intake air temperature sensor, engine coolant sensor, or fluid temperature sensor. Even between manufacturers, these test methods remain the same. The only difference between temperature sensors is that their temperature-resistance relationship may be different between manufacturers.

Testing Temperature Sensors with a Scan Tool

A scan tool can be used to quickly verify the accuracy of a temperature sensor. In addition, if the sensor reading is not correct, the scan tool can help in diagnosing the circuit and sensor. To test a temperature sensor with a scan tool, follow these steps:

1. Install a scan tool and choose the menu that shows data parameters. If the vehicle is at ambient air temperature you can compare the temperature sensor reading to other temperature sensor reading to determine if the reading is accurate, figure 6-10.

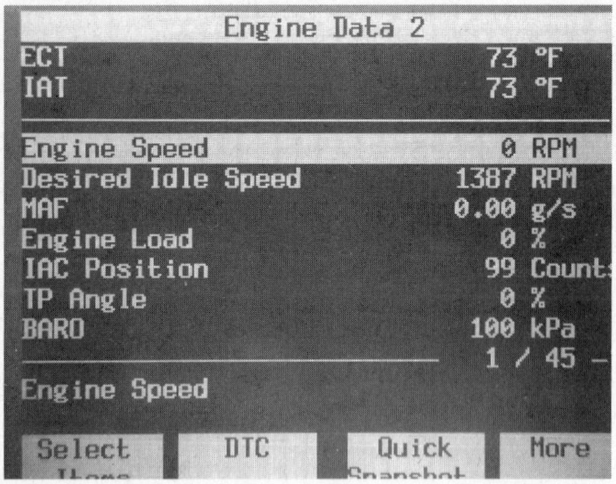

Figure 6-10. If the vehicle is at ambient air temperature, compare different temperature sensor readings to each other to determine their accuracy.

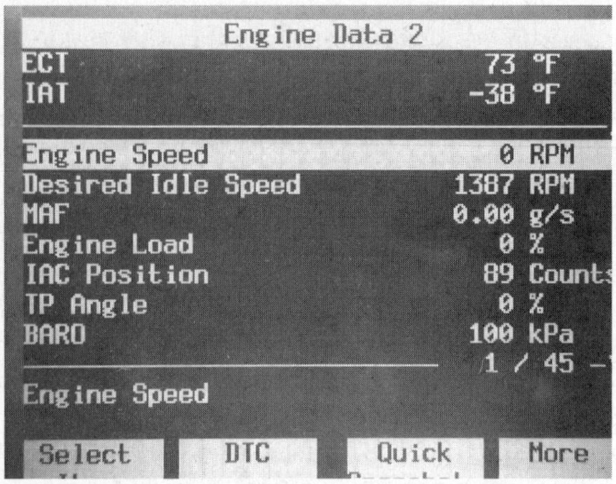

Figure 6-11. Disconnect the temperature sensor that you are testing. Observe the reading on the scan tool. It should show a negative reading.

2. If the temperature sensor reading is not accurate, disconnect the sensor. Observe the scan tool reading, figure 6-11. The reading on the scan tool should show a negative temperature reading. (How low the reading will go depends on the manufacturer.)
3. Short the 5-volt reference wire to the signal low (ground) using an appropriate jumper wire, figure 6-12.
4. Observe the scan tool reading. It should read 5 volts or, if the scan tool only shows temperature, it should show about 260°F or above, figure 6-13.
5. These steps check the integrity of the circuit and the ability of the PCM to interpret the signal. If both voltage and/or temperature reading reach

Figure 6-12. Jumper the temperature connector terminals together using the appropriate jumper wires.

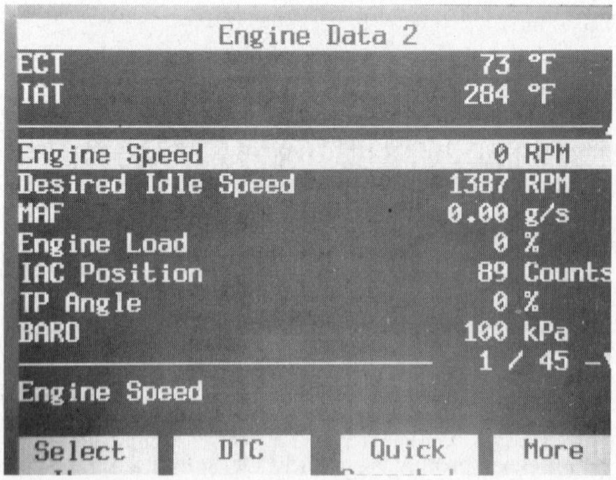

Figure 6-13. Observe the scan tool with the terminals jumped. The reading should be greater than 260°F.

their extreme limits, then the PCM and the circuit are OK. The temperature sensor should be replaced. If these results are not achieved then test the circuit for excessive resistance, shorts, or grounds. If no problems are found then replace the PCM.

Testing Temperature Sensors with a DVOM

A DVOM can be used to test temperature sensors and circuits. You can measure the resistance of the temperature sensor and compare its resistance value at a certain temperature to a temperature/resistance table provided by the manufacturer. In addition, the integrity of the circuit and the 5-volt reference can be verified with the DVOM. To test a sensor and circuit with a DVOM, do the following:

1. Disconnect the sensor. Connect a DVOM to the sensor and measure the resistance of the sensor, figure 6-14.
2. Measure the 5-volt reference voltage supplied to the sensor. Your measurement should be within a couple hundredths of 5 volts, figure 6-15.
3. Verify the ground to the sensor by measuring between the 5-volt reference and the ground terminal of the connector. The reading should be very close to the reading obtained in step 2, figure 6-16.
4. If the 5-volt reference measurement is too low, check for excessive resistance in the 5-volt circuit. Check for an open or shorted 5-volt circuit. If there are no opens, shorts to ground, or excessive resistance in the circuit, then replace the PCM.
5. If the resistance value of the sensor does not match the manufacturer's specifications, then replace the sensor.

Figure 6-14. Disconnect the temperature sensor and measure the resistance of the sensor. Compare the reading to a temperature/resistance table provided by the manufacturer.

TESTING THE THROTTLE-POSITION SENSOR

The throttle-position sensor is a potentiometer that connects to the throttle plates. It sends a variable voltage signal to the PCM to indicate throttle position and

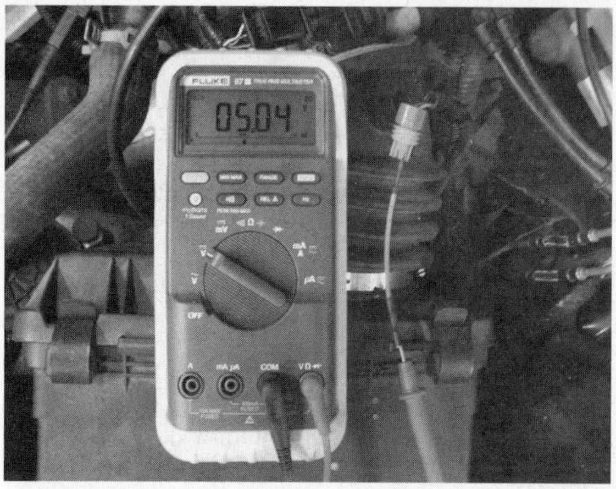

Figure 6-15. Measure the 5-volt reference circuit with the DVOM. The reading should be within a couple hundredths of a volt of 5 volts.

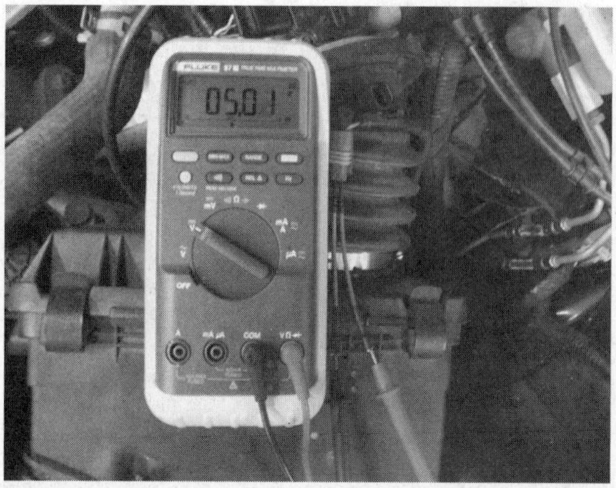

Figure 6-16. To verify the ground to the sensor, measure between the 5-volt reference and ground. This reading should be very close to the 5-volt reference reading. If it varies more than .02 volts, check the ground circuit.

also throttle movement. This signal varies between 0 and 5 volts; however, the voltage usually does not go below .3 volts or above 4.8 volts. If the voltage goes above or below these values, the PCM typically sets a code for high or low voltage.

Because the throttle-position sensor is mechanical in nature, it is not uncommon for it to fail at some point in the vehicle's life. Typical failures are often intermittent opens in the sensor as it moves. The TPS can be tested by various methods. The scan tool is not the best method to test a TPS because often a glitch in the TPS will not show on the scan tool. This is because a scan tool can only display samples based on the speed of the communications line. The best method for testing a TPS is either an oscilloscope or a DVOM with a MIN-MAX record feature. To test a TPS, follow these steps.

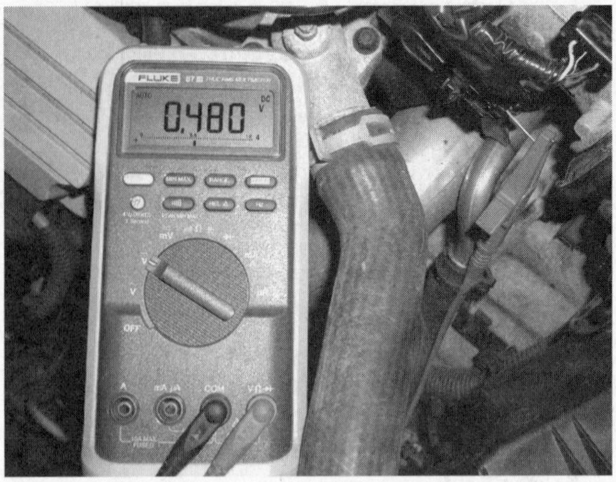

Figure 6-18. Turn the ignition on and set the DVOM to read DC voltage. This reading represents the closed-throttle voltage.

Testing the TPS Using a DVOM with MIN-MAX Record

1. Back probe the TPS connector signal wire with a suitable device (T-pins work very well for this), figure 6-17.
2. Turn the ignition on and set the DVOM to DC volts, figure 6-18.
3. Set up the DVOM in the MIN-MAX record mode, figure 6-19.
4. Slowly open the throttle until wide-open throttle is reached.
5. Play back the MIN-MAX record mode and observe the minimum voltage recorded by the meter, figure 6-20.
6. The minimum value recorded by the meter should not be lower than the closed throttle reading obtained in step 2. An open anywhere in the TPS range will result in the meter capturing a value less than the closed throttle reading.

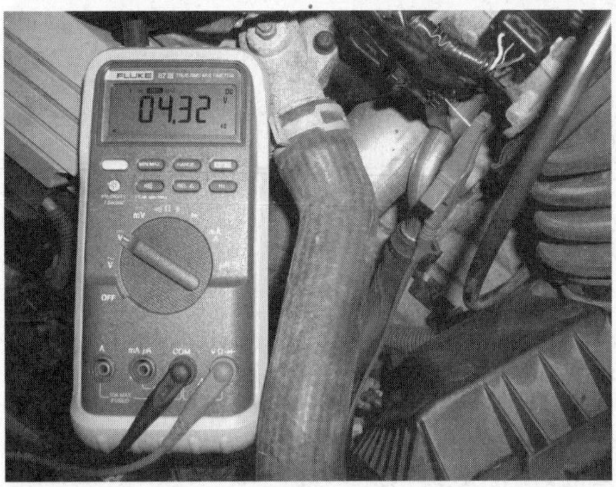

Figure 6-19. Set the DVOM to record the voltages on the MIN-MAX record mode.

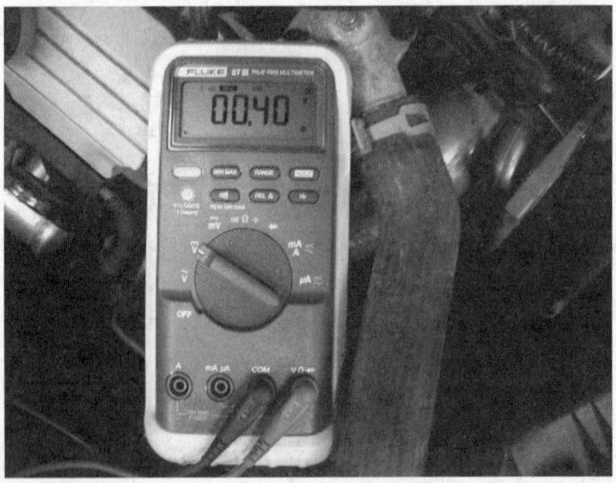

Figure 6-20. After opening the throttle by hand, play back the minimum and maximum values. The minimum should not be much lower than the closed throttle voltage recorded in step 2.

Figure 6-17. Back probe the TPS signal wire using a suitable back probe device. T-pins work very well for this.

TESTING MANIFOLD ABSOLUTE SENSORS

Manifold absolute sensors (MAP) are pressure sensors that sense engine vacuum. The analog type of MAP sensor produces a voltage signal that ranges from 0 to 5 volts. When no vacuum is present at the MAP sensor, the voltage output should be high, about 4.8 volts. At high vacuum levels the output of the MAP sensor should be low, about 1 volt. MAP sensor calibrations differ between manufacturers, so be sure to check the manufacturer's specification before testing the MAP sensor.

Ford MAP sensors differ from most other manufacturers in that they produce a digital voltage from the sensor. The Ford MAP sensor produces a digital signal that varies frequency as the vacuum changes. At 0 in. Hg the Ford MAP sensor produces a frequency of approximately 150 to 159 hertz. At 20 in. Hg the frequency output of the sensor decreases to approximately 102 to 104 hertz. These sensors should be checked by using a DVOM that can measure frequency. As vacuum is supplied with a hand-held vacuum pump, the frequency should decrease as the vacuum increases.

To test an analog-type MAP sensor using a DVOM and vacuum pump, do the following:

1. Disconnect the MAP sensor from the vacuum source.
2. Leaving the MAP sensor connector plugged in, back probe the signal wire at the MAP sensor, figure 6-21.
3. Connect a hand-held vacuum pump and set the DVOM to read DC volts, figure 6-22.
4. Apply vacuum in 2 in. Hg increments and record the voltage at each increment, figure 6-23.
5. Keep recording readings until you have applied between 20 to 25 in. Hg, figure 6-24. (This depends on your vacuum pump.)
6. Compare the readings to the manufacturer's specifications. If the readings are within specifications, then the MAP sensor is OK. Check the 5-volt reference and ground circuit for excessive resistance, grounded conditions, or open circuits. If no faults are found in the circuits and the MAP sensor meets specifications, then replace the PCM.

Figure 6-21. Disconnect the MAP sensor from the engine but leave the connector plugged in.

TESTING MASS AIR FLOW SENSORS

Two types of mass air flow sensors (MAF) are used on vehicles that are equipped with a MAF. The first and least commonly used is the vane airflow sensor. The

Figure 6-22. Connect the DVOM and record the MAP sensor reading. It should read close to 5 volts with no vacuum applied.

Figure 6-23. Apply vacuum in 2 in. Hg increments and record your results.

vane air flow sensor has a vane that attaches to a potentiometer. As the airflow moves the vane, the potentiometer moves and creates a variable voltage (analog) signal. These sensors were mostly used on 1980-era vehicles.

The most common type of MAF in use on current vehicles is the hot wire type of sensor. This sensor measures the amount of air entering the engine by heating a wire to an exact temperature, figure 6-25. As the air flows over the wire it cools the sensor wire. More current is needed to keep the wire at its desired temperature. The output of the MAF is converted to a digital signal that varies frequency as airflow changes. The PCM then converts this frequency to grams per second of airflow.

Figure 6-24. Continue applying vacuum until you reach 20 to 25 in. Hg. Compare the readings to manufacturer's specifications.

Testing a MAF with a DVOM

A MAF can be tested with a DVOM although this is not the best method. The DVOM must be able to measure frequency. Since the MAF produces a digital signal that varies in frequency, measuring the voltage of the signal wire will not indicate whether the MAF sensor is working correctly. However, if frequency is measured the technician should be able to observe the frequency change as the throttle plates are opened. The frequency should increase in a linear fashion with rpm, figure 6-26. To test a MAF with a DVOM, do the following:

1. Use a schematic to determine which wire is the signal wire. Most MAF sensors are three-or four-wire sensors. The three-wire sensor has power, ground, and signal wire. The four-wire sensor has power, ground, signal, and air temperature circuits.
2. Back probe the signal wire at the MAF sensor using the appropriate back probe device that will not damage the wiring, figure 6-27.
3. Connect the DVOM to the signal circuit and select frequency on the meter, figure 6-28.
4. Start the engine and allow the engine to idle, observing the frequency reading, figure 6-29.
5. Slowly increase the rpm and observe the frequency increase on the DVOM. It should increase proportionately with the increase of the rpm, figure 6-30.
6. If the frequency output is erratic or drops out as rpm increases, inspect the MAF sensor for contamination. If it is contaminated clean the wire element and retest. If the sensor is still erratic or the signal drops out, then replace the MAF sensor.

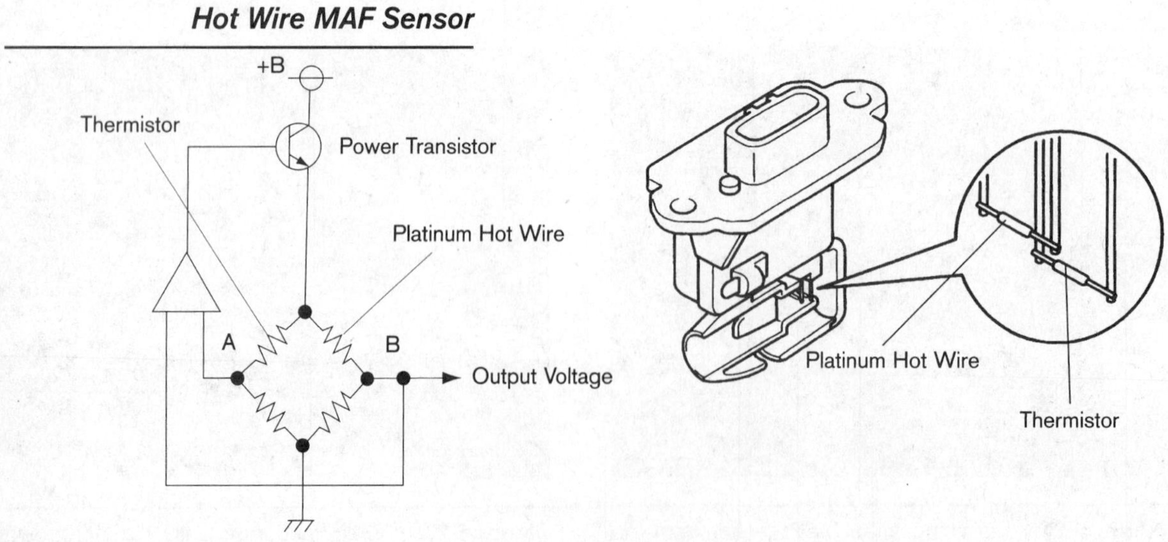

Figure 6-25. The hot wire MAF sensor. (Provided courtesy of Toyota Motor Sales U.S.A., Inc.)

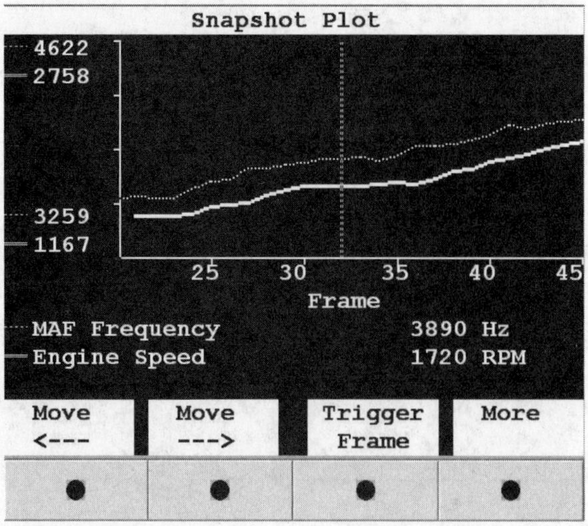

Figure 6-26. The frequency output of the MAF should increase as the rpm increases in a linear fashion.

Figure 6-27. Use a schematic to identify the signal wire of the MAF sensor. Back probe the signal wire using the appropriate back probe device.

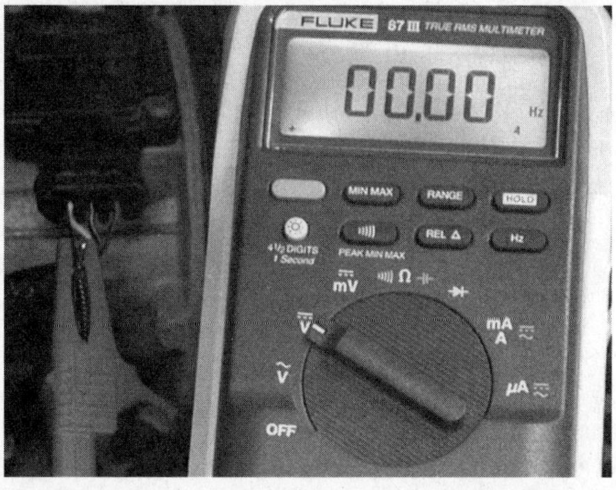

Figure 6-28. Set the DVOM to measure DC frequency.

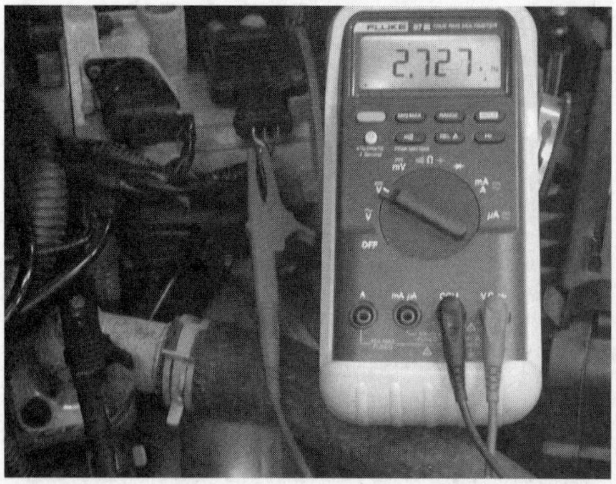

Figure 6-29. Start the engine and allow it to idle. Measure the frequency of the signal at idle.

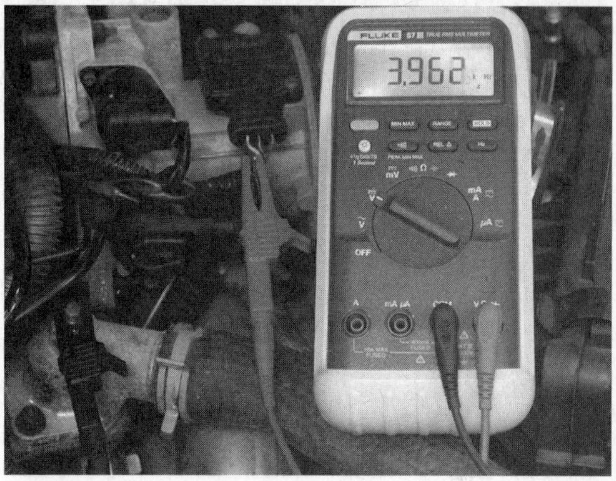

Figure 6-30. Slowly increase the engine rpms while observing the frequency on the meter.

7. If no signal is produced on the signal wire when the engine is running, check the power and grounds to the sensor and check the signal circuit for opens or grounded conditions.

Testing the MAF Sensor with an Oscilloscope

An oscilloscope can be used to test the MAF sensor. It actually is a better method than measuring frequency because the technician can see quick changes in the MAF sensor signal. It also is very effective in testing MAF sensors that are intermittent in nature. To test the MAF sensor using an oscilloscope, do the following:

1. Back probe the signal wire with a suitable back probe device, figure 6-31.
2. Connect the oscilloscope to the signal wire and ground, figure 6-32.

Figure 6-31. Back probe the signal wire at the MAF sensor.

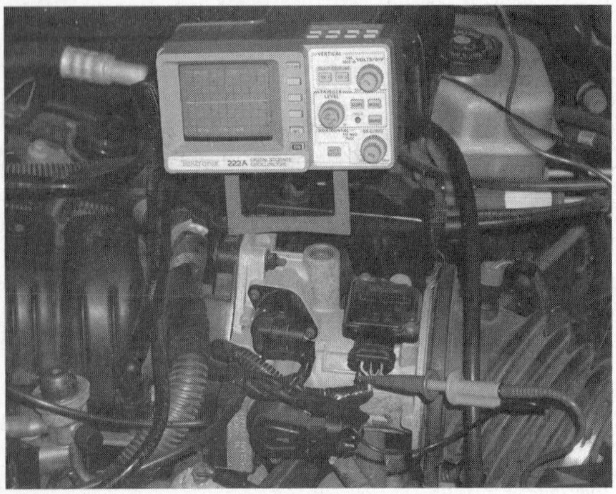

Figure 6-32. Connect an oscilloscope to the signal wire.

3. Start the engine and observe the signal on the oscilloscope, figure 6-33.
4. Raise the rpms and observe the signal frequency change on the oscilloscope, figure 6-34.
5. If no signal is present with the engine idling, test the power, ground, and signal circuits for opens, grounds, or excessive resistance.
6. If the signal is erratic or drops out when the throttle is moved, check the sensor for contamination. Clean the sensor and retest. If the signal is still erratic or drops out, replace the MAF sensor.

NOTE: Using aftermarket air filters that have elements that are oiled can cause contamination of the MAF sensor wire. In some cases the sensor can be damaged and must be replaced.

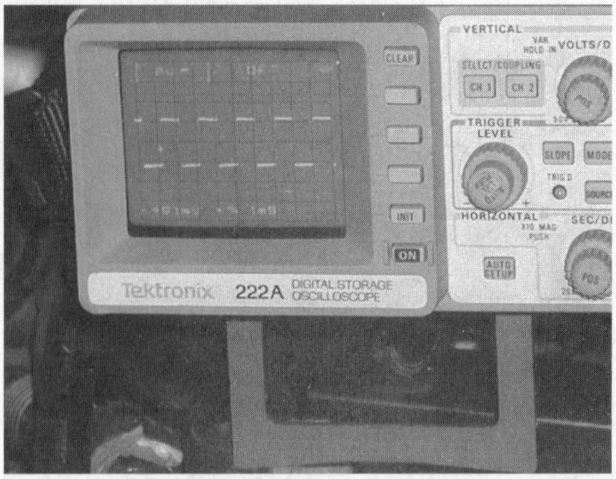

Figure 6-33. Start the engine and allow the idle to stabilize. The signal should be a smooth square wave.

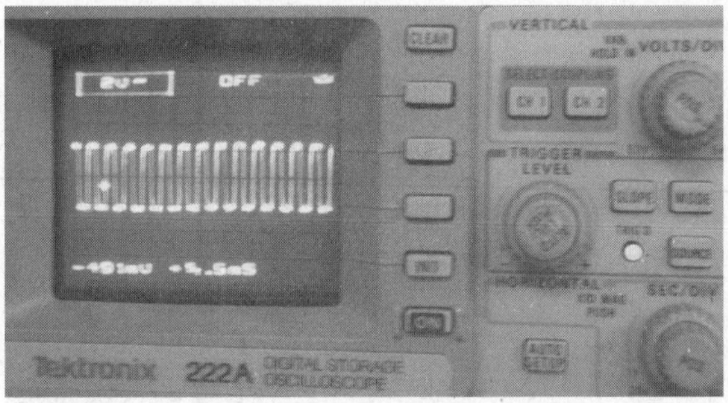

Figure 6-34. Slowly increase the engine rpms and observe the MAF sensor signal on the scope.

TESTING OXYGEN SENSORS

The zirconia oxygen sensor produces a voltage in proportion to the amount of oxygen in the exhaust stream. This voltage that is produced is between 0 and 1,000 mv. A rich exhaust stream has relatively little oxygen content which causes the zirconia type of oxygen sensor to produce a high voltage. When the exhaust stream is lean it has a high oxygen content and the oxygen sensor produces a low voltage. A properly operating oxygen sensor on a vehicle running properly should always be switching back and forth from lean to rich. This transition from rich to lean is referred to as oxygen sensor cross counts. A properly operating oxygen sensor should transition up to 10 times per second. As oxygen sensors degrade, their ability to switch decreases. These are often referred to as sluggish sensors. Many times a degraded oxygen sensor may not set a trouble code. However, a sluggish sensor may impact the emissions and/or driveability of the vehicle. Other oxygen sensor failures are usually detected by the PCM and a code is set. To test an oxygen sensor with a scan tool, do the following.

NOTE: In order to test the oxygen sensor the vehicle must be operating properly. If the vehicle is operating too lean or too rich, the oxygen sensor cannot be tested until proper operation of the vehicle is achieved.

1. Install a scan tool and start the vehicle. Operate the vehicle until the coolant temperature reaches 85° to 110°C, figure 6-35.
2. Set up the scan tool to record a snapshot (movie), figure 6-36.
3. Raise the engine rpm to 2,500 for one minute to precondition the oxygen sensor. Also ensure that the vehicle is in closed loop, figure 6-37.
4. Reduce the rpms to 2,000 and start the snapshot (record) on the scan tool, figure 6-38.
5. After the snapshot is complete, shut off the vehicle and play back the snapshot. Record the number of oxygen sensor voltage samples in the spaces below.

0–300 mV	______
301–450 mV	______
451–600 mV	______
601–1,000 mV	______

There should be more samples in the 0–300 mV and 601–1000 mV ranges. If more samples fall in the 301–450 mV and 451–600 mV ranges, then the oxygen sensor should be replaced.

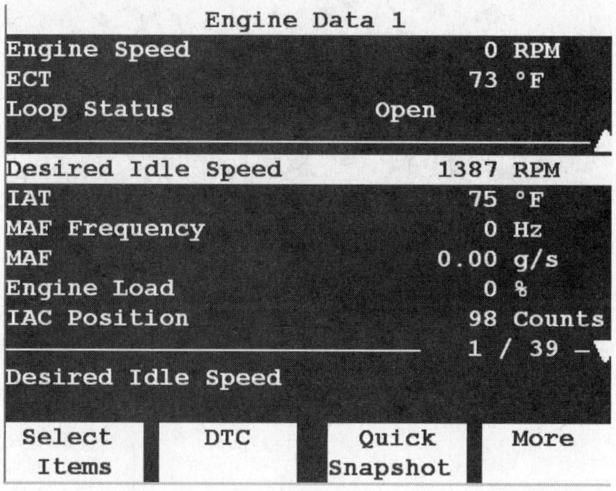

Figure 6-35. Operate the vehicle until the coolant temperature reaches 85° to 110°C.

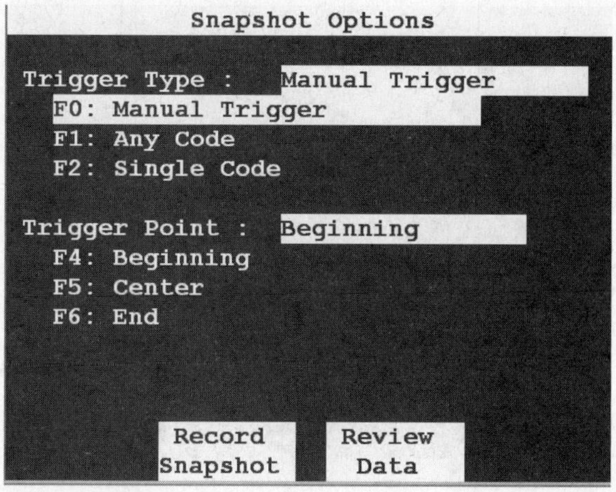

Figure 6-36. Set up the scan tool to record data.

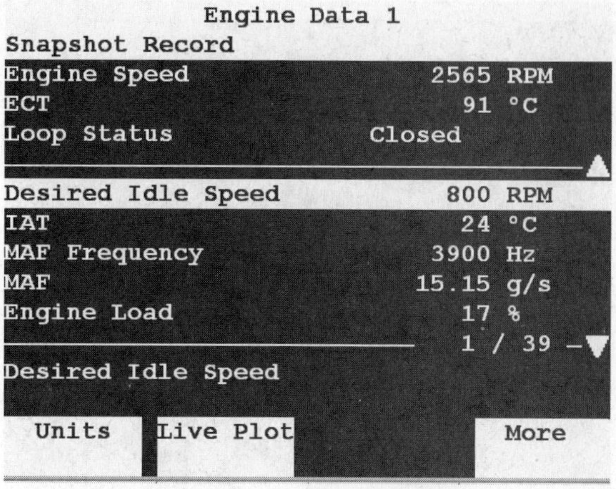

Figure 6-37. Run the engine at 2,500 rpms for 1 minute. Make sure the vehicle is operating in closed loop.

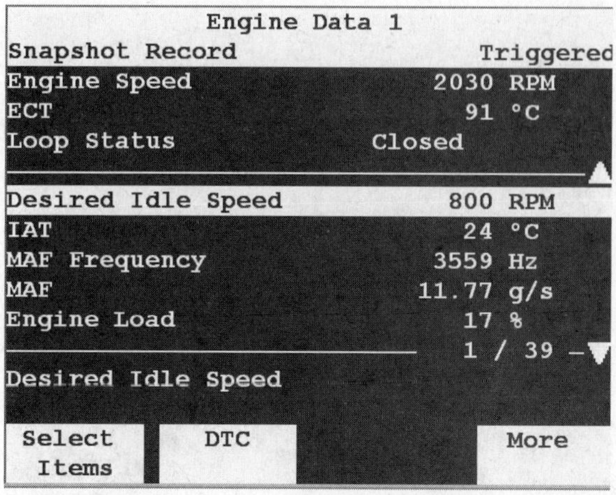

Figure 6-38. Maintain 2,000 rpms and trigger the scan tool to record the data.

Testing the Oxygen Sensor with a DVOM

1. Using a schematic, determine the signal wire from the oxygen sensor. Back probe the sensor signal wire with a T-pin or other suitable device, figure 6-39.
2. Set the DVOM to read DC volts.
3. Connect the red meter lead to the signal wire and a good ground.
4. Start the engine and run at 2,500 rpms for two minutes.
5. Select MIN/MAX mode on the DVOM and maintain engine rpm at 2,500 for two minutes, figure 6-40.
6. Record the minimum values, figure 6-41.
7. Record the maximum values, figure 6-42.
8. Record the average values, figure 6-43.

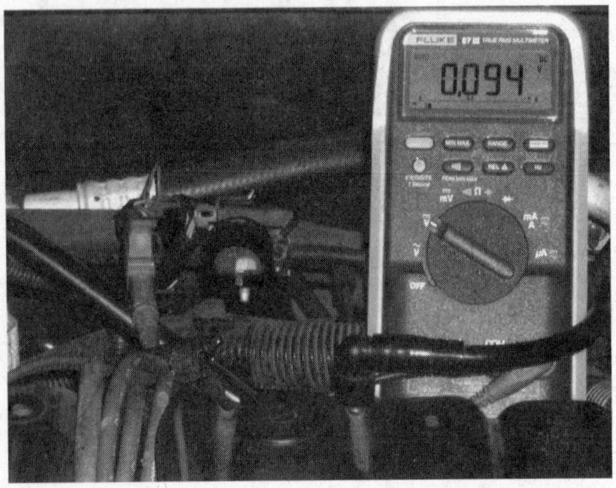

Figure 6-39. Back probe the signal wire from the oxygen sensor and connect the DVOM.

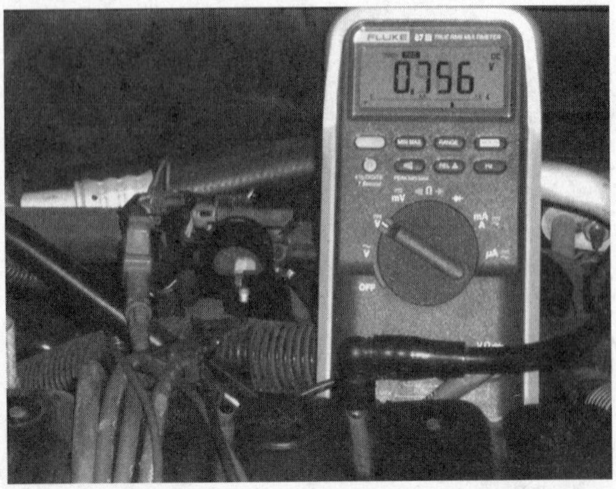

Figure 6-40. Set the DVOM to the MIN/MAX record mode and run vehicle for two minutes at 2,500 rpms.

9. The minimum value should be under 200 mV and the maximum value should be over 800 mV. The average should be about 450 mV. If the average is above 450 mV then the vehicle is operating on the rich side. If the average is under 450 mV then the vehicle is operating on the lean side or the oxygen sensor is defective. Check for lean conditions before replacing the oxygen sensor.

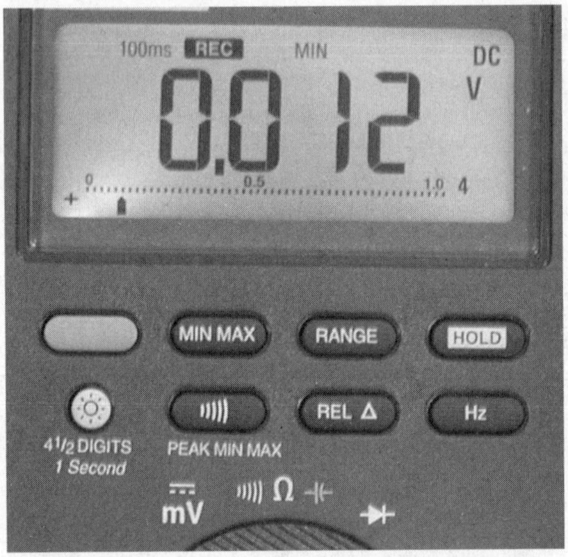

Figure 6-41. After two minutes, record the minimum voltage recorded. It should be under 200 mV.

Figure 6-42. View the maximum value recorded after two minutes. It should be above 800 mV.

Testing Oxygen Sensor Heaters

Most of the oxygen sensors used on current vehicles are of the heated type of sensors. This allows the sensor to start operating quicker and allows the vehicle to enter closed loop sooner. In addition, the post catalytic converter sensor is too far down the exhaust stream to stay hot enough to operate. OBD II systems require performance tests of the oxygen sensor heaters. However, some heated sensors are used on non-OBD II vehicles also. To test the oxygen sensor heater, do the following:

NOTE: The vehicle must be at ambient air temperature and the engine not started for at least two hours to perform this test.

1. Before turning the ignition on, connect a scan tool and program it for the vehicle, figure 6-44.
2. Select the data list that shows the oxygen sensor(s) voltages, figure 6-45.
3. Turn the ignition on but do not start the vehicle.
4. Observe the oxygen sensor(s) voltage. The voltage should be approximately 450 mV, figure 6-46.
5. Observe the oxygen sensor(s) voltage; within two minutes the voltage should have dropped to under 200 mV, figure 6-47.
6. If the sensor(s) voltage has dropped to under 200 mV in less than two minutes, then the oxygen sensor heater is working. If it has not, then the oxygen sensor heater is not working. Check the heater circuit for opens, grounds, or excessive resistance. If the circuit is OK, replace the oxygen sensor and retest.

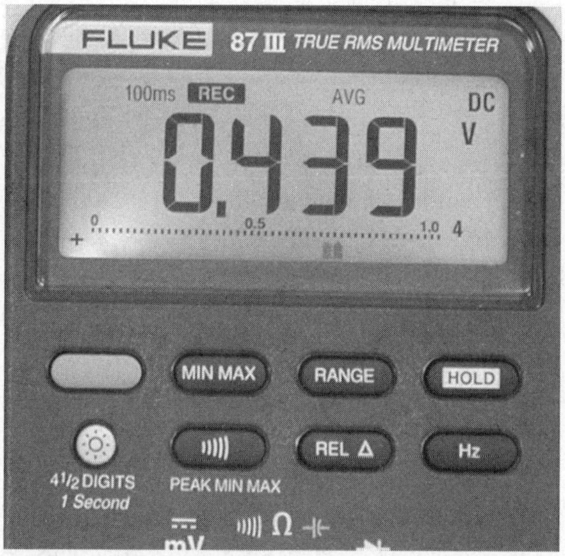

Figure 6-43. View the average voltage recorded after two minutes. It should be close to 450 mV.

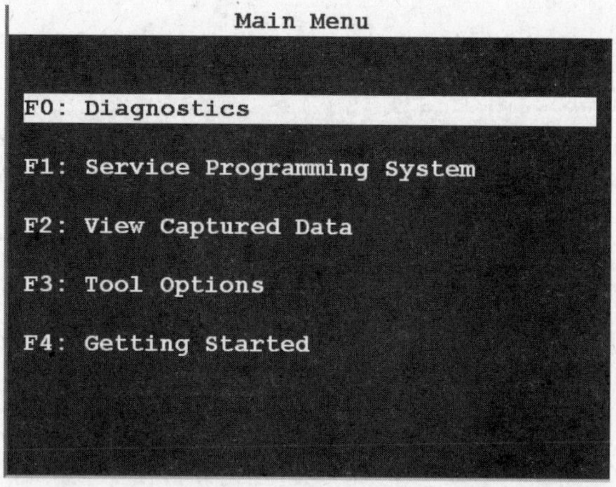

Figure 6-44. Before turning the ignition on, program the scan tool for the vehicle.

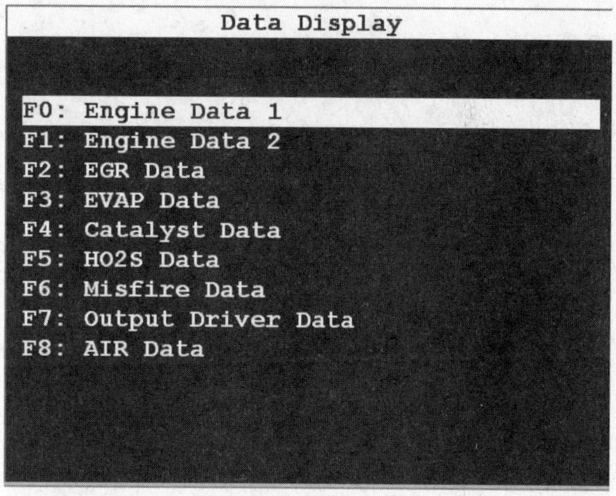

Figure 6-45. Select the data list that contains the oxygen sensor(s) voltage.

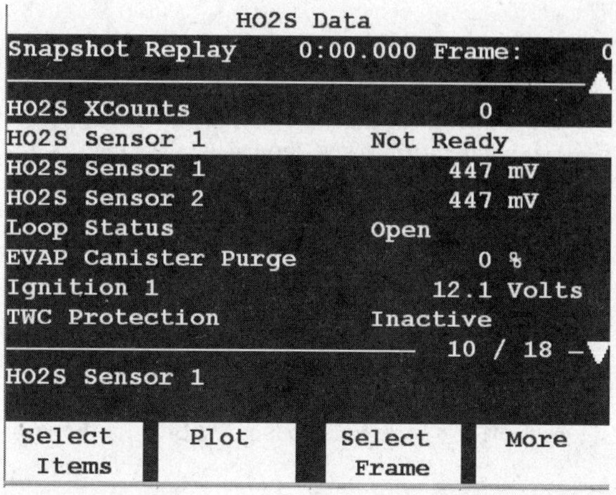

Figure 6-46. Observe the oxygen sensor(s) voltage on the scan tool. It should start at about 450 mV.

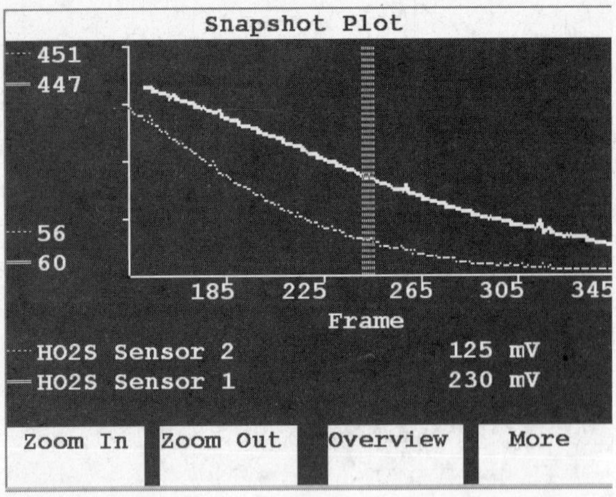

Figure 6-47. This plot shows that within two minutes the oxygen sensor(s) voltage dropped to under 200 mV, indicating that the oxygen sensor heater(s) is (are) working.

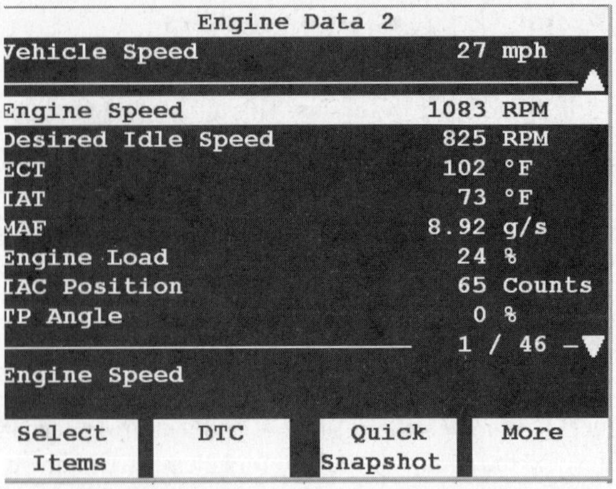

Figure 6-48. Verify the presence of the VSS signal to the PCM by using a scan tool.

TESTING OTHER INPUTS

Other input devices such as vehicle speed sensors, camshaft sensors, crankshaft sensors, and EGR pintle position sensors can be tested using a variety of methods such as testing with a DVOM, oscilloscope, or a scan tool. It is important for the technician to know what kind of sensor is being used on the vehicle before attempting to test the sensor. For instance, some crankshaft sensors are hall effect switches while other crankshaft sensors are magnetic. The technician must determine which type is being used before beginning testing. Testing crankshaft and camshaft sensors is covered in Chapter 12 of the *Shop Manual.*

Testing Vehicle Speed Sensors

Most vehicle speed sensors are of the magnetic type. They are usually located in the transmission. Magnetic speed sensors generate an AC voltage which changes frequency as the vehicle speed changes. Usually the vehicle speed sensor is a direct input to the PCM. On newer vehicles this speed input signal (which is received by the PCM) is then changed to a digital signal and sent to other systems such as the instrument panel cluster for operation of the speedometer and odometer. On other vehicles this speed information is transmitted via the serial data line to the various modules that need vehicle speed information. The quickest way to verify if the speed sensor is working is by using a scan tool, figure 6-48. If the vehicle speed is shown on the scan tool when the vehicle is being driven, then the vehicle speed sensor must be operating correctly. If no speed is indicated on the scan tool, then the vehicle speed sensor and circuit must be tested using a DVOM or oscilloscope. To test the vehicle speed sensor and circuit, do the following:

Figure 6-49. Carefully back probe at the PCM connector or the vehicle speed sensor.

1. Using a schematic, determine the circuits from the speed sensor to the PCM.
2. Back probe the circuit from the speed sensor either at the sensor or the PCM, whichever is easiest to access, figure 6-49.
3. If using a DVOM, connect the red meter lead to one circuit from the speed sensor and the black lead to the other circuit from the speed sensor. Select AC volts on the meter, figure 6-50.
4. With the wheels off the ground and the vehicle suitably supported, start the engine and place the transmission in drive.
5. Measure the voltage from the vehicle speed sensor, figure 6-51. You should obtain a minimum output of about one volt.

Figure 6-50. Connect the DVOM and select AC volts.

Figure 6-51. With the vehicle raised and properly supported, start the engine and place the transmission in "Drive."

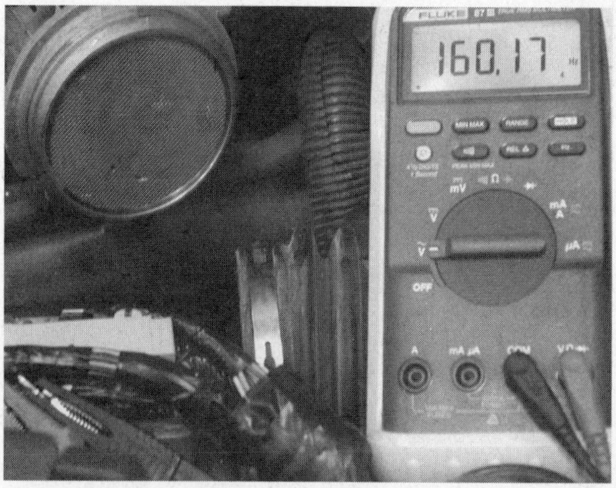

Figure 6-52. Select hertz on the DVOM and watch the frequency. It should increase with speed.

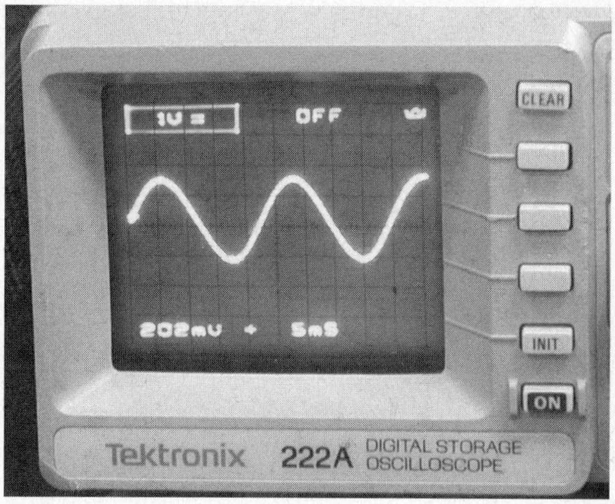

Figure 6-53. If an oscilloscope is available the signal can be checked with the scope.

6. Select AC frequency of the DVOM and observe the frequency of the vehicle speed sensor. The frequency should increase with speed, figure 6-52.
7. If you have the availability of an oscilloscope, connect the scope to the speed sensor circuit and observe the waveform on the scope. It should look like the pattern shown in figure 6-53.
8. If the correct readings were obtained with the DVOM or oscilloscope, check the circuit between the PCM and speed sensor for opens.
9. If no signal was obtained from the speed sensor, verify that the circuits are not grounded and replace the vehicle speed sensor.

7

Testing Engine Management Output Devices

OBJECTIVES

Upon completion and review of this chapter, you will be able to:

- Determine low-side and high-side drivers.
- Test fuel injector circuits.
- Test and or replace idle air control devices.
- Perform a functional EGR systems test.
- Test electronic throttle actuators and circuits.

INTRODUCTION

The majority of engine management output devices electrically work one of three ways. Some output devices are simply switched on or off. This is usually the case for relays that are controlled by the PCM. The second way outputs are controlled is by using pulse width modulation (PWM) to control the operation of a device. PWM-controlled devices can be controlled by the PCM at varying percentages of on and off time. These percentages can vary between 0 percent or off to 100 percent or full on. By varying the on and off time of a digital signal, very precise control can be achieved. An example would be a vacuum solenoid that controls the vacuum to an EGR valve. By varying the on and off rate (duty cycle) of the signal to the vacuum solenoid, the vacuum flow through the solenoid can be precisely controlled. Pulse width modulation can also control the injector on time or the movement of a linear motor such as an electronic EGR valve. The third type of output signal is the digital output. Digital output signals are often used to control stepper motors such as an idle air control motor used to control engine idle speed.

LOW-SIDE AND HIGH-SIDE DRIVERS

The term *driver* refers to the transistors inside the PCM that control output devices. Transistors used in this fashion are electronic switches that turn the output circuit on and off. A low-side driver is a transistor that controls the ground circuit to an output device. When the driver is switched on, the ground path to the output device is completed and current flows in the circuit energizing the output device. A high-side driver is a transistor that controls the voltage to an output device. When the driver is switched on voltage is applied to the output device which is grounded. This allows current to flow through the driver and output device to ground, and the output device is energized.

TESTING PCM OUTPUTS

Testing PCM outputs may be accomplished using a variety of tools and methods. The technician needs to know how the output device is controlled. This can be determined by using service information that provides a description of the device, or, in many cases, by studying a schematic for that device. Often the schematic will identify whether the output device is a switched output or a PWM-controlled output. In addition, if the output device is a switched output the schematic will usually show the technician if the circuit is a low-side or high-side driver. This will help the technician decide which are the most appropriate methods and tools needed to test the output device and circuit.

Testing most PCM outputs is relatively straightforward. The circuits to the output device can be tested for continuity and grounds using a DVOM. Many output devices can be tested by measuring resistance or directly jumping the device with power and ground. The main difficulty in testing outputs has been the ability of the technician to observe the PCM's ability to actually switch the output on and off during garage conditions. This is because many outputs are only actuated under specific operating conditions that can't always be replicated in the shop. For instance, in order to get the PCM to energize the evaporative purge solenoid, certain operating conditions must exist such as vehicle speed or a percentage of throttle opening that may not be able to be replicated in the shop.

To get the PCM to energize an output, the PCM and scan tools need bidirectional capability. This simply means that the PCM has the capability to execute commands from the scan tool to operate an output even if the conditions for operating that output are not being met. In addition, the scan tool must be capable of communicating these commands to the PCM. While most early engine management systems and scan tools did not have this bidirectional capability, most current systems and scan tools now have this function.

NOTE: This bidirectional capability is dependent on both the scan tool and the PCM. While manufacturer's scan tools have a full range of bidirectional support for testing PCM outputs, many non-OEM scan tools are limited in their bidirectional capabilities. Always consult the scan tool's reference guide to determine the capability of the scan tool that is being used.

Having a scan tool with bidirectional testing capabilities aids the technician in testing output devices and circuits. Using the scan tool, the technician can quickly command the PCM to energize a particular output device and verify that the output device either operates or does not operate correctly. If the output device does not operate when commanded on using the scan tool, the technician can perform circuit testing using a DVOM and other appropriate tools. Many scan tools also have specific test routines that allow the technician to perform complete system tests. For instance, some scan tools allow the technician to perform a functional test of the evaporative system in the service bay. This allows the technician to confirm a repair without having to drive the vehicle and meet the enabling criteria for the evaporative system test to run. This saves time and helps the technician validate the repair. In many instances the manufacturer's diagnostic troubleshooting procedures are written for and require the use of a scan tool to diagnose output circuits and devices.

When diagnosing inoperative PCM outputs, always follow the manufacturer's diagnostic routines whenever possible. In addition, many PCM output drivers are fault protected. This means that if the PCM detects a circuit that is shorted or in some cases open, it may disable the driver to protect it from damage due to excessive current flow. If the output device's resistance decreases due to an internal short, the current in the circuit will increase. This would cause the driver in the PCM to overheat and burn up. To prevent damage to the PCM, the driver will disable under these conditions. In some cases the PCM may set a code indicating the driver has been disabled. Because many PCM output drivers are packaged in a single module of four drivers (quad drivers), up to four individual outputs may be disabled. This can create some confusion on the technician's part. The technician may have to test four individual output devices and circuits to determine which output circuit or device is causing the problem.

NOTE: There have been many instances where the technician has replaced the PCM due to a fault driver code. This rarely will fix the complaint. The fault driver code simply indicates that the PCM is performing as designed in protecting itself from damage due to excessive current flow.

TESTING FUEL-INJECTOR CIRCUITS

The PCM controls the fuel injector(s) by using a low-side driver. The injector receives voltage from a switched ignition circuit and the PCM driver grounds the injector circuit to energize the fuel injector. The PCM calculates the length of injector on time based on inputs from various sensors. If the vehicle is a sequentially fuel-injected engine, then each fuel injector has its own driver and circuit. If the vehicle does not use sequential fuel injection, then a driver may control multiple injectors at one time.

In Chapter 8, specific fuel-injector testing such as pressure drop and injector balance testing will be covered. For now, testing of the injector driver circuit will be covered. One of the most effective tools for testing injector driver circuits are fuel-injector "noid" lights. These are low current test lamps that are designed to plug directly into the fuel-injector connector, figure 7-1. They flash brightly whenever the fuel-injector driver is working. This verifies the presence of both voltage to the fuel injector and the PCM driver grounding the circuit. They are very bright and allow the technician to visually confirm the operation of the fuel-injector circuit. To test the fuel-injector driver circuit, perform the following:

1. Disconnect the affected injector circuit or, in the case of a no-start condition, disconnect one injector on each bank, figure 7-2.
2. Connect the "noid" light to the injector connector, figure 7-3.
3. Crank or start the engine and observe the noid light; it should flash. If it does not flash, proceed to step 4.
4. Disconnect the "noid" light and check for voltage to the injector connector with a test light, figure 7-4.

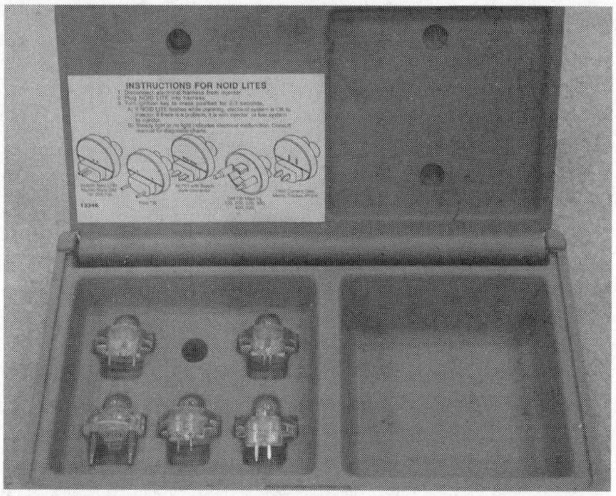

Figure 7-1. Injector "noid" lights are available in sets that fit most injector connectors.

Figure 7-2. Disconnect the connector from the injector driver that you want to test.

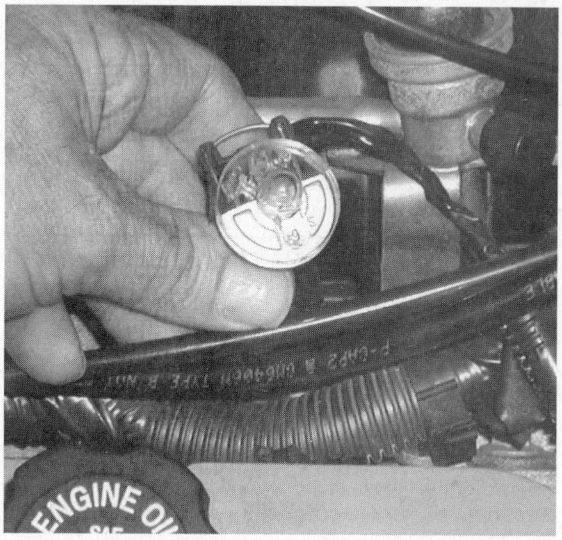

Figure 7-3. Install the "noid" light onto the connector. It should not light until the engine is started or cranked. If the driver is working correctly, it will flash.

Figure 7-4. If the "noid" light did not flash, test for the presence of voltage at the injector connector.

Stepper Motor IAC Operation

The ECM commands a change in IAC position by sequentially turning on the stepper motor coils.

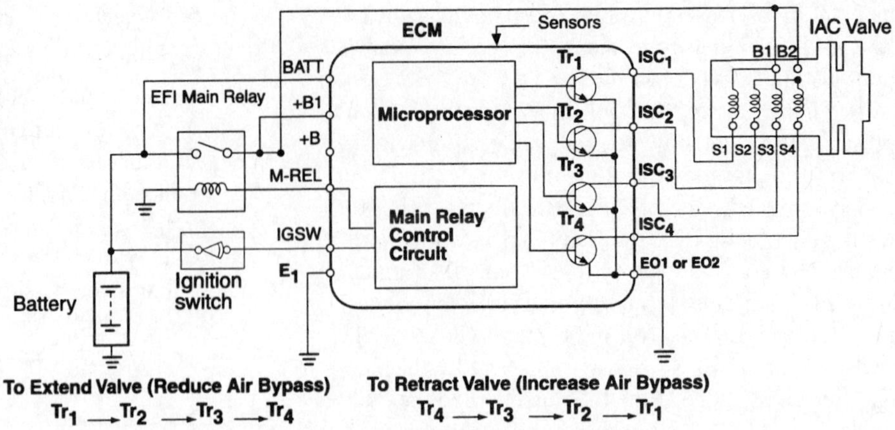

Figure 7-5. Stepper motor IAC operation. (Provided courtesy of Toyota Motor Sales U.S.A., Inc.)

5. If voltage is present, check for an open in the circuit going to the PCM. If no opens are present, replace the PCM.

NOTE: This test assumes that only one injector circuit is not working. If none of the injectors are working, then the technician must test the ignition reference signal to the PCM. If the PCM fails to detect an ignition reference signal then it will not enable the fuel injectors. Other faults that could cause the PCM not to enable any of the injectors are the PCM itself or a failure in the vehicle theft system, if equipped. Anytime there is a no-start condition, be sure to check multiple injectors with the "noid" light.

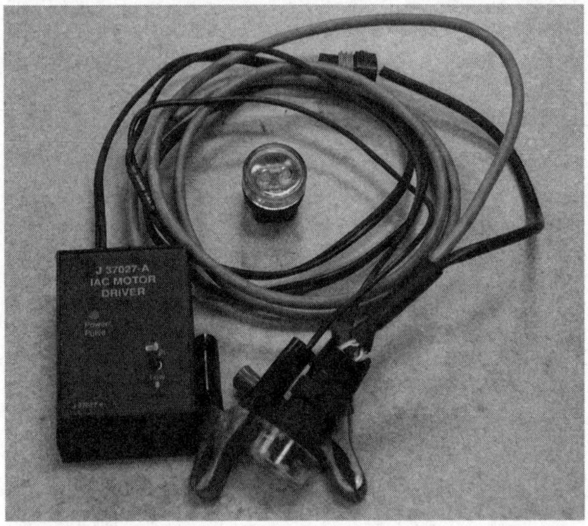

Figure 7-6. An IAC driver and test lights are available for some vehicles.

TESTING THE IDLE AIR CONTROL OUTPUT

Most idle air control valves are the stepper motor design. The PCM pulses the coils alternately inside the idle air control valve which causes the idle air control valve to move in steps (usually referred to as counts), figure 7-5. The PCM controls the position of the idle air control valve based on the desired idle speed as calculated by the PCM. The only time that the idle air control valve controls engine speed is when the driver is not controlling the engine speed with the throttle. The difficulty in testing the IAC circuit is that the PCM only pulses the control circuits as needed to maintain idle speed. In addition, these pulses occur very quickly and are hard to observe using a test lamp or DVOM. Some tool manufacturers have developed a "noid" lamp tester that allows the technician to observe the pulses from the PCM. In addition, a tool is provided that allows the technician to pulse the IAC in and out independent of the PCM, figure 7-6. In addition, the scan tool can often be used to verify the correct operation of the IAC valve. To test the operation of the IAC valve and circuit, do the following:

1. Connect a scan tool and select engine outputs from the menu, figure 7-7.
2. Control the idle speed up and down using the scan tool, figure 7-8. Does the commanded engine speed match the actual engine speed and change with the scan tool? If so, then the IAC is working correctly. If no, then proceed to step 3.

3. Disconnect the IAC and install the "noid" light to the IAC harness, figure 7-9.
4. Command the IAC to move using the scan tool and observe the "noid" lights. Each of the four LEDs should cycle from red to green, figure 7-10.
5. If all of the LEDs are cycling from red to green, then the PCM and circuits are OK. Replace the IAC valve.
6. If any of the LEDs did not cycle, then check each of the four circuits from the PCM to the IAC for opens or shorts to ground. If no opens or shorts are found, then replace the PCM.

TESTING PCM-CONTROLLED RELAYS

Many switched outputs are relay controlled. This is due to the fact that many outputs that are PCM controlled draw too much current for the PCM to control directly. Figure 7-11 shows a switched PCM relay that controls the fuel-pump circuit. Many fuel pumps draw between 5 to 10 amps, which is more current than the driver could handle. However, the relay coil usually only draws 250 to 500 mA of current. Many PCM-controlled outputs are relay switched. Some examples

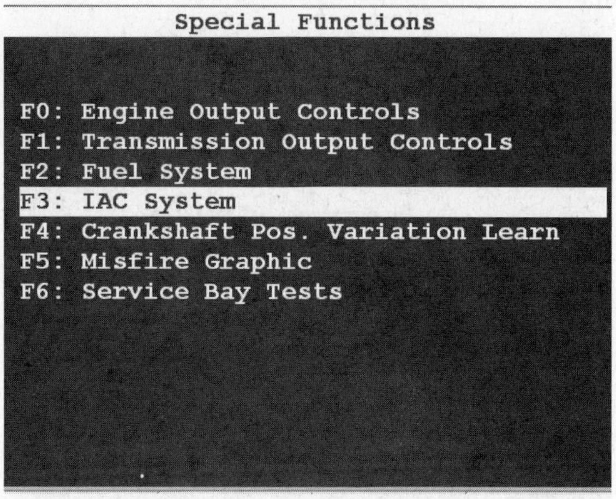

Figure 7-7. Select the IAC test to perform diagnostics on the IAC system.

Figure 7-9. Use the IAC "noid" lights to test the IAC circuits and IAC drivers.

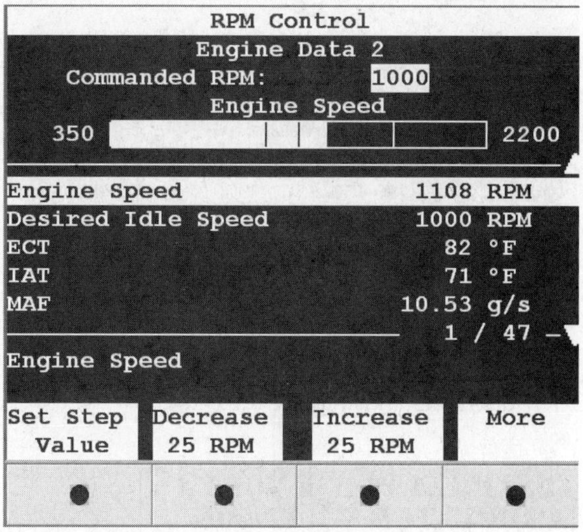

Figure 7-8. Command increases and decreases in the engine speed to determine if the IAC responds.

Figure 7-10. Using the scan tool to command IAC movement, watch the "noid" lights. They should flash from red to green.

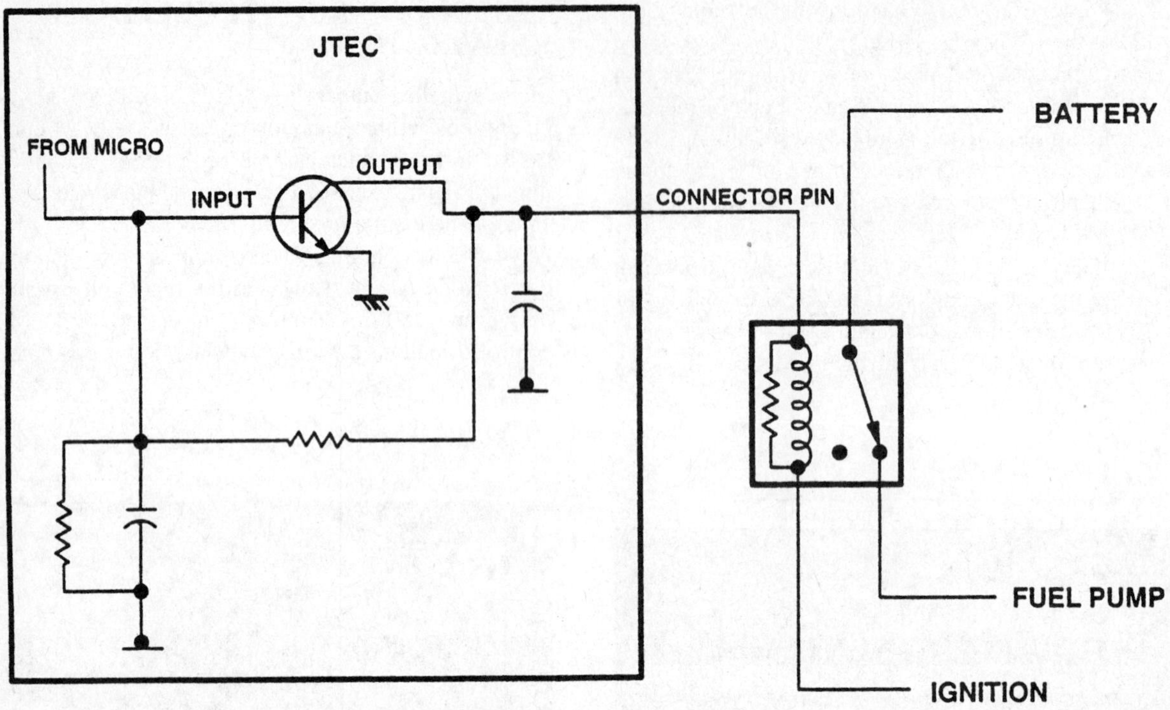

Figure 7-11. Fuel pump relay circuit. (Courtesy of DaimlerChrysler Corporation)

of switched PCM outputs that are relay controlled are cooling fans, air conditioning compressor, electric air pumps, fuel pumps, main power relays, and starter relays. As mentioned previously in this text, these relays could be switched by low-side or high-side drivers. The most common method is for the relay to be switched by a low-side driver as shown in figure 7-11. To test a relay that is PCM switched using a scan tool, do the following:

1. Connect the scan tool and select the output menu, figure 7-12.
2. Select the controlled device to be tested, figure 7-13.
3. Locate the relay on the vehicle, figure 7-14.
4. Using the scan tool, command the device on while touching the relay with your hand, figure 7-15. You should feel it click. If not, proceed to step 5.
5. Using a schematic, check for voltage to the appropriate relay terminals. The coil and the switched side of the relay should both have voltage. If there is no voltage at either one of these terminals, diagnose the loss of voltage. If there is voltage at both terminals, proceed to step 6.
6. Remove the relay and connect a 12-volt test light to battery power. Probe the relay terminal connector that goes to the PCM, figure 7-16.
7. Command the device on with the scan tool. The test light should illuminate if the PCM driver is working, figure 7-17. If the test light does not illuminate, check the circuit for opens. If no opens are found, then replace the PCM.

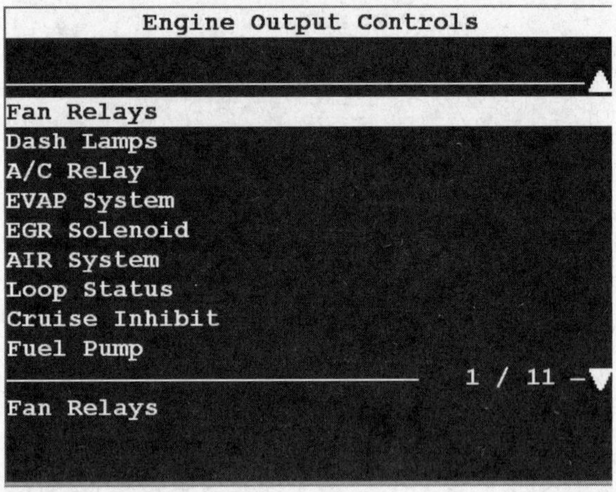

Figure 7-12. Use the scan tool to select the relay-controlled output that you want to test.

TESTING A PULSE WIDTH MODULATED SOLENOID

Many solenoids and other devices are controlled by the PCM using pulse width modulation (PWM). Pulse width modulation control is a digital signal generated by the

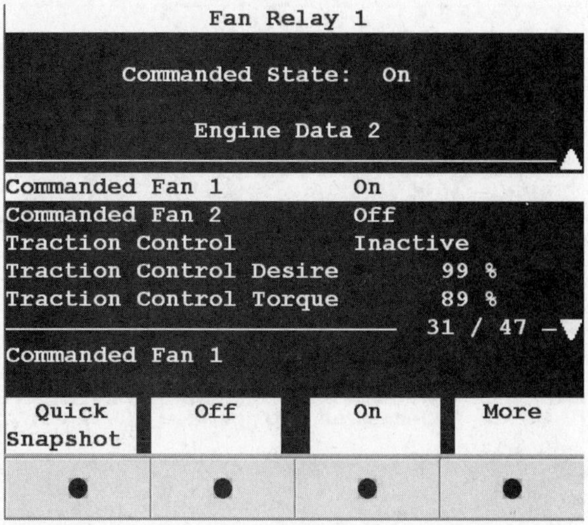

Figure 7-13. Select the controlled output on the scan tool.

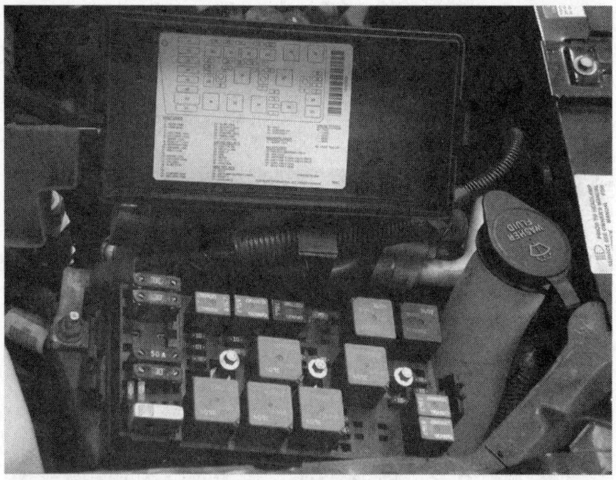

Figure 7-14. Locate the relay for that controlled output on the vehicle.

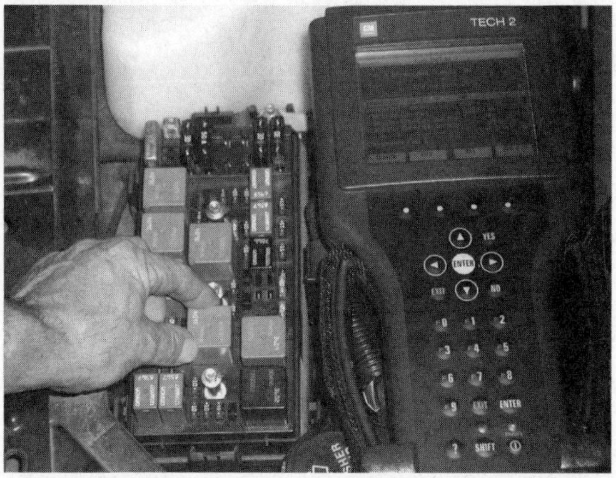

Figure 7-15. Touch the relay with your hand as you actuate the controlled device with the scan tool. You should feel it click.

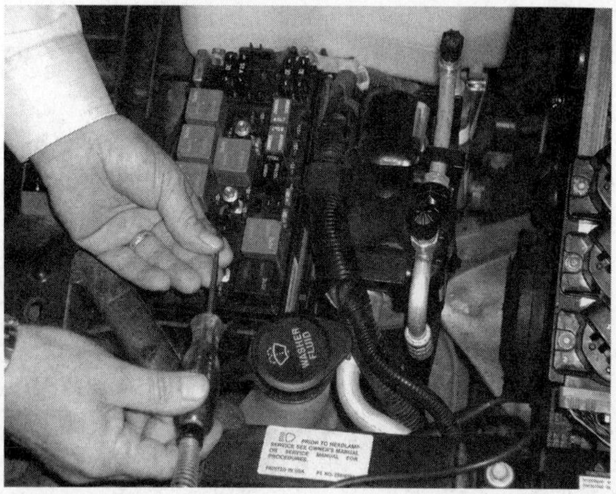

Figure 7-16. To check the driver, install a test light from B+ to the driver circuit on at the relay connector (for low-side drivers). The test light should not light.

Figure 7-17. Activate the driver with the scan tool. The test light should light if the driver and circuit are OK.

PCM that can vary the length of the on and off time of the signal. This provides very precise control of solenoids. A pulse width modulated signal may be verified by using a test light, DVOM, or oscilloscope. A scan tool is often required in order to command the PCM to actuate the device. To test a pulse width modulated signal of a solenoid using a scan tool and a test light, perform the following:

1. Connect a scan tool and select the output menu, figure 7-18.
2. Disconnect the solenoid and install a test light across the connector terminals in series, figure 7-19. The test light should not light.
3. Using the scan tool, command the solenoid on to a percentage of duty cycle (not 100 percent), figure 7-20.

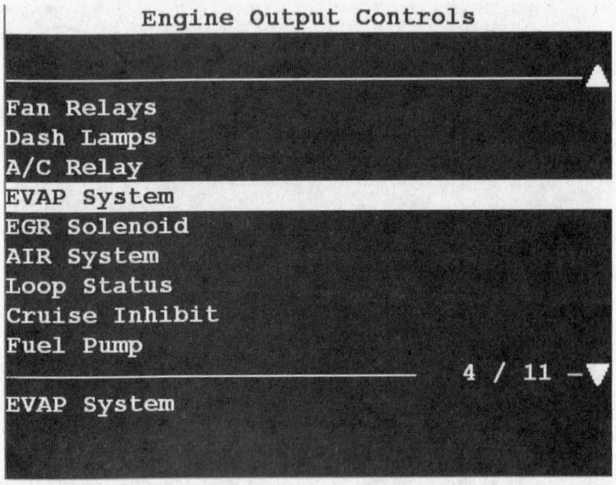

Figure 7-18. Select engine outputs and select the pulse width modulated controlled device.

Figure 7-19. Disconnect the device and install a test light in series across the connector.

4. The test light should blink on and off if the PWM signal is present.
5. If the test light does not blink, verify that voltage is present at one terminal of the connector. If voltage is not present, find the cause of no voltage. If voltage is present, check the circuit to the PCM for opens.
6. If there are no opens in the circuit to the PCM, then replace the PCM.

TESTING AN ELECTRONIC EXHAUST GAS RECIRCULATION VALVE

Electronic exhaust gas recirculation (EGR) valves are controlled by the PCM. They usually incorporate a pintle position feedback sensor that provides the PCM with pintle position information. This provides diagnostic capabilities of the EGR valve and circuitry. A scan tool is helpful in diagnosing electronic EGR valves. Using the output control function of the scan tool, the valve can be commanded to move to various positions while the technician monitors the pintle position sensor feedback signal. Correct operation of the EGR valve can be determined using this method. To test an electronic EGR valve using a scan tool, perform the following:

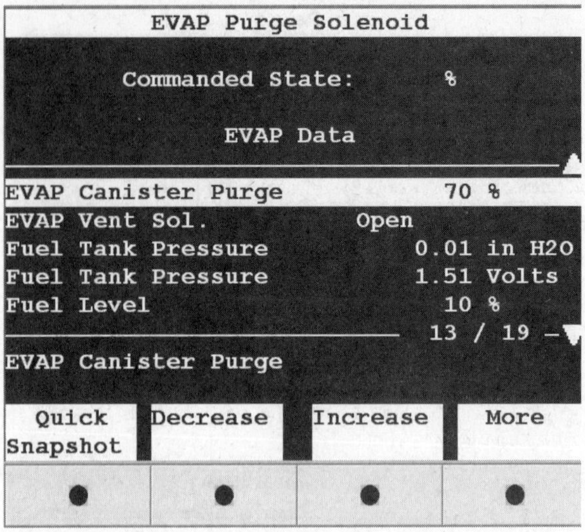

Figure 7-20. Using the scan tool, command the device on to a certain percentage duty cycle. The test light should flash.

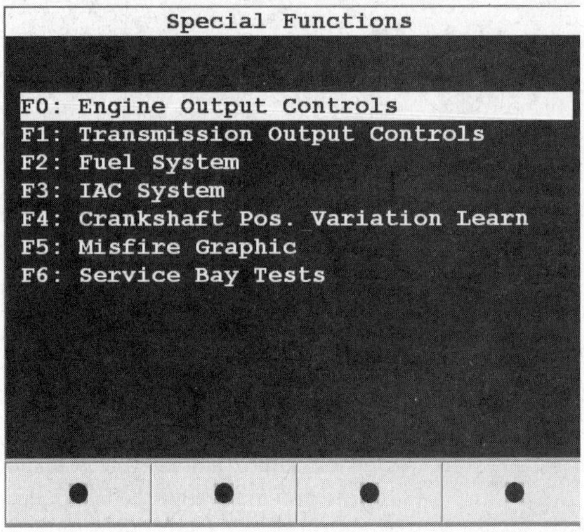

Figure 7-21. Connect the scan tool and select engine output controls.

1. Connect the scan tool and select the engine output menu, figure 7-21.
2. Select EGR valve output, figure 7-22.

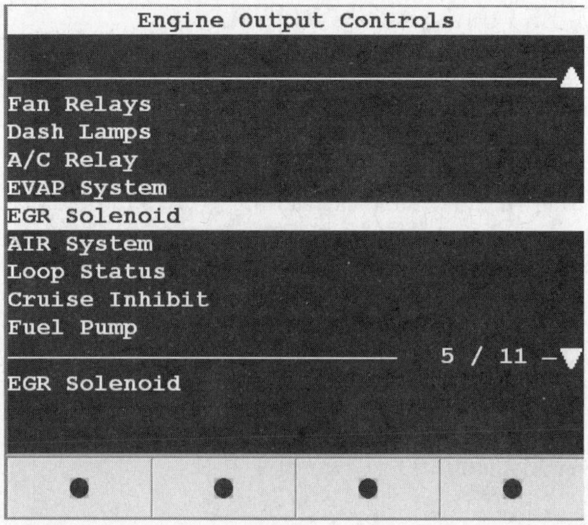

Figure 7-22. Select the EGR output.

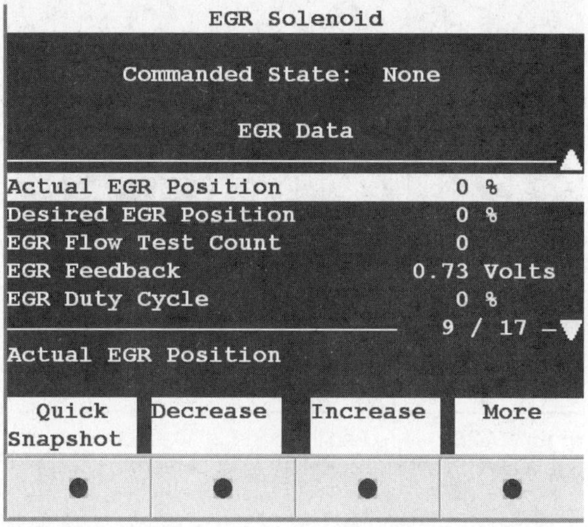

Figure 7-23. Observe the desired EGR position against the actual EGR position. Both should show 0 percent.

3. Monitor the desired EGR position with the actual EGR position on the scan tool, figure 7-23.
4. Command the EGR valve to various positions while monitoring the desired and actual position, figure 7-24. These should match as the valve is commanded to various positions. If they do not match, proceed to step 5.
5. Remove the valve and check for physical binding or a stuck valve.
6. If the valve did not move, check the circuit by placing a test light across the EGR control circuit and the EGR ground circuit. Command the valve to move using the scan tool. The test light should blink if the circuit and PCM are working correctly, figure 7-25. If the test light does not blink, check the circuits for opens and grounds. If no problems are found, replace the PCM.

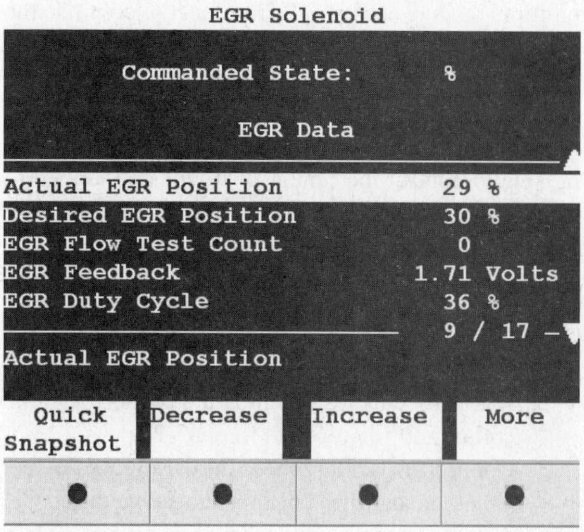

Figure 7-24. While commanding the EGR position to various positions, observe the actual position to determine if it matches.

Figure 7-25. To test the EGR driver, install a test light in series with the EGR control circuit and the EGR ground circuit. Using the scan tool, command the EGR to move. The test light should illuminate.

TESTING THE IGNITION CONTROL OUTPUT

The ignition control (IC) output is the timing output signal from the PCM. Pre-OBD II vehicles used different names for this circuit such as SPOUT or EST. All OBD II vehicles refer to this as IC. If the vehicle is equipped with a distributor system or distributorless ignition system, this signal usually goes from the PCM to the ignition module. However, if the system has no ignition module and the PCM directly controls the coil(s), then this signal is the signal that controls the coil(s). In all cases, however, this signal is going to be a digital output from the PCM to the coil(s) or ignition module. If this signal is not present the PCM will not be able to control the timing or it will not be able to trigger the ignition coil(s). To test for the presence of the ignition control signal, do the following:

1. Use a schematic and determine the circuit to the coil or the ignition module coming from the PCM that is the IC circuit.
2. Back probe the IC circuit at the connector of the coil or ignition module, figure 7-26.
3. Connect an oscilloscope to the back probe device. Crank or start the vehicle and observe the scope. A digital signal should be present, figure 7-27.
4. If a scope is not available, the signal can be tested with a DVOM set to read DC frequency, figure 7-28.

Figure 7-26. In this coil-over-plug ignition system, each coil has an IC circuit. Back probe the IC circuit at the coil.

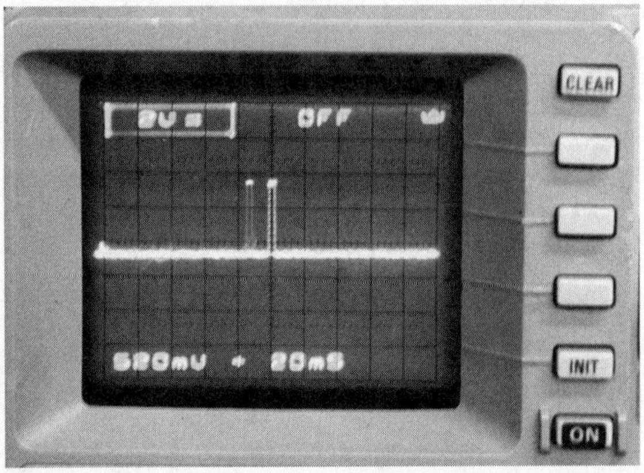

Figure 7-27. If the IC signal is present, a digital signal should be observed on the oscilloscope.

TESTING ELECTRONIC THROTTLE ACTUATORS

The electronic throttle actuator uses a motor to control the position of the throttle plates. There is no cable that attaches the accelerator pedal to the throttle plates. Many manufacturers are starting to use throttle actuator systems. A scan tool is essential to diagnose throttle actuator systems. Due to the safety aspects of throttle actuator systems, they must have redundant systems to ensure accurate input data. Most throttle actuator systems incorporate dual throttle-position sensors and dual accelerator pedal position sensors. A scan tool is used to view the data of the throttle actuator systems, figure 7-29.

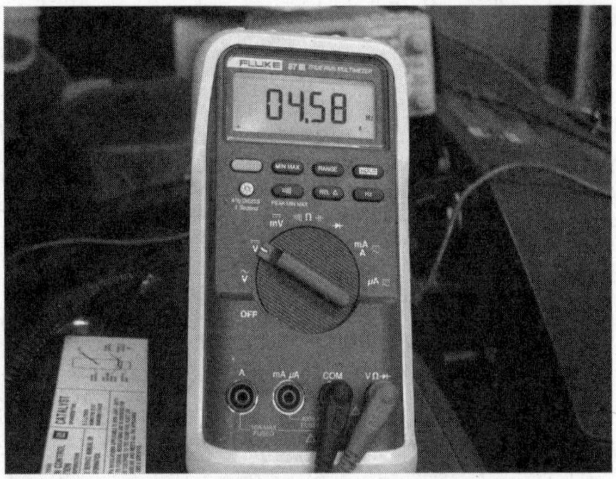

Figure 7-28. If no oscilloscope is available, you can use a DVOM set to frequency to observe the IC signal.

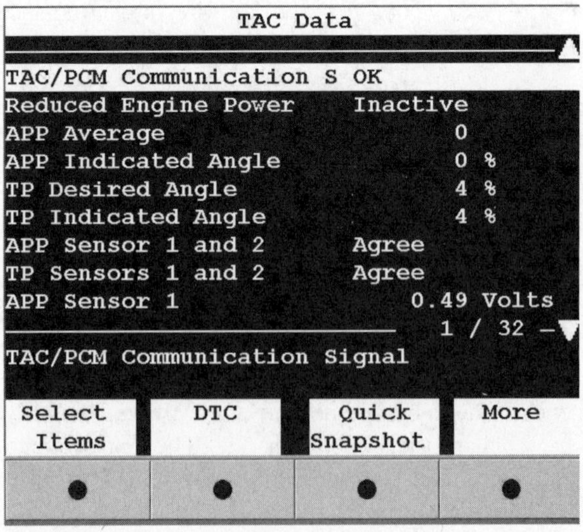

Figure 7-29. The scan tool is required to view the extensive data list for the throttle actuator system.

The throttle actuator cannot be moved manually. If the technician needs to increase engine speed during diagnostics, either a helper will be required to raise the engine speed using the accelerator pedal or a scan tool is required to increase the throttle opening to the desired engine speed. A scan tool will also be required to diagnose problems involving the throttle actuator system. To test the operation of a throttle actuator system using a scan tool on a GM vehicle, do the following:

1. Connect the scan tool and choose special functions, figure 7-30.
2. Under the menu special functions, choose the submenu for throttle actuator control, figure 7-31.
3. Using the scan tool, increase or decrease the throttle to the desired rpms, figure 7-32.
4. If the throttle actuator fails to achieve the desired rpms, then you must follow the published diagnostics to diagnose the condition.

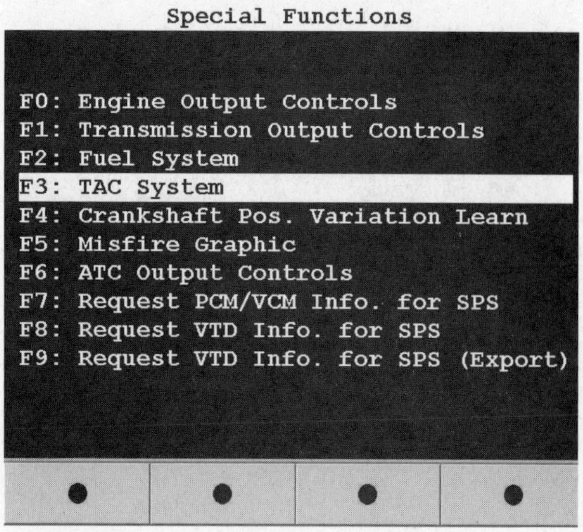

Figure 7-31. Choose the TAC submenu.

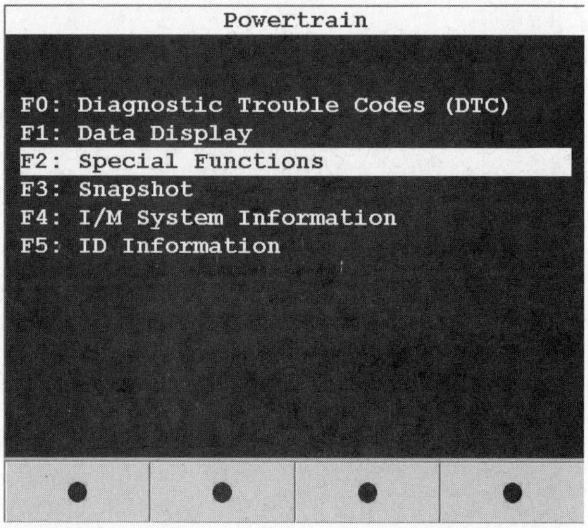

Figure 7-30. To control the throttle actuator using the scan tool, choose the special functions menu.

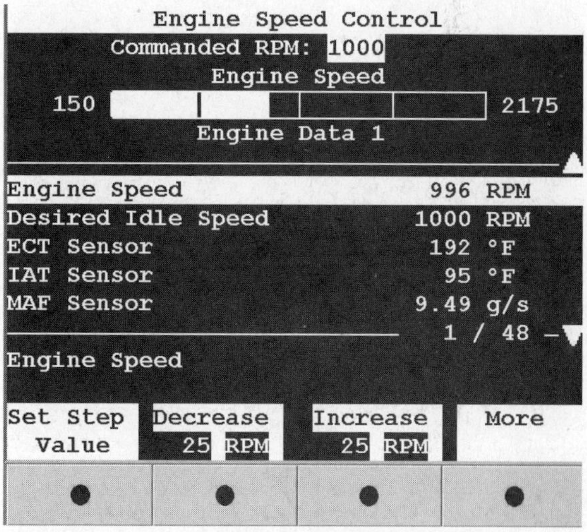

Figure 7-32. The engine speed can then be controlled using the scan tool.

8

Gasoline Fuel-Injection Testing and Service

OBJECTIVES

Upon completion and review of this chapter, you will be able to:

- Perform static and dynamic fuel pressure tests.
- Test and/or replace fuel injectors.
- Test fuel pressure regulators.
- Perform fuel injector cleaning procedures.

INTRODUCTION

All modern electronic fuel-injection (EFI) systems are based on original designs invented by the Bendix Corporation. Research and testing efforts, led in large part by Robert Bosch GmbH, are responsible for vast improvements to today's EFI systems. Most current EFI systems, regardless of the name by which they are known, are based on Bosch components and have evolved from Bosch research and development.

This chapter covers basic procedures and techniques associated with EFI diagnosis and service. These procedures and techniques are based on operating principles common to all EFI systems regardless of make or model. However, it is important to note that manufacturer system information, procedures, specifications, and special test equipment are often required to properly diagnose and service EFI systems. Also, since EFI systems are part of the computer-controlled engine management system, an overall understanding of electronic engine control strategies, functions, and operation is essential.

FUEL-INJECTION SAFETY

In addition to the safety procedures specified in the *Safety Summary* at the front of this volume, there are safety precautions to observe whenever working with an EFI system:

- High-pressure EFI systems generally use fuel pressures between 35 psi (241 kPa) and 50 psi (344 kPa), while low-pressure EFI systems carry about 15 psi (103 kPa). Some low-pressure systems contain a built-in bleed to reduce fuel-line pressure after engine shutdown and to prevent vapor lock. Regardless of the system, however, there is far more residual pressure in an EFI system fuel line than in a carbureted fuel system. For this reason, fuel system pressure must be relieved before opening any fuel line on a fuel-injected engine. Specific procedures for releasing residual fuel pressure are provided in Chapter 3 of this *Shop Manual.*
- Never disconnect or connect any electrical circuits or components when the ignition switch is

on, unless directed to do so by the manufacturer's testing or service procedures. Breaking or making such a connection can result in a high-voltage surge through the system, which can damage electronic components.

ELECTRONIC SYSTEM SAFETY

Solid-state components are sensitive to electrostatic discharge (ESD). Some manufacturers use the symbol shown in figure 8-1 on wiring diagrams and service replacement components to warn the user that the circuit and/or part is subject to ESD damage. This symbol should serve as a reminder that static charges and discharges must be avoided when working on engine control or EFI systems. Even if the symbol is not present, assume that a solid-state component is ESD sensitive. To avoid possible damage to solid-state components, practice the following safety precautions:

- Avoid touching any exposed electrical terminals on connectors or components. If using a tool to separate a connector, do not let it touch the exposed terminals.
- Do not jumper, ground, or probe any components or connectors with test equipment unless specified by the manufacturer's diagnostic procedure. When test equipment is used, connect the ground lead first.
- Before opening a package containing a solid-state component, always touch the package to a good ground. Keep the component in its package prior to installation.

Figure 8-1. Any time that this figure is shown on a schematic, it indicates that the circuit or component is sensitive to damage from ESD. Take particular care when handling or testing the component. (Courtesy of ESD Association)

EFI TROUBLESHOOTING STRATEGY

Driveability complaints can result from one problem in several systems or multiple problems in a single system. Do NOT immediately turn to the EFI system as the cause of a driveability complaint. The following list identifies components and areas of operation that should be checked before proceeding with detailed EFI system tests:

- Battery and charging system performance
- Ignition components, spark plugs, distributor cap and rotor, ignition wires
- Ignition timing and timing advance capability
- Engine compression
- Engine power balance
- Fuel-pump pressure and volume
- Air cleaner and crankcase vent filters
- PCV, EGR, EVAP, and secondary air-injection system operation
- Diagnostic trouble codes (DTCs) in PCM memory

In addition to these systems, also visually inspect electronic system wiring and connectors for conditions that can cause high resistance. High resistance in engine control system circuits can cause driveability symptoms. Remember that effective driveability diagnosis requires a consideration of the functional and operational relationships between ALL related systems and subsystems.

Driveability diagnosis starts by carefully noting specific complaints by the vehicle's driver. The driver knows his/her vehicle best and is usually aware of subtle changes in performance. Many diagnostic troubleshooting charts begin with a clearly defined symptom, figure 8-2.

One final note about EFI diagnosis and service—all EFI systems are essentially the same between car lines. While there are minor operational and component differences, the necessary diagnostic and service information is provided in the appropriate manufacturer shop manuals, or in service guides developed by independent publishers.

ELECTRONICS

Input sensors provide the powertrain control module (PCM) with information about various engine conditions such as engine coolant temperature, engine speed, and exhaust gas oxygen content. The PCM controls fuel injector on-time based on this and other sensor information.

Effective engine performance diagnosis requires a thorough understanding of the relationships between input sensors, the related data provided by these sensors, and the operation of the fuel injectors. While all input data is important for good driveability, *engine load* is particularly important for proper fuel-injection control.

A variety of inputs is used by the PCM to calculate load, many of which vary depending on the specific sys-

Symptom	OASIS Number
Poor Fuel Economy	622000

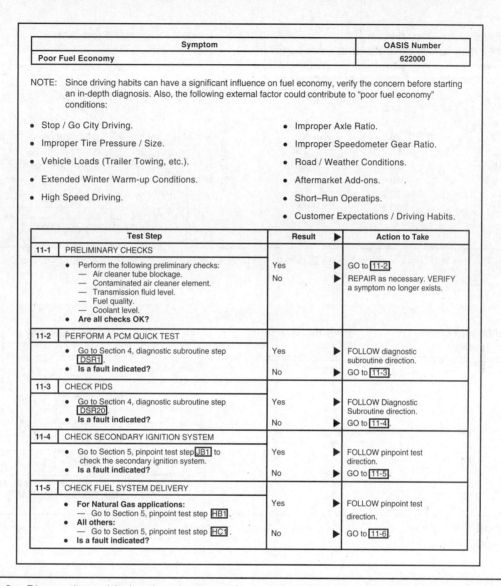

NOTE: Since driving habits can have a significant influence on fuel economy, verify the concern before starting an in-depth diagnosis. Also, the following external factor could contribute to "poor fuel economy" conditions:

- Stop / Go City Driving.
- Improper Tire Pressure / Size.
- Vehicle Loads (Trailer Towing, etc.).
- Extended Winter Warm-up Conditions.
- High Speed Driving.
- Improper Axle Ratio.
- Improper Speedometer Gear Ratio.
- Road / Weather Conditions.
- Aftermarket Add-ons.
- Short–Run Operatips.
- Customer Expectations / Driving Habits.

	Test Step	Result ▶	Action to Take
11-1	PRELIMINARY CHECKS		
	• Perform the following preliminary checks: — Air cleaner tube blockage. — Contaminated air cleaner element. — Transmission fluid level. — Fuel quality. — Coolant level. • **Are all checks OK?**	Yes ▶ No ▶	GO to 11-2. REPAIR as necessary. VERIFY a symptom no longer exists.
11-2	PERFORM A PCM QUICK TEST		
	• Go to Section 4, diagnostic subroutine step DSR1. • **Is a fault indicated?**	Yes ▶ No ▶	FOLLOW diagnostic subroutine direction. GO to 11-3.
11-3	CHECK PIDS		
	• Go to Section 4, diagnostic subroutine step DSR20. • **Is a fault indicated?**	Yes ▶ No ▶	FOLLOW Diagnostic Subroutine direction. GO to 11-4.
11-4	CHECK SECONDARY IGNITION SYSTEM		
	• Go to Section 5, pinpoint test step JB1 to check the secondary ignition system. • **Is a fault indicated?**	Yes ▶ No ▶	FOLLOW pinpoint test direction. GO to 11-5.
11-5	CHECK FUEL SYSTEM DELIVERY		
	• **For Natural Gas applications:** — Go to Section 5, pinpoint test step HB1. • **All others:** — Go to Section 5, pinpoint test step HC1. • **Is a fault indicated?**	Yes ▶ No ▶	FOLLOW pinpoint test direction. GO to 11-6.

Figure 8-2. Diagnostic troubleshooting charts match symptoms to recommended actions. (Courtesy of Ford Motor Company)

tem, make, and model. The most common load inputs are the manifold absolute pressure (MAP) sensor, mass air flow (MAF) sensor, and vane air flow (VAF) sensor. Remember, none of these sensors work alone. All systems use a throttle-position sensor (TPS) to tell the computer how far the throttle is open and how *fast* it is opening or closing. Many systems also use an intake air temperature (IAT) sensor that communicates the temperature of the air charge in the intake manifold to the PCM.

MAP Sensor

The MAP sensor is connected to the intake manifold, usually by a vacuum hose. The MAP sensor provides the PCM with an indication of engine load by sending a voltage or frequency signal that varies with changes in manifold pressure. The more the throttle plate is open, the higher the manifold pressure, figure 8-3. When the engine is not working hard, at idle for instance, manifold pressure is low (strong manifold vacuum). The MAP sensor sends a low-voltage or low-frequency signal to the PCM. When the engine is working harder, such as during acceleration, intake manifold pressure increases (manifold vacuum drops) and the MAP sensor sends a higher voltage or frequency signal.

MAP sensors are most commonly found in "speed density" systems in which the PCM uses engine load, engine speed, and intake air temperature information to calculate engine load. Be aware that MAP sensors are sometimes used as an auxiliary sensor in mass airflow (MAF) systems.

A MAP sensor is similar to other sensors in that it has power and ground circuits that can be checked for opens or shorts with a DVOM. The signal can be measured

VACUUM AT SEA LEVEL (in. Hg)	MANIFOLD ABSOLUTE PRESSURE (kPa)	SENSOR VOLTAGE
0	101.3	4.5
3	91.2	4.0
6	81.0	3.5
9	70.8	3.0
12	60.7	2.5
15	50.5	2.0
18	40.4	1.5
21	30.2	1.0
24	20.1	0.5

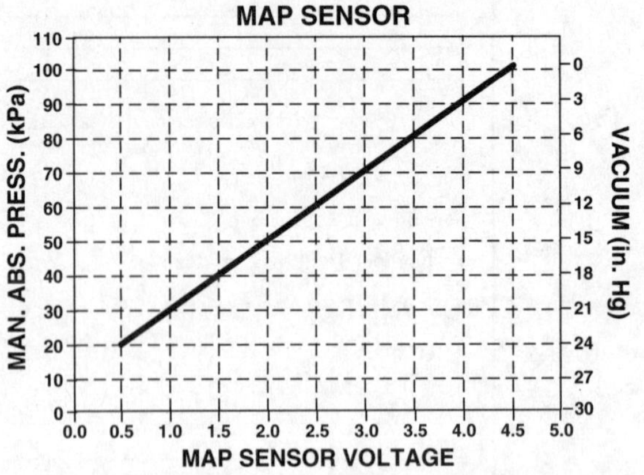

Figure 8-3. The MAP sensor provides the PCM with an indication of engine load. (Courtesy of Ford Motor Company)

MASS AIRFLOW (gm/sec)	SENSOR VOLTAGE
0	0.2
2	0.7
4	1.0
8	1.5
15	2.0
30	2.5
50	3.0
80	3.5
110	4.0
150	4.5
175	4.8

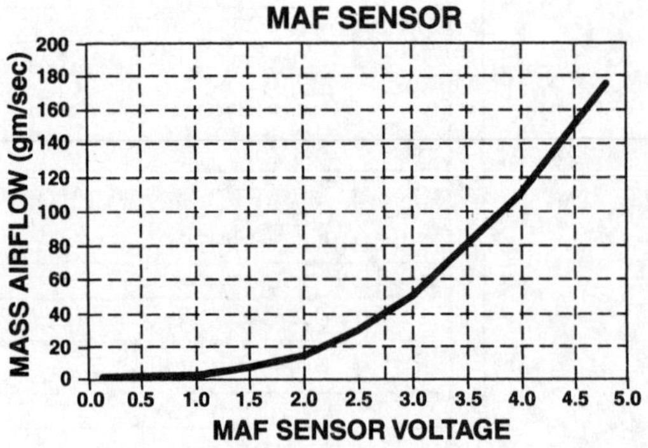

Figure 8-4. The MAF sensor's air mass measurements help the PCM estimate air density. (Courtesy of Ford Motor Company)

when the engine is running to see if the sensor is operating properly. Be sure to inspect the vacuum hose carefully. Vacuum hoses which are cracked, kinked, or otherwise damaged will cause driveability complaints ranging from hesitation to poor fuel economy. Refer to the appropriate service publications for specific testing techniques and voltage/frequency specifications.

MAF Sensor

The mass air flow (MAF) sensor measures air mass as intake air flows over a sensing element in the air inlet. The PCM uses the signal from the MAF sensor to estimate air density, which is an indicator of engine load, figure 8-4.

MAF sensors are located in the air intake, between the air cleaner and the throttle body. Depending on the manufacturer, they send a voltage or frequency signal that increases in proportion to the volume of air flowing into the engine. The condition of connecting air ducts and clamps is critical to proper MAF operation. If air enters the engine without being measured, the PCM cannot add fuel to compensate for the extra air, so the vehicle runs lean. Check that the MAF sensor is mounted firmly and squarely, the air cleaner is properly installed, and all clamps are tight.

The component tests contained in most service publications provide tests to check the integrity of the power, ground, and signal circuits. Some tests involve monitoring the MAF signal while operating the engine at various speeds to see if the signal voltage or frequency properly corresponds to the change in rpm.

VAF Sensor

A vane air flow (VAF) sensor, figure 8-5, is similar to a MAF sensor because it also measures intake air. However, a VAF sensor measures the volume of air rather than the density or mass of the air. A VAF sen-

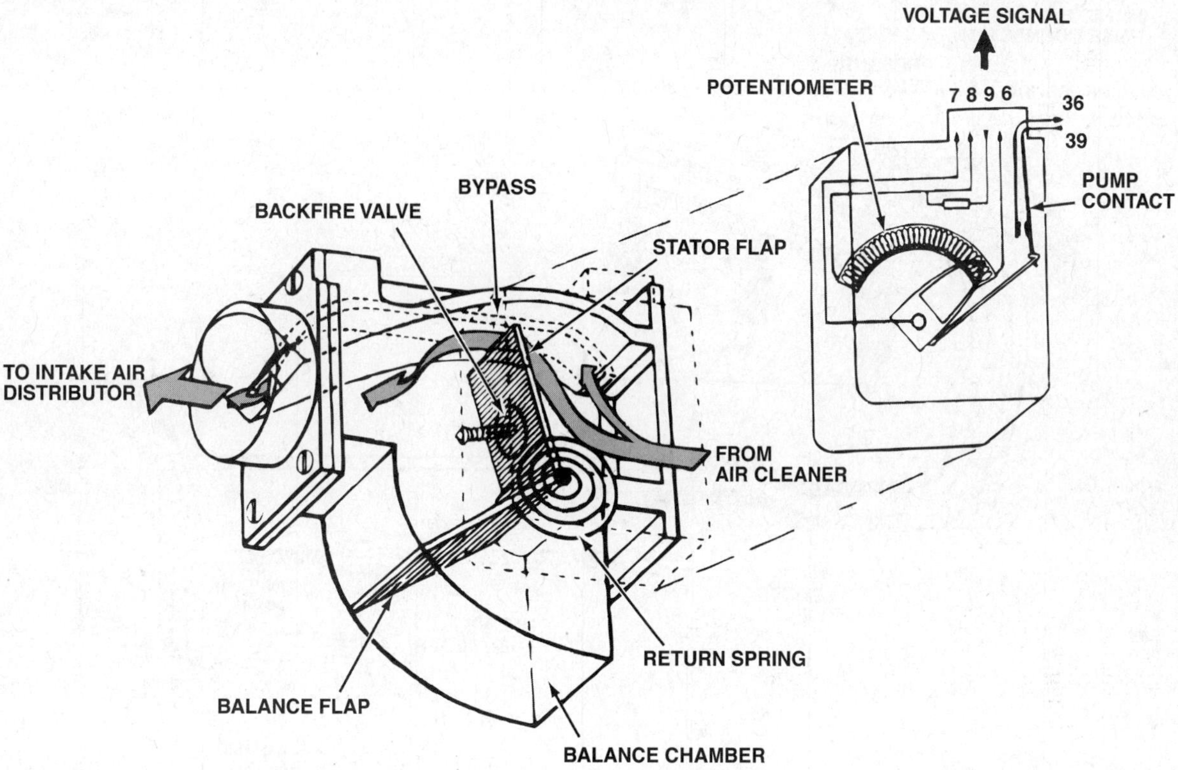

Figure 8-5. The vane air flow (VAF) meter uses a movable vane and potentiometer to measure intake air volume.

sor is constructed with a potentiometer operated by a mechanical vane device located in the air inlet. The vane moves in relation to how much air is flowing into the intake manifold. The VAF potentiometer converts mechanical movement into an electrical signal, sending a voltage signal that increases as airflow increases. The PCM uses the signal to calculate engine load.

An airflow sensor that does not work properly can cause hard starting, erratic performance, and poor fuel economy. You can check this type of sensor with the following procedure and the manufacturer's specifications:

1. Turn on the ignition key but do not start the engine.
2. Remove the sensor air intake duct or air filter as required to reach the sensor.
3. Slowly push the sensor flap wide open with gentle pressure. As the flap opens slightly, the electric fuel pump should come on. Make sure the flap moves in both directions without binding.
4. Turn off the ignition key, then unplug the sensor harness connector.
5. Refer to the manufacturer's specifications to determine the correct terminals and resistance values, then check resistance with an ohmmeter. Move the flap by hand and compare the readings to specifications.

FUEL SYSTEMS

The fuel system supplies the fuel injectors with clean fuel at a controlled pressure. Most modern fuel-injection systems are sequential fuel-injection (SFI) systems, meaning there is one injector for each cylinder and each injector is switched on, in sequence, at the beginning of its cylinder's intake stroke. The PCM controls the fuel pump, monitors the fuel pump circuit, and controls injector operation. There are three main types of fuel systems in use today:

- Returnable fuel systems
- Mechanical-returnless fuel systems
- Electronic-returnless fuel systems

Returnable Fuel System

Fuel systems using a fuel-return line to return unused fuel to the tank are the most common systems in service today. A typical returnable fuel system, figure 8-6, consists of a fuel tank, fuel-pump unit, fuel supply and return lines, filters, a pressure test point, fuel rail, injectors, and a fuel-pressure regulator.

On port-type fuel-injection systems, the fuel-pressure regulator is attached to the return side of the fuel rail, downstream of the injectors. The fuel-pressure regulator is easily identified by the vacuum line attached

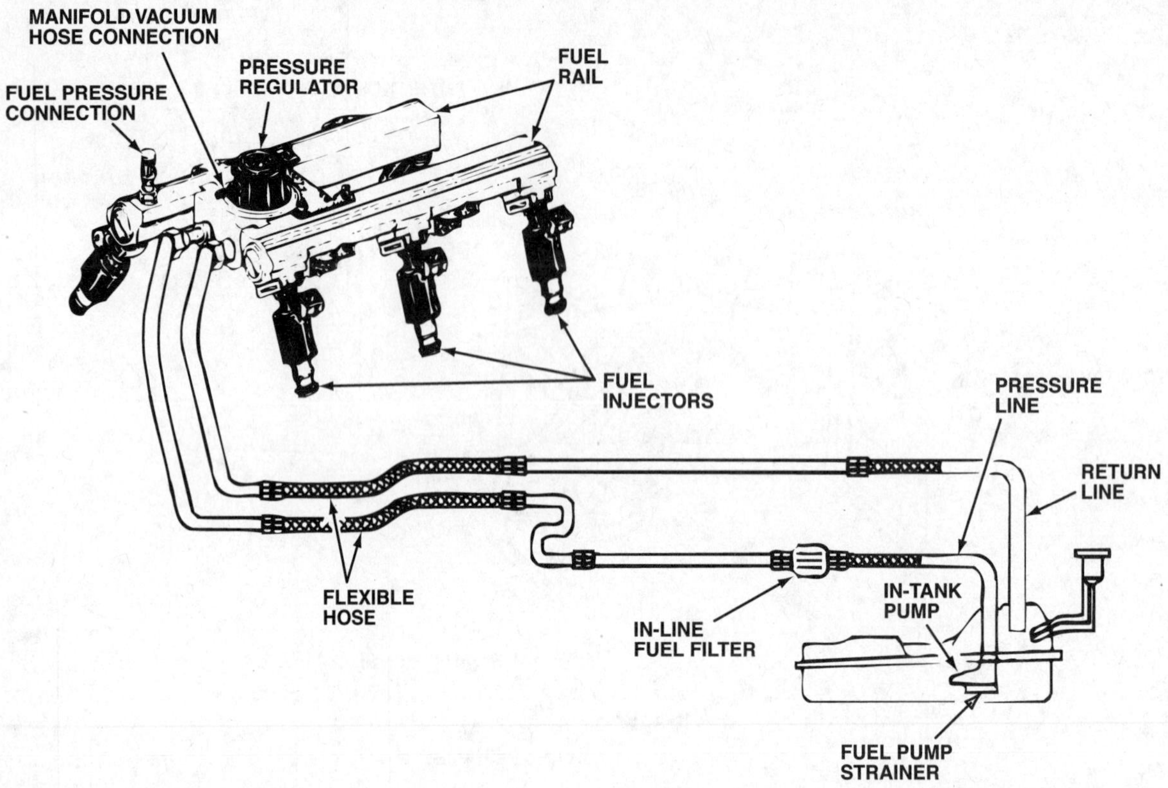

Figure 8-6. A typical returnable fuel system.

to the regulator. Intake manifold vacuum is supplied to the regulator to promote a steady pressure drop across the injectors during varying engine load situations. At idle, strong manifold vacuum pulls the regulator open. Fuel not used by the injectors passes through the regulator and returns to the tank. As engine load increases, manifold vacuum drops (manifold pressure increases) and spring pressure closes the regulator, blocking the return line to the tank. It is important to check the vacuum line during visual inspection. A leaking or kinked vacuum line will cause high fuel pressure at idle and result in a rich condition.

Throttle-body injection systems use a pressure regulator, but no intake manifold vacuum source is needed. The pressure regulator is built into the throttle-body assembly. The pressure regulator may sometimes be changed as an assembly. However, never attempt to open the regulator itself. It contains a strong spring that can cause injury if the regulator is opened.

Mechanical-Returnless Fuel System

Mechanical-returnless fuel systems first appeared on Chrysler products beginning in 1993. A typical mechanical-returnless fuel system consists of a fuel tank, fuel-pump module, fuel-pressure regulator, filter, supply line, fuel rail with a pulse damper and pressure test point, and injectors, figure 8-7.

This system is easy to recognize because there is no fuel-return line coming off the fuel rail and no pressure regulator visible. The fuel-pressure regulator is part of the fuel pump module located in the fuel tank. Fuel pressure is regulated to 49 psi (338 kPa) by a calibrated spring in the regulator. Excess fuel drains directly into the fuel tank. The regulator also contains a filter assembly and a check valve to maintain residual fuel pressure when the engine is off.

Some, but not all, fuel-pressure regulators are replaceable. The fuel tank must be removed to gain access to the fuel pump module and the integral fuel-pressure regulator. Fuel filters on many applications are part of the fuel-pump module. There may be no external filter or regular filter service interval. Fuel-pressure testing procedures are different for these systems; be sure to check the service manual before testing.

Electronic-Returnless Fuel System

Electronic-returnless fuel systems are found on many newer vehicles. A typical electronic-returnless fuel system, figure 8-8, consists of a fuel tank, fuel pump, fuel pump driver module, fuel rail pressure sensor, fil-

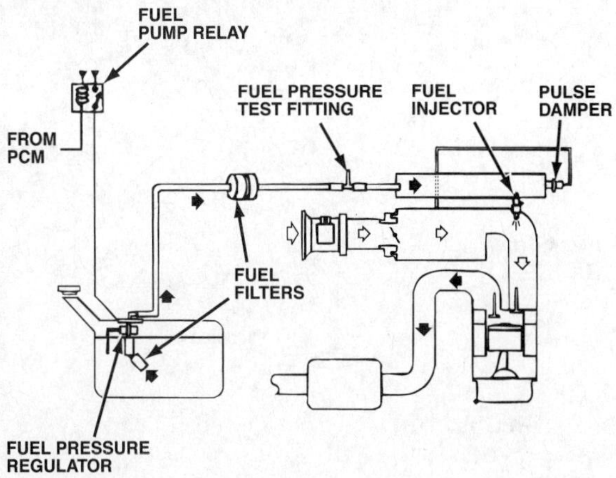

Figure 8-7. A typical mechanical-returnless fuel system.

ter, supply line, fuel temperature sensor, fuel rail with pressure test point, and the injectors.

Identify this system by looking for the fuel-pressure sensor on the fuel rail. The sensor has a three-wire connector and a vacuum line connection. Not all systems use a fuel-temperature sensor. Again, there is no fuel-return line from the fuel rail to the fuel tank.

The fuel-rail pressure sensor provides the PCM with fuel-pressure data, actually, the pressure difference between fuel-rail pressure and intake manifold pressure. The PCM uses this data to vary the duty cycle to the fuel-pump driver module, which modulates voltage to the fuel pump to maintain proper fuel pressure. Input from the engine fuel temperature sensor (when applicable) is used to vary fuel pressure in order to avoid fuel system vaporization (vapor lock).

Electronic-returnless injection systems normally use an external fuel filter and have a specified service interval. Fuel-pressure test procedures vary, so use the service manual to determine the best way to test and service the system.

PRESSURE TESTING

The first diagnostic steps with any EFI system are to make sure that:

- Fuel pump delivery pressure is satisfactory
- Fuel pressure holds for a period of time without bleeding down
- All lines are secure and there are no external leaks
- The filter(s) are replaced if there is any indication of obstruction

Before tapping into a fuel system with a fuel-pressure gauge, it is very important that any remaining fuel pressure in the system is relieved. Follow the manufacturer's procedure for relieving fuel pressure carefully to avoid sprayed fuel, possible personal injury, or damage to the vehicle. Be sure the gauge set will withstand the expected fuel pressure of the system you are testing.

A fuel-pressure gauge is connected to the system in one of two ways:

1. To the pressure test port on the fuel rail for vehicles equipped with a pressure test/relief valve. Remove the cap from the fitting, figure 8-9, and connect a high pressure gauge with a suitable hose and connector.
2. If there is no test/relief valve, connect the gauge into the inlet line of the fuel rail with a tee fitting, figure 8-10.

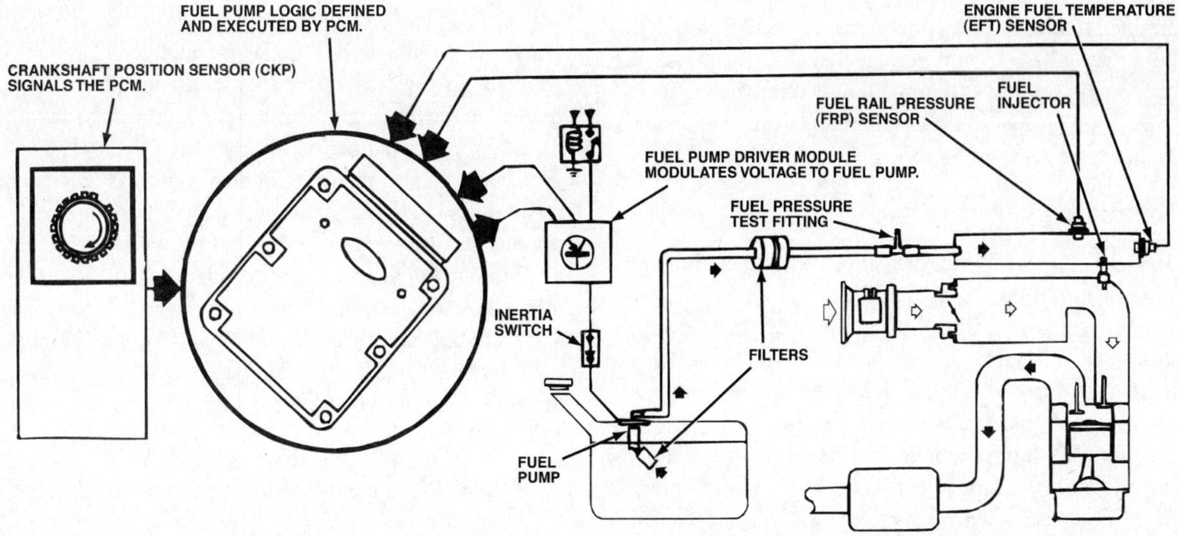

Figure 8-8. A typical electronic-returnless fuel system.

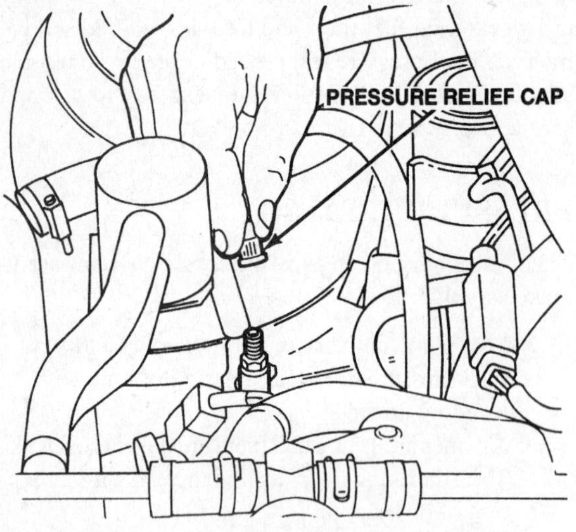

Figure 8-9. Some vehicles have a pressure test or pressure relief fitting for attaching the fuel-pressure gauge.

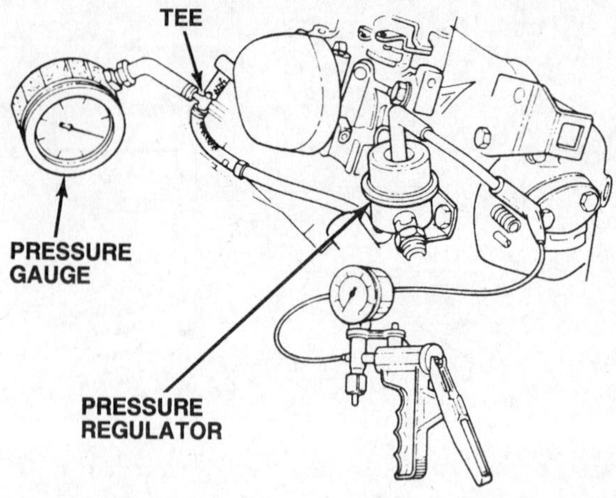

Figure 8-10. Tee the pressure gauge into the inlet line if there is no test or relief fitting. A hand vacuum pump may be needed to do the test.

There are three fuel-pressure readings manufacturers may require during diagnosis:

- *System or static pressure:* fuel pressure with the ignition on and the engine off.
- *Regulated or running pressure:* fuel pressure with the engine running, usually at curb idle speed.
- *Residual or rest pressure:* fuel pressure that remains in the fuel rail after a specified amount of time with the ignition and engine off.

System Pressure

System pressure is the easiest to check. Turn the ignition switch to the "run" position without starting the engine. If the system is working normally, you will hear the fuel pump run for a few seconds. At the same time, fuel pressure will immediately climb to the specified pressure. Some manufacturers refer to system pressure as static pressure.

Regulated Pressure

To check regulated fuel pressure, start the engine and run at a specified rpm (generally curb idle) while noting the gauge reading. Some manufacturers' procedures on returnable fuel systems call for installing a hand vacuum pump to the fuel-pressure regulator as shown in figure 8-10. Once installed, a specified amount of vacuum is applied and fuel pressure is monitored with respect to the amount of vacuum applied. The procedure may involve applying differing amounts of vacuum to obtain a range of readings. Pinch or clamp off the vacuum line and watch the pressure gauge. If pressure drops, the vacuum diaphragm is leaking and the pressure regulator must be replaced. Then, with the engine idling, pull the vacuum line from the pressure regulator while observing the pressure gauge. Fuel pressure should rapidly increase as the regulator closes and blocks the fuel-return line.

Regulated fuel pressure below specifications indicates a defective regulator, defective fuel pump, restricted fuel lines, or restricted fuel filter. If fuel pressure is above specifications, check the vacuum signal to the fuel-pressure regulator. Also check the regulator for a leaking diaphragm, and the fuel-return line for restrictions.

Residual Pressure

Residual fuel pressure helps ensure fast restarts and prevents vapor lock by maintaining pressure on the fuel when the engine is off. Some manufacturers refer to residual pressure as rest pressure. Check the service manual to see how much fuel pressure the system should maintain over what period of time. Note that as the engine cools, fuel in the fuel rail also cools and contracts, resulting in lower fuel pressure. Fuel pressure of 0 psi is not unusual after a vehicle has completely cooled.

When residual fuel pressure is below specification, a leak is indicated somewhere in the system. External leaks should have been found during the preliminary visual inspection. If there are no external leaks, there are only two or three places fuel can escape from the fuel rail:

- Into the engine through one or more leaking fuel injectors
- Into the fuel tank through a defective fuel-pressure regulator (return type systems only)

- Into the fuel tank through a leaking check valve in the fuel pump or tank-mounted pressure regulator

To test a return type system for internal leaks, locate the two flexible fuel lines that connect hard fuel lines to the fuel rail. Cycle the ignition on to build system pressure, then use hose pinching pliers to clamp both flexible fuel lines. If residual pressure drops below specification, one or more fuel injectors are leaking. Remove the spark plugs and use an emissions analyzer to check for high HC concentrations above each spark plug hole. Very high HC readings indicate a leaking injector.

If residual fuel pressure does not drop with both flexible lines blocked, remove the hose pinch pliers from the return line. A drop in pressure now indicates a leaking fuel-pressure regulator. If pressure drops when the hose pinch pliers are removed from the inlet line, the check valve in the fuel pump is leaking. Some pressure drop is normal over time, so it may be necessary to cycle the ignition switch on during the procedure to restore residual fuel pressure.

Returnless systems are tested in a similar manner. Some manufacturers require a hose and adapter be installed between the fuel line and the fuel rail, figure 8-11. The hose pinch pliers are used to block the fuel fail side of the adapter setup. A drop in residual pressure means the check valve is leaking in the tank-mounted fuel-pressure regulator. Next, the fuel line side of the adapter setup is blocked. If residual pressure drops now, one or more injectors are leaking.

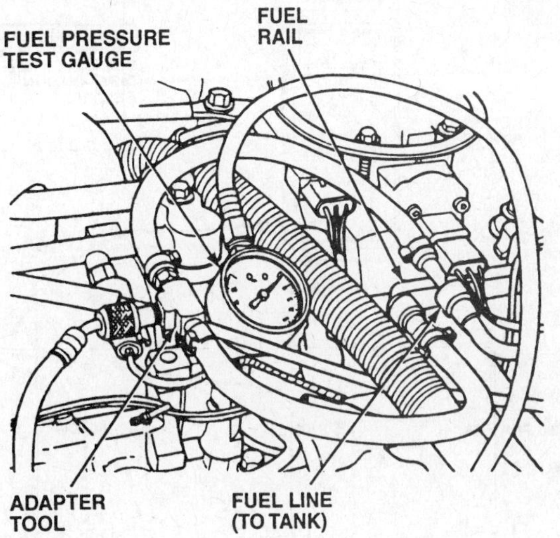

Figure 8-11. Test adapter used to test residual pressure on returnless fuel supply systems.

FUEL-INJECTOR TESTING

Electrical Operation

Fuel injectors are solenoid-operated actuators. This means they can be tested using the same basic techniques used with any other solenoid-operated device. However, before testing fuel injectors, make sure:

- The fuel pump delivery pressure is correct
- The battery is fully charged
- The charging system is working properly
- The PCM is receiving the supply voltage it requires
- The PCM is properly grounded

Refer to the schematic in figure 8-12 while reading the following text. On most systems, the fuel injectors are always "enabled," meaning battery/charging system voltage is available when the key is in the RUN position. With operating voltage always present, all that is needed to actuate the injectors is a completed circuit to ground. This is the job of the PCM; it provides the switch and the path to ground to complete the circuit.

As discussed previously, the PCM controls the on-time of the injectors in relation to input data from various sensors. The amount of time the PCM turns the fuel injectors *on* is known as the pulse width. The injector pulse width varies depending on the demand for fuel. The on-time of an injector can be as short as 2 milliseconds (.002 second) when the demand for fuel is low, such as during deceleration. Or, on-time can be as long as 30 milliseconds (.030 second) when the demand for fuel is great, such as during acceleration. Pulse width specifications for various operating conditions can be found in most service manuals.

Referring to the schematic in figure 8-12, battery/charging system voltage should be present on circuit 639 (the power feed circuit to each injector) any time the key is in the RUN position. Available voltage can be easily checked using a DVOM. Don't forget to test for unwanted circuit resistance by measuring voltage drop on the positive side of the circuit. If voltage drop exceeds specifications, typically 500–600 mV, test for voltage drop across each electrical connector in the circuit.

A DVOM does not provide an accurate means of checking the ground side of the fuel-injector circuit because of the rapid cycling (pulse width) of the injector by the PCM. A scan tool provides the best method to accurately check the electrical operation of the injectors. All OBD II electronic engine control systems have a diagnostic data link connection where pulse width data can be retrieved using a properly programmed scan tool (refer to Chapter 5, *Electronic Engine Control System Testing and Service*).

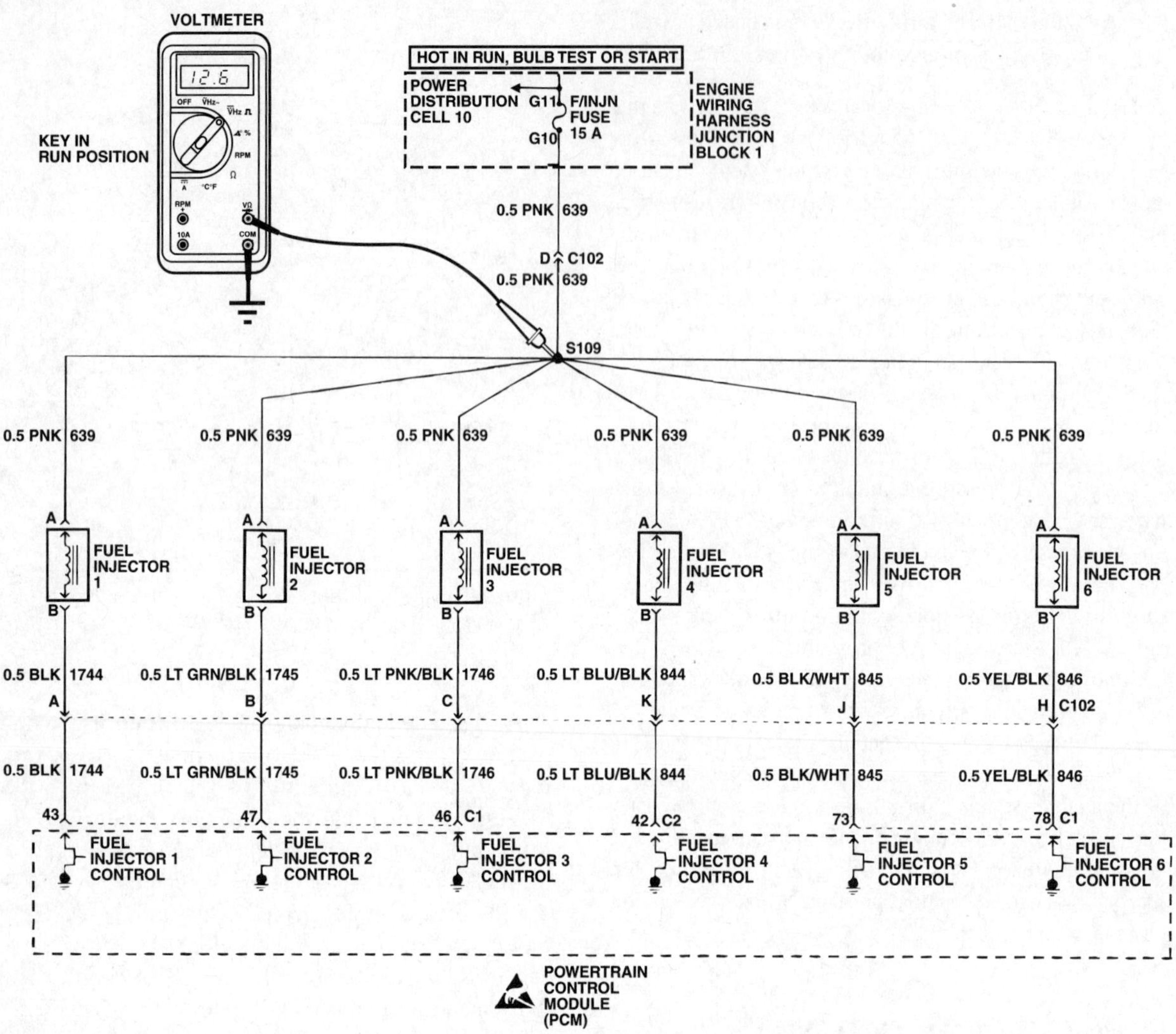

Figure 8-12. Typical fuel-injector schematic. Available voltage can be easily checked using a DVOM. (Courtesy of General Motors Corporation)

Oscilloscope testing is another effective method of checking the ground side of a fuel-injector circuit. A digital storage oscilloscope (DSO) is preferred because problem patterns can be recorded and played back at a later time. The positive scope lead is attached to the ground side of the injector connector. When the vehicle is running, the scope displays a pattern of voltage (trace) as it travels through the injector coil windings. It is important to remember that the injector is open and delivering fuel when the PCM completes the ground circuit, so the injector is on when voltage is low or zero and off when voltage is high.

The quality of ground-side circuit connections can be determined by examining the injector pattern as it parallels the 0 volt line. A good ground circuit produces a flat, smooth line that is very close to 0 volt. Unwanted circuit resistance causes the trace to stay above the 0 volt line on the scope. Loose connections cause the scope trace to look rough along the low-voltage section of the pattern. Due to different injector circuit designs, all injector patterns do not look the same, figure 8-13. Be sure you know what a normal pattern looks like before condemning a particular injector. If necessary, look at patterns for several injectors on the same vehicle to determine which pattern looks different from the rest.

Additional Fuel-Injector Tests

To determine if the injectors are functioning properly, manufacturers usually specify one or more of the following checks:

- Check multiport injectors by listening to each one with a stethoscope while the engine is running at

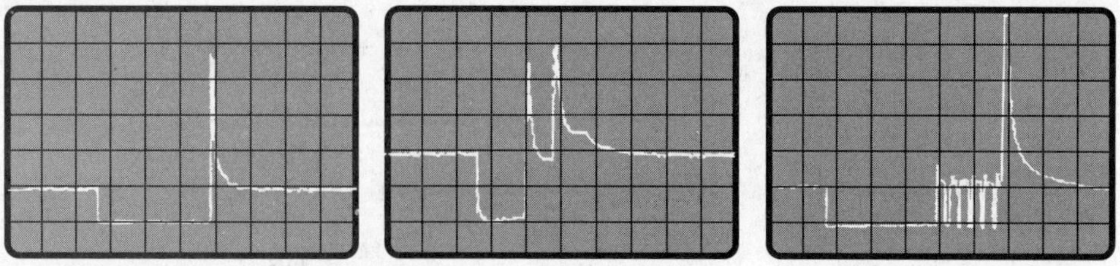

Figure 8-13. Typical fuel-injector oscilloscope patterns.

idle. A steady clicking indicates solenoid activity. All injectors should sound about the same.

- Throttle-body style injector operation is checked by removing the air cleaner cover or air intake duct, running the engine at idle, and observing the spray pattern from the injector tip. The spray pattern should be steady, uniform, and conical in shape.
- Use an ohmmeter to measure the injector solenoid coil resistance. Turn the ignition off. Unplug the injector electrical connector, connect the ohmmeter leads to the two injector pins, and note the meter reading. Then move one test lead to the side of the injector to check for a short. If the resistance reading is not within specifications, or if the injector is shorted, replace it with a new one. Remember that electrical resistance increases at higher temperatures. Whenever possible, test injector resistance at normal operating temperatures.
- Check the injector harness for shorts or opens by unplugging it from the PCM (ignition off) and connecting the ohmmeter between the supply and return terminals for each injector or group of injectors, figure 8-14. When the injectors are energized in groups, the resistance of the harness and injector group will be less than that of a single injector. A resistance reading higher than specified by the manufacturer results from an open in one or more injectors or circuits. A short causes a lower resistance reading in one or more injectors or circuits.
- Some systems are designed so a 12-volt test lamp can be used to check the signal to the injector. Never apply battery voltage directly to an injector to test its operation unless specifically directed by the manufacturer's test procedure. If the system can be tested with a 12-volt test lamp, connect your lamp in series between the supply wire and injector in the harness connector, or in series between the injector and ground, figure 8-15. When the engine is cranked, the lamp will light each time an injector or group of injectors is energized if they are functioning properly.

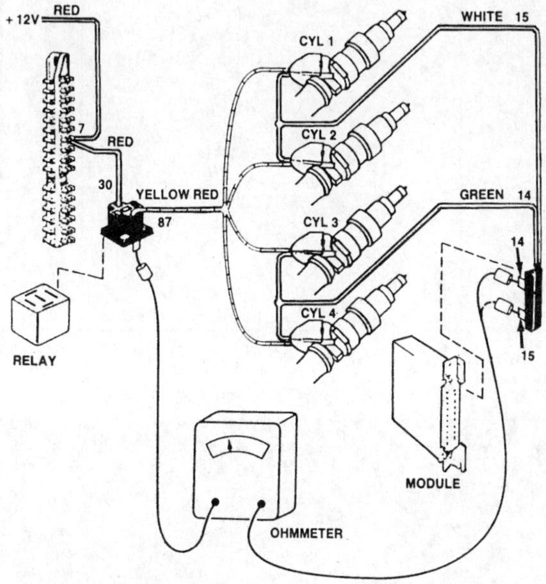

Figure 8-14. Injector wiring harness, complete with injectors, can be checked with an ohmmeter.

- Noid lights, figure 8-16, can always be used safely to check the electrical signal to the injector. Noid lights have enough internal resistance to prevent PCM damage. The lights are plugged into the injector connector and blink when the engine is cranked or started.
- The injector pulse width can be checked with an oscilloscope. As previously discussed, pulse width is the duration in milliseconds that the injector is energized, or ON. Refer to the equipment manufacturer's instructions and specifications to check injectors in this way.
- Don't forget the O-rings. All port type fuel injectors use rubber O-rings to seal the injector port. The O-rings become hard and brittle with age and cause vacuum leaks. Symptoms include rough idle, stalling, and hesitation. To test for leaks around injector O-rings, slowly flow propane around each injector with the engine at

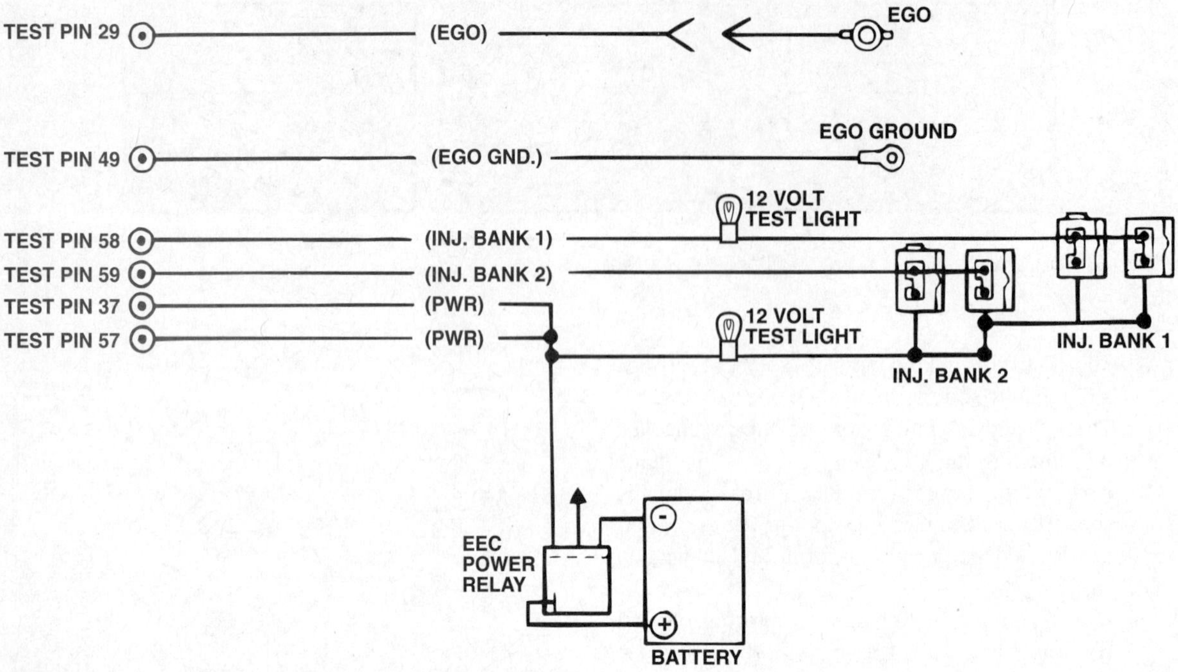

Figure 8-15. Testing injector signals with a 12-volt test light.

Figure 8-16. Testing electrical signal to the injector with a noid light.

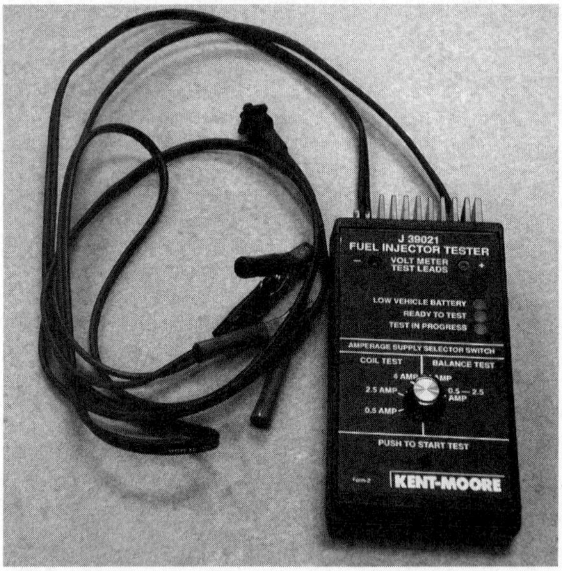

Figure 8-17. This tool performs an injector pressure drop test and an injector coil test on various fuel injectors.

idle. Listen to the engine. When propane enters the intake manifold through a leaking O-ring, the engine will run smoother and the rpms will increase. Be extremely careful when using propane. It is very flammable and will ignite easily when exposed to sparks or flame.

Injector Pressure Drop Testing

This test is sometimes referred to as an injector flow balance test. The test is performed with the engine off. A special balance tester is required to turn the injector on for a precise time period, figure 8-17. This allows the injector to spray a measured amount of fuel into the manifold, causing a drop in fuel-rail pressure that can be recorded and compared.

All injectors have a pressure drop specified by the manufacturer. Any injector whose pressure drop exceeds the specified average drop of the other injections

Figure 8-18. Connect the fuel-pressure gauge to the fuel rail.

Figure 8-19. Cycle the ignition so that the fuel pump runs. Record this pressure reading.

is leaking and should be replaced. Injectors whose pressure drop is less than specified are dirty or shorted and should be cleaned or replaced as necessary.

The following is a typical procedure to measure injector pressure drop. It requires the use of a special tool to pulse the fuel injectors a precise amount of time:

1. Make sure the engine is off and cool. Relieve residual fuel pressure. Connect a suitable pressure gauge to the fuel pressure tap on the fuel rail, figure 8-18.
2. Unplug all injector connectors. Connect the special tester to one injector.
3. Turn on the ignition and bleed all air from the pressure gauge according to the equipment manufacturer's procedure.
4. Turn OFF the ignition for 10 seconds, then turn the ignition back ON to bring fuel pressure to its maximum. Note and record the initial pressure reading, figure 8-19.
5. Press the tester button. Generally, longer pulse duration will yield more accurate test results, figure 8-20. Note and record the lowest pressure reading, figure 8-21. Subtract this reading from the initial reading obtained in step 4 and record the difference. This is the amount of injector pressure drop.
6. Connect the tester to the next injector to be checked and repeat steps 4 and 5. Record the readings.
7. Repeat step 6 for all injectors. Compare the results to specifications and retest any injector that shows a pressure difference of 1.5 psi (10 kPa) in either direction from the average of the other injectors, figure 8-22.

Figure 8-20. Press the button on the fuel-injector tester and wait for the injector to stop clicking.

Figure 8-21. When the injector stops clicking, immediately record the pressure reading on the pressure gauge.

Cylinder	#1	#2	#3	#4	#5	#6	#7	#8
Initial Pressure Reading	315	314	315	316	315	314		
Pressure Reading After Drop	150	155	160	156	140	153		
Difference	165	159	**155**	160	**175**	161		

Note: No injector drop should vary by more than 10 kPa.
Injector #3 shows too little of a drop.
Injector #5 shows too much of a drop.
Clean and retest the injectors.

Figure 8-22. Record the pressure readings for each injector. Subtract the initial reading from the reading after activating the injector. All injector drops should be within 10 kPa of each other.

8. Do not repeat this test more than once (including retesting of injectors) without running the engine to clear the unburned fuel and prevent possible flooding. If any injector fails the retest, replace it with a new one.
9. Pressure drop readings that are uneven may also indicate dirt and varnish build-up around the injector pintle. Clean the injectors, if recommended, using the manufacturer's recommended procedures and retest, before replacing a set of injectors.

Figure 8-23. Relieve the fuel pressure and disable the fuel pump.

Injector-Coil Testing

To test the coil of injectors, technicians typically have measured the resistance of the fuel-injector coil and compared the resistance to specifications. While this method will detect an injector whose coil is open and in some cases will detect an injector coil that is shorted, it won't always detect shorts in injector coils.

As many technicians are aware, electrical shorts do not always present themselves until there is current flowing in the circuit and a component heats up. Some injectors suffer from shorted coils but only when hot or during operation. A tool has been developed that will supply the correct amount of injector current flow for a duration of time that allows the technician to detect shorts in the injector coil. This tool is often referred to as an injector balance tester. This tester has various current settings for different injectors. In addition, it has a built-in timer that flows current for a specified duration; long enough to heat the coil of the injector but not long enough to damage the fuel-injector coil. When the tool is supplying current to the injector coil, the technician uses a DVOM to measure a voltage drop across the injector. This voltage drop should be within a specific range if the injector coil is good. If the injector coil has a short, the voltage drop measured will not be within the specified range and the technician knows that the injector coil is shorted. To test injectors using the injector balance tester, do the following:

1. Disconnect all the injectors.
2. Relieve the fuel pressure and disable the fuel pump, figure 8-23.
3. Connect the tester to the battery and to the first injector, figure 8-24.
4. Select the correct current setting on the balance tester, figure 8-25.
5. Connect the DVOM to the tester and select DC volts, figure 8-26.
6. Push the tester button and monitor and record the voltage reading on the DVOM, figure 8-27.
7. Compare the DVOM reading to the specifications that are provided with the balance tester, figure 8-28.
8. Replace any injectors that fall outside of the specifications.

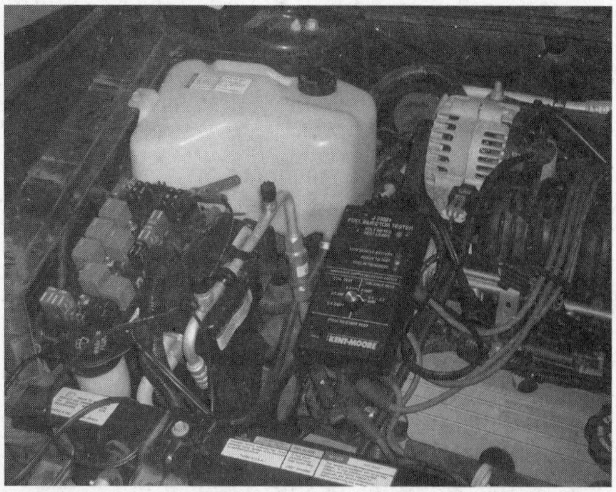

Figure 8-24. Connect the tester to the battery and to the injector.

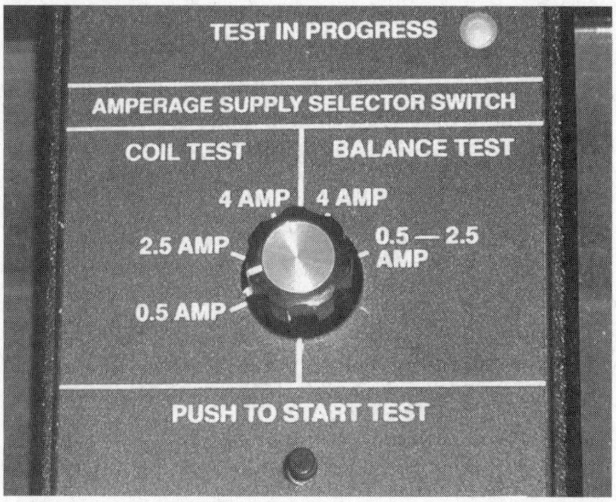

Figure 8-25. Select the correct current setting for the type of injector being tested.

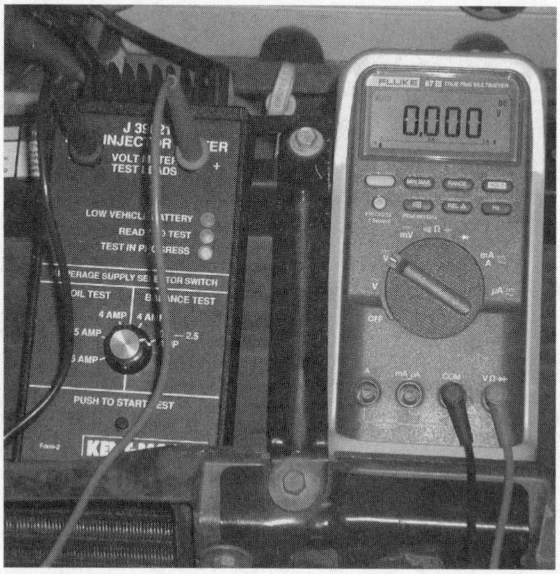

Figure 8-26. Connect the DVOM to the tester and select DC volts.

Figure 8-27. Push the button on the tester and observe the voltage on the DVOM.

FUEL-INJECTOR CLEANING

Check the manufacturer's recommendations for injector service before attempting to clean injectors. Many late-model injectors are a deposit-resistant design, and should NOT be cleaned according to some manufacturers. Most older pintle valve injectors can be cleaned if varnish or carbon builds up on the tip.

Various cleaning devices and special solutions are available to clean fuel injectors. Some require the injectors to be removed from the fuel rail or throttle body. Others, figure 8-29, can be used to clean the injectors without removing them.

When cleaning injectors, it is very important to follow the manufacturer's directions *exactly.* The cleaning solvents used are very powerful, and can:

- Affect the injector coil wire insulation, resulting in low resistance or a shorted injector
- Increase friction at critical mating surfaces and cause premature wear
- Corrode plated parts if allowed to remain in the system

To ensure that all traces of solvent are purged from the fuel system, start and run the engine on premium unleaded gasoline after the injectors have been cleaned. *Do not* check system operation with an infrared analyzer immediately after cleaning the injectors. Any solvent

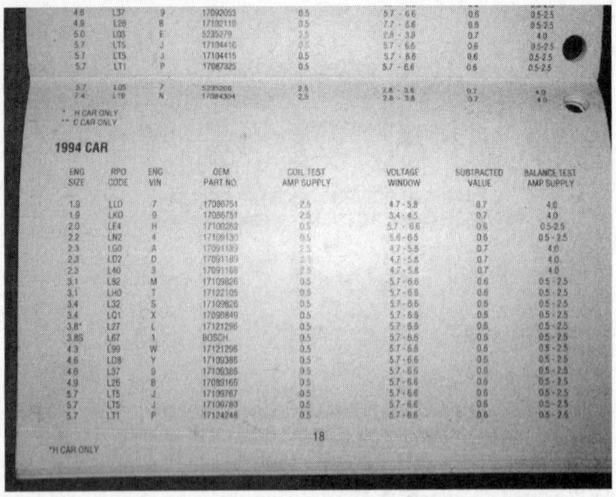

4.8	L37	9	17092053	0.5	5.7 - 6.6	0.6	0.5-2.5
4.9	L26	B	17102119	0.5	7.7 - 8.6	0.6	0.5-2.5
5.0	L03	E	5235279	2.5	2.8 - 3.8	0.7	4.0
5.7	LT5	J	17104416	0.5	5.7 - 6.6	0.6	0.5-2.5
5.7	LT5	J	17104415	0.5	5.7 - 6.6	0.6	0.5-2.5
5.7	LT1	P	17087325	0.5	5.7 - 6.6	0.6	0.5-2.5
5.7	L05	7	5235206	2.5	2.8 - 3.8	0.7	4.0
7.4	L19	N	17084304	2.5	2.8 - 3.8	0.7	4.0

* H CAR ONLY
** C CAR ONLY

1994 CAR

ENG SIZE	RPO CODE	ENG VIN	OEM PART NO.	COIL TEST AMP SUPPLY	VOLTAGE WINDOW	SUBTRACTED VALUE	BALANCE TEST AMP SUPPLY
1.9	LL0	7	17086751	2.5	4.7 - 5.8	0.7	4.0
1.9	LK0	9	17086751	2.5	3.4 - 4.5	0.7	4.0
2.0	LE4	H	17100262	0.5	5.7 - 6.6	0.6	0.5-2.5
2.2	LN2	4	17109130	0.5	5.6 - 6.5	0.6	0.5 - 2.5
2.3	LG0	A	17091189	2.5	4.7 - 5.8	0.7	4.0
2.3	LD2	D	17091189	2.5	4.7 - 5.8	0.7	4.0
2.3	L40	3	17091168	2.5	4.7 - 5.8	0.7	4.0
3.1	L82	M	17109826	0.5	5.7 - 6.6	0.6	0.5 - 2.5
3.1	LH0	T	17122105	0.5	5.7 - 6.6	0.6	0.5 - 2.5
3.4	L32	S	17109826	0.5	5.7 - 6.6	0.6	0.5 - 2.5
3.4	LQ1	X	17090849	0.5	5.7 - 6.6	0.6	0.5 - 2.5
3.8*	L27	L	17121296	0.5	5.7 - 6.6	0.6	0.5 - 2.5
3.8S	L67	1	BOSCH	0.5	5.7 - 6.6	0.6	0.5 - 2.5
4.3	L99	W	17121296	0.5	5.7 - 6.6	0.6	0.5 - 2.5
4.6	LD8	Y	17109386	0.5	5.7 - 6.6	0.6	0.5 - 2.5
4.6	L37	9	17109386	0.5	5.7 - 6.6	0.6	0.5 - 2.5
4.9	L26	B	17089166	0.5	5.7 - 6.6	0.6	0.5 - 2.5
5.7	LT5	J	17109767	0.5	5.7 - 6.6	0.6	0.5 - 2.5
5.7	LT5	J	17109780	0.5	5.7 - 6.6	0.6	0.5 - 2.5
5.7	LT1	P	17124248	0.5	5.7 - 6.6	0.6	0.5 - 2.5

18

*H CAR ONLY

Figure 8-28. Compare the voltage reading obtained to specifications.

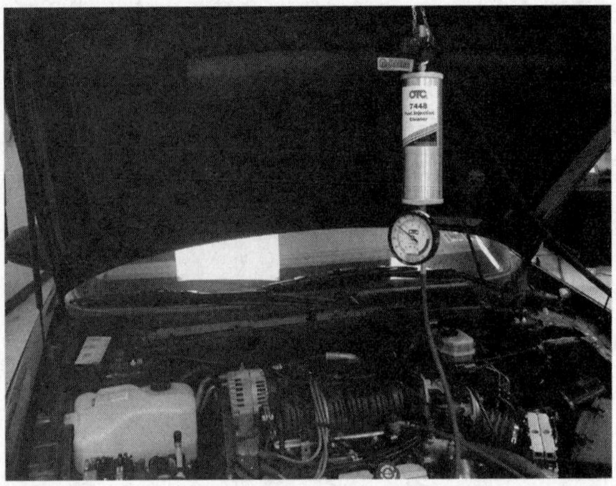

Figure 8-29. This fuel-injector cleaner is pressurized with air. The cleaner can be mixed or purchased premixed.

that remains in the system will pass through the exhaust and may damage analyzer components.

BASIC IDLE ADJUSTMENTS

Fuel-injection systems, like carburetor fuel systems, are responsible for maintaining the correct air-fuel ratio for all operating conditions, and maintaining proper engine speed at idle. Idle speed adjustment is a routine procedure required on a regular basis with carburetor systems, but it is not always as routine with fuel-injection systems. Late-model electronic engine management systems control idle speed based on sensor inputs and PCM programming. The following segments cover the systems commonly in use, point out which adjustments can be made, and which operations are computer-controlled. Systems that allow idle adjustment require certain preliminary steps:

1. Set the parking brake and securely block the drive wheels.
2. Connect an engine tachometer.
3. Start the engine and warm to normal operating temperature.
4. Check the vehicle emission control information (VECI) decal in the engine compartment for specifications and specific adjustment procedures.
5. If an infrared exhaust analyzer is being used, install the probe in the tailpipe.
6. Check ignition timing and adjust as necessary before adjusting the idle speed.

Bosch L-Jetronic Adjustments

Idle mixture and idle speed adjustments are possible on early L-Jetronic and similar airflow-controlled port injection systems. Idle mixture is adjusted with an air-bypass screw on the airflow meter, figure 8-30. Air that bypasses the sensor does not affect the electrical signal sent to the PCM, so increasing the volume of air bypassing the throttle causes a leaner air/fuel ratio. Turn the screw clockwise (inward) to richen the mixture or counterclockwise (outward) to lean the mixture. Service manuals often refer to this procedure as a "CO adjustment."

Idle speed is adjusted with an idle air bypass screw located on the throttle body, figure 8-30. Turning this screw counterclockwise (outward) allows air to bypass the throttle plate and increases idle speed. Turning the same screw clockwise reduces idle speed. You may also notice an adjustment screw for the throttle plate angle, but it is set at the factory and should not be adjusted.

No idle speed adjustment is possible on late-model L-Jetronic systems, because they use comptuer-controlled idle air control valves. Idle mixture adjustment is not part of normal tune-up procedures. The adjustment screw is normally covered by a plug to prevent tampering.

Other EFI Systems

Minimum idle speed or minimum air rate is adjustable on some fuel-injected engines. The adjustment controls the minimum speed at which an engine idles when the idle speed control (ISC) motor is fully retracted, or the idle air control (IAC) valve/motor is closed. Generally, the term *idle speed control (ISC)* refers to systems and adjustment procedures that adjust the actual angle of the throttle plate. The term *idle air control (IAC)* refers to systems and adjustment procedures that adjust idle

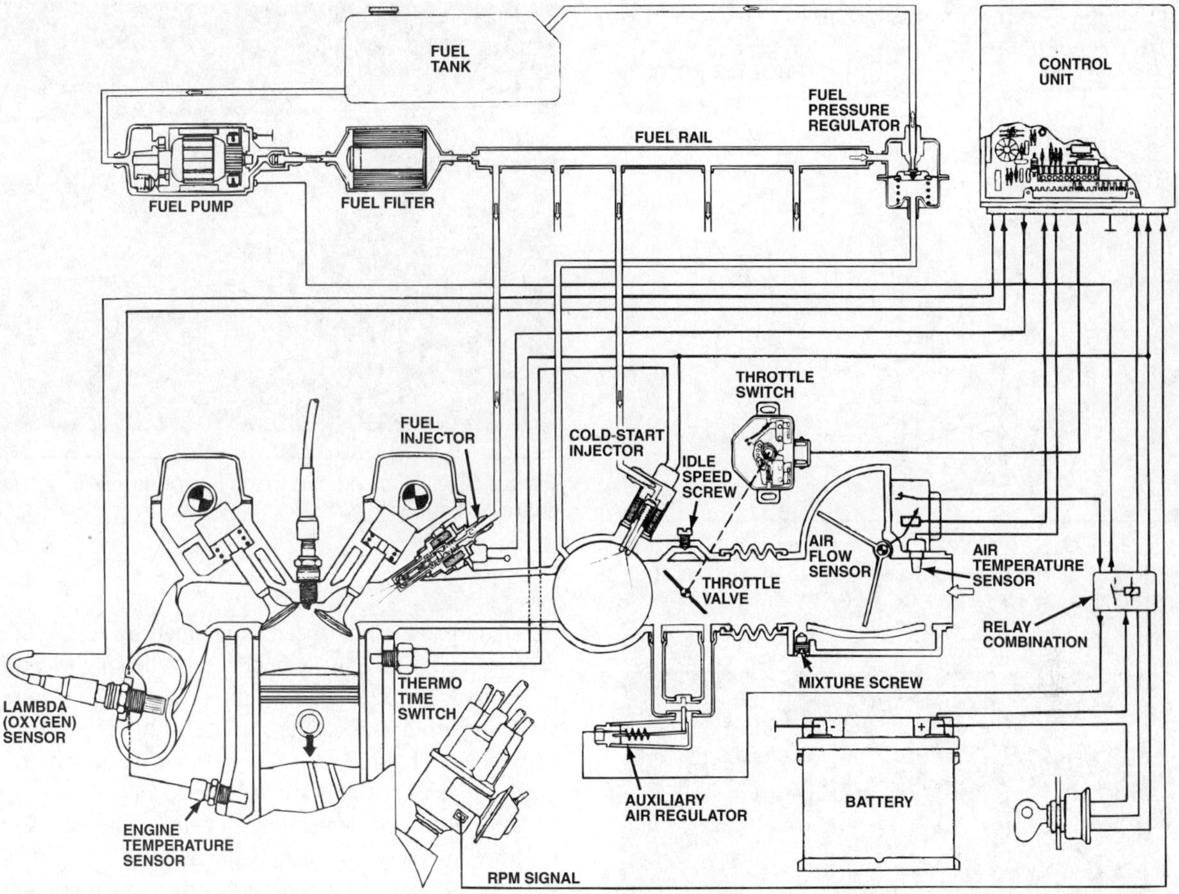

Figure 8-30. Idle speed and mixture on early airflow-controlled multipoint fuel-injection systems can be adjusted. (Courtesy of Volkswagen of America)

speed by controlling air that bypasses the throttle plate. The manufacturer's procedure and specifications for checking and adjusting the minimum idle speed of an engine must be followed *exactly.* There are more than 50 different idle adjusting procedures, which cannot be covered in detail in this publication. However, there are basic principles that apply to all vehicles.

Check minimum idle speed and adjust, if necessary, on any engine that exhibits a rough or rolling idle, a tendency to stall, or whenever the throttle body or TPS sensor has been serviced. If the engine has a port-injection system, check the throttle body for sludge buildup before adjusting the minimum idle speed, as a poor idle or stalling condition can result from sludge buildup. Check the service manual to be sure throttle-body cleaning is recommended by the manufacturer; some manufacturers use a coating on the throttle-body that is damaged by chemical cleaners.

Before making any adjustments, the engine must be warmed to normal operating temperature. Ignition timing should be checked and adjusted as necessary. Accessories such as air conditioning should be off unless the manufacturer's procedures specify otherwise.

Minimum idle speed cannot be adjusted by changing the position of the IAC valve or ISC motor. The throttle or idle stop screw in the throttle body is used for adjustment. The stop screw looks and acts much like the idle speed screw on a carburetor. On most Ford systems, the stop screw is located on the outside of the throttle body, figure 8-31, contacts the throttle linkage, and is usually sealed with paint. GM TBI and PFI throttle bodies locate the stop screw under a concealment plug, figure 8-32, to discourage unauthorized adjustment. Access to the screw requires plug removal. On many engines, this can be done without throttle-body removal. The hardened steel plug used with early GM throttle bodies requires use of a drill and punch for removal. Throttle bodies after mid-1984 use a soft plug that can be pried out with the help of an awl or small punch, figure 8-33. After making adjustments on a GM throttle body, cover the adjustment screw with a small quantity of silicone sealant or equivalent.

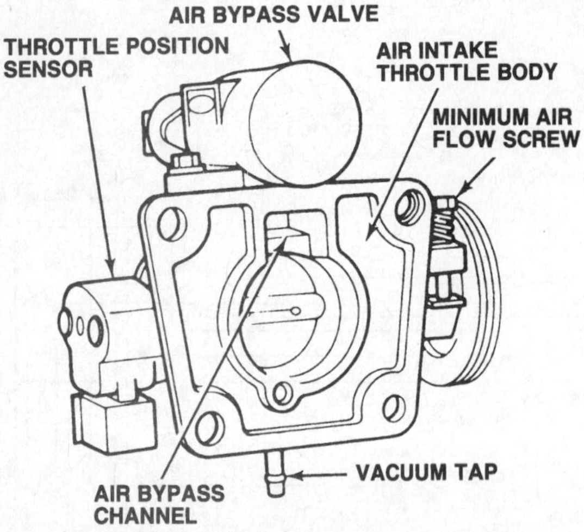

Figure 8-31. Typical minimum airflow or throttle stop screw used by Ford EFI systems for idle speed adjustment. (Courtesy of Ford Motor Company)

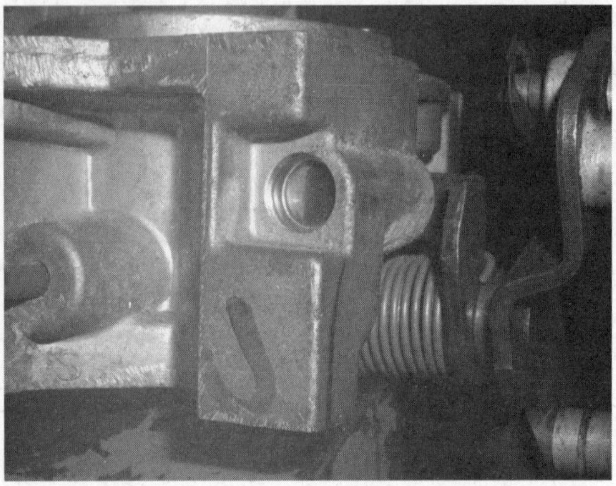

Figure 8-32. This photo shows the plug that seals the minimum air rate adjustment screw.

To maintain and adjust idle speed during normal operation, most systems use a throttle air bypass valve solenoid, figure 8-34, or an idle air control (IAC) motor, figure 8-35 controlled by the PCM. The air bypass valve, common on Ford Motor Company products, is usually located on the throttle body, but may be located elsewhere and connected to the throttle body with hoses. IAC motors, found on General Motors and Chrysler products, are mounted on the throttle body. Both air bypass valves and IAC motors open or close an air bypass

Figure 8-33. To remove the plug, puncture it with a sharp awl and pry out.

passage to adjust idle speed as required by engine accessory loads or engine temperature. When higher rpm is required, during engine warm-up for example, more air is bypassed around the throttle plate. The bypass valve solenoid also acts as an electronic dashpot by reducing engine rpm slowly during deceleration.

Late-model fuel-injection systems generally use adaptive memory to change the idle speed control strategy as engine wear occurs. After any minimum idle speed adjustment, the computer's strategy must be reset according to the manufacturer's procedure. With some engines, it may be necessary to drive the vehicle to allow the computer to relearn the idle speed strategy.

THROTTLE-BODY INJECTOR UNIT OVERHAUL

This next segment of photo sequences covers the important points in disassembling and assembling five TBI units. They are:

- Rochester Model 200
- Rochester Model 700
- Motorcraft CFI Single-Injector unit
- Motorcraft CFI Dual-Injector unit
- Chrysler Low-Pressure TBI unit

The study of this segment will reveal how simple these units are in comparison with carburetors. Once the components are removed from the casting, they can be cleaned with a soft-bristle brush and aerosol cleaner, or submerged in a carburetor cleaning tank if necessary.

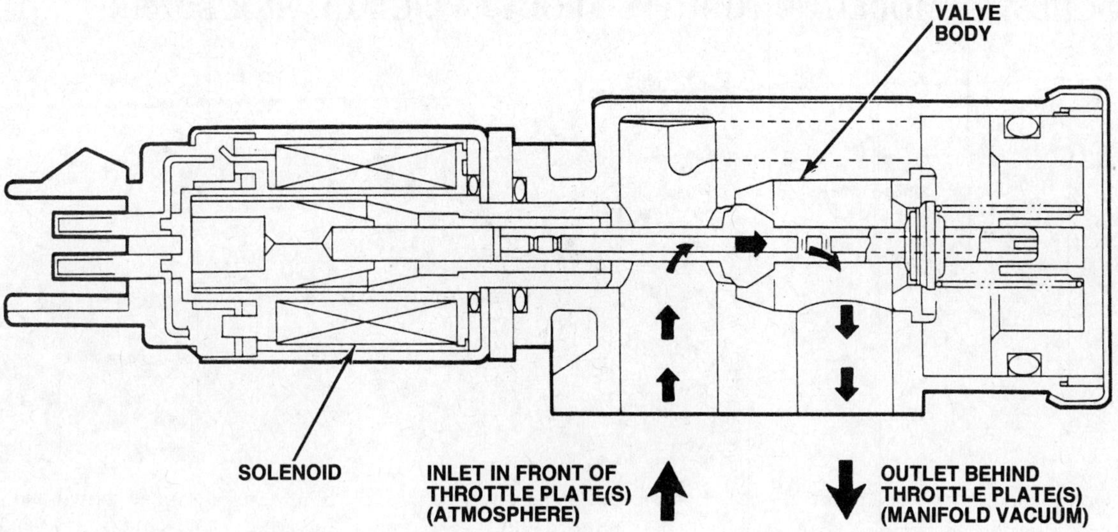

Figure 8-34. The throttle air bypass solenoid controls idle speed on some Ford multipoint injection systems. (Courtesy of Ford Motor Company)

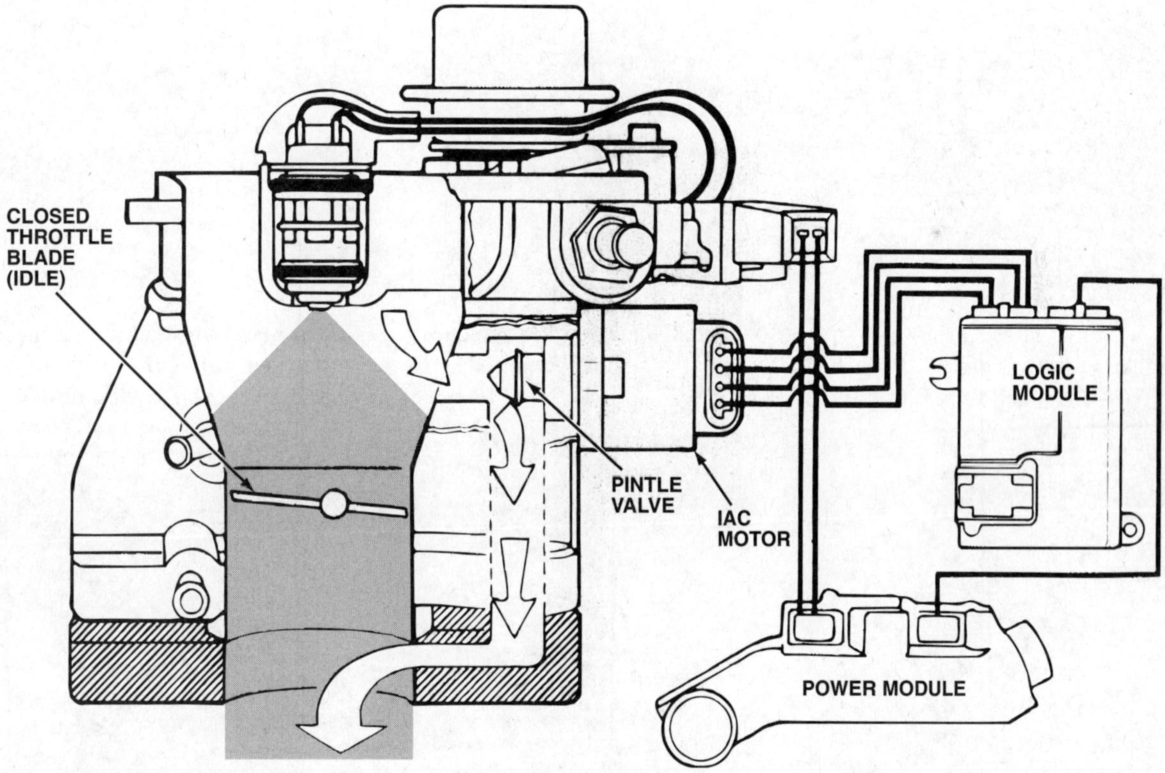

Figure 8-35. Chrysler's IAC motor controls airflow through an air bypass in the throttle body. (Courtesy of DaimlerChrysler Corporation)

ROCHESTER MODEL 200 THROTTLE-BODY OVERHAUL PROCEDURE

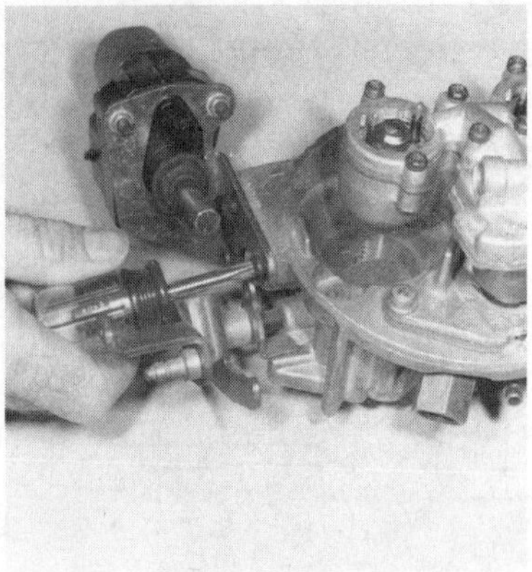

1. Unbolt and remove idle speed control motor assembly from throttle-body casting.

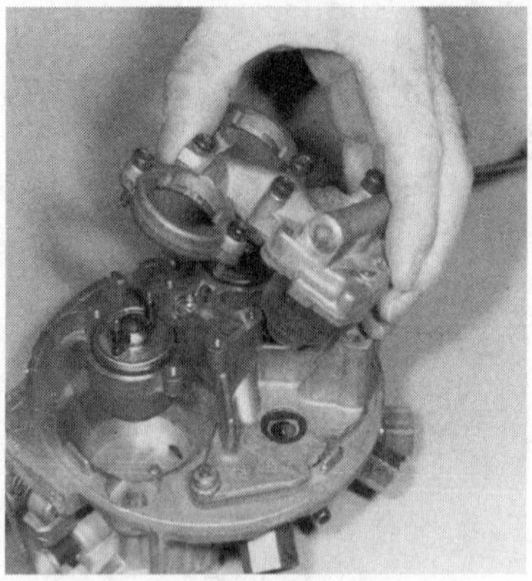

2. Loosen fuel meter cover screws, then lift cover with pressure-regulator assembly off of fuel meter body.

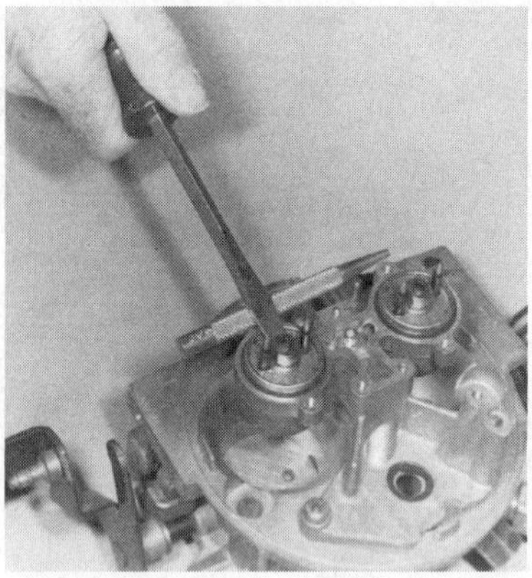

3. Use a fulcrum as shown and pry injectors from casting. A pry slot is provided in injector housing.

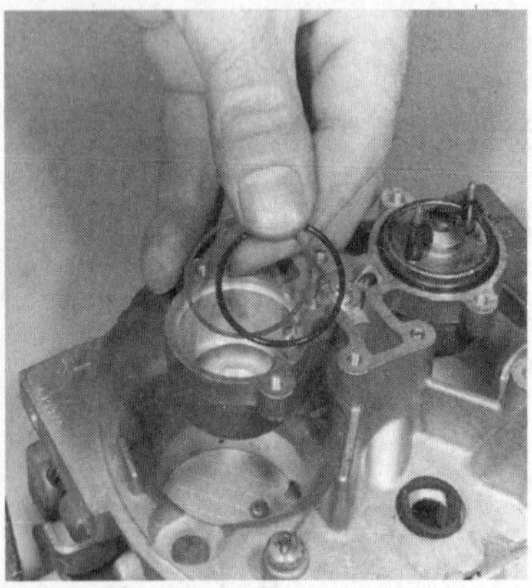

4. Remove large O-ring and steel backup washer from top of injector cavity, and small O-ring from bottom of cavity.

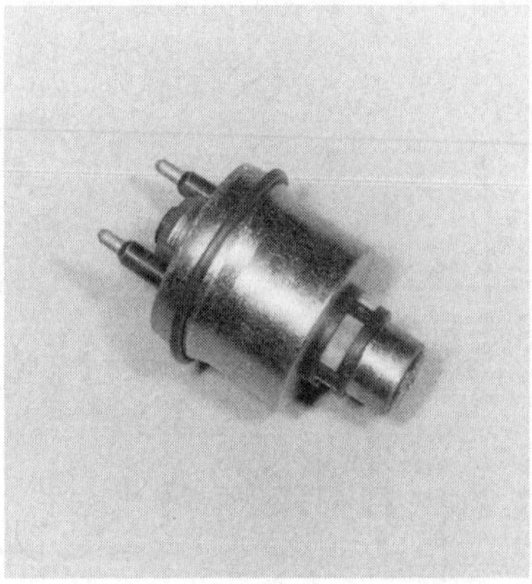

5. Check injector filter screen for signs of clogging. Replace all O-rings.

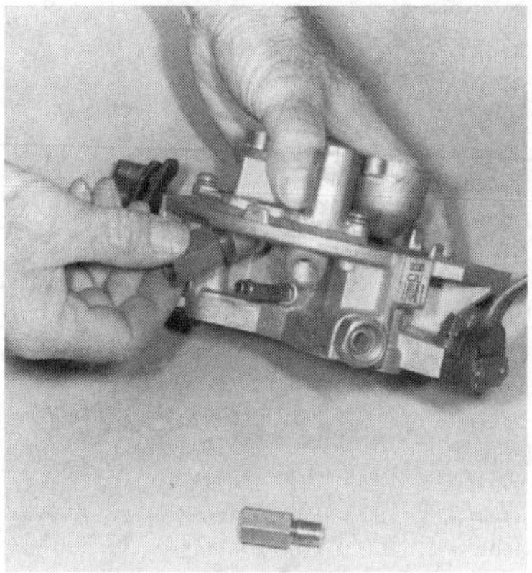

6. The inlet and outlet fittings must be removed before fuel meter body can be separated from throttle body.

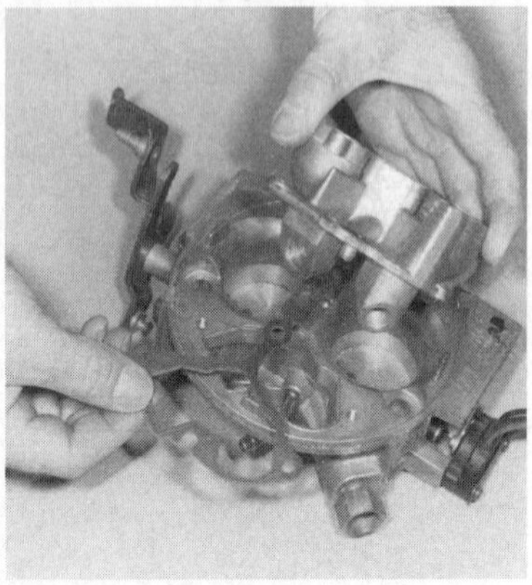

7. Remove fuel meter body screws. Separate fuel meter and throttle bodies. Remove and discard gasket.

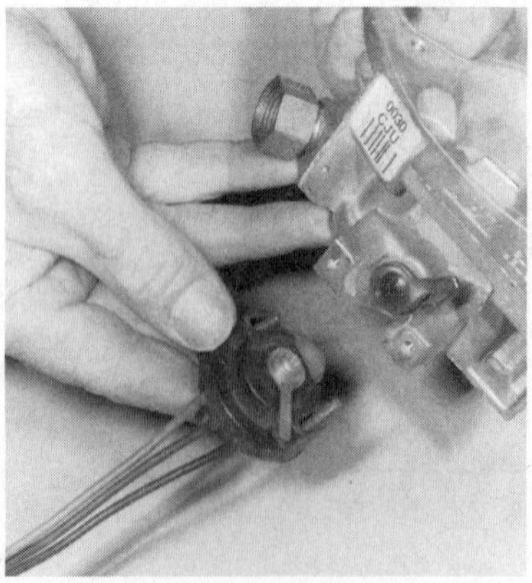

8. Remove the two screws holding the throttle-position sensor to the throttle-body casting. Remove sensor.

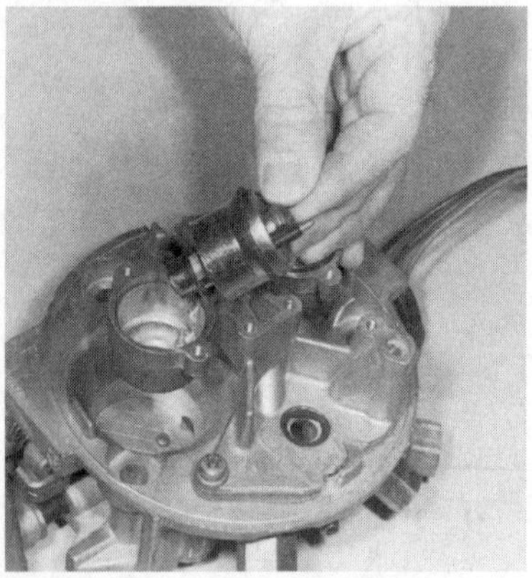

9. When reinstalling injectors, pin on injector housing must align with slot in casting for injector to fully seat.

ROCHESTER MODEL 700 THROTTLE-BODY OVERHAUL PROCEDURE

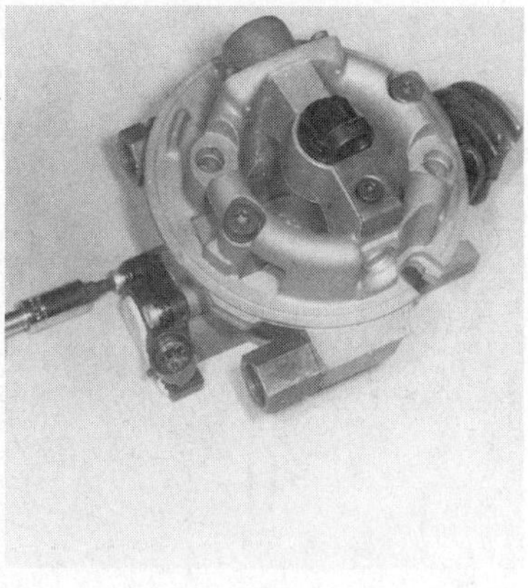

1. Unbolt throttle-position sensor fasteners. All fasteners are a Torx head design.

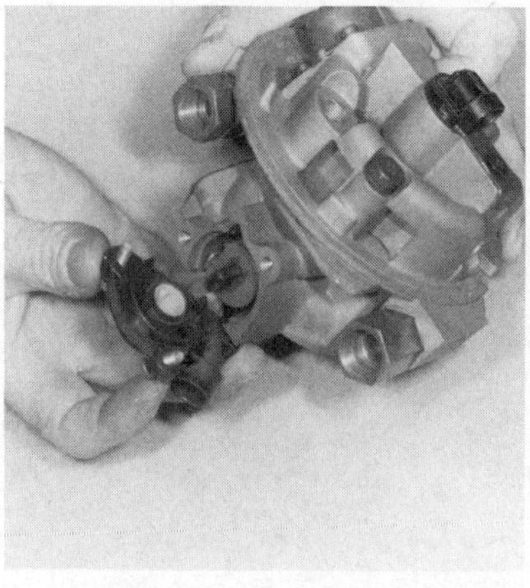

2. Remove the throttle-position sensor. Check shaft seal and replace if damaged.

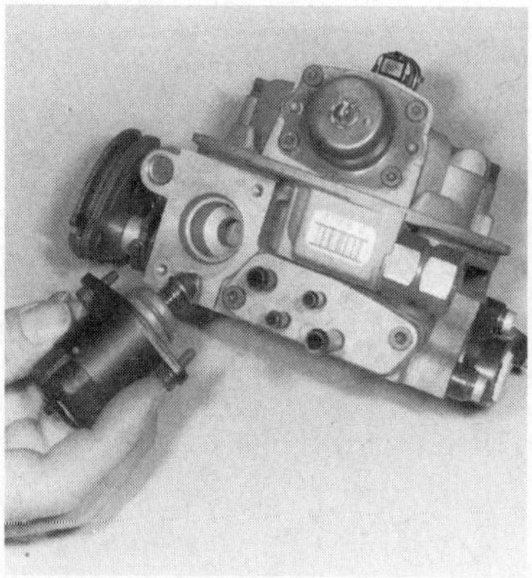

3. Unbolt and remove idle air control assembly from throttle-body casting.

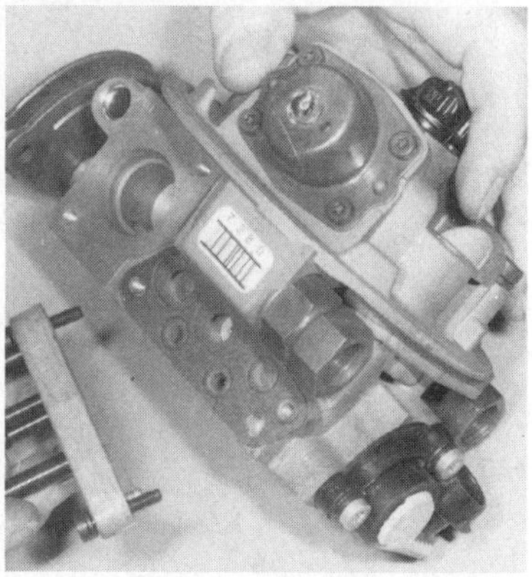

4. Unbolt and remove vacuum tube module assembly. Remove and discard the module assembly gasket.

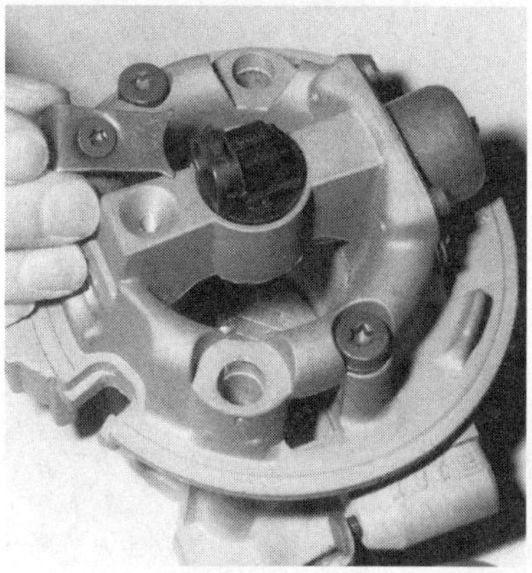

5. Remove the screw holding the injector retainer clamp. Remove screw and retainer.

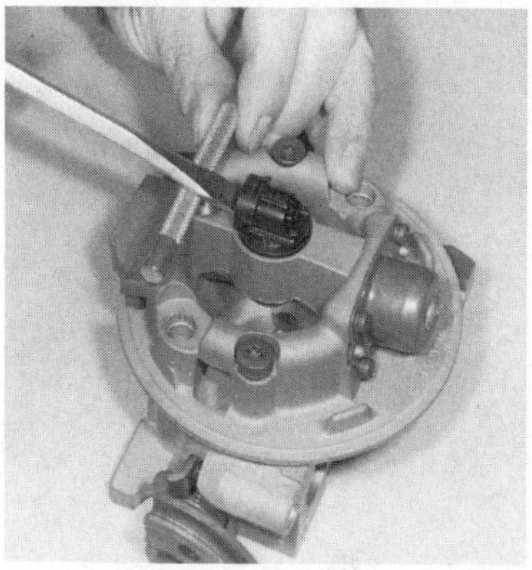

6. Use a fulcrum and flat-blade screwdriver as shown to pry injector from fuel meter body.

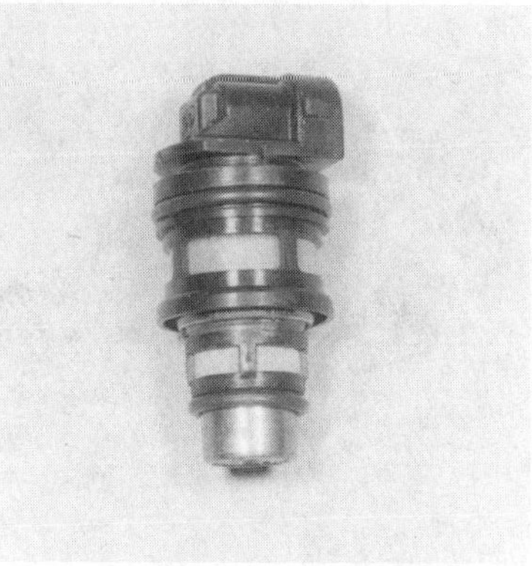

7. Multec injector uses two filter screen assemblies. Fuel inlet is behind lower filter; upper one is a purge filter.

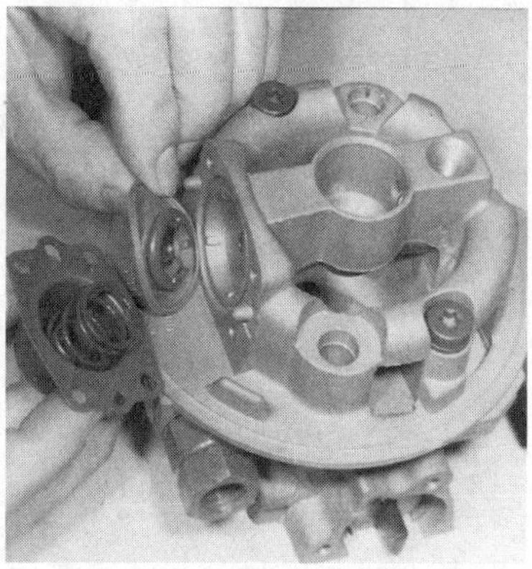

8. Remove pressure-regulator assembly. Design allows replacement of regulator diaphragm.

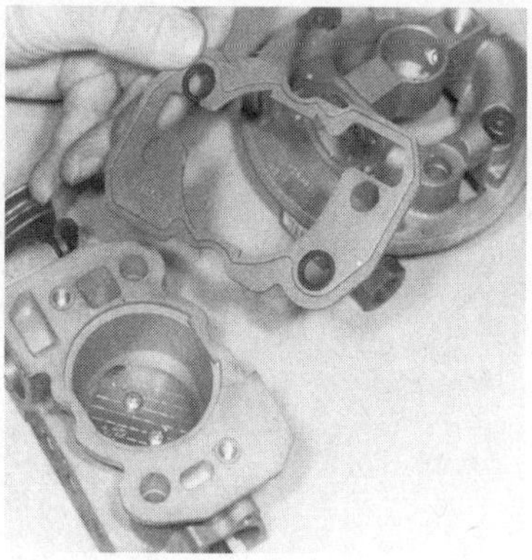

9. Remove fasteners holding fuel meter body to throttle body. Separate bodies; remove and discard gasket.

MOTORCRAFT CFI SINGLE-INJECTOR OVERHAUL PROCEDURE

1. Hold fuel-pressure regulator cover in place and remove the four cover screws, then carefully lift off cover.

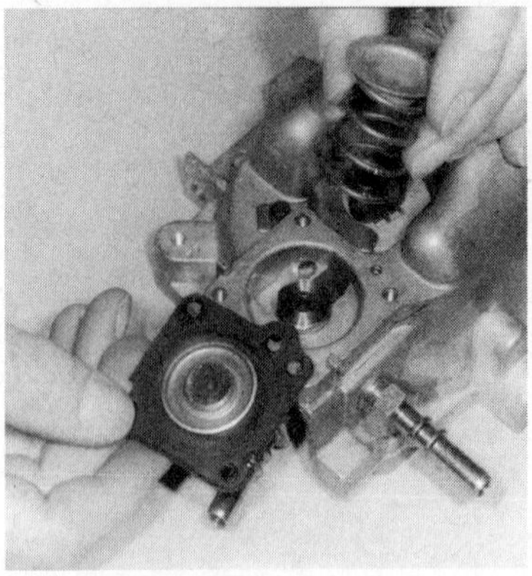

2. Remove pressure-regulator cup, spring, and diaphragm. Check diaphragm carefully for defects and replace if necessary.

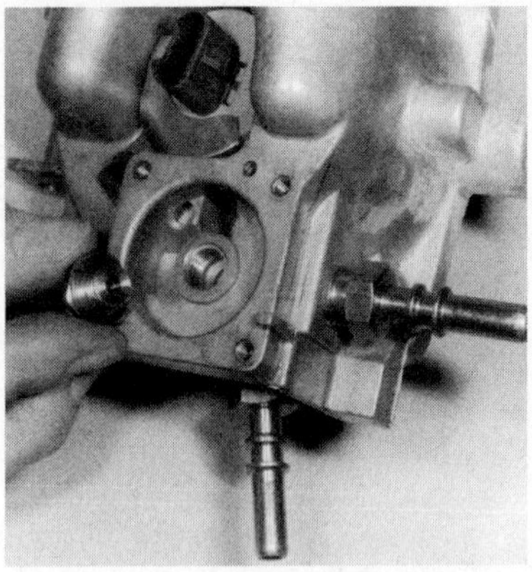

3. Remove the pressure-regulator valve seat from the regulator bore with a suitable socket wrench.

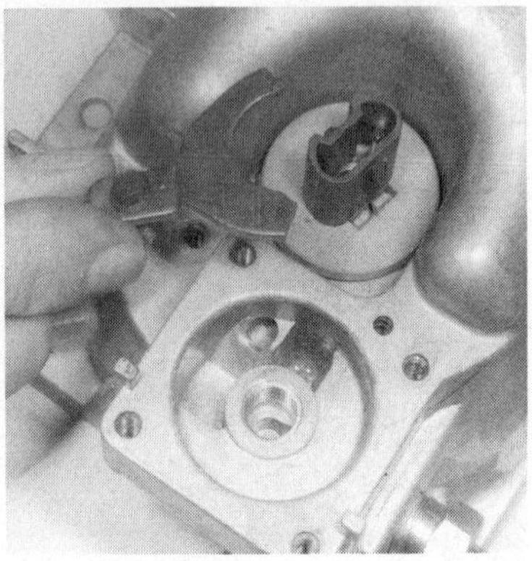

4. Loosen injector retainer capscrew. Remove retainer and capscrew from throttle-body housing.

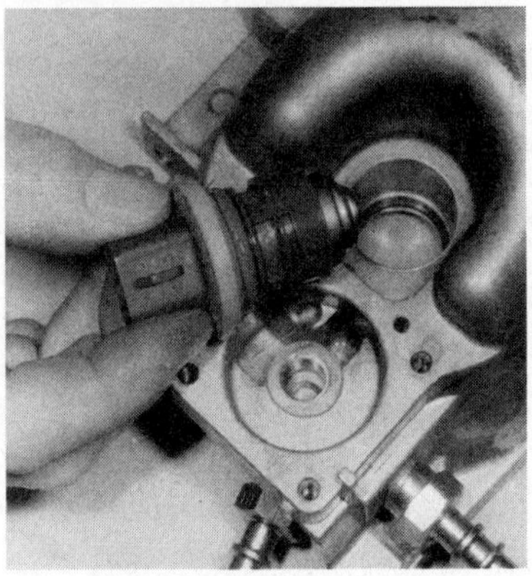

5. Apply finger pressure from beneath the injector nozzle to remove the injector. Discard O-ring in injector cavity.

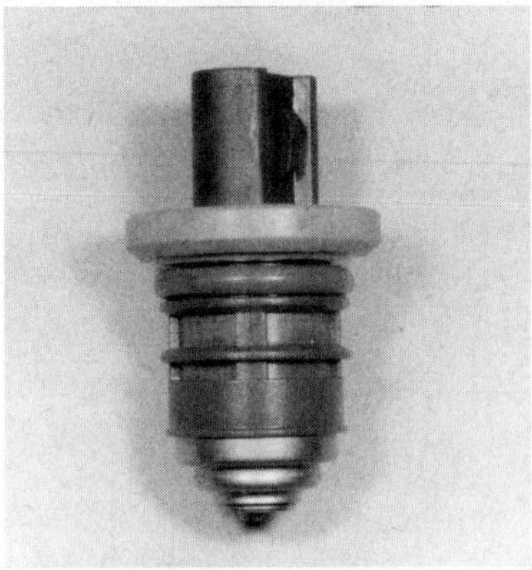

6. Check injector filter screen for signs of clogging. Replace all O-rings.

7. Remove the fuel inlet fitting. Reach inside with needle-nose pliers. Then, carefully extract filter screen. Replace if required.

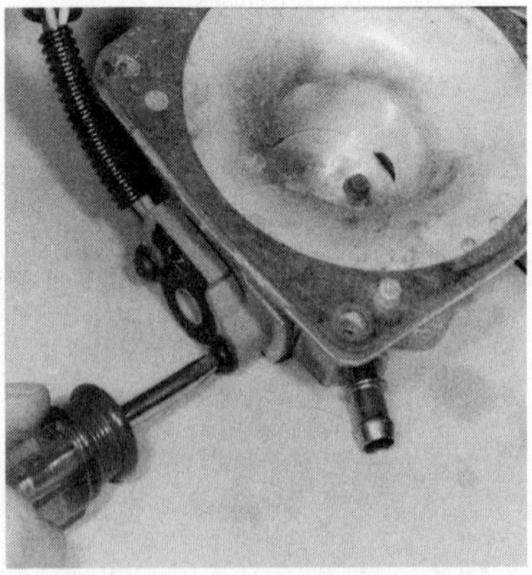

8. Remove four screws holding throttle body to main body. Separate bodies and discard gasket. Remove throttle-position sensor.

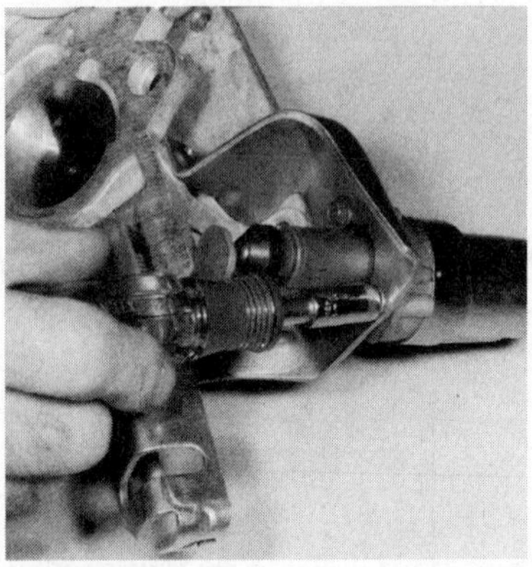

9. Remove throttle actuator from bracket. Bracket does not have to be removed for cleaning.

MOTORCRAFT CFI DUAL-INJECTOR OVERHAUL PROCEDURE

1. Remove air cleaner stud, then invert fuel charging assembly and remove four throttle-body attaching screws.

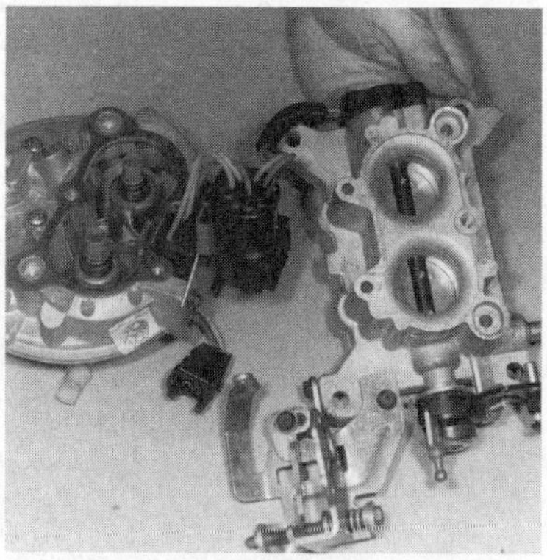

2. Separate throttle body and upper body assemblies. Remove and discard upper body gasket.

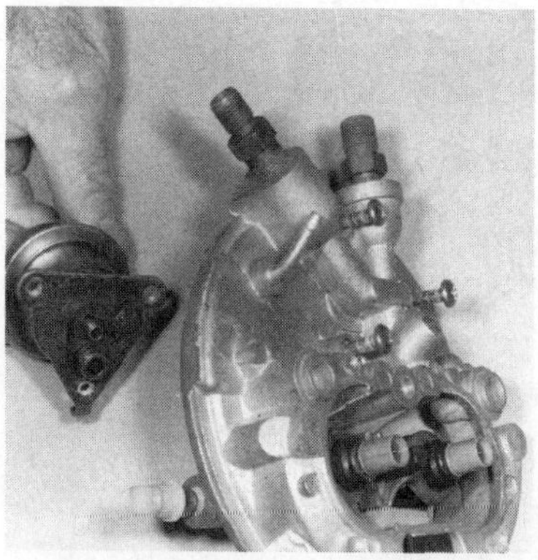

3. Remove three pressure-regulator screws. Pull regulator off from underside of upper body. Check gasket and O-ring condition.

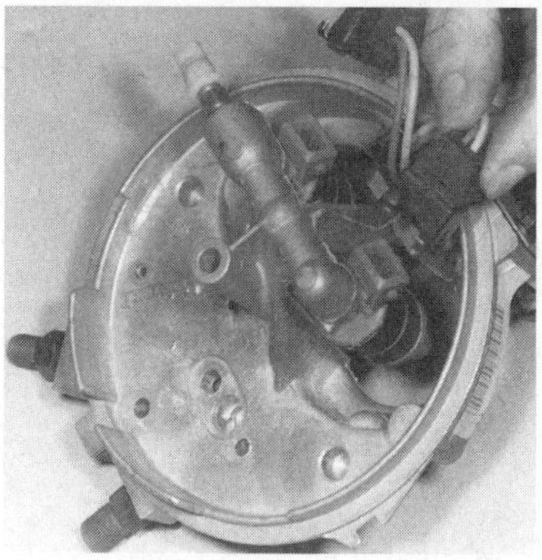

4. Carefully disengage injector connectors by pulling straight out and off of injectors.

5. Remove 10-pin connector. On earlier models with a wiring harness connector, simply loosen (do not remove) connector screw.

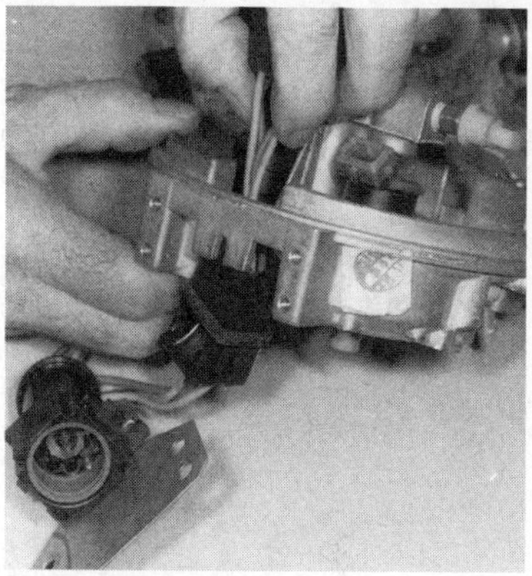

6. Depress tabs on harness connector and remove from upper body, feeding injector connectors through casting slot one at a time.

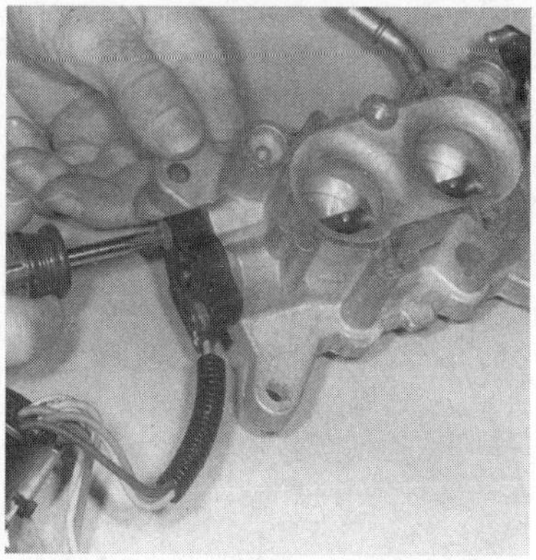

7. Unscrew and remove throttle-position sensor from upper body. Set harness and sensor assembly to one side.

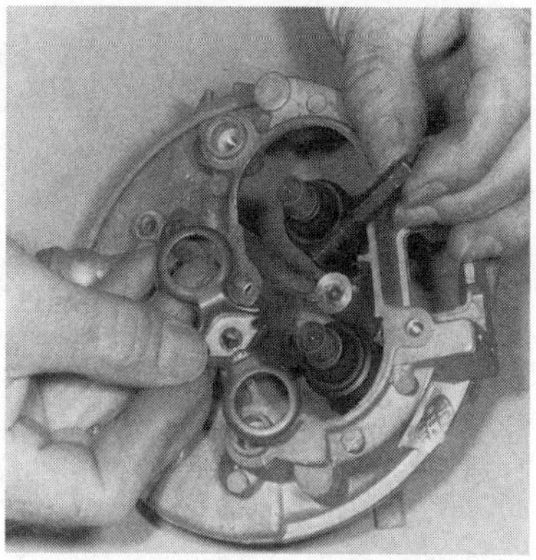

8. Remove screw holding injector retainer. Remove retainer from upper body.

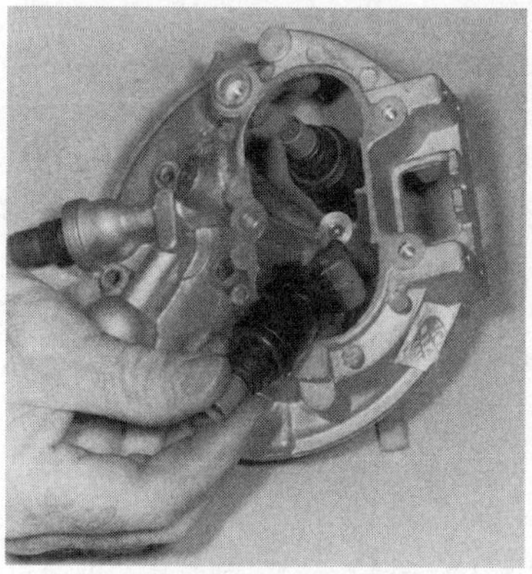

9. Carefully pull injectors from upper body. Mark injector as choke or throttle side. Replace O-rings.

CHRYSLER LOW-PRESSURE THROTTLE-BODY OVERHAUL PROCEDURE

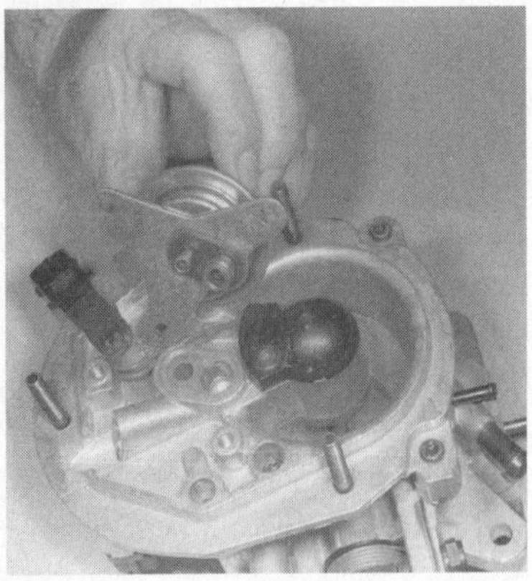

1. Unbolt and remove fuel-pressure regulator. Remove and discard gasket and O-ring.

2. Remove injector cap screw, then insert screwdriver blades in cap pry slots and pry the cap loose.

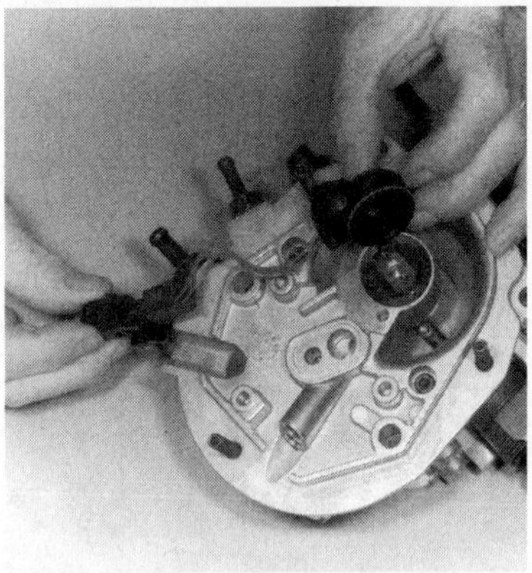

3. Disengage wiring harness connector from throttle body and remove cap assembly. Replace cap O-ring.

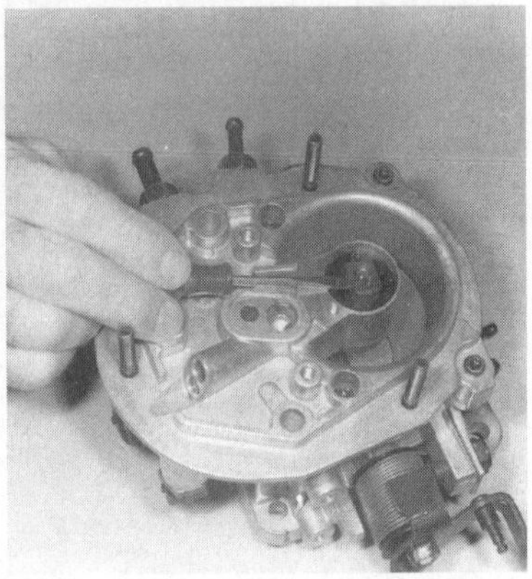

4. Insert screwdriver tip in injector slot and lift injector from pod with downward leverage.

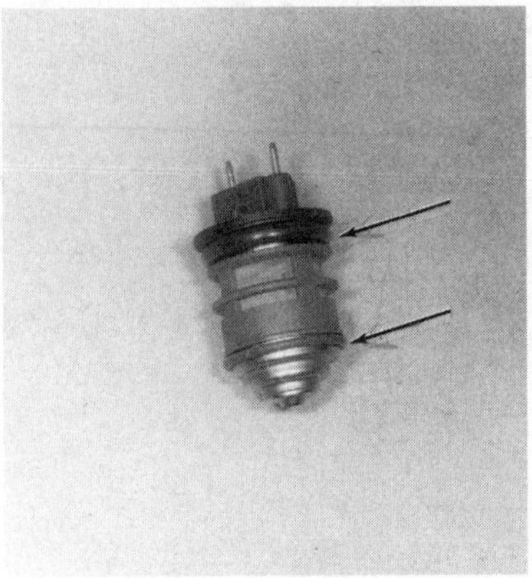

5. Check injector filter screen for signs of clogging. Replace all O-rings.

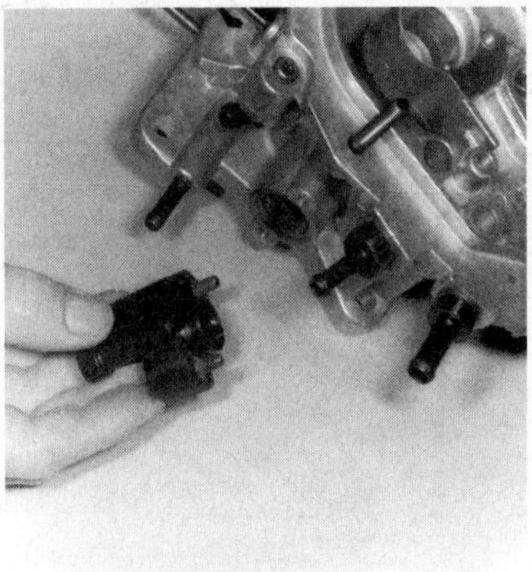

6. Remove the two throttle-position sensor screws. Remove sensor from throttle shaft. Replace O-ring, if used.

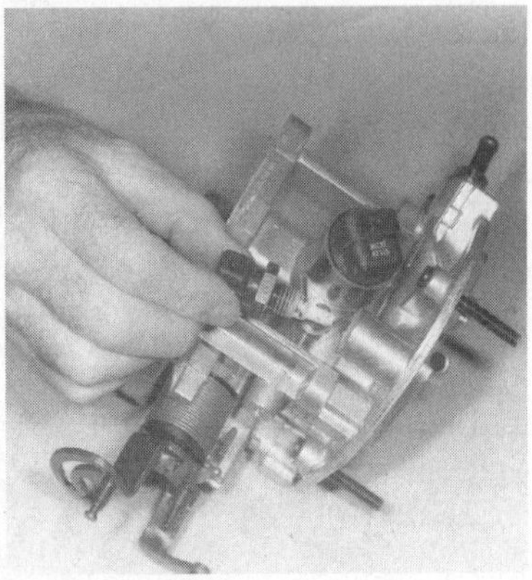

7. Remove temperature sensor. Do not wipe off heat-transfer grease from sensor tip.

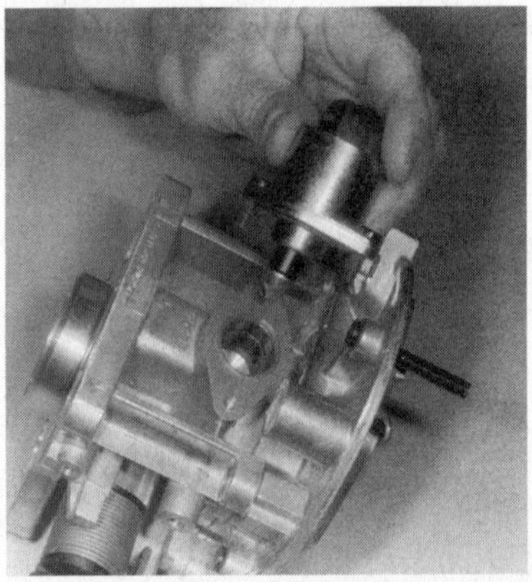

8. Remove the two automatic idle speed motor screws. Remove motor from throttle body. Discard O-ring.

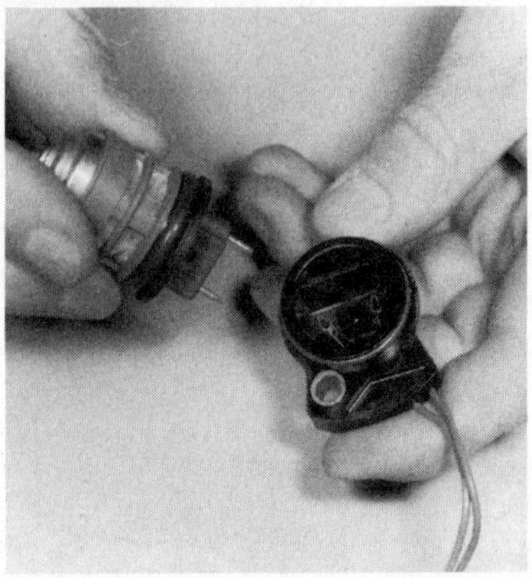

9. To reinstall injector easily, align protrusion on injector to match slot in cap. Press cap and injector together, then install in pod as an assembly.

FUEL-RAIL AND THROTTLE-BODY OVERHAUL

These photo sequences cover the basic points of disassembling and assembling typical PFI fuel rails and air intake throttle bodies. All are relatively simple units, particularly when compared to carburetors and TBI units. There are no adjustments to be made during reassembly. Once the components have been removed from the fuel rail. it can be submerged in a carburetor cleaner tank, if necessary. Be careful not to immerse any components that contain plastic or rubber in carburetor cleaning solutions.

The twin, or split, fuel rail disassembled in this sequence is a Rochester Model F8A used on V-8 engines. The Model T2A air intake throttle body also is used on V-8 engines but has evolved into the Model B210. The adjustable TP sensor used on T2A and early B210 throttle bodies was replaced by a nonadjustable sensor on 1990 and later models. In 1992, the coolant cover was redesigned to incorporate an air vent valve and the outer edges of the throttle valves are coated with a sealing compound to prevent air bypass at closed throttle. To prevent damage to this sealing compound on 1992 and later models, *do not* immerse this throttle body in a cold immersion cleaner.

ROCHESTER MODEL F8A FUEL-RAIL DISASSEMBLY

1. Use screwdriver blade to properly position injector retaining clips or clip may be damaged when pulling injector from fuel rail.

2. Proper clip position. Use rocking motion to counter injector O-ring sealing action. Clip will come out with injector. *Do not lose clip.*

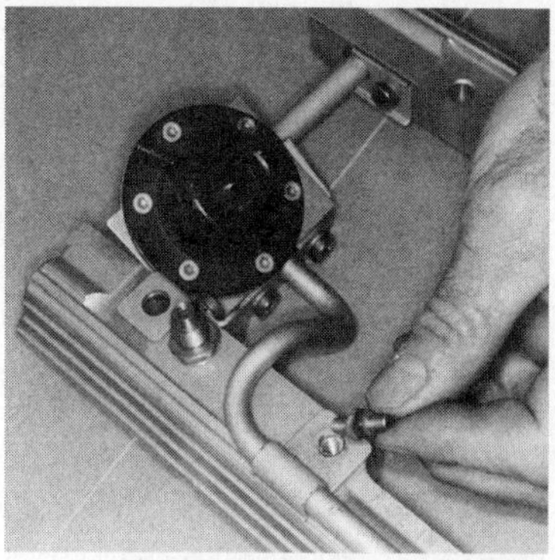

3. Unscrew and remove rear bracket attaching stud from fuel-rail tubing bracket near pressure-regulator assembly.

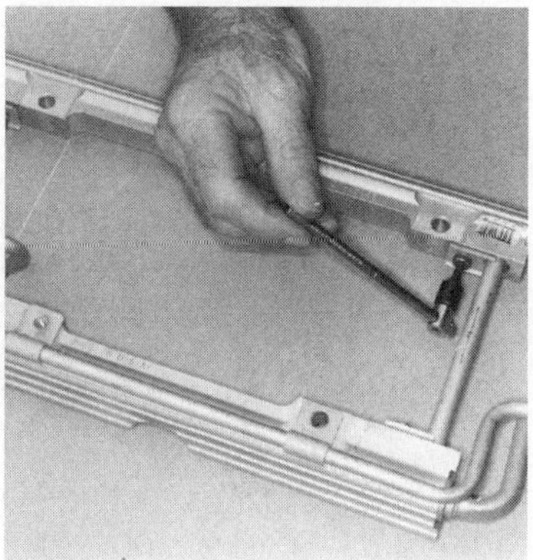

4. Crossover tubes use O-ring seals and are held in place by retainer clips and Torx-head screws. Remove the clips on one side of the fuel rail.

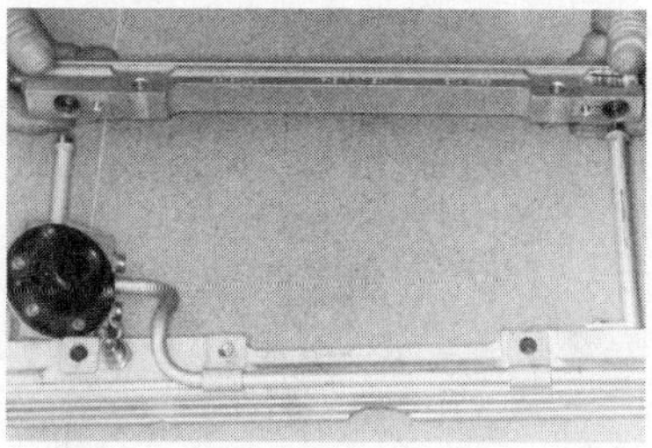

5. Once the two clips have been removed, separate the fuel rail from the crossover tubes. Although the O-rings should stay on the tubes, they may remain in the fuel rail and must be removed. Discard all O-rings and install new ones during reassembly.

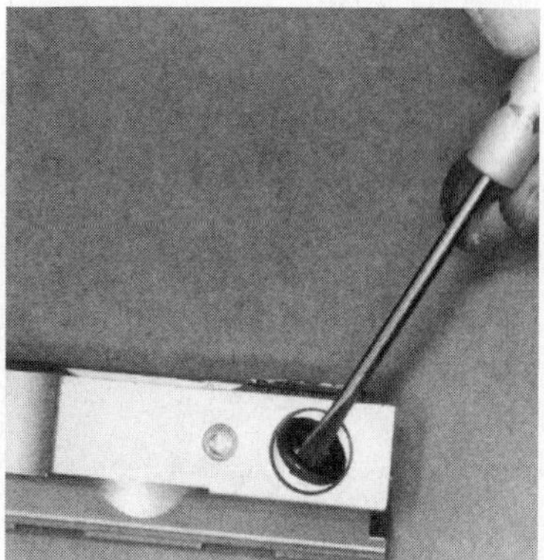

6. *Carefully* remove O-ring seals from fuel rail. Damage to the aluminum fuel rail may prevent the new O-rings from sealing properly.

ROCHESTER MODEL F8A FUEL-RAIL DISASSEMBLY

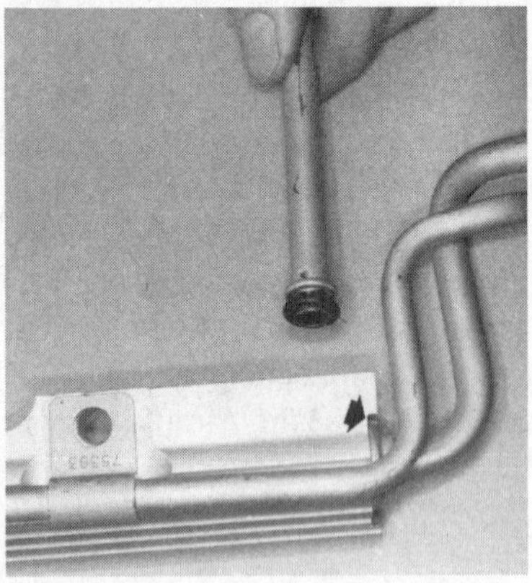

7. Remove the retainer clip on opposite rail, separate the crossover tube, and discard the O-ring. Fuel inlet line (arrow) is swedged into the rail. *Do not remove.* Replace rail if damaged.

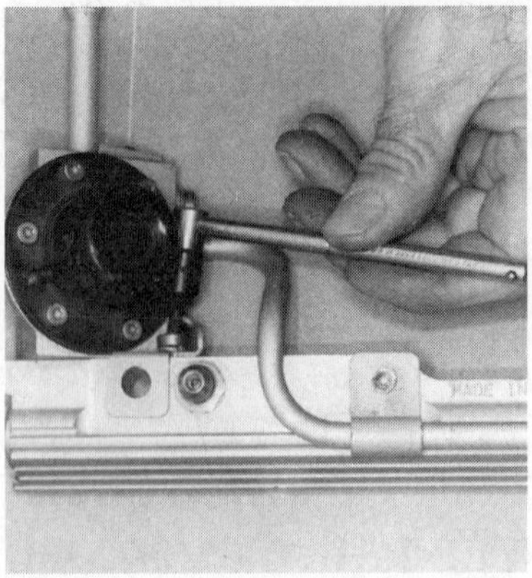

8. Remove two screws holding U-shaped retainer at pressure-regulator assembly. Space limitation requires special tool.

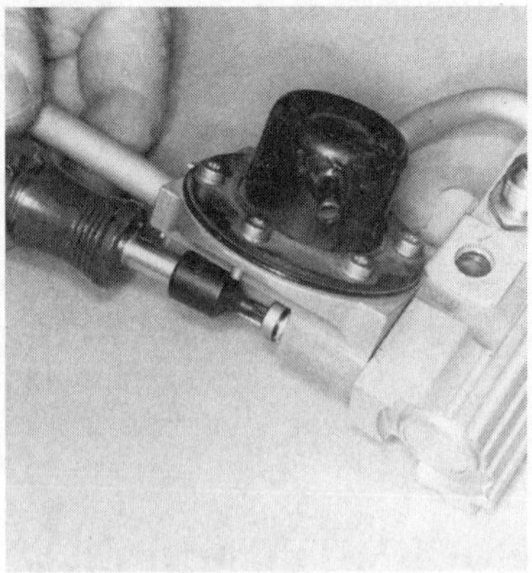

9. Be sure to remove semi-hidden third screw on opposite side of regulator housing or regulator will not separate from fuel rail.

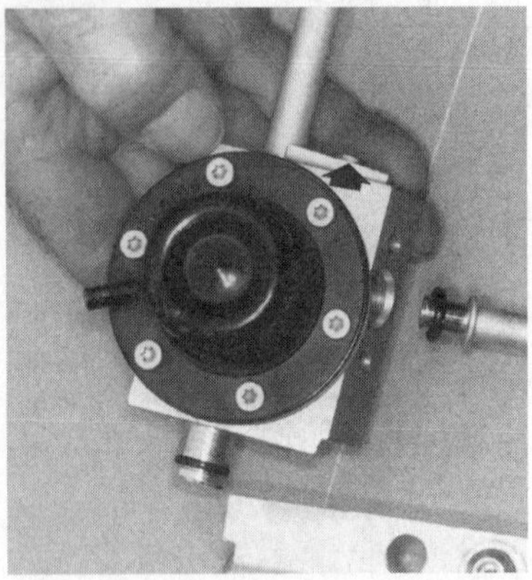

10. Remove the other retainer clip (arrow) and separate the crossover tube from the regulator housing.

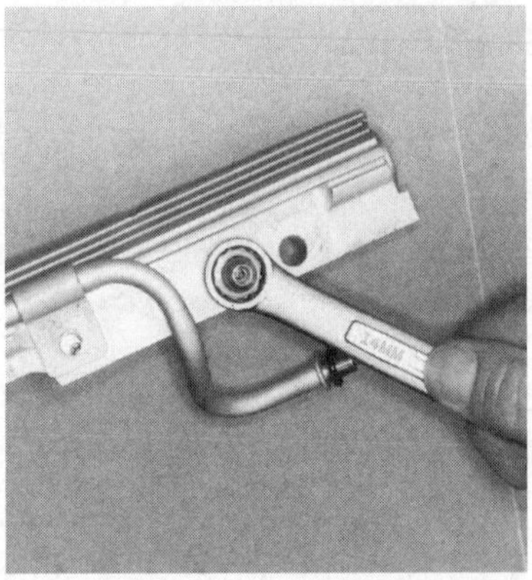

11. Use a 14-mm box-end wrench to loosen and remove fuel-pressure valve and O-ring from fuel rail. Clean fuel-rail components.

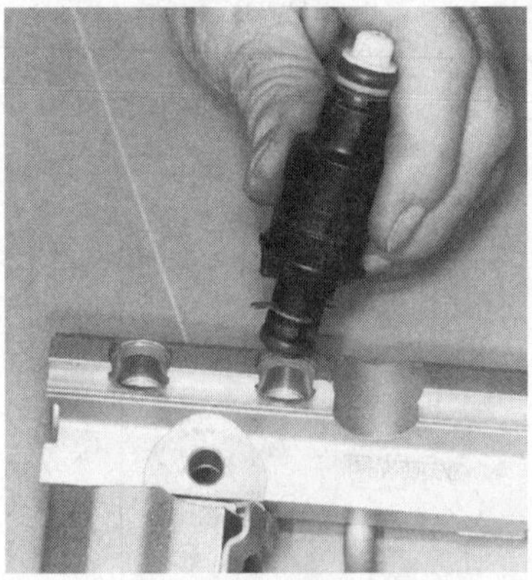

12. Use new O-rings in assembly. Position injector retaining clips and install injectors with a downward motion until fully seated. Snap the clips in place with screwdriver blade.

ROCHESTER MODEL T2A AIR INTAKE THROTTLE-BODY DISASSEMBLY

1. The T2A throttle body uses an adjustable TP sensor held in place by two Torx-head screws. Loosen and remove the two sensor retaining screws.

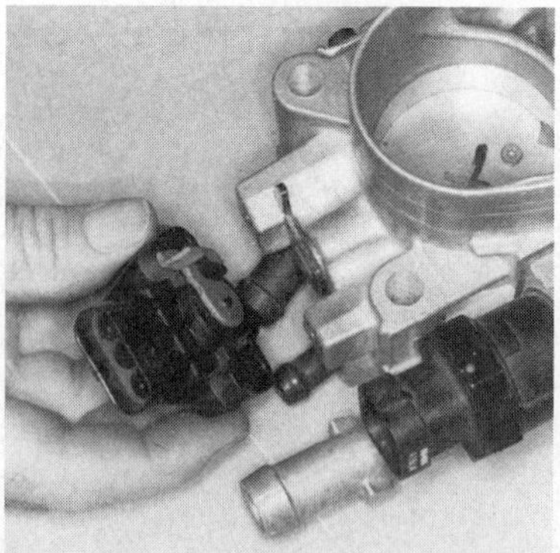

2. Disengage the sensor arm from throttle valve shaft and remove it from the throttle body. After reassembly, it will have to be adjusted.

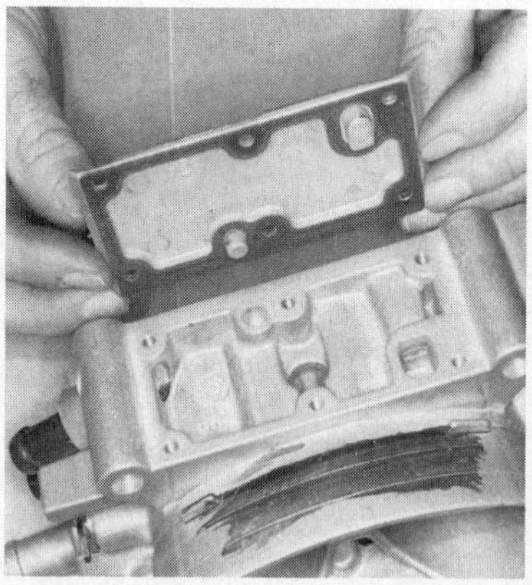

3. Remove clean air cover screws. Tap cover with a soft-faced hammer and remove from throttle body. Remove and discard gasket.

4. Idle air control (IAC) valve threads use sealant, making it difficult to remove with the coolant cover in place. To avoid damage, remove cover screws first.

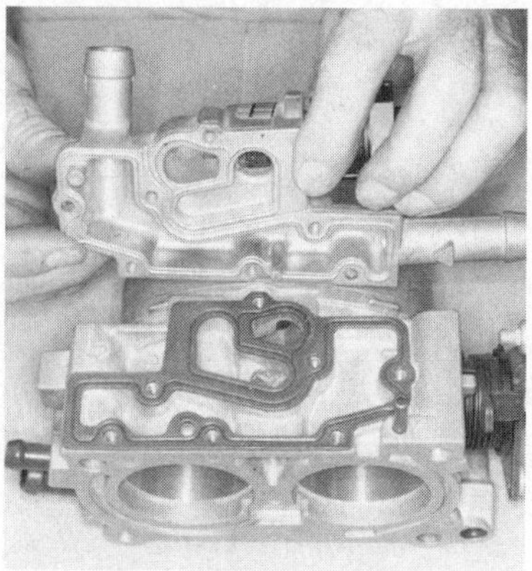

5. Tap the coolant cover with a soft-faced hammer and remove it from throttle body. Remove and discard gasket.

6. Clamp coolant cover between protective jaws of a vise and remove IAC valve with a suitable open-end wrench. Replace gasket.

9

Turbocharger and Supercharger Testing and Service

OBJECTIVES

Upon completion and review of this chapter, you will be able to:

- Troubleshoot and repair turbochargers and controls.
- Perform maintenance services required on turbochargers and superchargers.
- Troubleshoot and repair superchargers and controls.

INTRODUCTION

Turbochargers have been used on a great many 4-cylinder and some V-6 engines, with most manufacturers offering one or more turbocharged models. Turbocharger use on new models is in decline, but there are still many turbocharger units in service. Although turbochargers are not complicated devices, they do require troubleshooting and maintenance. This chapter contains the basic service and troubleshooting procedures you need to know.

Use of superchargers began to increase in the 1980s. While they are not as common as turbochargers, several manufacturers have current models with superchargers. Later in this chapter, basic supercharger service and troubleshooting are covered.

TURBOCHARGER SAFETY

Working with turbochargers requires that you observe certain specific precautions in addition to the safety precautions already given for fuel systems:

- The high speeds and temperatures at which a turbocharger operates make lubrication a critical factor in service life and performance of the unit. Lubrication is provided by the engine oil system, figure 9-1. Most manufacturers recommend that the oil and filter be changed more frequently on a turbocharged engine than on a normally aspirated engine. Lubrication problems should be taken care of as soon as they occur, and you should always make sure that the turbocharger oil supply system, figure 9-2, is working properly.
- Whenever you change the oil on a turbocharged engine, or have installed a new turbocharger, disable the ignition and crank the engine (not more than 15 seconds at a time) to prime the unit. Continue until the oil pressure warning light or gauge indicates that oil pressure is stable.
- Priming time is reduced by filling the new oil filter with clean engine oil before installation. It also is a good idea to disconnect the oil drain line

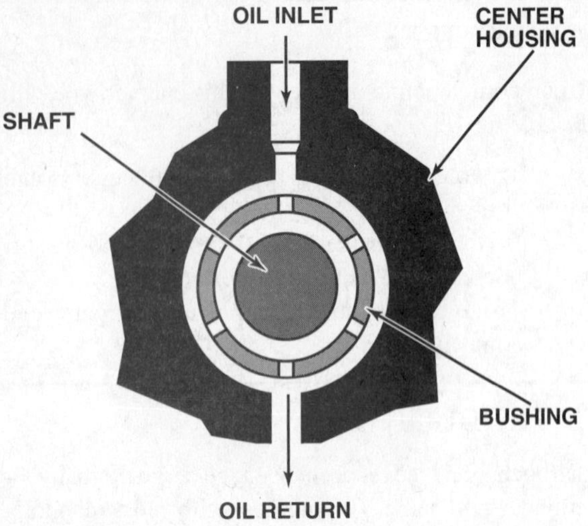

Figure 9-1. Engine oil passes through the turbocharger center housing to lubricate the shaft bearings.

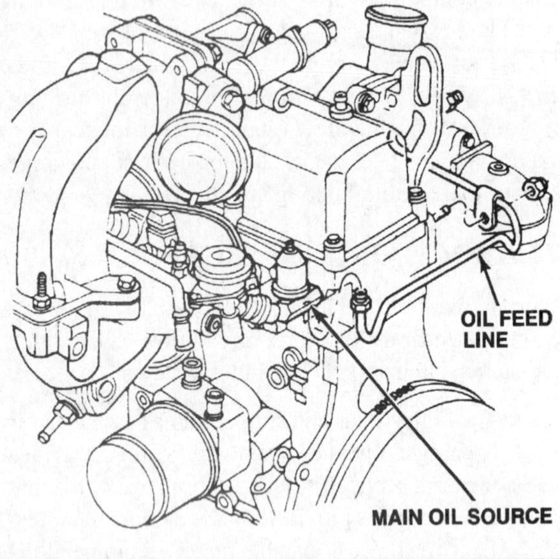

Figure 9-2. Engine oil reaches the turbocharger through a feed line to the center housing. (Courtesy of Ford Motor Company)

at the turbocharger to bleed any air from the center housing, and reconnect it after the turbo is primed. On some turbochargers, you may be able to disconnect the oil inlet at the turbocharger and prime the unit by adding oil directly to the center housing, figure 9-3.

- Let a turbocharged engine idle for 20 to 30 seconds after starting before you accelerate it. This ensures that the turbocharger is receiving a constant supply of oil.
- Let a hot turbocharged engine idle for one to two minutes before shutting it off. As explained in

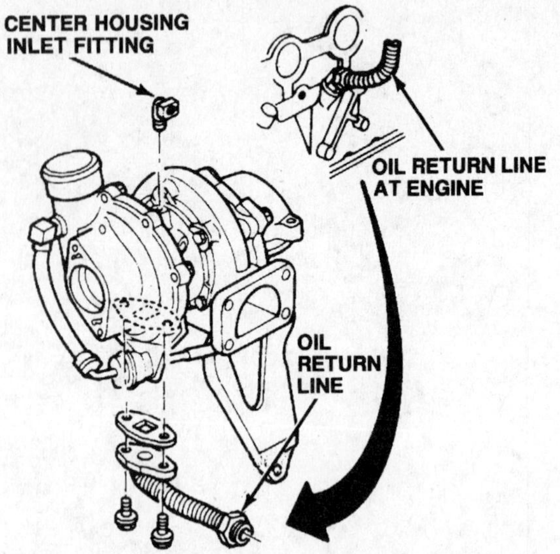

Figure 9-3. Some turbochargers can be primed by adding oil through the center housing inlet. (Courtesy of Ford Motor Company)

Chapter 10 of the *Classroom Manual,* the turbocharger rotates at speeds up to 100,000 rpm and cannot decelerate as rapidly as the engine. If you do not allow it to "unwind" properly, the oil supply to the bearings is interrupted, which causes damage.

- Always make sure the air cleaner is in place and the air intake duct is properly connected to a turbocharger before starting and running the engine. Turbochargers operate with extremely close tolerances and any dirt or contamination drawn into the unit increases wear and may damage the internal parts.
- Whenever you remove or disconnect a turbocharger, clean the area around it with a noncaustic solution and cover any openings to the engine or manifold to prevent any foreign material or contamination from entering while the unit is off the engine.
- Make sure that you do not bend or nick the blades of the turbine or compressor. Even slight damage creates an imbalance when the shaft rotates, resulting in damage to the bearings or turbine or compressor wheels. Check for binding by rotating the turbine and compressor by hand before reconnecting an inlet or outlet duct. Do not work on or around the turbocharger unless it is cold. Turbochargers tend to stay hot long after the engine has been shut off and, like catalytic converters, they cause severe burns.

WHY TURBOCHARGERS FAIL

Turbocharger failure is generally caused by improper lubrication or dirt or other contamination in the system. Before you can identify and correct a turbocharger problem, you must first determine exactly what is causing it. Most problems fall into one of the following categories:

- Lack of lubrication
- Contaminated oil
- Intake air contamination

Lack of Lubrication

This is caused by using the wrong type or viscosity of oil, a restriction in the oil supply line to the turbocharger, or a worn engine that produces low oil pressure.

Contaminated Oil

This usually results from not changing the oil often enough. Infrequent oil and filter changes let dirt and acid accumulate in the oil and allow sludge to form. The oil system can also be contaminated by the failure of an engine bearing, piston ring, or other internal component. Other possible sources are the use of the wrong type or viscosity of oil, a ruptured oil filter, or an oil filter bypass valve that sticks open.

If a dirty or contaminated oil system is not corrected soon enough, it causes breakdown or oxidation of the oil. Since the oil passages in the turbocharger center housing are very small, oxidation of the oil is extremely damaging. With the turbine running at very high speeds, the dirty or contaminated oil is flung against the hot walls of the center housing where it turns to sludge that eventually blocks lubrication to the bearing and the oil drain line. If the oil cannot flow through its normal channels, it leaks past the compressor and turbine seals, figure 9-4. Such oil also may "coke," or form hard deposits that further restrict oil passages and cause bearing and shaft wear.

Oil breakdown is also caused by external factors, such as engine overheating, a coolant leak into the engine, or excessive blowby in the crankcase.

Intake Air Contamination

This generally results from contamination being drawn into the compressor inlet due to a leaky air duct, or from running the engine without an air filter. In extreme cases, it may be caused by improper placement of small tools or fasteners while working on the turbocharger system.

Exhaust system debris such as hard carbon particles or particles from a damaged catalytic converter (when installed between the turbocharger and exhaust manifold) may enter the turbine. If so, damage results.

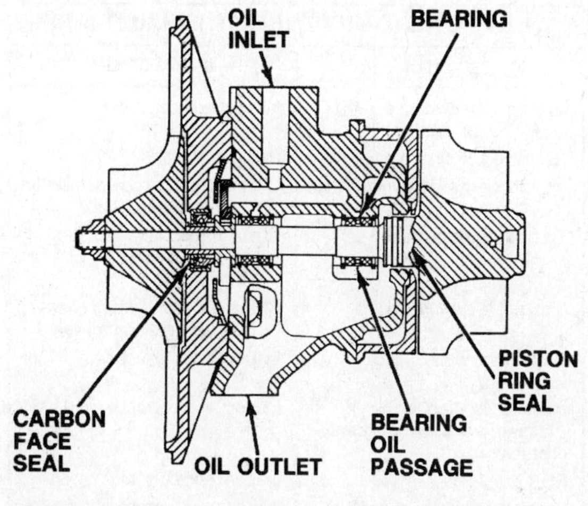

Figure 9-4. Any restriction in the lubrication system will cause oil to leak past the compressor and turbine seals. (Courtesy of Ford Motor Company)

TROUBLESHOOTING A TURBOCHARGER

Remember that a turbocharger does nothing more than supply an increased volume of air for combustion. Just as with carburetors or fuel injectors, problems often are caused by the ignition, cooling, lubrication, or emission systems that affect the operation of the turbocharger. These systems, as well as the fuel system, must all be working properly or replacing the turbocharger may not correct the customer complaint.

Turbocharger Performance Complaints

Troubleshooting a turbocharger starts with an evaluation of the customer complaint or the symptoms exhibited by the vehicle. Figures 9-5 and 9-6 contain common complaints associated with turbocharged engines and their possible causes. Since many of these complaints are identical to those encountered on a normally aspirated engine, they are often caused by other engine systems and not the turbocharger. The troubleshooting of problems that seem to be caused by the turbocharger is covered in the following sections.

Turbocharger Noises and Leakage

The first step in troubleshooting a turbocharger system is similar to that with any other engine system: Look over the system. Make sure that the air cleaner filter is in place and that there are no loose connections or obvious

LACK OF POWER OR EMITS BLACK EXHAUST SMOKE	
POSSIBLE CAUSE	**CORRECTION**
Damaged or disconnected air cleaner ducting	Inspect and correct
Restricted air filter element	Replace air filter
Intake or exhaust manifold leak	Check turbocharger installation for air or exhaust leak
Turbocharger damage	Check turbocharger rotating assembly for binding or dragging
Exhaust system restriction	Check exhaust system check manifold heat control valve
Carburetor or fuel injection problem	Inspect and correct
Incorrect ignition timing, advance or other ignition problem	Inspect and correct
EGR problem	Inspect and correct
Lack of compression or other engine wear	Inspect and correct
Low boost pressure	Check boost pressure and wastegate operation; adjust or replace as required

Figure 9-5. If the engine lacks power or emits black exhaust smoke, check these possible causes.

HIGH OIL CONSUMPTION OR EMITS BLUE EXHAUST SMOKE	
POSSIBLE CAUSE	**CORRECTION**
Excessive blowby or PCV problem	Check for engine wear; service PCV system
Engine oil leakage	Inspect and correct
Worn engine rings, cylinders, valves, valve guides	Check for engine wear; check for low compression and cylinder leakage
Leaking turbocharger oil seals†	Replace turbocharger
Restricted turbocharger oil return or too much oil in center housing	Check oil return line for restrictions and blow out with compressed air; check center housing for sludge; clean as required
Carburetor or fuel injection problem	Inspect and correct
Incorrect ignition timing, advance, or other ignition problem	Inspect and correct
EGR problem	Inspect and correct
† Smoke and detonation indicates a leak on the compressor side; smoke alone indicates a leak on the turbine side.	

Figure 9-6. If oil consumption is too high or the engine emits blue exhaust smoke, check other engine systems in addition to the turbocharger.

sources of leaks, restrictions, or contamination in the ducting or hoses.

Inspect the exhaust system. Look for burned areas caused by exhaust leakage. Make sure the manifolds, exhaust pipes, and tubing are in good condition and that their fasteners and connections are not loose. It is sometimes possible to correct such a problem simply by tightening the fasteners, if it is caught soon enough.

TURBOCHARGER PROBLEMS	
SOUND	**CAUSE**
Louder than normal noise which includes hissing	Exhaust leak
Uneven sound that changes in pitch	Restricted air intake from a clogged air cleaner filter, bent air ducting or dirt on the compressor blades
Higher than normal pitch sound	Intake air leak
Sudden noise reduction, with smoke and oil leakage	Turbocharger failure
Uneven noise or vibration	Possible shaft damage, damaged compressor or turbine wheel blades
Grinding or rubbing sounds	Shaft or bearing damage, misaligned compressor or turbine wheel
Rattling sound	Loose exhaust pipe or outlet elbow, damaged wastegate

Figure 9-7. Listen to the sound of a turbocharger when diagnosing problems.

Normal turbochargers make a slight, uniform whistling noise the steadily increases in frequency as engine speed climbs. Although each turbocharger version has a different sound, you should be able to tell the difference between the sound of a normal turbocharger and one that has problems once your ear has become accustomed to working around them. Figure 9-7 contains various turbocharger sounds and the possible causes.

To check for intake leaks, spray a soapy water solution or a nonflammable solvent on the suspect areas with the engine running. If there is an intake leak, the liquid will be drawn into the leak and engine speed changes. The same procedure can be used to check for an exhaust leak, but bubbles form if there is a leak. A mechanic's stethoscope is also useful for locating an exhaust leak.

Inspecting a Turbocharger

Once you have done a visual inspection and corrected any obvious defects, disconnect the air inlet and exhaust outlet at the turbocharger. A flashlight and inspection mirror are handy tools to look inside the turbine and compressor housings to determine if there are any mechanical problems. Broken or bent blades indicate physical damage caused by foreign objects. Wear marks on the blades or housing indicate that the wheel is misaligned and has been rubbing. This generally is caused by high shaft or bearing wear.

Rotate the wheel by hand while listening and feeling for any rubbing or binding. Rotate the wheel while trying to move it from side to side to check the clearance by hand. There should be no rubbing or binding when you apply this pressure. Move both ends of the shaft up and down at the same time while rotating the assembly to check radial clearance. There should be little, if any, movement and the wheels should not contact the housing.

Check the shaft seal areas, figure 9-4, for signs of oil leakage. Because compressor pressure has a tendency to force the oil back into the center housing, you are more likey to find leakage at the exhaust turbine seal.

If the turbocharger passes inspection to this point, it is probably satisfactory, but you should check actual bearing clearances with a dial indicator.

Checking Bearing Clearances

Turbocharger design requires the use of full-floating bushings. This means that labrication flows between the shaft and bearings, and between the bearings and their bores, figure 9-4. A thrust bearing or washer is also used at one end of the shaft to control shaft endplay. All bearing clearances must be kept within specifications to maintain the close tolerances between the turbine and compressor wheels and the housing.

Before checking clearances, there are certain steps that you must do:

- Look up the clearance specifications provided by the manufacturer. Although they vary slightly, most turbos use a radial clearance of 0.003 to 0.006 inch (0.076 to 0.152 mm) and an axial clearance of 0.001 to 0.003 inch (0.025 to 0.076 mm).
- Obtain a dial indicator kit with the necessary adapters to perform the bearing clearance check. Some tool manufacturers offer kits with adapters for use with various turbocharger units.
- It is possible to check bearing clearances on some installations with the turbocharger on the engine, but most manufacturers recommend (and it is a good idea) that you remove the unit. Follow the service manual procedure for removal and installation.
- Disconnect and move the intake and exhaust tubes or ducting away from the turbocharger housing. If necessary, remove them from the engine to prevent any possible damage.
- Remove the fastener used to connect the wastegate actuator linkage rod to the wastegate assembly, figure 9-8.
- Unbolt and remove the exhaust outlet housing or elbow from the turbine housing, figure 9-9, to access the end of the shaft for an axial clearance check.
- Place a suitable container under the unit and disconnect the oil drain line from the center housing to check radial clearance. If necessary, remove the fitting from the center housing. On some water-cooled turbos, you also may have to disconnect one or both of the coolant fittings.

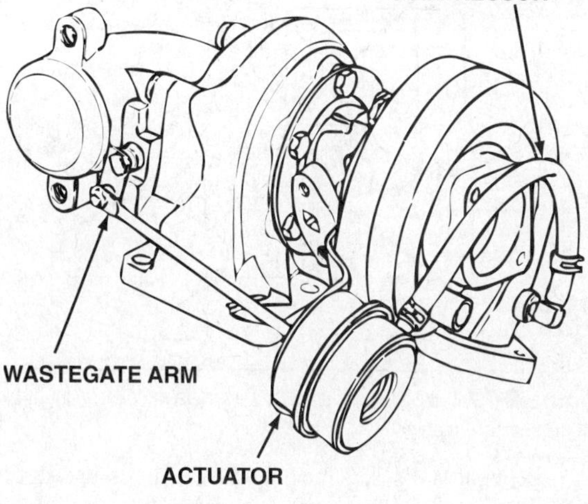

Figure 9-8. A pin or clip usually holds the actuator rod to the wastegate arm. (Courtesy of Ford Motor Company)

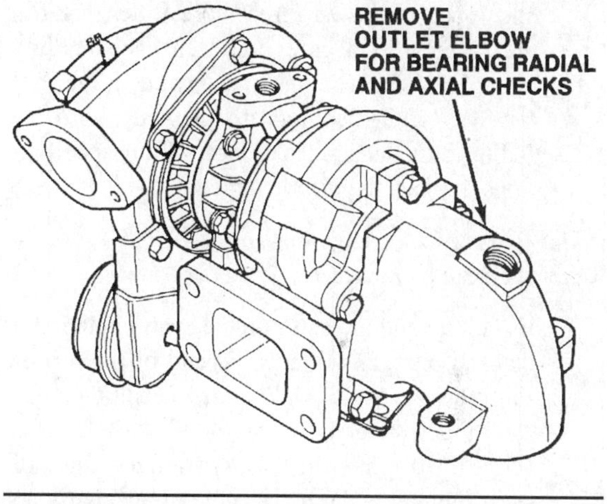

Figure 9-9. Remove the exhaust outlet elbow to provide access for bearing checks. (Courtesy of Ford Motor Company)

Radial Clearance

To check radial clearance:

1. Fit the necessary adapter on the dial indicator plunger and install the dial indicator on the center housing so that the plunger and adapter enter the oil outlet port, figure 9-10, and contact the shaft.
2. Move the shaft away from the dial indicator plunger as far as possible by manually applying pressure equally and simultaneously at point A, figure 9-10. The amount of movement should be very slight.
3. Hold the shaft away from the plunger and set the dial indicator gauge to zero.
4. Repeat step 2, applying pressure at point B, figure 9-10, to move the shaft as far as possible toward the dial indicator. Note the gauge reading.

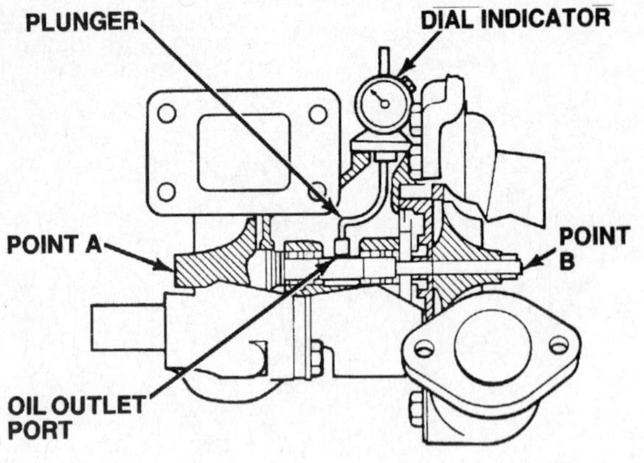

Figure 9-10. Bearing radial clearance check. (Courtesy of Ford Motor Company)

5. Repeat steps 2, 3, and 4 several times to verify the measurement. Make sure that the gauge needle returns to zero when the shaft is released.
6. Rotate the compressor and turbine wheels 90 degrees and repeat the procedure. Note the reading and repeat this step three more times, rotating the wheels 90 degrees between each reading. Compare these readings to determine if the shaft is bent or if the bearings are unevenly worn.
7. If the radial clearance is not within specifications, replace the turbocharger.

Axial Clearance

To check axial clearance:

1. Install the dial indicator on the center housing so that the plunger contacts the end of the turbine shaft, figure 9-11. Make sure the plunger and shaft touch squarely.
2. Move the turbine wheel away from the dial indicator plunger as far as possible by manually applying pressure equally and simultaneously at point. A, figure 9-11. The amount of movement should be very slight.
3. Hold the shaft away from the plunger and set the dial indicator gauge to zero.
4. Repeat step 2, applying pressure at point B, figure 9-11, to move the shaft as far as possible toward the dial indicator. Note the gauge reading.
5. Repeat steps 2, 3, and 4 several times to verify the measurement. Make sure that the gauge needle returns to zero when the shaft is released.
6. If the axial clearance is not within the manufacturer's specifications, replace the turbocharger.

Testing Boost Pressure and Wastegate Operation

Boost pressure and wastegate operation are both tested with a dial indicator and pressure gauge. In most cases, both items are checked at the same time, since they are interrelated. Not enough boost pressure generally indicates that the wastegate leaks or is opening too much or too soon. Too much boost pressure usually indicates a stuck or broken wastegate actuator diaphragm or link.

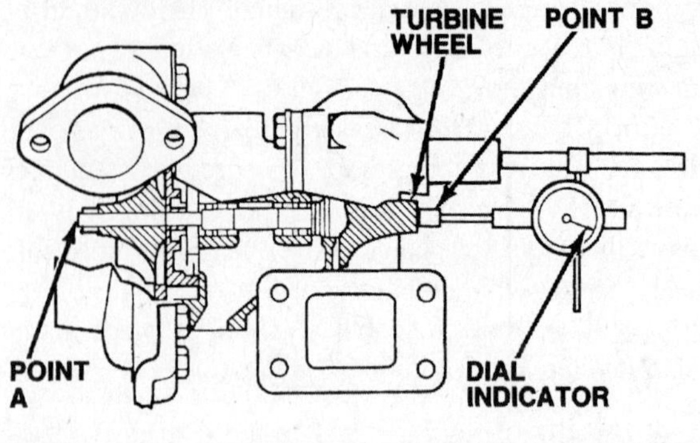

Figure 9-11. Bearing axial clearance check. (Courtesy of Ford Motor Company)

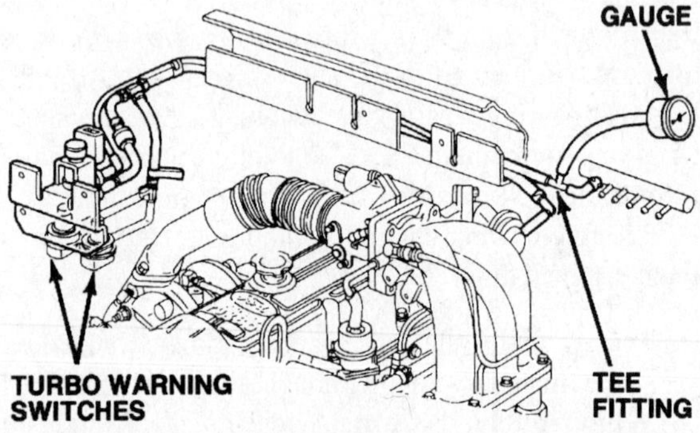

Figure 9-12. With some turbos, you can connect the test gauge to the vacuum lines for the boost gauge pressure switches. (Courtesy of Ford Motor Company)

Boost Pressure Test

The most useful and accurate test of maximum boost pressure is made by applying a driving load on the vehicle with a chassis dynamometer or during a road test. The only test equipment required is a pressure gauge. Connect the pressure gauge to a pressure port on the compressor side of the turbocharger. Some turbos allow you to tee the test gauge into the line running to a warning lamp pressure switch, figure 9-12, or an instrument panel boost gauge. Regardless of where you connect the test gauge, use a length of hose that reaches from the engine compartment into the driver's compartment so that you are able to read it easily.

With the engine at normal operating temperature, accelerate at full throttle from zero to 40 or 50 mph (64 or 80 kmh), noting the maximum test gauge reading under load. Slow the vehicle and repeat the test several times to make sure the reading is correct. If the maxi-

Figure 9-13. Wastegate operation can be checked with a hand pressure pump and test gauge.

mum pressure reading you get during this sequence is within specifications, check the wastegate and actuator. Some wastegates can be adjusted; others must be replaced.

Wastegate Actuator Test

An external rod connects the wastegate actuator to the wastegate arm or link, figure 9-8. On some models, the rod length is adjustable, figure 9-13. However, many manufacturers seal the adjustment and require that the wastegate actuator be replaced if either the boost pressure or wastegate operation is not within specifications. Check wastegate operation with the engine off. You need a source of air pressure, a pressure gauge, and a dial indicator for this test. The same pressure tester used for checking cooling system operation can be used to test the actuator. Although there may be some test variations between models, most include the following steps:

1. Connect a hand-operated pressure pump with gauge, figure 9-13, or a compressor air line equipped with a pressure regulator and gauge, to the wastegate actuator.
2. Apply approximately 5 psi (34 kPa) of pressure to the actuator diaphragm and maintain the pressure level for 60 seconds. If the pressure reading falls below 2 psi (14 kPa) in that time, the actuator diaphragm is defective.
3. Clamp or mount a dial indicator on the turbocharger housing so that its plunger contacts the actuator rod, figure 9-14. Remove all pressure from the actuator and set the indicator gauge to zero.
4. Apply the specified test or maximum boost pressure to the actuator diaphragm, noting the amount of rod movement shown on the indicator gauge. Watch the gauge closely, because the rod

Figure 9-14. Mount a dial indicator as shown to measure wastegate actuator rod travel.

Figure 9-15. Disconnection points to remove the wastegate actuator. Make sure the wastegate arm will swing through a 45-degree arc when disconnected.

travel is generally quite small (usually about 0.015 inch or 0.385 mm). If rod travel is not within specifications, either replace or adjust the actuator as recommended by the manufacturer.
5. Remove the fastener holding the actuator rod to the wastegate arm or link, then move the arm to make sure that it travels freely through a 45-degree arc, figure 9-15. If the arm does not move freely, the wastegate is sticking or binding from exhaust deposits. Install a new turbocharger outlet elbow and wastegate assembly.

REPLACING TURBOCHARGER COMPONENTS

Some turbocharger can be repaired or rebuilt if they are damaged or worn. In most cases, however, the manufacturer does not sell the necessary replacement

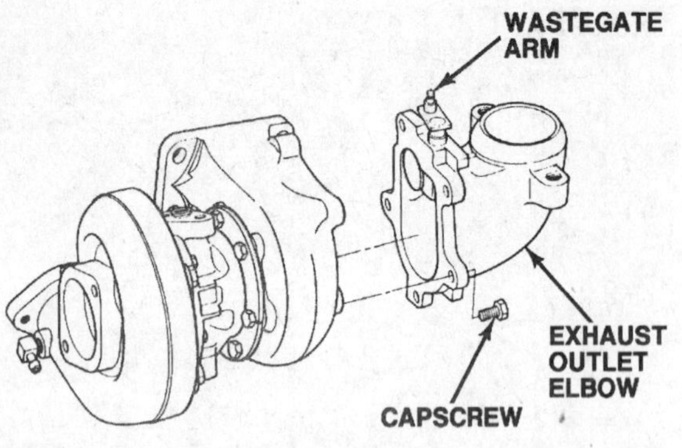

Figure 9-16. Exhaust elbow outlet replacement. (Courtesy of Ford Motor Company)

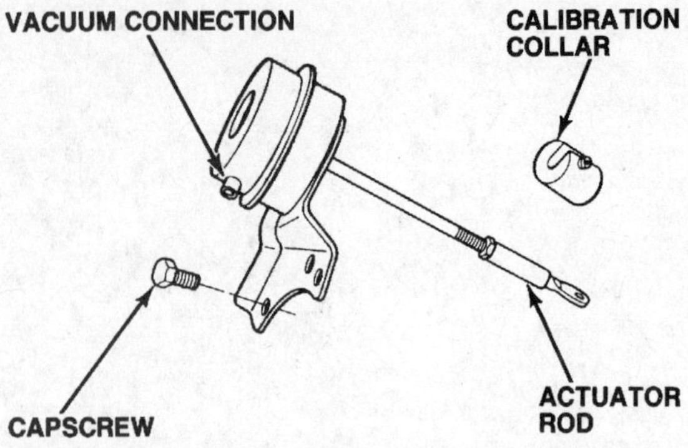

Figure 9-17. Replacement actuators use a horseshoe calibration clamp that is discarded after the actuator is installed and adjusted. (Courtesy of Ford Motor Company)

parts and recommends that the turbocharger unit be replaced with a new or factory-rebuilt one. You can generally purchase replacement exhaust outlet elbows, wastegate assemblies, and actuators.

Turbocharger replacement procedures vary according to the turbo model so you must have the manufacturer's procedure. It is not a difficult procedure, but is generally a time-consuming one.

You must remove the turbocharger from the engine to replace the exhaust outlet elbow, figure 9-16. The actuator can often be replaced without removing the turbocharger. Some actuators are mounted to a bracket on the turbocharger, figure 9-15; others are a part of the bracket, figure 9-17.

To replace an actuator:

1. Disconnect the vacuum hose or hoses at the actuator.
2. Remove the pin, clip, snapring, or other fastener holding the actuator rod to the wastegate link.
3. Remove the fasteners holding the actuator to the bracket, figure 9-15, or the bracket to the turbocharger, figure 9-17. Remove the actuator from the turbocharger.
4. Install the new actuator to the turbocharger or bracket. Replacement actuators generally have a horseshoe calibration clamp, figure 9-17, on the rod.
5. Loosen the actuator rod locknut and turn the end of the rod in or out as required until it aligns with the wastegate arm pin or hole.
6. Install the pin, clip, snapring, or other fastener to connect the end of the rod to the wastegate arm, then tighten the locknut securely. Some manufacturers specify using Loctite or staking the exposed threads on the rod.
7. Reconnect the vacuum hose or hoses to the actuator.
8. Remove the horseshoe calibration clamp from the actuator rod and discard it.

SUPERCHARGER SERVICE AND DIAGNOSIS

Ford and General Motors have made no provision for servicing the Roots-type blower currently in use. Any component that fails in the supercharger is serviced by replacing the entire unit. The gears and ball bearings are lubricated by a synthetic oil stored in an oil reservoir inside the supercharger housing. On, earlier GM supercharger units, this reservoir is sealed and requires no service; the synthetic oil cannot be checked, added to, or changed. Ford, however, recommends checking the oil level every 30,000 miles (48,000 km) and provides a plug for this purpose at the lower front of the supercharger housing. Ford also makes available a special 90-weight synthetic lubricant for topping up the oil reservoir.

Other than chekcing the blower drive belt(s), along with other belts, at the specified maintenance interval, the major service to be performed on these superchargers is to ensure that no air leaks exist on either side of the unit. The Ford V-6 installation contains many possible points, figure 9-18, where leakage may develop. Any leak on the intake side draws dirt and dust into the blower where it causes internal damage. Leakage between the Mass Air Flow (MAF) Sensor and the blower inlet allows the blower to draw in quantities of unmeasured air. Since the engine computer regulates fuel metering according to the MAF sensor signal, this additional air affects the air-fuel ratio and causes the engine to run lean. The opposite is true if a leak develops on the outlet side of the blower. In this case, the mixture is rich because air is leaking out of the system before it mixes with the fuel provided by the computer. A lean exhaust trouble code may result from a blower intake leak; an outlet leak may cause a rich exhaust code.

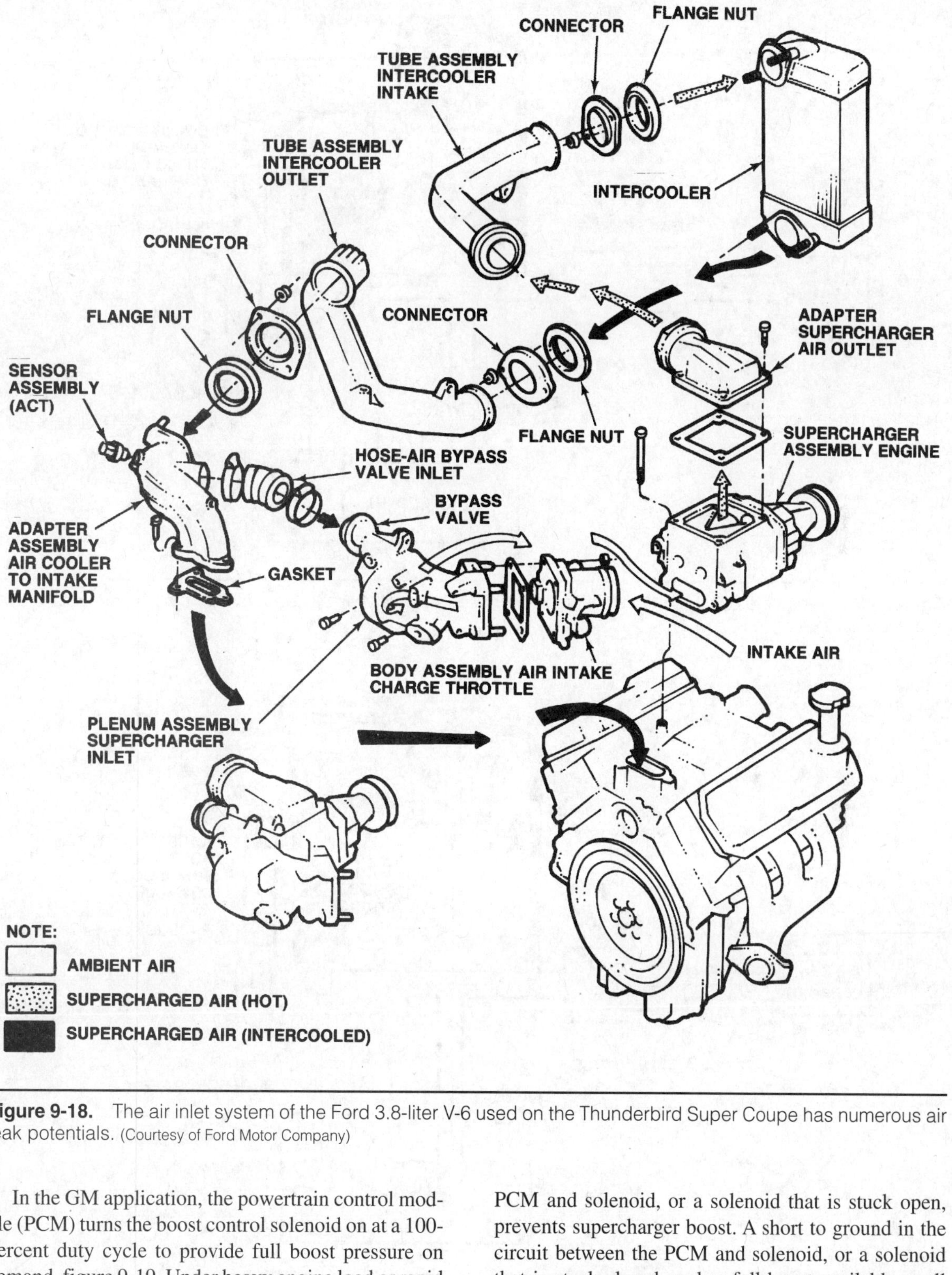

Figure 9-18. The air inlet system of the Ford 3.8-liter V-6 used on the Thunderbird Super Coupe has numerous air leak potentials. (Courtesy of Ford Motor Company)

In the GM application, the powertrain control module (PCM) turns the boost control solenoid on at a 100-percent duty cycle to provide full boost pressure on demand, figure 9-19. Under heavy engine load or rapid deceleration conditions (and in reverse gear), the PCM turns the solenoid off. This opens the bypass valve, recirculating boost pressure back through the supercharger inlet, figure 9-20. An open circuit between the PCM and solenoid, or a solenoid that is stuck open, prevents supercharger boost. A short to ground in the circuit between the PCM and solenoid, or a solenoid that is stuck closed, makes full boost available at all times, resulting in a possible overboost condition during periods of high engine load. Any restriction in the vacuum line to the bypass valve actuator causes a rough idle condition and poor fuel economy. The boost

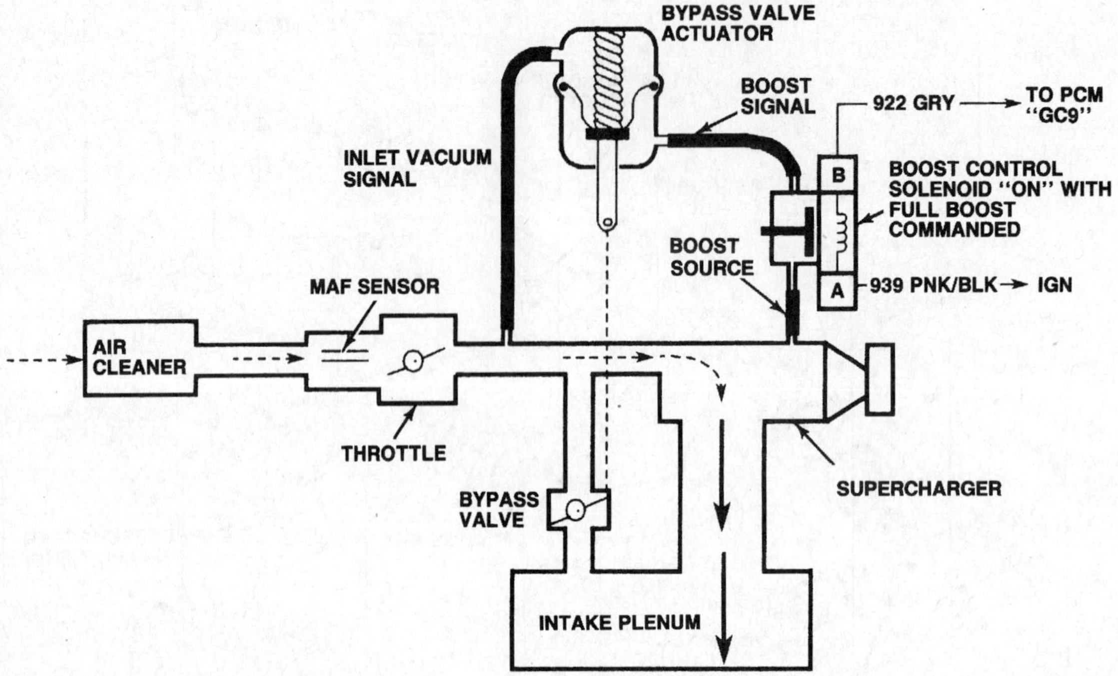

Figure 9-19. When the boost control solenoid is on, the bypass valve is closed for full boost. (Courtesy of General Motors Corporation)

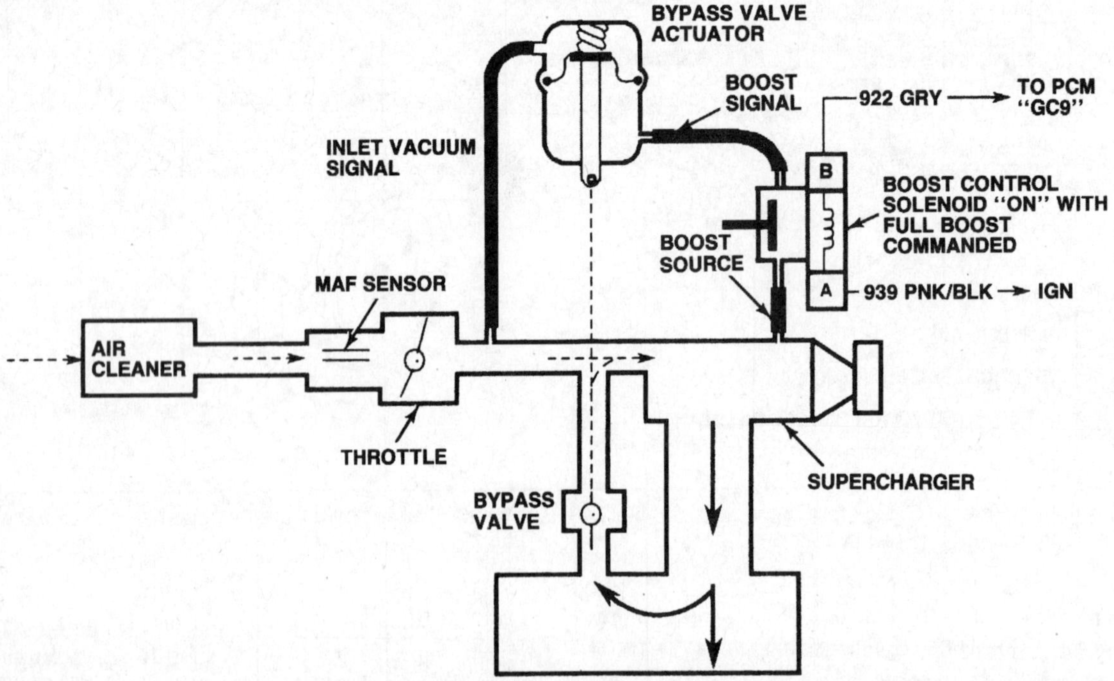

Figure 9-20. If the boost solenoid is off, the bypass valve opens to recirculate boost pressure. (Courtesy of General Motors Corporation)

Figure 9-21. Locate the oil plug on the supercharger.

Figure 9-22. Clean the area around the plug with a clean towel to ensure no debris or contamination gets into the supercharger.

solenoid can be checked with a test light or operated by a scan tool during a boost solenoid output test with a vacuum gauge.

Supercharger Oil Level Inspection

Most current superchargers use synthetic oil for lubrication. Typically this oil should last the life of the supercharger. The oil level should be checked periodically to ensure an adequate oil supply. Always check the manufacturer's recommendations for the correct type and amount of oil to use when servicing the supercharger. To check the oil, do the following:

1. Allow the supercharger to cool approximately two to three hours if the vehicle has been warmed up.
2. Locate the oil plug on the supercharger, figure 9-21.
3. Clean any dirt or debris from around the plug to prevent oil contamination, figure 9-22.
4. Remove the oil plug, figure 9-23.
5. Inspect the oil level. It should be to the bottom of the threads of the inspection hole.
6. Reinstall the inspection plug and torque to specifications.

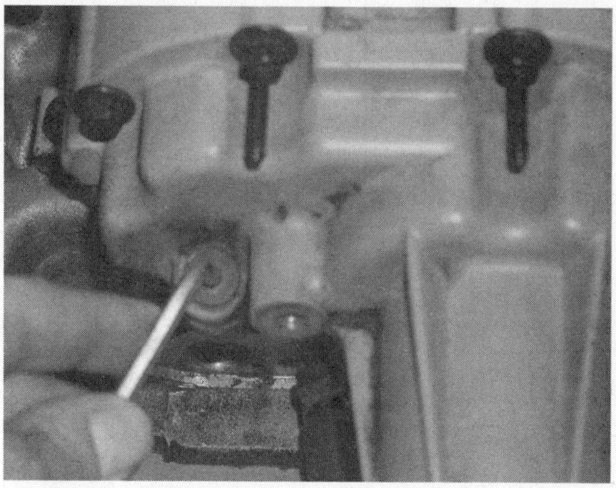

Figure 9-23. Remove the plug and check the oil level. It should be to the bottom of the threads.

10

Emission Testing

OBJECTIVES

Upon completion and review of this chapter, you will be able to:

- Calibrate emission analyzers.
- Perform an emission test as required by the state if applicable.
- Test a vehicle's emission and determine necessary repairs.
- Perform an OBD II test.
- Determine causes of excessive emissions on a vehicle.

INTRODUCTION

Emission testing is usually performed in conjunction with a state emission inspection program. The type of emission testing equipment varies depending on what type of testing program that particular region has implemented. For a review of various test programs, review Chapter 12 of the *Fuel Systems and Emission Controls Classroom Manual.* In addition to performing state-required emission testing, many of these emissions testers may also be used for diagnostic testing or to validate emissions repairs after repairs have been made to a vehicle.

THE EMISSION ANALYZER

The emission gas analyzer, figure 10-1, is used to measure the exhaust gas concentrations in the exhaust stream of a vehicle. The most common type of exhaust gas analyzer is the infrared gas analyzer. The most common infrared analyzers available are the 4-gas analyzer that measures HC, CO, O_2, and CO_2 and the 5-gas analyzer that measures nitric oxide (NO) in addition to the other four gases previously mentioned. Nitric oxide is a major component of oxides of nitrogen (NO_x).

The basic operation of an infrared exhaust gas analyzer consists of shooting a beam of infrared light through the exhaust sample. This beam is split into three parts that pass through optical filters. HC, CO, and CO_2 each absorb different wavelengths of light. After the infrared beam is passed through the gases, the infrared beams are electronically analyzed to determine how much of the infrared beam has been absorbed by the different gases. The amount of absorption determines the amount of HCs, COs, and CO_2 that is present in the exhaust stream. O_2 is measured using an electrochemical sensor that is somewhat similar to the O_2 sensor used on the vehicle, figure 10-2.

Figure 10-1. Emission gas analyzers are produced by a variety of different manufacturers. Only certain analyzers are approved by each state to operate as a state-approved emissions analyzer.

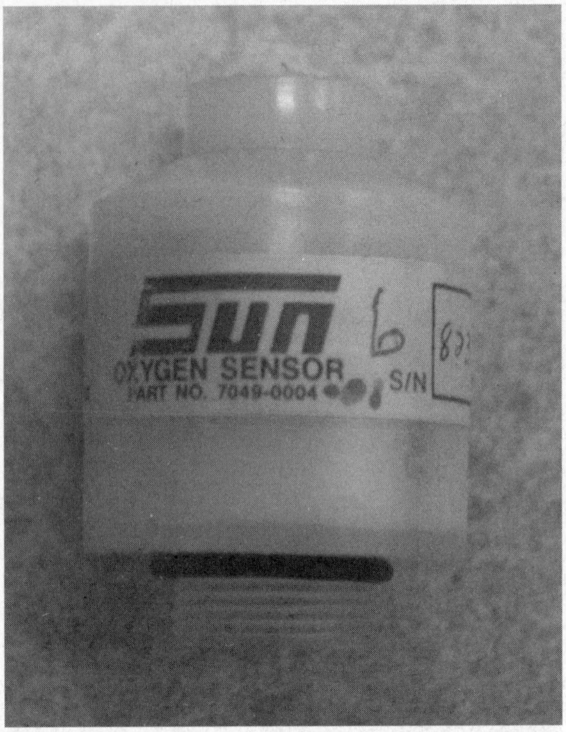

Figure 10-2. This photo shows an example of an oxygen sensor used by emissions analyzers.

NO_x gas concentrations cannot be directly measured with a 5-gas analyzer. The analyzer actually measures the concentrations of nitric oxide (NO), which is one component of NO_x. Therefore, the actual amount of NO_xs present can only be assumed by the analyzer based on the percent of NO present. If NO is high then NO_x will also be high.

UNITS OF MEASUREMENTS

The five gases are measured differently depending on the type of equipment used. The most common measurements used by 4- or 5-gas analyzers are as follows:

- HC in parts per million (ppm)
- NO_x in parts per million (ppm)
- CO in percentage (%)
- CO_2 in percentage (%)
- O_2 in percentage (%)

Constant volume sampling equipment, which is the type of equipment used during the Federal Test Procedure (FTP) or the IM240 test, measures exhaust levels by mass in grams per mile (gms/mi). This equipment is only used in areas that have centralized test centers where vehicles are tested but usually not repaired. Due to the cost of the equipment it is not practical for repair facilities to have this type of equipment to test or validate repairs. The two different types of measurements cannot be compared to each other. This creates a difficult situation when a repair facility is presented with a vehicle that has failed as IM240 type of test and the test results are given in grams per mile measurements. After repairs are performed and the repair facility measures the exhaust concentrations in ppm or percentages, there is no sure way to determine if these percentages are low enough for the vehicle to pass the IM240 type of test. In addition, there is no way for the repair facility to simulate the same amount of load that the vehicle was tested under without having a vehicle dynamometer. NO_x formation is virtually non-existent at idle or at low load levels. Therefore, if a vehicle has failed an IM test due to excessive NO_x there is no practical way to test for NO_xs in the repair facility.

PORTABLE 5-GAS ANALYZERS

The portable 5-gas analyzer is a lightweight test unit, figure 10-3, that can be used to help technicians validate emissions-related repairs. It allows the technician to actually drive the subject vehicle and observe or capture data as the vehicle is being driven. This tool is very useful in areas that perform ASM or IM240 tests

Figure 10-3. A portable 5-gas analyzer is useful in testing vehicles when no loaded-mode test equipment is available.

using dynamometers. While this type of emission tester still measures gases in ppm and percentages it gives the technician the ability to baseline a subject vehicle. Baselining a vehicle refers to driving the vehicle and recording the emissions prior to performing a repair. For instance, if the vehicle failed an ASM or IM240 test for high NO_x levels the technician would drive the vehicle and record the NO_x levels first using a portable 5-gas analyzer. After repairs are made to the vehicle another test drive is performed and the NO_x levels are again recorded. If the repairs were successful the technician should see a drop in NO_x levels between the baseline test and the second test after the repairs. While this method will not definitely indicate to the technician that the vehicle will pass the IM240 or ASM test, it at least serves as a good indicator that the repairs made to the vehicle were appropriate for the failure. Portable 5-gas analyzers are often used by repair facilities in areas of the country that have centralized emissions testing programs.

GAS ANALYZER CALIBRATION

Gas analyzer calibration is crucial in order to obtain accurate gas sample readings. The calibration method varies by manufacturer; however, most gas analyzers use similar calibration methods. The first type of calibration is zeroing. This process often occurs periodically as the analyzer is running. This process involves purging the analyzer with fresh air and then adjusting all of the readings to zero except O_2. O_2 readings should be around 21 percent whenever the analyzer is sampling outside air. This is because that is the percentage of oxygen in the air. If the analyzer does not read close to 21 percent, then service needs to be performed on the analyzer.

The next part of the calibration procedure is to connect a bottle of calibration gas, figure 10-4, to the analyzer. This calibration gas typically contains fixed percentages of HC, CO, and CO_2. For example the HC may be 300 ppm, CO 1 percent, and CO_2 6 percent. The analyzer will then sample this "tri blend" calibration gas and compare its reading to the gas mixture. If they do not agree, then the analyzer will require service before being used. If the analyzer is being used for a state emission testing program then this gas calibration is required every certain number of days. If the calibration is not performed or the gas calibration fails then the emission analyzer will lock out the technician to prevent any further test being performed. If the analyzer is not being used for state emission testing programs the gas calibration check should be performed on a periodic basis as routine maintenance.

The HC hang up check is usually performed before the start of emissions testing. This term refers to making sure that the sample probe is at least two feet off of the shop floor and sampling the amount of HCs in the surrounding air. Since the analyzer relies on fresh air, if there is a large concentration of HCs in the shop air the analyzer would not be accurate.

CAUTION: Running vehicles without exhaust extraction hoses or spraying certain chemicals can cause higher than normal HC concentrations. Always ensure that the area around the emission analyzer is well vented.

A leak test should be performed periodically or every three days if the analyzer is a state-certified unit. The leak check tests the integrity of the sample hose and connections to ensure no leaks are present that would allow outside air to be drawn into the sampling equipment. Follow the equipment manufacturer's recommendations for performing a leak test. On state-certified emissions analyzers, if this test is not performed as required the software will lock the machine so that it may not be used until the leak test is performed.

Figure 10-4. Shown are two bottles of calibration gas used to calibrate this state-certified emission tester.

PRECONDITIONING THE VEHICLE

Before performing an emissions test, either for a state-mandated emissions test or a diagnostic test, the vehicle must be preconditioned. Preconditioning involves operating the vehicle until closed loop operation has been achieved. This ensures that the O_2 sensor(s) are operating and providing feedback to the PCM. In addition, preconditioning ensures that the catalytic converter has reached operating temperature to effectively treat the exhaust gases. During this preconditioning it might be helpful to connect a scan tool and verify engine operating temperature and closed loop fuel control, figure 10-5.

Some state-inspection procedures may specify specific preconditioning procedures. Be sure to follow the state requirements when performing a state-certified test procedure. Another problem that sometimes occurs during state testing is that vehicles sit and idle for extended periods of time while waiting for an emissions test. This usually occurs at centralized testing centers where long lines are present. Some of these vehicles may drop out of closed loop operation because the O_2 sensor cools down. If this happens the vehicle may fail the emissions test. Some vehicles may require an engine shutdown, restart, and operation at 2,000 rpm for two minutes before testing. Be sure to follow the manufacturer's or the equipment manufacturer's recommendations when preconditioning the vehicle.

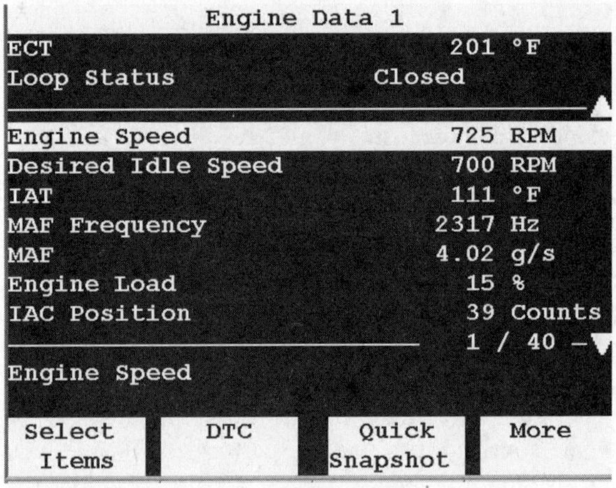

Figure 10-5. A scan tool is useful to determine if the vehicle is properly preconditioned.

PERFORMING A TWO-SPEED TAILPIPE TEST

Many vehicles are tested with a two-speed tailpipe test as part of a state-mandated emissions testing program. In addition, this two-speed tailpipe test is often performed as a diagnostic test either prior to or after repairs. The procedure may vary slightly in different programs; however, the procedure should be very sim-

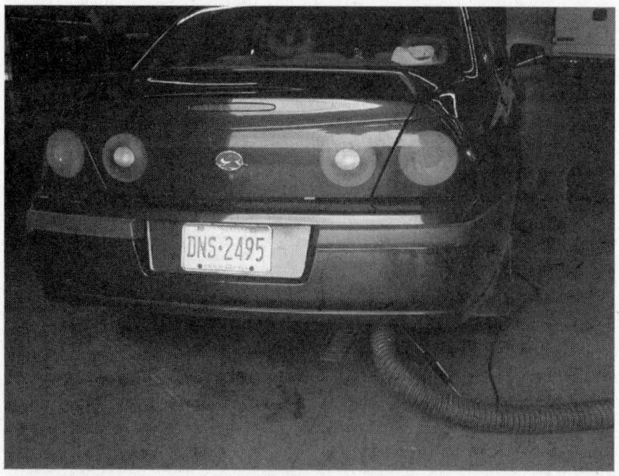

Figure 10-6. Insert the exhaust probe at least 10 inches into the tailpipe and install exhaust removal equipment.

Figure 10-7. Install the rpm hookup to the engine.

ilar to the one outlined in the following text. To perform a two-speed tailpipe test, do the following:

1. Set the parking brake, block the wheels, and insert the analyzer probe at least 10 inches into the tailpipe, figure 10-6.

CAUTION: Connect exhaust extraction equipment to remove dangerous exhaust fumes.

2. Connect the analyzer rpm hookup to the engine, figure 10-7.
3. Select the correct menu on the emissions analyzer and input the required information, figure 10-8.
4. Start the engine and precondition until operating temperature and closed loop operation is achieved.
5. Raise the engine speed to 2,500 rpms and maintain this speed until the exhaust readings stabilize. Record these readings, figure 10-9.
6. Return to idle speed and allow the exhaust readings to stabilize. This may take up to 30 seconds. Record these readings, figure 10-10.
7. Compare the readings obtained to figure 10-11, or compare the readings to the standards set by the state.

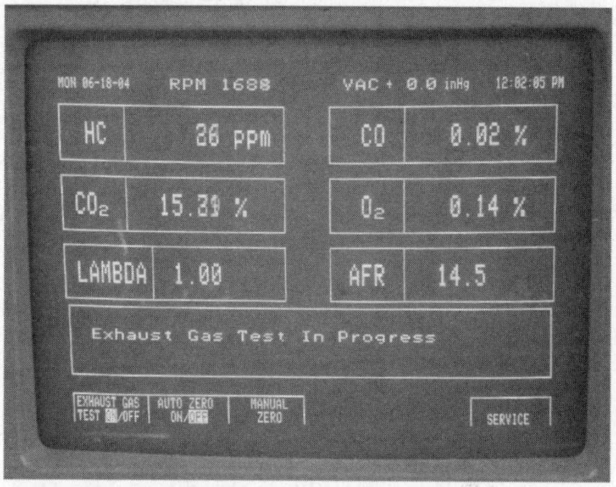

Figure 10-8. Select the correct menu on the analyzer and input the necessary information.

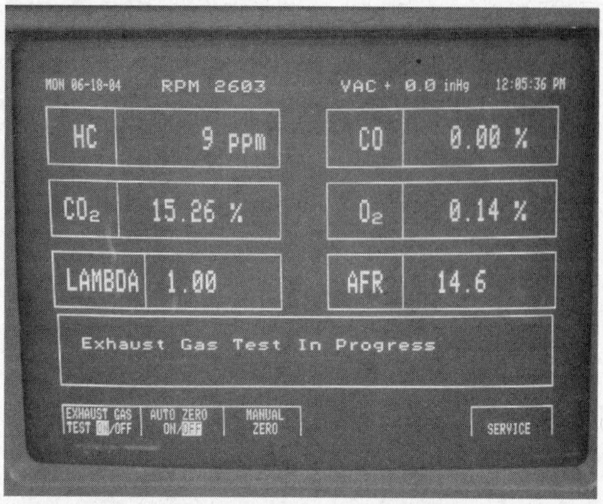

Figure 10-9. Raise the idle to 2,500 rpms and allow the readings to stabilize.

PERFORMING AN ASM 5015 TEST OR SIMILAR TEST

An ASM 5015 test and other similar tests such as the ASM 2525 are performed with the vehicle on a dynamometer. The vehicle is then operated under load on the dynamometer as specified by the test type, and the exhaust readings are obtained while the vehicle is operating under load. This type of test is often referred to

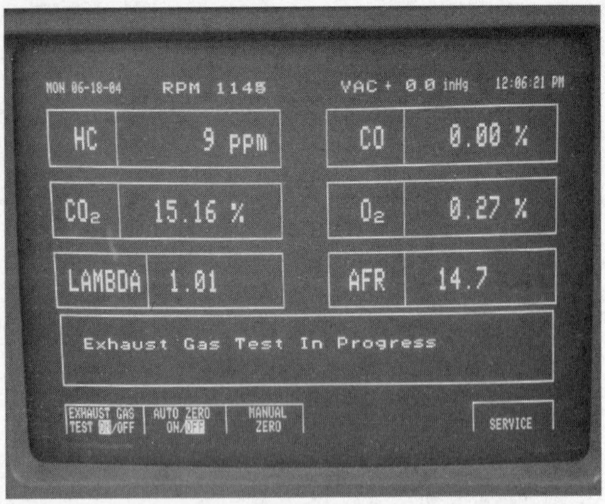

Figure 10-10. Lower the rpms and let the vehicle idle. Allow the readings to stabilize.

as a loaded-mode test. There are different versions of this loaded-mode test; however, the test methods are similar. An important part of a loaded-mode test is the vehicle preinspection safety evaluation. This determines whether the vehicle is in a safe operating mode to perform the test. This preinspection safety evaluation is especially critical for vehicles that are subject to dynamometer testing. To perform the preinspection safety evaluation, do the following:

1. Check for excessive crankcase oil leakage, excessive transmission oil leakage, and excessive coolant leakage, figure 10-12. If excessive leakage is noted, do not test the vehicle.
2. Check for visible smoke coming from any source other than the tailpipe. This may indicate excessive fluid leakage.
3. Check for leaking or missing exhaust systems. If the exhaust system parts are leaking or missing, do not test the vehicle. Testing the vehicle with a leaking exhaust can lead to an invalid test due to an inadequate exhaust sample.
4. Listen for any obvious mechanical engine problems that indicate that the vehicle should not be tested (such as excessive engine knock).
5. Check tires for excessive wear, different tire sizes on the same axle, tire pressure, tread separation, or any other defect that would make it unsafe to test the vehicle on the dynamometer, figure 10-13. ***Do not test any vehicle on a dynamometer that has unsafe tires.***
6. If the vehicle has none of the previously mentioned obviously unsafe items, place the vehicle on the dynamometer and run the vehicle on the dynamometer to center the vehicle. If the vehicle cannot be stabilized on the dynamometer, do not test the vehicle.

Automobiles and Light-Duty Trucks Less than 6000 Pounds GVWR		
Model Year	**CO %**	**HC ppm**
1975 – 1979	4.0	400
1980	3.0	300
1981 – 1992	1.2	220
1993 & Newer	1.0	130

Automobiles and Light-Duty Trucks Less than 6000 – 9000 Pounds GVWR		
Model Year	**CO %**	**HC ppm**
1975 – 1978	6.0	650
1979	4.0	400
1980	3.0	300
1981 - 1992	1.2	220
1993 & Newer	1.0	180

Figure 10-11. Each state may set their own standards for emissions testing. The standards shown in this figure should be similar.

Procedures may vary from state to state when performing an ASM 5015 or similar test. Be sure to follow the correct procedures as required by that testing program. The following represents a typical test procedure for performing an ASM 5015 test on a vehicle. To perform the ASM 5015, do the following:

1. Drive the vehicle onto the dynamometer, figure 10-14. Ensure that if the vehicle is equipped with traction control or all-wheel drive, it is disabled.

NOTE: If you cannot disable the traction control or all-wheel drive, then do not attempt to test the vehicle.

2. Lower the dynamometer lift.
3. Start the engine, place the selector in "drive," and slowly increase the vehicle speed until the wheels are stabilized, figure 10-15.

Figure 10-12. Before testing the vehicle, perform a safety evaluation to determine if the vehicle should be tested.

Figure 10-13. Carefully inspect the tires before testing the vehicle on the dynamometer.

Figure 10-14. Drive the vehicle onto the dynamometer after it is determined that the vehicle is safe to test.

NOTE: If the vehicle cannot be stabilized, abort the test.

4. Apply the parking brake and install the appropriate vehicle restraint devices to the vehicle, figure 10-16.
5. Input the required information into the emissions analyzer, figure 10-17.
6. Insert the exhaust probe into the tailpipe at least 10 inches, figure 10-18. If the vehicle is equipped with dual exhaust, two probes must be used.
7. Attach the exhaust extraction equipment, figure 10-19.
8. Attach the engine rpm pickup to the vehicle, figure 10-20.
9. If the ambient air temperature is above 72°F in the shop, an auxiliary fan should be placed in front of the vehicle to aid in cooling and prevent engine overheating, figure 10-21.
10. Start the engine and place the selector in drive for automatic transmissions and second gear

Figure 10-15. Drive the vehicle on the dynamometer to stabilize the vehicle before applying the restraints.

Figure 10-16. Apply the proper restraints to the vehicle.

Figure 10-17. Input the required information into the analyzer. Many state units incorporate a bar code scanner.

for manual transmissions. Follow the prompts on the emission analyzer to perform the test, figure 10-22.

11. Maintain 15 mph ± 1 mph until the analyzer indicates that the test is complete.
12. After completion of the test, shut off the engine and remove the exhaust probe, vehicle restraints, and tachometer lead.
13. Print and review the test results, figure 10-23.
14. A printout is usually given to the customer and in some cases another copy is made for the station records, figure 10-24.

PERFORMING AN OBD II I/M TEST

In many areas of the country, OBD II tests are being performed on 1996 and newer vehicles. This test ensures that the OBD II is operating correctly and the check engine light is not illuminated. In addition, many areas include a gas cap test as part of this procedure. The procedure may vary in different areas; however, the procedure is similar.

To perform an OBD II I/M check, do the following:

1. Enter the vehicle information into the analyzer and select the test type, figure 10-25.

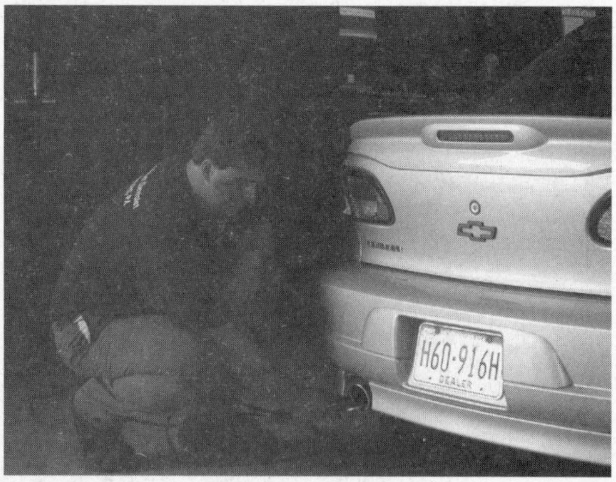

Figure 10-18. Insert the tailpipe probe at least 10 inches into the tailpipe.

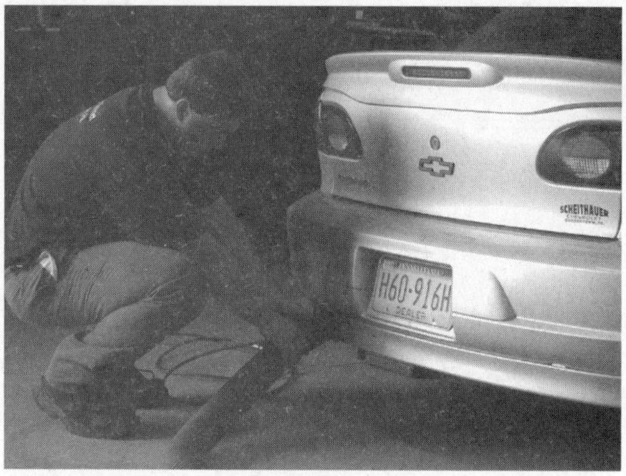

Figure 10-19. Install exhaust extraction equipment over the tailpipe and exhaust probe.

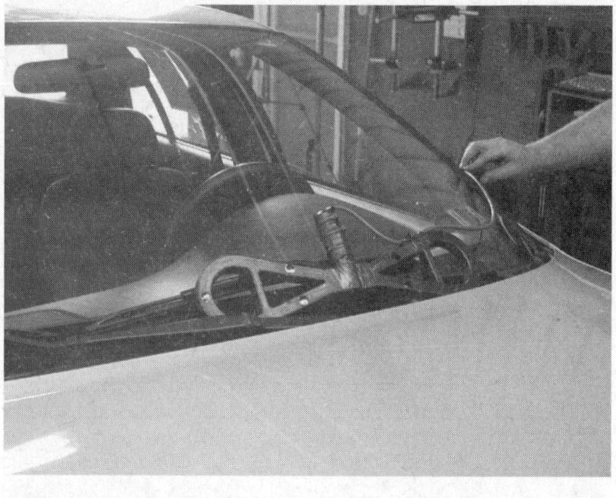

Figure 10-20. Install the rpm pickup to record engine rpm. This is an inductive version that does not require the hood to be open.

Figure 10-21. If the ambient air temperature is above 72°F, use an auxiliary fan to prevent overheating.

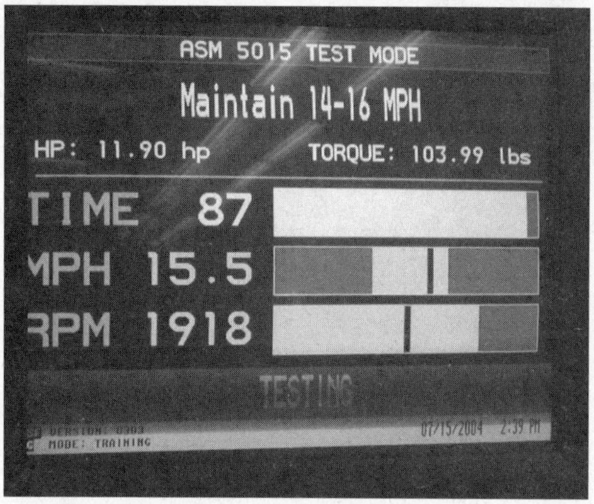

Figure 10-22. Follow the emissions analyzer prompts for driving the vehicle during the test.

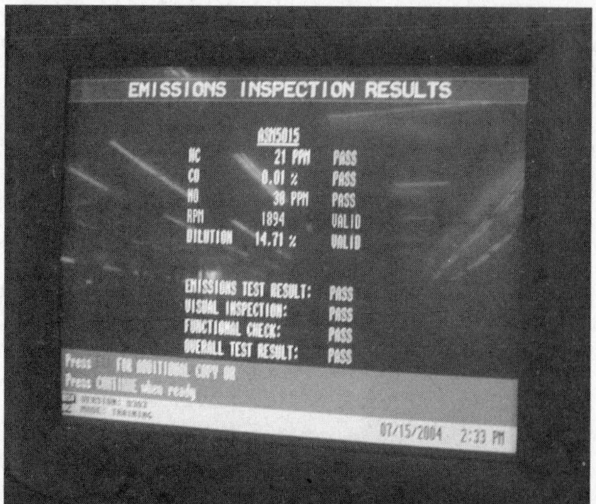

Figure 10-23. Review the test results at the conclusion of the test.

COMMONWEALTH OF PENNSYLVANIA
VEHICLE EMISSIONS INSPECTION REPORT

3GNGK26U8YG180552

9999999999

Test Date/Time: 07/13/2004 @ 16:30:17

VEHICLE INFORMATION

Year:	2000	Engine Size:	4300	Odometer Reading:	049001
Make:	CHEVROLET	Cylinders:	08	Plate #:	H60915H
Model:	C3500 PICKUP	GVWR:	08600	Title:	9999999999
VIN:	3GNGK26U8YG180552	Estimated Test Weight:	UNKNOWN	Inspection Type:	Initial
County:	BUCKS			Record #:	002028

EMISSIONS CONTROL SYSTEMS INSPECTION

Air Inj. System:	PASS	Catalytic Converter:	PASS	Gas Cap Integrity:	PASS
EGR System:	PASS	Evaporative Control System:	PASS		
PCV System:	PASS	Fuel Inlet Restrictor:	PASS		

TAILPIPE EMISSIONS INSPECTION

MODE	CO %			HC ppm			RPM		DILUTION	
2 Speed Idle	Limit	Reading	Result	Limit	Reading	Result	Reading	Result	Reading	Result
IDLE	1.00	0.13	PASS	180	117	PASS	606	PASS	14.5%	PASS
2500 RPM	1.00	0.36	PASS	180	68	PASS	2399	PASS	14.5%	PASS

OVERALL TEST RESULTS: **PASSED**

Emissions Control Systems Visual/Functional Inspection: **PASS**

Tailpipe Inspection: **PASS**

TIN: **137513057262** Sticker Number: **IM45298332**

PLEASE RETAIN THIS DOCUMENT FOR YOUR RECORDS.

Vehicle tested in accordance with federal regulations and Pa. Title 67, Chapter 177

EMISSIONS INSPECTION STATION

STATION NAME:	SCHEITHAUER CHEVROLET INC	STATION #:	2592
ADDRESS:	780 W END BLVD, QUAKERTOWN PA 18951	ANALYZER #:	SE270076
PHONE:	1-215-536-2100	SOFTWARE VERSION:	0303
INSPECTOR NAME:	JASON A. DEVINE	INSPECTOR ID:	

VEHICLE EMISSIONS INSPECTION QUESTIONS

For additional information, please contact the Customer Hotline at

(800)265-0921

Inspector's Signature: JASON A. DEVINE

Figure 10-24. A printout of the test results.

Figure 10-25. Enter the required information into the OBD II analyzer.

Figure 10-26. Most OBD II tests require a gas cap test also.

NOTE: An initial test is entered for the first test and a retest is selected after the vehicle has failed the first test.

2. If required, perform the gas cap test when instructed by the analyzer, figure 10-26.
3. Turn the ignition on, engine not running, and observe the malfunction indicator lamp (MIL). It should illuminate, figure 10-27.

NOTE: On some vehicles the malfunction indicator lamp may go out after several seconds. This is acceptable as long as the light initially illuminated with the key on, engine off.

4. Start the engine and observe the malfunction indicator lamp (MIL). It should remain off when the engine is running.

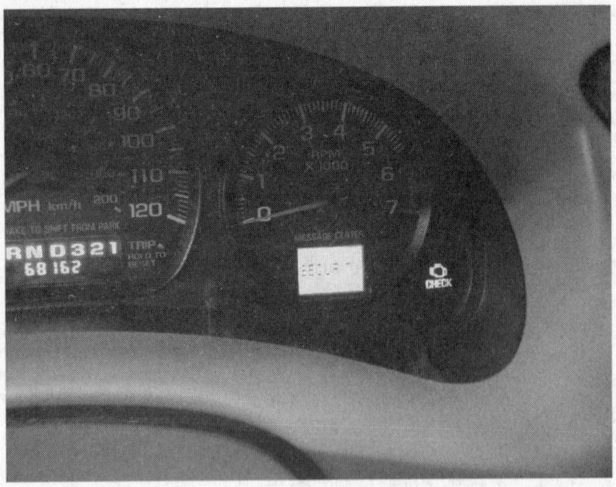

Figure 10-27. With the ignition on and the engine off, verify that the MIL is illuminated.

Figure 10-28. Connect the test cable to the vehicle's DLC connector.

5. Turn the engine off and connect the OBD II cable to the vehicle's data link connector (DLC), figure 10-28.

NOTE: Wait 15 seconds before turning the ignition on.

6. Start the engine and allow the analyzer to obtain the readiness status of the OBD II system. In addition, the analyzer will look at the MIL request and any stored trouble codes if the MIL is requested on, figure 10-29.
7. Shut off the engine and disconnect the DLC cable from the vehicle.
8. Print and review the test results. The vehicle will pass the test if the MIL illuminated with the key on, engine off; if the MIL was not on with the engine running; if the majority of the readiness flags have passed (varies with programs); and there is no MIL request.

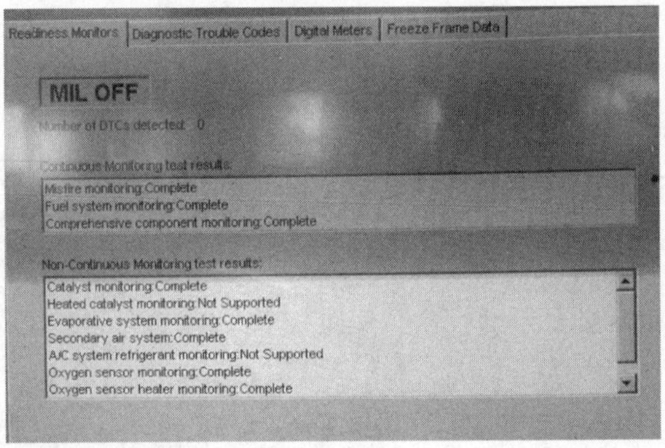

Figure 10-29. The OBD II test will display the results of the continuous monitors and trouble codes if there are any set.

DIAGNOSING EMISSIONS FAILURES

A gas analyzer is essential for diagnosing emissions failure on pre-OBD II vehicles. If the vehicle has failed a state-mandated test, the cause of the failure should be indicated on the test report. However, after any repairs have been made to the vehicle an emissions diagnostic test should be performed to determine if the emissions levels have been reduced. This will help the technician determine if the vehicle will likely pass a retest. Higher-than-normal HC and CO levels can easily be measured using a 4-gas analyzer. The most difficult gas failure to verify will be NO_x failures since these almost always occur when the vehicle is under load. Therefore, if the repair facility does not have a portable 5-gas analyzer or a dynamometer to test the vehicle, it will be impossible for the repair facility to verify any repairs performed on the vehicle.

Common Causes of High HC

Following is a list of possible causes for high HC failures. All the causes for high HC failures listed may not apply to all vehicles. It will depend on which fuel-delivery system the vehicle is equipped with and what type of airflow system is on the vehicle. Hich HC failures may be caused by:

- Excessively rich or lean fuel mixture
- Misfire
- Leaking fuel injectors
- Improper ignition timing
- Vacuum leaks
- Mass airflow or manifold absolute pressure sensor
- Evaporative system malfunction

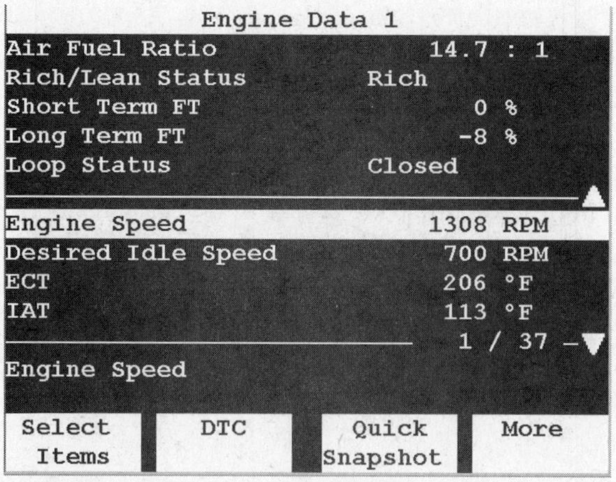

Figure 10-30. A scan tool is useful in determining if the vehicle is running too rich or lean.

- O_2 sensor contamination
- EGR valve stuck
- Fuel pressure
- Coolant temperature sensor
- Engine mechanical condition
  - Leaking head gaskets
  - Defective rings, pistons, cylinders
  - Worn camshaft
  - Low compression
  - Defective valves or guides

Remember that high HC can be caused by either an overly rich condition or lean condition. A scan tool is very useful for looking at engine data parameters that could be causing either a rich or lean condition, figure 10-30. Pay particular attention to the O_2 sensor voltages and the fuel trim numbers. This will help the technician determine if the high HCs are being caused by an overly rich or lean condition. Also always consider engine mechanical conditions and perform the appropriate test such as compression testing and cylinder leakage tests. Non-OBD II vehicles may fail an emissions test for rich or lean conditions without setting any diagnostic trouble codes.

Common Causes for High CO

High CO is caused by excessively rich conditions. The most common causes for high CO are:

- Rich fuel mixture
- Defective PCV system
- Restriction in the air cleaner of air-induction system
- Leaking fuel injectors
- Over-advanced ignition timing
- Air-injection system inoperative
- Coolant temperature sensor
- Mass air flow or manifold absolute sensor
- Excessive fuel pressure

Common Causes for Low CO_2

Remember that CO_2 is an efficiency gas. CO_2 levels should typically be higher than 8 percent Low CO_2 levels may be caused by:

- Exhaust gas dilution due to leaking exhaust
- Rich mixture
- Restricted air-induction system or air filter
- Leak in exhaust sampling equipment

Typically as CO_2 levels decrease, CO increases. Therefore, whenever CO_2 is below normal always look at the other gas levels. Unless there is a dilution of the sample due to an exhaust leak the technician should note abnormal readings in the other gas values.

Common Causes for Low O_2 and High CO

Common causes for low O_2 and high CO levels are:

- Rich mixture
- Leaking fuel injectors
- Restricted air filter or air-induction system
- Evaporative system malfunction
- Fuel-contaminated crankcase
- Defective coolant temperature sensor

Common Causes for High O_2 and Low CO

High O_2 with corresponding low CO levels are most commonly caused by a lean mixture. If the mixture is excessively lean, HC will also start to rise. The most common causes of high O_2 and low CO are:

- Lean mixture
- Vacuum leaks
- Defective fuel injector
- Improper air-injection reaction operation
- O_2 sensor defective or circuit malfunction

Common Causes for High NO_x

Many loaded-mode exhaust emissions failures are excessive NO_x failures. Since NO_x formation does not occur at idle or low-load conditions, the only effective way to detect high NO_x conditions is a loaded-mode test or by using a portable 5-gas analyzer to measure NO_x on a road test.

Excessive NO_x levels are usually caused by one of three things: excessive heat, defective EGR system, or a defective catalytic converter. The most common causes for high NO_x levels are conditions that cause higher heat in the cylinder(s). Whenever diagnosing causes for high NO_x levels, always consider and check for problems that could cause higher heat levels in the cylinder or the engine. Below is a list of possible causes for higher-than-normal NO_x emissions that relate to heat:

- Carbon buildup on pistons
- Over-advanced ignition timing
- Lean fuel mixture
- Restricted EGR passages or inoperative EGR valve
- Cooling system malfunction
- Detonation
- Excessive coolant concentraton
- Inoperative spark control system
- Stuck or inoperative thermostatic air inlet door

Another cause for high NO_x emissions could be a defective three-way catalytic converter. However, there is no effective test method to determine the capability of the converter to reduce NO_x formation. Therefore, before replacing the catalytic converter, make sure that all other possible causes for excessive NO_x production have been eliminated. If the catalytic converter is suspected an increase in CO and HC will also likely be evident.

11

Testing and Servicing PCV, AIR, EGR, and Catalytic Converters

OBJECTIVES

Upon completion and review of this chapter, you will be able to:

- Perform a PCV systems functional test and determine necessary repairs.
- Perform AIR systems functional test and determine necessary repairs.
- Perform EGR systems functional test and determine necessary repairs.
- Determine if the OBD II systems monitors have run and passed on an OBD II equipped vehicle.

POSITIVE CRANKCASE VENTILATION SERVICE

There are three requirements for proper positive crankcase ventilation (PCV) system operation:

1. The control valve must be operating properly (not restricted or stuck open).
2. Air and vapor must flow freely through the system (no blockage or collapse of the crankcase breather or manifold vacuum hose).
3. A properly operating intake manifold vacuum supply with no leaks in the connecting hoses or engine gaskets and seals.

A plugged system, figure 11-1, or a valve stuck in the closed position causes an increase of blowby pressure in the crankcase. If the pressure becomes too great, the following symptoms may occur:

- Blue tailpipe smoke from increased oil consumption
- Engine oil leaks caused by blown seals and gaskets
- A pool of engine oil in the air cleaner. This occurs when liquid oil instead of just vapor is forced back through the crankcase breather hose.

If there is a vacuum leak in the hose between the valve and the manifold port, figure 11-2, too much air will enter the intake manifold and may cause a lean air-fuel mixture. If the vacuum leak is not too great, the engine's electronic control system may compensate for the extra air. If not, fuel economy and tailpipe emissions suffer. If the vacuum leak is too large, driveability symptoms may occur such as a rough idle, stumble, or hesitation.

The PCV system is checked and serviced by:

- Visually inspecting the components.
- Observing the system's performance under operation.
- Replacing worn, damaged, and inoperative components.

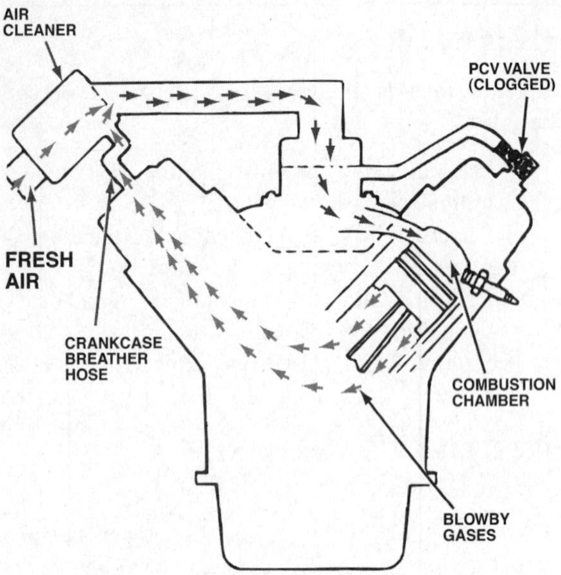

Figure 11-1. A clogged PCV system can cause serious problems.

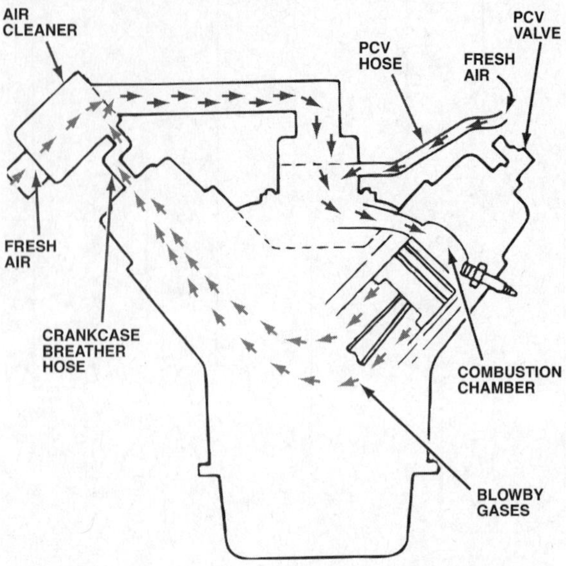

Figure 11-2. A vacuum leak between the PCV valve and the intake manifold port can cause driveability, fuel economy, and tailpipe emissions problems.

NOTE: There are no adjustments to any components in the system.

PCV VISUAL INSPECTION

As with all system diagnostics, visual inspection is the first step in checking the PCV system. The following is a typical PCV-system visual inspection procedure. Be sure to refer to the appropriate service publication for the vehicle being serviced.

1. Check all hoses for proper routing and tight connections.
2. Inspect PCV hoses for cracks, brittleness, and clogging. Replace any that are bad.
3. Remove the air cleaner cover and check the air cleaner box and crankcase filter for oil deposits.

NOTE: Oil deposits in the air cleaner assembly of an engine with high mileage may be due to worn rings or a clogged PCV valve. Functionally test the PCV system before suspecting piston ring wear.

4. Check the crankcase ventilation filter (if equipped) for clogging. It may be located in:
   a. The intake air cleaner, figure 11-3
   b. The crankcase breather air hose, figure 11-4
   c. The valve cover oil filler cap, figure 11-5
   d. The crankcase breather hose between the air cleaner and oil filler cap on certain Ford 2.3-liter and 3.8-liter engines, figure 11-6
5. Replace the crankcase breather filter in the air cleaner housing, figure 11-3, or the hose, figure 11-4, if it is dirty. Foam and wire gauze filters should not be cleaned. On Ford 2.3-liter and 3.8-liter engines, replace the entire breather hose and oil fill cap assembly at specified intervals.

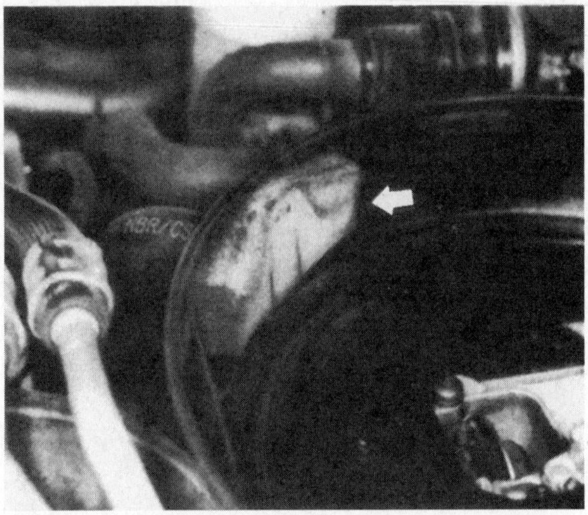

Figure 11-3. A ventilation filter in the air cleaner may become clogged with oil.

6. Clean the wire mesh filter installed in the valve cover oil filler or breather cap, figure 11-5, by washing the cap in solvent and letting it air dry.

FUNCTIONAL TESTS

Though not in common use today, various testers are available to check overall PCV system operation. These testers generally measure crankcase vacuum

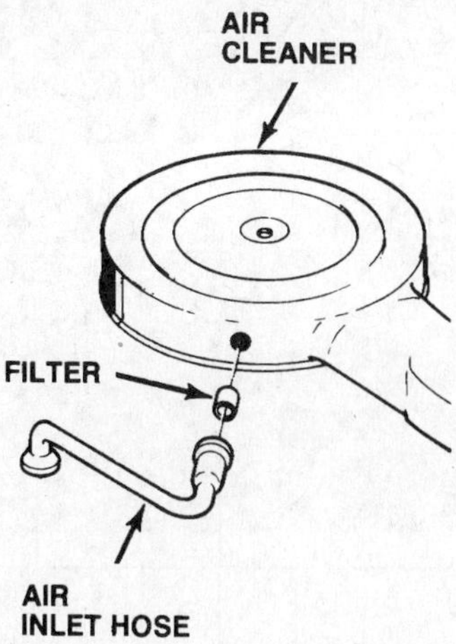

Figure 11-4. Replace the ventilation filter in the PCV inlet hose if it is dirty.

Figure 11-5. The wire mesh filter in the oil fill cap should be washed in solvent and air dried.

while the engine is idling. Install the tester in place of the oil filler cap on the valve cover, figure 11-7 and note the reading with the engine running.

NOTE: These testers are not reliable when checking fixed orifice PCV systems that don't use a PCV valve.

Crankcase Vacuum Test

When a PCV tester is not available, hold a small piece of stiff paper over the oil filler opening or closure hose with the engine idling, figure 11-8. In a properly operating system, crankcase vacuum will pull the paper down against the opening. If the paper is not pulled down, or if it is blown upward, the system is not working properly or crankcase blowby pressure is excessive. An improperly operating system may be caused

Figure 11-6. Some Ford 2.3-liter and 3.8-liter engines use breather hose and filler cap ventilation filters.

Figure 11-7. A typical PCV system tester.

by too much blowby, which happens to older engines, or by a clogged PCV valve. The previously noted paper test will not work on fixed orifice systems.

Intake Vacuum Test

The following is another typical method for testing the PCV system when results of the crankcase vacuum test are not conclusive.

1. Pull the PCV valve out of the valve cover.
2. Listen for a hissing noise as air passes through the valve.
3. Place a finger over the valve, figure 11-9. A strong vacuum should be felt and engine idle speed should drop slightly.

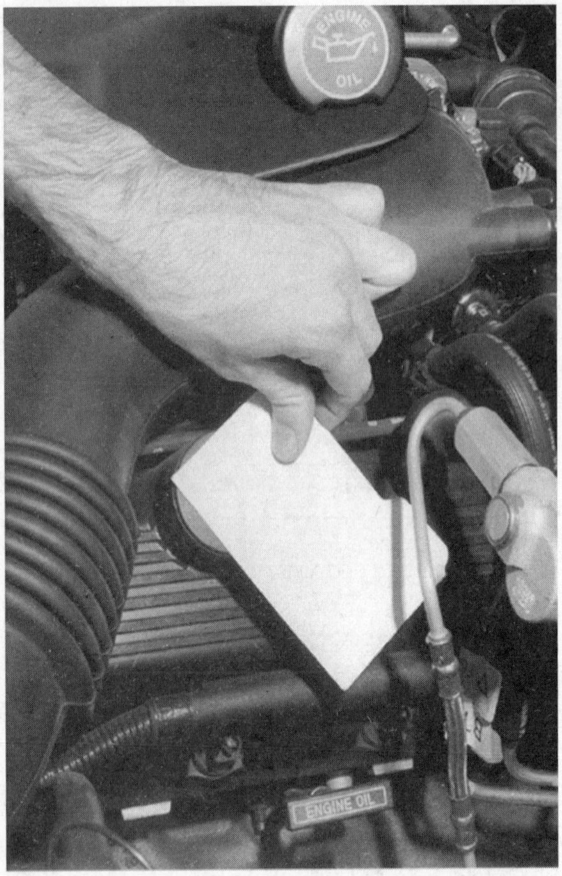

Figure 11-8. If a PCV tester is not available, a piece of paper can be used to test for vacuum.

Engine speed may be affected only momentarily by this test. The engine's electronic control system is capable of adjusting the air-fuel mixture to compensate for the lean conditions created by this procedure. If no hissing noise is heard when the valve is initially removed and if no vacuum can be felt when placing a finger over the valve, suspect a clogged valve, vacuum hose, or vacuum port.

It is also important to make sure the vehicle has the correct PCV valve installed. PCV valves are manufactured with calibrated flow rates based on the engine in which they are used. The wrong valve might increase or decrease flow. Match the part number on the PCV valve with the manufacturer's specifications to determine if the correct valve is being used.

On some vehicles, the evaporative emission vapor canister is purged through the PCV line. If the PCV system does not pass the crankcase vacuum or intake vacuum tests, disconnect and plug the canister purge hose, and repeat the tests. If the PCV system passes with the purge hose plugged, the purge hose or connection is leaking. If the PCV system is found to be

Figure 11-9. Test a PCV valve by placing a finger over the end of the valve with the engine running.

clogged, check the evaporative system. An evaporative system that has not been able to purge may have a saturated charcoal canister.

Orifice Flow Control System

An orifice flow control system operates the same as a valve system. However, a fixed orifice is used in place of a variable orifice PCV valve.

Refer to the manufacturer's specifications to determine the size of the orifice. To check the orifice for plugging, insert a drill bit of the same size into the orifice, figure 11-10. If the drill bit does not insert, or inserts with difficulty, the orifice is clogged and must be cleaned. Cleaning is accomplished using smaller drill bits and solvent.

To check the dual orifice valve design, disconnect the vacuum line at the bottom of the valve and connect a hand vacuum pump in its place. With the engine running at a fast idle, apply 10 to 12 inches (255 to 305 mm) of vacuum and listen for a change in engine rpm. If there is no change in engine speed, replace the dual orifice valve. If engine rpm changes, the valve is good. Next, perform the test shown in figure 11-11 to complete the system check.

PCV VALVE SERVICE

The first step is to be sure the correct valve is installed. Refer to the manufacturer's specifications for the correct part number. PCV valves are not a serviceable item and must be replaced at regular intervals. The vehicle emission control (VEC) label on some vehicles, figure 11-12, identifies the interval for PCV valve replacement. The decal may also identify the valve part number. The same information is usually found in the service section of the owner's manual.

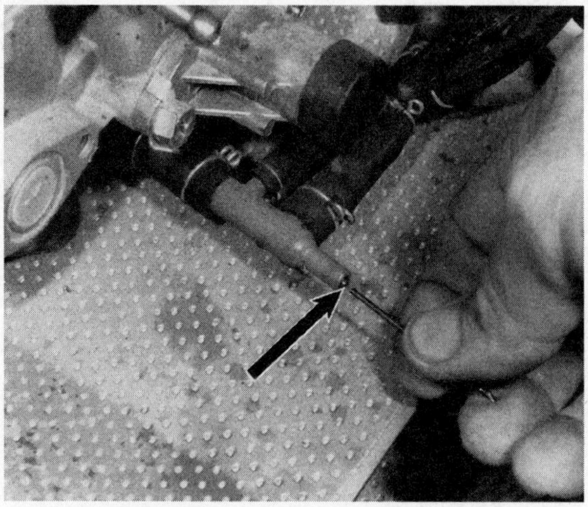

Figure 11-10. A fixed orifice can be cleaned with solvent and a suitable drill bit.

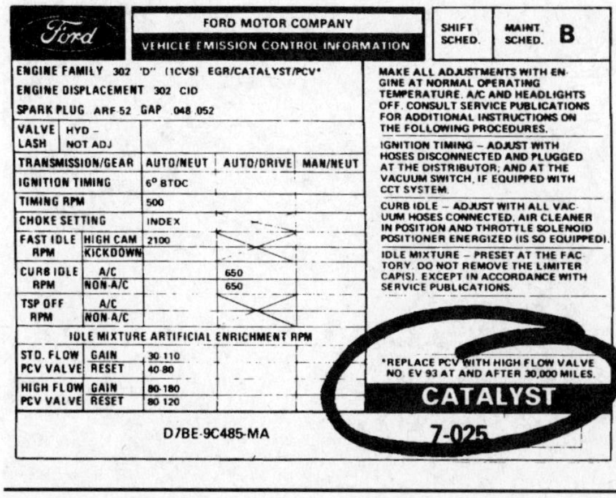

Figure 11-12. Manufacturer's PCV valve specifications are posted on the VEC label. (Courtesy of Ford Motor Company)

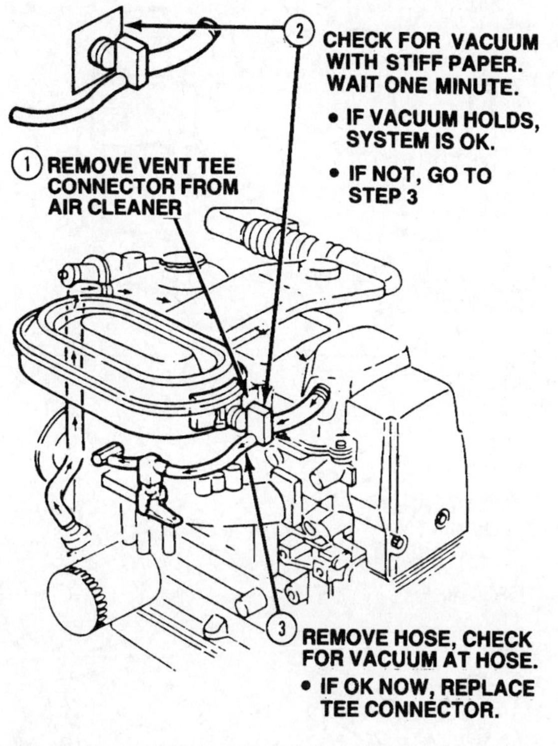

Figure 11-11. A fixed orifice system check. (Courtesy of Ford Motor Company)

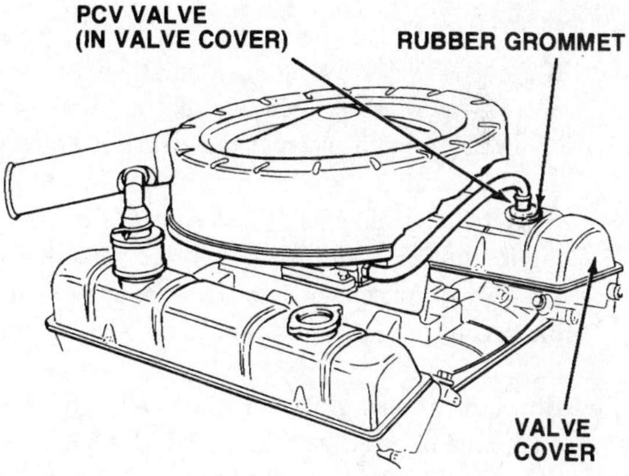

Figure 11-13. A valve cover PCV valve location.

Location

PCV valves are usually found in a crankcase vent hose that leads to manifold vacuum on an intake manifold port. The valve may be installed at either end of the hose. Sometimes it fits into a rubber grommet in the valve cover, figure 11-13. In somes cases, the PCV valve is installed in the side of the engine block or in the intake manifold, figure 11-14, where it picks up crankcase vapors from under the intake manifold.

Replacement

The following is a typical procedure for replacing the PCV valve. Check the service manual for the vehicle being serviced for specific directions.

1. Disconnect the PCV vent hose from the valve cover or intake manifold. Separate the hose pieces if the valve is installed in a 2-piece hose.
2. If the valve is held in the hose with a clamp, squeeze the clamp ends together with pliers and slide it back on the hose. Remove the valve from the vent hose by twisting and pulling it.
3. Disconnect the other end of the vent hose. Blow into the hose with compressed air to make sure there are no restrictions. Connect the vacuum port end of the hose.

Figure 11-14. Some PCV valves are located in the intake manifold underneath a cover that must be removed to access the valve.

4. If the PCV valve uses a push-on elbow or connector, figure 11-15, remove the old elbow or connector from the valve end and install it on the new PCV valve.
5. Push the new valve into the hose with a twisting motion until it is fully seated. If the valve has a flow arrow, make sure the arrow points toward the vacuum source. If used, slide the hose clamp back in position.
6. Check the rubber grommet at the valve cover or intake manifold. If deteriorated or damaged, remove it and install a new one.
7. Connect the crankcase hose to install the PCV valve in 2-piece hoses. With 1-piece hoses, wipe the inside edge of the rubber grommet at the valve cover or intake manifold with silicone lubricant and seat the valve in the grommet.

Filter Service

Some PCV systems draw crankcase inlet air from the filtered side of the air cleaner. These systems require no special filter service. However, if the system uses a flame arrestor, clean the screen with solvent and allow to air dry. Other PCV systems use a polyurethane foam or gauze filter pack (also called an air inlet filter) installed in the side of the air cleaner housing, figure 11-3. The filter is replaced as follows:

1. Remove the air cleaner cover.
2. Disconnect the crankcase breather hose from the filter pack fitting in the air cleaner housing, figure 11-16.
3. Slide the retainer clip from the filter pack nipple and remove the filter pack from the air cleaner housing.

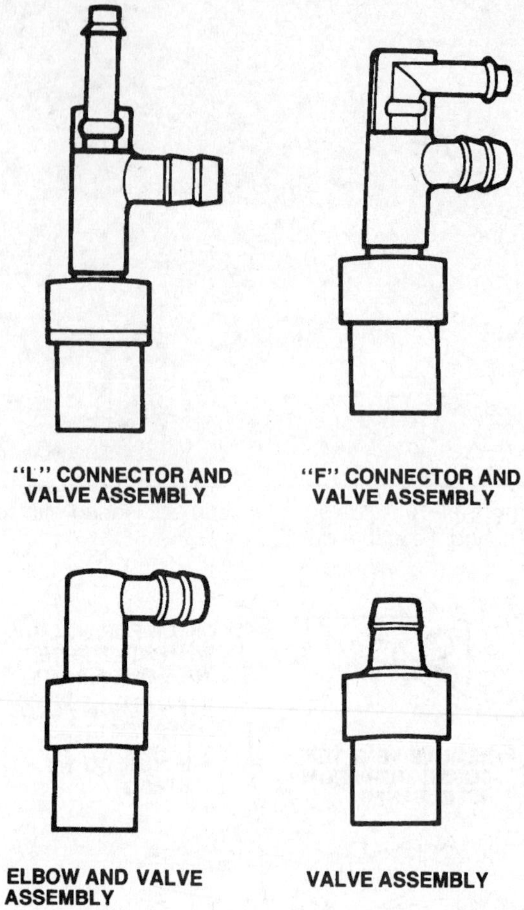

Figure 11-15. PCV valves use various plastic snap-on connectors. (Courtesy of Ford Motor Company)

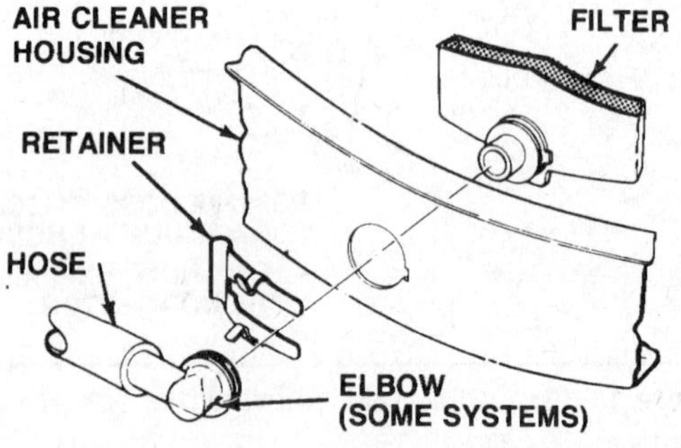

Figure 11-16. Typical PCV filter installation in an air cleaner housing.

4. Insert a new filter pack in the side of the air cleaner housing. Secure the filter pack in the air cleaner with the retainer clip.
5. Connect the hose to the filter pack and reinstall the air cleaner cover.

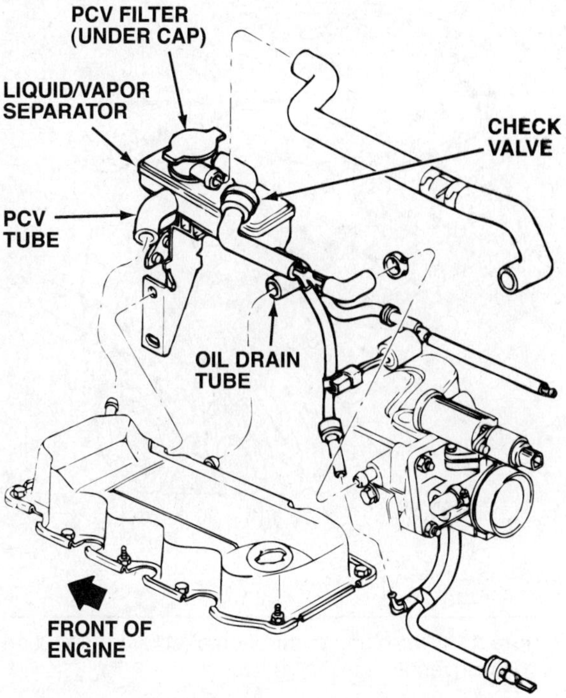

Figure 11-17. A PCV filter located under a liquid/vapor separator cap. (Courtesy of Ford Motor Company)

Sometimes air inlet filters are sealed in oil filler caps. The filter is usually a wire mesh. To service this type of filter, remove the filler cap and soak the complete cap and filter in parts cleaning solvent. Allow it to drain and dry. Do not use compressed air to dry this type of filter; the air pressure will damage the wire mesh. If the filter in an oil filler cap is damaged or permanently clogged, the entire cap and filter must be replaced.

Manufacturers continue to find new places to install the PCV filter. Ford 4-cylinder EFI turbo systems use a liquid/vapor separator, figure 11-17. The PCV filter is serviced by lifting the separator cap and replacing the filter.

Orifice Flow Control System

The orifice is usually located at the valve cover, intake manifold, or throttle body. Service the hoses the same as for a valve system, remembering that the orifice cannot be removed.

As previously mentioned, the orifice is cleaned using a drill bit the same size as the orifice opening. However, care must be taken not to use a drill bit that is too large. The orifice is a calibrated opening. If the orifice is enlarged while cleaning, it will allow too much flow and affect engine operation. Whenever cleaning an orifice, work carefully and slowly to avoid damaging the orifice walls. If the orifice is damaged, it may mean replacing the valve cover, intake manifold, or throttle body. Import vehicles with orifice flow control systems each require slightly different procedures, depending on the system. Refer to the service manual for specific procedures.

DIAGNOSING AND SERVICING AIR, EGR, AND CATALYTIC CONVERTERS

This section provides basic diagnostic and service techniques associated with late-model secondary air injection (AIR), exhaust gas recirculation (EGR), and catalytic converter systems. The basic components of these systems have not changed much over the years. However, air injection and EGR system controls have significantly changed. Early model air injection and EGR systems are mechanically controlled using vacuum and temperature operated valves, switches, and diaphragms. In contrast, late-model systems are computer-controlled and monitored by the vehicle's onboard diagnostic system. As a result, a total systems approach to diagnosis and testing is required.

SECONDARY AIR-INJECTION (AIR) SYSTEMS

The AIR system supports the catalytic converter in oxidizing exhaust pollutants by forcing fresh air into the exhaust system during engine warmup. There are two major types of AIR systems in use today:

- Electric air pump
- Belt-driven air pump

While the electric air pump design is more complex than the belt-driven air pump design, the function is the same.

Electric Air Pump System

The electric air pump AIR system, figure 11-18, consists of an electric air pump (EAP), single or dual combination diverter valve(s), bypass solenoid, solid-state relay, powertrain control module (PCM), and related wiring and vacuum hoses. The PCM uses various electronic engine control system inputs to initiate AIR operation. When the engine is started, the strategy determines when to enable the EAP. The EAP is enabled when the PCM signals the solid-state relay and the bypass solenoid to begin operation. Once the catalyst is *lit-off* by the fresh air forced into the exhaust stream, the PCM signals the solid-state relay to stop EAP operation and the bypass solenoid to stop the vacuum supply to the diverter valve(s). The PCM evaluates AIR system effectiveness using the oxygen sensors.

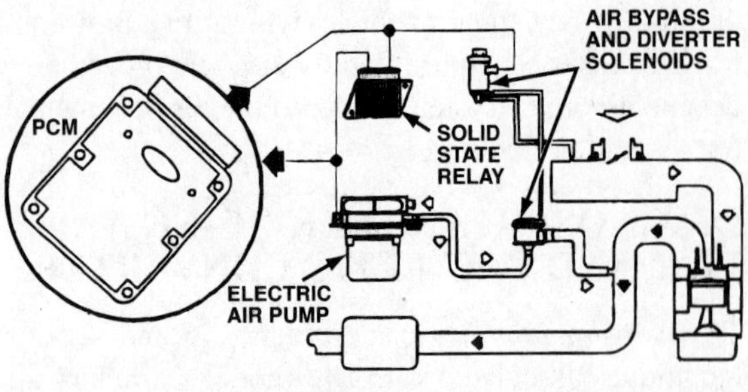

Figure 11-18. The PCM turns the electric air pump system on and off as needed.

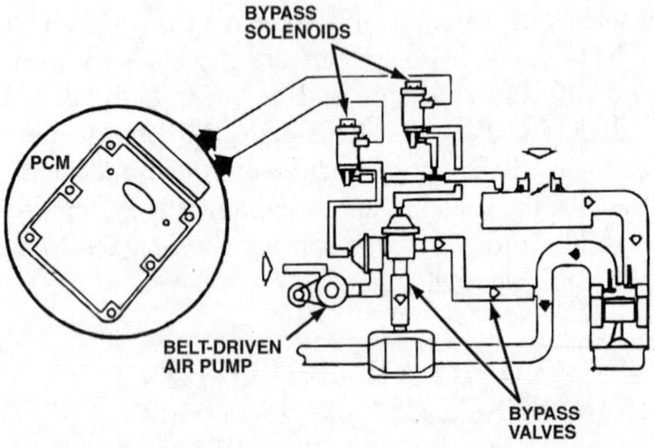

Figure 11-19. The PCM controls airflow in belt-driven air-injection systems.

Belt-Driven Air Pump System

The belt-driven air pump system, figure 11-19, consists of a belt-driven air pump, single or dual diverter valve(s), bypass valve, solenoids, PCM, and related wiring and vacuum hoses. Some systems contain an air filter/silencer. There are many different system configurations based on specific vehicle designs. In general, the PCM provides one or more signals that enable one or more bypass or diverter solenoids. The solenoids control one or more bypass or diverter valves to route secondary airflow. Some belt-driven air pumps operate whenever the engine is running. Some are controlled by the PCM through the use of a magnetic clutch, like the clutch used on the AC compressor.

Diagnosis and Testing

The following are basic diagnostic and testing procedures that apply to all makes and types of AIR systems. It is important to note that the AIR system is like any other computer-controlled system in that many concerns set diagnostic trouble codes (DTCs). Always refer to the manufacturer's service publications for specific pinpoint test instructions.

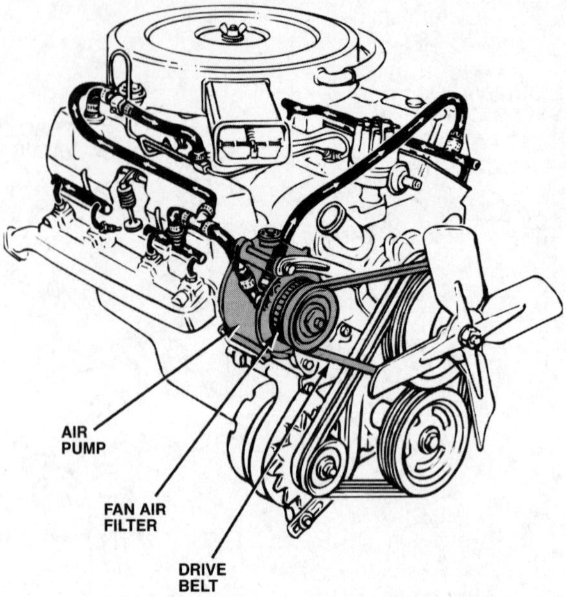

Figure 11-20. Many components require inspection on belt-driven air-injection systems.

General Inspection

The following is a general inspection procedure for all makes and types of AIR systems. Always perform a thorough general inspection before checking individual components.

1. On belt-driven systems, figure 11-20, check the air pump drive belt for wear, deterioration, or other damage. Make sure the pump turns freely if the belt is missing or broken. Replace the belt or adjust the tension as required. If the drive belt also operates the air conditioning compressor, set the belt tension to compressor tension requirements. Do not pry on the pump housing when adjusting belt tension.
2. Inspect all air system hoses, vacuum lines, check valves, and system tubing, figure 11-21. Inspect hoses for loose connections, cracking, brittleness, or burning. Inspect tubing to see if it is crimped, pinched, or corroded. Tighten loose connections and replace any defective components.
3. Inspect electrical connections and wiring for looseness and chafing. Repair or replace as required.

OBD II Air System Monitor Diagnostics—Electric Air Pump (EAP)

The OBD II AIR system monitors are a function of the PCM. One is called a dedicated or system monitor that tests AIR system operation. The system monitor is enabled during AIR system operation only after specific enable criteria (operating conditions) are met.

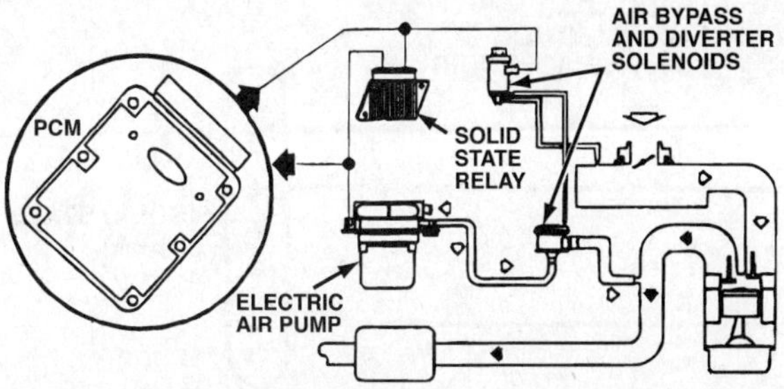

Figure 11-21. This PCM-controlled air-injection system uses a control valve to direct airflow.

A second monitor, called a comprehensive component monitor, constantly monitors two AIR system circuits, the solid-state relay primary circuit, and the solid-state relay secondary circuit, figure 11-22. Circuit voltage is monitored and compared with the PCM's programming during AIR system operation. Voltage values that are too high or too low for any given operating condition signify potential open or short concerns, and cause the PCM to set a diagnostic trouble code (DTC).

An example of an OBD II SAE standard DTC that can be set in conjunction with a solid-state relay circuit failure is P0412. The pinpoint test for this DTC, figure 11-23, involves using a DMM to check circuit voltage and resistance to determine if there is a circuit fault, solid-state relay fault, bypass solenoid fault, EPA fault, or PCM fault.

The OBD II AIR system monitor performs a functional check that tests the ability of the AIR system to inject fresh air into the exhaust. The functional check relies on feedback from the oxygen sensors to determine the presence of airflow. The functional check may be conducted in two parts: at startup when the EAP is normally commanded ON, or during hot idle if the startup test was not able to be performed. Failures detected during a functional test also cause the PCM to set a DTC. The malfunction indicator lamp (MIL) is activated if this monitor test fails on two consecutive drive cycles.

An example of an OBD II SAE standard DTC that can be set in conjunction with a failed functional check is P0411. DTC P0411 indicates secondary air was not detected during the functional test. Possible causes for this concern include:

- Blocked or leaking EAP hose
- Stuck, blocked, or leaking bypass solenoid
- EAP fault

OBD II Air System Monitor Diagnostics—Belt-driven Air Pump

The AIR monitor for belt-driven applications consists of two output state monitor configurations in the PCM: the bypass solenoid electrical circuit, and the diverter solenoid electrical circuit. OBD II SAE standard DTCs associated with a bypass solenoid circuit fault are P0413 and P0414—no change in voltage output when the solenoids are activated. Possible causes for this concern include:

- Open or shorted solenoid circuits
- Solenoid faults
- Faulty PCM

The pinpoint test for this concern, figure 11-24, involves testing circuit voltage and resistance values using a digital multimeter (DMM).

A functional check, or system monitor that tests the ability of the system to inject fresh air into the exhaust, is also performed. As with EAP applications, the functional check on belt-driven air pump systems relies on oxygen sensor feedback to determine airflow presence. The DTC and related pinpoint test associated with this concern are the same as for EAP applications—P0411, secondary air not detected during the functional test.

Basic Component Tests

The previous segments briefly outlined the characteristics of a few DTC-related tests. These tests check the integrity of electrical circuits as well as physical operating conditions of the air pump and the various solenoids. This segment outlines *basic* test procedures for air pumps, check valves, and solenoid valves. Remember that system designs vary between manufacturers. Always consult manufacturer publications for specific test instructions.

Belt-driven Air Pumps

Air pump output pressure can be measured with a pressure gauge and adapter as follows:

1. Use an adapter as shown in figure 11-25. The hole in the end of the pipe plug relieves pump output pressure during the test to prevent pump damage.
2. On magnetic clutch type systems, prepare the vehicle so that operating conditions allow clutch engagement or use a scanner to activate the clutch through bi-directional control (if applicable).
3. Disconnect the air outlet hose at the air management valve and connect the hose to the pressure gauge/adapter with a hose clamp.
4. Ensure that the hose and gauge are positioned away from moving parts such as the fan and drive belts.
5. Start and run the engine at a steady 1,000 rpm.
6. Watch the gauge reading. If output pressure is 1 psi (7 kPa) or more, the pump is satisfactory. If the pressure is erratic or below specification, the air pump should be replaced.

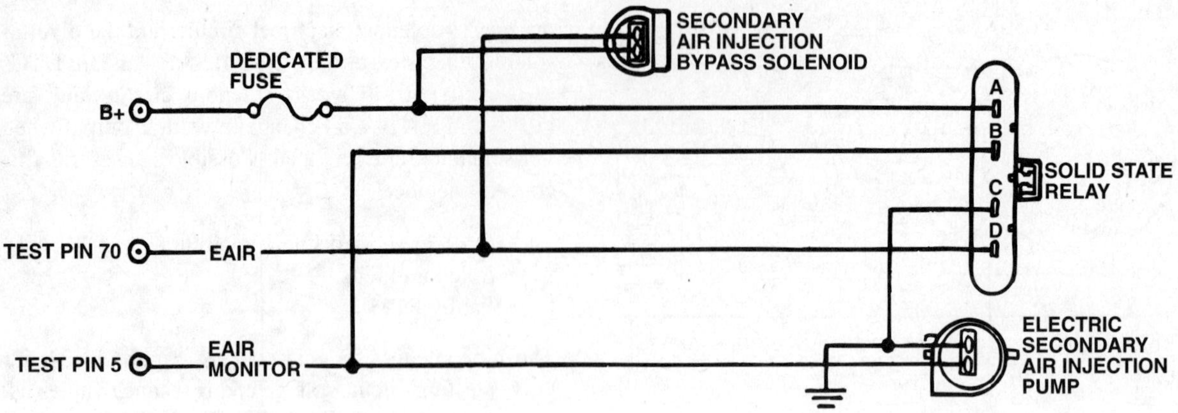

Figure 11-22. The PCM continuously monitors electronic AIR system circuits.

Test Step		Result	Action to Take
HM1	CHECK B+ VOLTAGE TO SOLID STATE RELAY		
	Diagnostic Trouble Code (DTC) P0412 indicates EAIR primary circuit fault. Possible causes: Fuel system — EAIR circuit open. — EAIR circuit short to power. — AIR bypass solenoid fault. — Solid state relay fault. — Damaged PCM. — Damaged electric AIR pump. • Disconnect SSR. • Key on, engine off. • Measure voltage of B+ circuit at SSR harness connector and battery negative post. • Key off. • **Was voltage greater than 10.5 volts?**	Yes ▶ No ▶	Supplied voltage is OK. GO to HM2. GO to HM6.
HM2	CHECK EAIR CIRCUIT FOR OPEN IN HARNESS		
	• Disconnect AIR bypass solenoid. • Remove Secondary Air Injection System dedicated fuse temporarily. • Install breakout box, leave PCM disconnected. • Measure resistance of EAIR circuit between PCM test pin 70 and SSR harness connector and AIR bypass harness connector. • **Is resistance greater than 5.0 ohms?**	Yes ▶ No ▶	GO to HM3. REPAIR open circuit. COMPLETE PCM Reset to clear DTCs. RESTORE vehicle. RERUN Quick Test.
HM3	CHECK EAIR CIRCUIT FOR SHORT TO POWER AND GROUND WITH DISCONNECT AIR BYPASS SOLENOID		
	• Measure resistance between PCM test pin 70 and PCM test pins 51, 71, 90, 97 and 103. • **Is each resistance greater than 10,000 ohms?**	Yes ▶ No ▶	The EAIR harness is OK. GO to HM4. REPAIR short circuit. COMPLETE PCM Reset to clear DTCs. RESTORE vehicle. RERUN Quick Test.
HM4	CHECK EAIR CIRCUIT FOR SHORT TO POWER AND GROUND		
	• Reconnect AIR bypass solenoid. • Measure resistance between PCM test pin 70 and PCM test pins 51, 71, 90, 97 and 103. • **Is each resistance greater than 10,000 ohms?**	Yes ▶ No ▶	The EAIR circuit with AIR bypass solenoid is OK. GO to HM5. REPLACE AIR bypass solenoid. COMPLETE PCM Reset to clear DTCs. RESTORE vehicle. RERUN Quick Test.

Figure 11-23. Factory test procedures are necessary when diagnosing trouble codes.

Test Step		Result ▶	Action to Take
HM73	CHECK SILENCER / FILTER FOR OBSTRUCTION		
	• Remove inlet hose (if equipped). • Inspect inlet of silencer / filter for blockage (bugs, leaves, debris). • **Is inlet open?**	Yes ▶ No ▶	REPLACE AIR pump and VERIFY a symptom no longer exists. RESTORE vehicle. REMOVE all debris and VERIFY a symptom no longer exists. RESTORE vehicle.
HM75	DIAGNOSTIC TROUBLE CODES (DTCS) P0413, P0414, P0416 AND P0417: CHECK VPWR CIRCUIT FOR OPEN HARNESS		
	DTCs P0413, P0414, P0416 and P0417 indicate that voltage output for Secondary Air Injection solenoid(s) did not change when activated. Possible causes: — AIRB/AIRD circuit(s) shorted to power. — AIRB/AIRD circuit(s) open or shorted to ground. — AIR bypass / AIR diverter solenoid(s) resistance out of range. — Damaged PCM. • Disconnect AIR bypass / AIR diverter solenoid connecter. • Key on, engine off. • Measure voltage of VPWR circuit between AIR bypass solenoid and battery negative post. • Repeat for AIRO diverter solenoid if equipped. • Key off. • **Was voltage greater than 10.5 volts?**	Yes ▶ No ▶	GO to HM76. REPAIR open circuit. RESTORE vehicle. RERUN Quick Test.
HM76	CHECK BOTH AIR BYPASS / AIR DIVERTER SOLENOID RESISTANCE		
	• Measure both AIR bypass / AIR diverter solenoid for resistances. • **Is each resistance between 50 and 100 ohms?**	Yes ▶ No ▶	GO to HM77. REPLACE solenoid assembly. RESTORE vehicle. RERUN Quick Test.
HM77	CHECK AIRB / AIRD CIRCUIT FOR OPEN IN HARNESS		
	• Install breakout box, leave PCM disconnected. • Measure resistance of AIRB circuit between AIRB circuit at breakout box and AIRB circuit at harness connector. • Measure resistance of AIRD circuit between AIRD circuit at the breakout box and AIRD circuit at harness connector. • **Is each resistance less than 5.0 ohms?**	Yes ▶ No ▶	GO to HM78. SERVICE open harness circuit. RESTORE vehicle. RERUN Quick Test.

Figure 11-24. Circuit testing involves visual inspection as well as measuring voltage and resistance values.

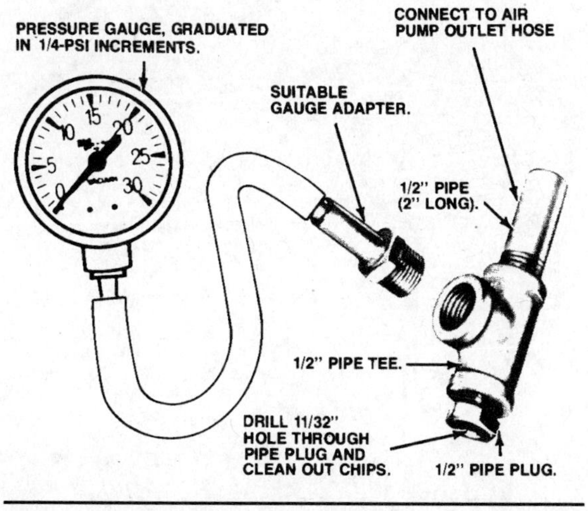

Figure 11-25. A pressure gauge and adapter are needed to test air pump pressure.

Electric Air Pumps

In terms of *air pumping,* there is no functional difference between electric and belt-driven air pumps. However, there is a major difference in pump control, which significantly impacts diagnosis and testing. As stated previously, belt-driven pumps operate any time the engine is running. As a result, the pump pressure test outlined in the preceding segment requires no special set-up—the test *stands alone.* In contrast, magnetic clutches and electric pumps do not operate anytime the engine is running. The electric pump runs only when turned ON by the system's electronic controls. Without proper consideration of the system's electronics, a misdiagnosis is likely.

Electric air pump operation is checked as follows:

1. Disconnect the air hose from the diverter valve.
2. Start the engine.

3. Carefully place a hand over the outlet of the hose.
4. After a 5-second delay, air pressure should be present for 30 to 90 seconds. There may be no air pressure specifications associated with this test. The pump is considered functional if air pressure is felt. Refer to the manufacturer's test procedures for specific instructions and pressure specifications—if provided.
5. The pump is working if airflow is present. However, the pump should be checked for water contamination before releasing the vehicle to the customer. Water contamination reduces the life of the pump. The check the pump for water, disconnect the air hoses and the electrical connector and tilt the pump to see if any water leaks out, figure 11-26. Replace the pump if it is contaminated with water. Use RTV sealer to seal the hose connections during reassembly.
6. If no air is present, it is necessary to ensure the pump-to-diverter air hose is not blocked or leaking.
7. If the hose is OK, the next step is to check the voltage and continuity of the various air pump, solid-state relay, and PCM electrical circuits, figure 11-22. The air pump is replaced *only after* all electrical circuits and components check OK. Refer to the manufacturer's pinpoint tests for specific test instructions.

In an actual diagnostic situation, electric air pump tests would be performed in conjunction with a pinpoint test. Other checks may be performed before a direct check of the pump itself. Remember, it is important to properly follow pinpoint test instructions to avoid misdiagnosis.

Check Valves

A check valve, figure 11-27, allows airflow in only one direction. A defective check valve can let hot exhaust gases enter the AIR system, which can damage hoses, other valves, and the air pump.

A simple test is to remove the check valve hose and carefully feel for exhaust escaping to the non-exhaust side of the valve. To bench test a check valve, use a hand-operated vacuum pump with an adapter on the valve. A check valve on its exhaust side should hold 15 to 20 inches (380 to 580 mm) Hg of vacuum for about 30 seconds if it is working properly. The valve is defective if air passes in both directions, or does not pass in either direction.

Air Control Valves

Early model vehicles used simple thermal vacuum switches to control air bypass and diverter functions. However, today's designs use computer-controlled solenoids to control air bypass and diverter functions. These solenoids are either attached directly to the

HM15 | CHECK AIR PUMP FOR WATER

NOTE: Water ingested in the electric AIR pump will reduce the life of the pump.

- **Disconnect electric AIR pump connector and air hoses.**
- **Carefully tilt electric AIR pump in various positions to verify if any water is present.**
- **Is any water present?**

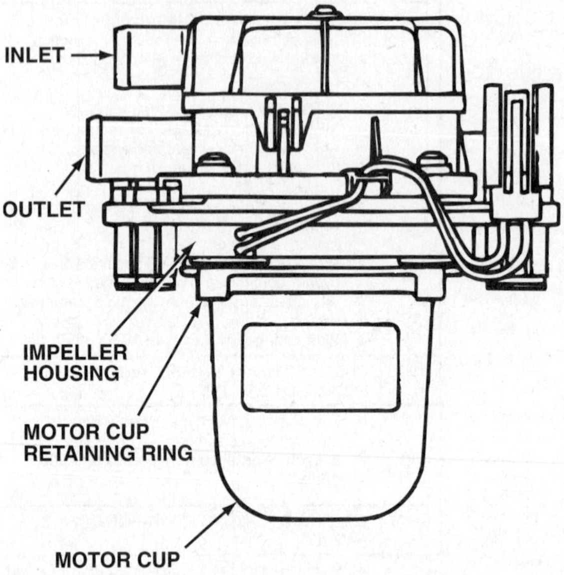

Figure 11-26. Checking for water contamination is an important part of electric air pump inspection.

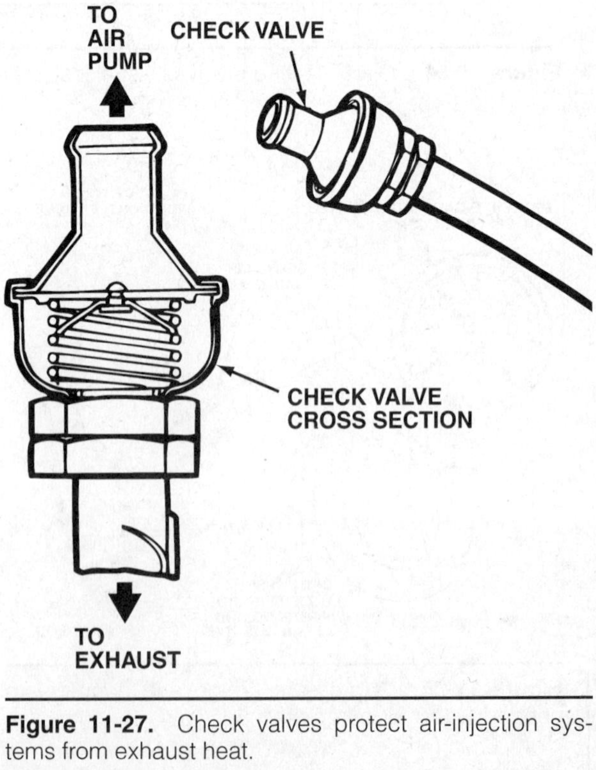

Figure 11-27. Check valves protect air-injection systems from exhaust heat.

valve, or they may be placed in the vacuum lines to the various diaphragms. A single solenoid/valve or a combination of solenoids/valves may be used to redirect fresh air as follows:

- To the exhaust ports or manifold when the engine is first started to help bring the oxygen sensor and catalytic converter to operating temperature, and to help control HC and CO emissions.
- Downstream to the second bed of a dual bed converter after engine warmup to help with the oxidation process or to the atmosphere on a system equipped with a 3-way catalyst.
- Away from the exhaust during deceleration to prevent backfiring and protect the converter from an overheat condition.
- Away from the exhaust during wide-open throttle or heavy load conditions to protect the converter from overheating.
- Away from the exhaust if a control system defect is detected by the PCM.

The following are basic procedures that apply to most solenoid air control valve tests. Be sure to refer to the manufacturer's service information when testing components for specific instructions and specifications.

- Test solenoids for continuity and resistance with an ohmmeter, figure 11-28. Note that computer-operated solenoids have a minimum and a maximum coil resistance.
- Use a voltmeter of 12-volt test lamp to check output voltage from the computer to the air control solenoids.

Air-injection System Service

This segment covers basic procedures and techniques associated with the service of air-injection system components. Remember to always refer to the manufacturer's service information for specific instructions and specifications.

In general, air-injection components do not require periodic maintenance, and cannot be serviced. Defective parts are replaced. However, system longevity and performance can be maintained by periodic inspection of the system's belt, hoses, check valves, tubing, and wiring.

Hose and Tubing Service

If system hoses need replacement, make sure the new hose is suitable for air-injection system use. Other types of hose will deteriorate rapidly because of excessive heat. Install new clamps and tighten the connections securely.

Air-injection tubing and air nozzle manifolds are typically made of steel, and are usually replaced with preformed parts. To prevent seizing, use an an-

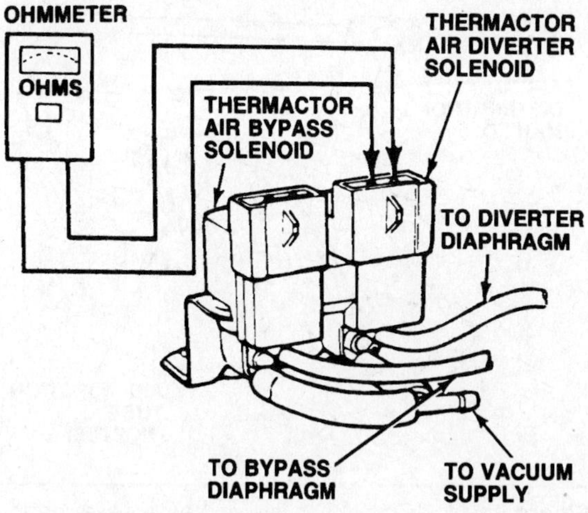

Figure 11-28. Use an ohmmeter to test solenoids for continuity and resistance.

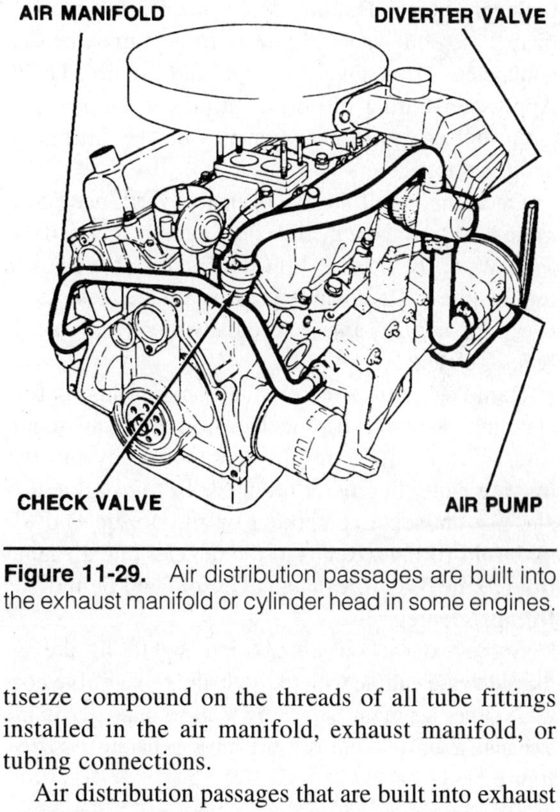

Figure 11-29. Air distribution passages are built into the exhaust manifold or cylinder head in some engines.

tiseize compound on the threads of all tube fittings installed in the air manifold, exhaust manifold, or tubing connections.

Air distribution passages that are built into exhaust manifolds or cylinder heads, figure 11-29, usually require no periodic inspection or service beyond the check valves. The internal air distribution passages can be inspected for carbon buildup and cleaned whenever a cylinder head or exhaust manifold is removed. The carbon can be softened with solvent or carburetor cleaner and removed by running a stiff, small-diameter wire brush through the passages.

External air-injection manifolds and injection tubes or nozzles can be removed from the engine for

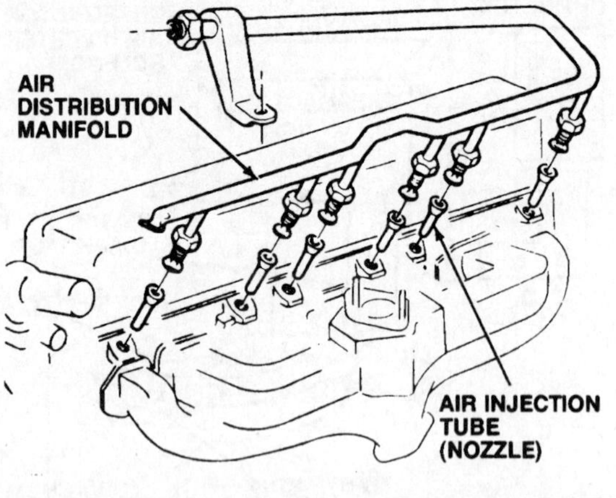

Figure 11-30. This AIR system uses an external air distribution manifold with injection nozzles.

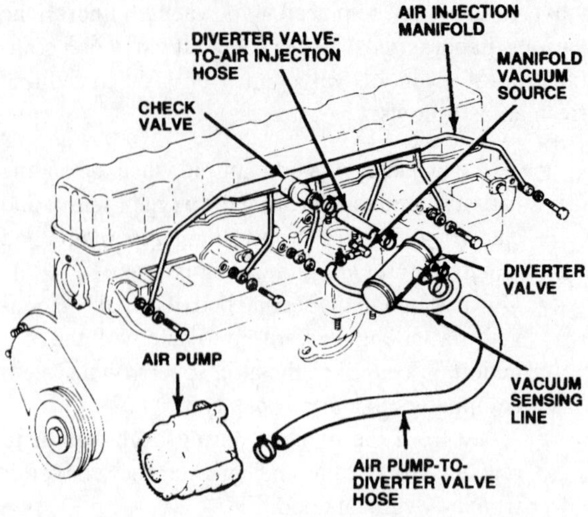

Figure 11-31. This air-injection manifold uses banjo-type fittings and hollow bolts to deliver pump air to the exhaust manifold.

replacement or cleaning. Many manifolds are one-piece assemblies and all nozzle fittings must be disconnected to remove the manifold, figure 11-30. Apply penetrating oil to the fittings to help loosen them and use flare nut wrenches on the fittings to avoid damage.

Once the manifold has been removed, use penetrating oil to loosen the nozzles in the head or manifold. Work the nozzle from the head with a hooked tool or pliers. Removable nozzles are stainless steel. Clean the nozzles and manifolds with solvent or carburetor cleaner.

Remove check valves from manifolds before cleaning. Remove carbon from nozzles and manifolds with a stiff wire brush. Replace any nozzles that are badly burned or plugged. To reinstall a nozzle, use antiseize compound on the nozzle and the manifold fittings. Reinstall the nozzles and air manifold to the head or exhaust manifold and tighten the fittings securely.

Some external air-distribution manifolds are attached to the engine with individually replaceable nozzles. Other air manifolds deliver air to passages in the exhaust manifold but do not have separate nozzles, figure 11-31.

Air Pump Service

Removing a belt-driven air pump is similar to removing an alternator. It may be necessary to remove other accessory units and their drive belts to gain access to the air pump and its belt. If the pump is mounted low on the engine, it may be easier to remove the pump from underneath the vehicle. On some models, it may be necessary to remove the drive pulley from the air pump shaft in order to get enough clearance to remove the pump. Note that some truck applications have dual air pumps, figure 11-32.

Electric air pump removal and installation is typically simpler than removing and installing belt-driven pumps because there is no crankshaft-driven belt to contend with, figure 11-33. In general, removing an electric air pump involves disconnecting the attaching hoses, the electrical connector, and the pump housing fasteners. Depending on the application, other components may need to be moved to gain access. Refer to the manufacturer's service information for specific procedures.

Outlet Tube Replacement

Some pumps have an outlet tube pressed into the rear of the pump body. The outlet tube on these applications can be removed from the pump by grasping it in a vise and pulling it out with a twisting motion. Do not clamp the pump body in the vise because the aluminum casting may crack.

Install the outlet tube by tapping it in place with a wooden block and hammer. Support the pump when tapping the tube in place.

EXHAUST GAS RECIRCULATION (EGR) SYSTEM

Recirculating exhaust gases dilute the air-fuel mixture and limit the formation of NO_x emissions. NO_x emission control is especially important during high combustion temperature and lean air-fuel mixture conditions such as cruising or moderate acceleration. EGR is not required at idle speed, low operating temperatures, or very rich mixture conditions.

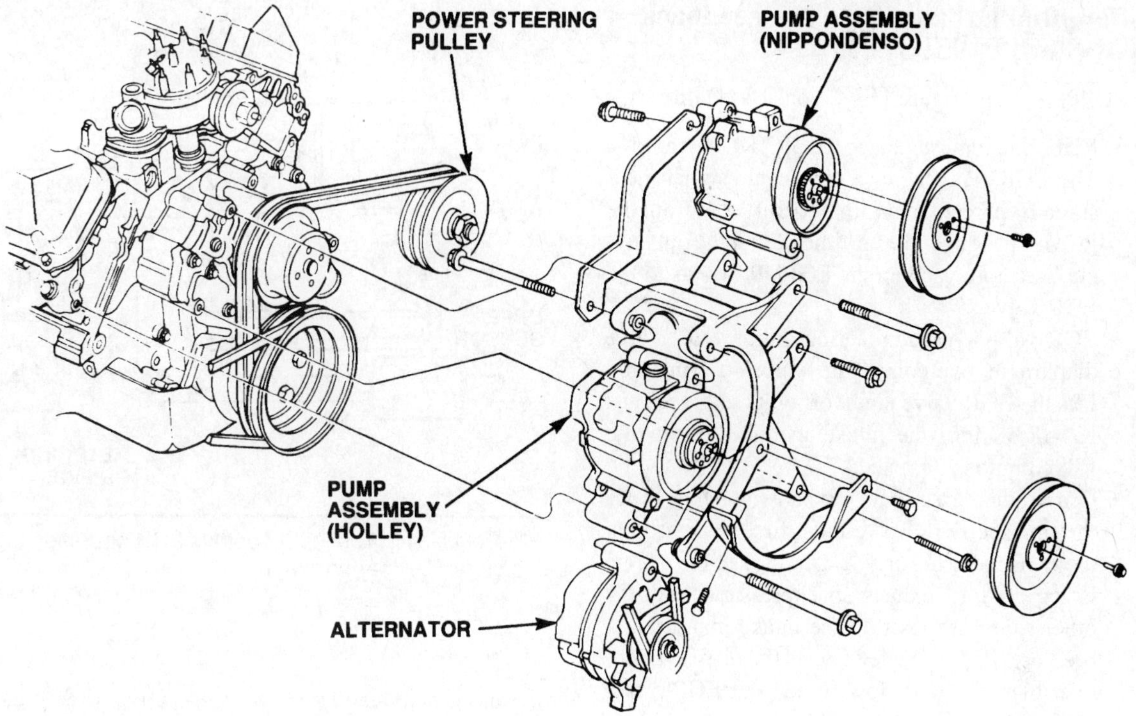

Figure 11-32. Some truck applications have dual air pumps.

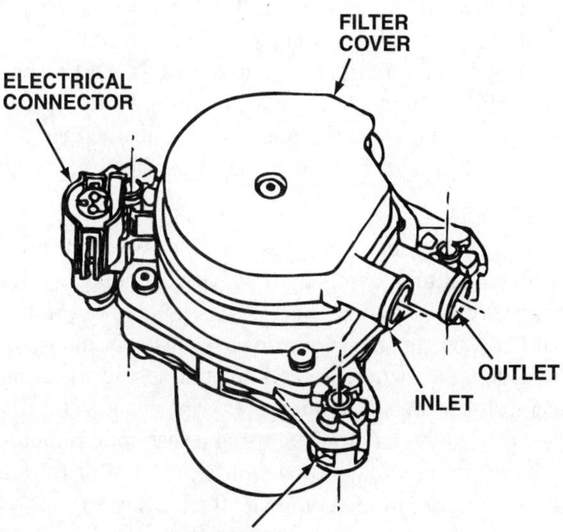

Figure 11-33. To remove an electric air pump, disconnect electrical connectors, hoses, and remove mounting bolts.

An improperly operating EGR system can cause several driveability complaints such as:

- Rough idle
- Stalling
- Stumbling during acceleration
- Detonation or pinging during acceleration (spark knock)
- Surging at cruising or high speeds
- Lack of power
- Poor fuel economy

This segment briefly discusses basic EGR operating strategies and the three main types of EGR systems commonly used by today's domestic manufacturers. Due to the variations in system types and designs, be sure to use the manufacturer's diagnostic and service resources for specific test and service procedures.

Basic Operating Strategy

Like almost every other engine system on today's vehicles, the EGR system is computer controlled. While there are several different EGR types and designs, the operating strategies are essentially the same. The PCM receives input signals that represent engine coolant temperature, intake air temperature, throttle position, engine load, engine speed, and vehicle speed to determine if operating conditions are appropriate for EGR flow. For the PCM to initiate EGR flow, the engine must be warm, above idle speed, lightly loaded, and either under light acceleration or at a steady speed. The PCM deactivates the EGR during cold engine operation, idle, wide-open throttle, or whenever an electronic control system malfunction is detected.

Differential Exhaust Pressure Feedback EGR (Ford DPFE System)

The DPFE system, figure 11-34, consists of a(n):

- EGR vacuum regulator (EGRVR) solenoid—The EGRVR receives a duty cycle signal measured by percent of on-time (0 to 100%) from the PCM. The longer the on-time during the duty cycle, the more vacuum the EGRVR diverts to the EGR valve.
- EGR valve—vacuum acting on the EGR valve diaphragm overcomes the valve's spring and lifts the EGR valve pintle off of its seat. Exhaust gas flows into the intake manifold when the valve pintle is unseated.
- Orifice tube assembly—Exhaust gas that flows through the EGR valve must first pass through the EGR metering orifice. One side of the orifice is exposed to exhaust backpressure, and the other side is exposed to the intake manifold. A pressure drop is created across the orifice whenever there is EGR flow. When the EGR valve closes, there is no longer flow across the metering orifice and pressure on both sides of the orifice is the same.
- DPFE sensor—The DPFE sensor measures the pressure drop across the metering orifice and relays a proportional voltage signal (0.0 to 5.0 volts) to the PCM. The PCM uses this signal as a report of EGR flow.

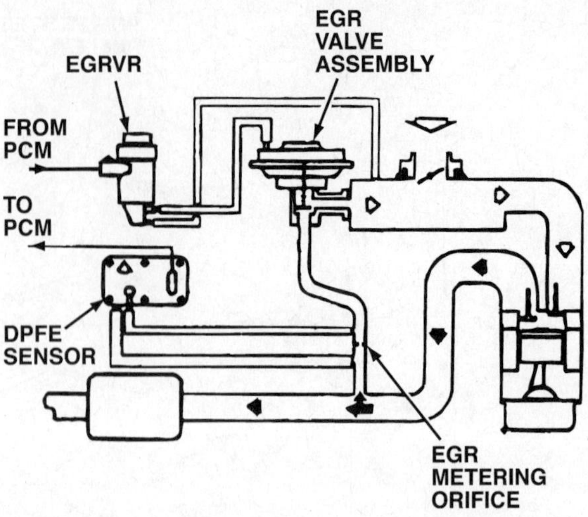

Figure 11-34. The PCM controls EGR operation and monitors EGR performance.

OBD II System Monitor

The DPFE system monitor is a PCM function that tests the integrity and flow characteristics of the EGR system. The monitor is activated during EGR operation after certain base engine conditions are satisfied. Once activated, the monitor performs tests during various engine operating modes as follows:

- The DPFE sensor and circuit are continuously tested for opens and shorts.
- The monitor tests for a stuck-open EGR valve or EGR valve flow occurring at idle.
- The DPFE sensor upstream hose is tested for a disconnected or plugging condition once per drive cycle.
- An EGR flow rate test is performed when engine speed and load are moderate and the EGRVR vacuum duty cycle is high.

The EGRVR circuit is continuously tested during vehicle operation for opens and shorts by the comprehensive component monitor.

The malfunction indicator lamp (MIL) is activated immediately by the comprehensive component monitor and is activated after two consecutive drive cycles by the system monitor.

Basic Testing

As with other computer-controlled systems, most EGR problems will set diagnostic trouble codes (DTCs). Following are basic examples of how to check the EGR valve, the EGRVR solenoid, and the DPFE sensor. The manufacturer's diagnostic and service resources contain specific pinpoint test instructions for each of the major components of the system.

EGR Valve

There are various DTCs that can be set in connection with a malfunctioning EGR valve depending on the nature of the malfunction. As was pointed out in the earlier segment on air-injection systems, it is important to follow the manufacturer's pinpoint test instructions carefully to avoid misdiagnosis. The manufacturer's test procedure may involve using a scan tool to monitor related input/output data during the test or for bi-directional control to command the EGR open.

Assuming for the purpose of this discussion that other tests have been performed, a typical DPFE system EGR valve test, figure 11-35, is performed as follows:

- Disconnect the vacuum hose at the EGR valve and plug the hose.
- Connect a hand vacuum pump to the EGR valve.
- Start the engine and allow to idle.
- Slowly apply 8 to 10 in-Hg (27 to 34 kPa) vacuum to the valve.
- If rpm changes and/or if the engine starts to stall, the valve is opening and allowing EGR flow. If

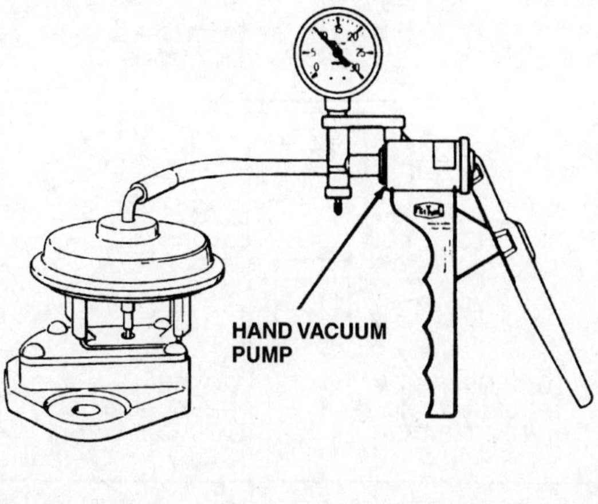

Figure 11-35. Test the EGR valve with a hand-operated vacuum pump.

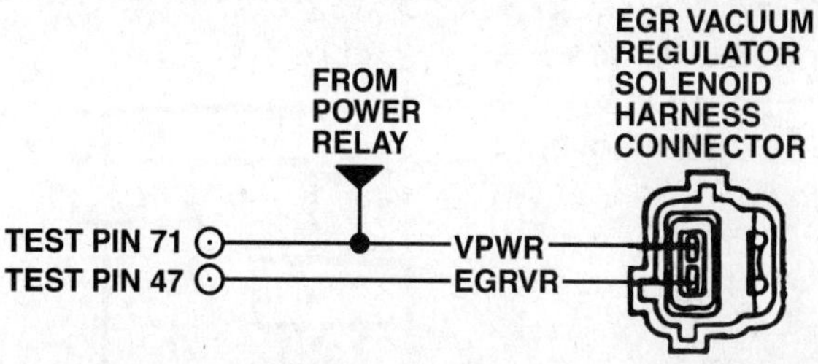

Figure 11-36. Test the EGRVR solenoid using an electrical schematic, service information, and a digital multimeter.

there is no change in the way the engine runs, the valve is not functioning or the EGR ports are clogged.

EGRVR Solenoid

Many tests can be performed on the EGRVR solenoid depending on the type of problem and what DTCs, if any, are set. Basic electrical checks, figure 11-36, include measuring the resistance across the terminals of the solenoid, checking for opens or shorts in the EGRVR, and checking for continuity in the circuit between the EGRVR and the PCM. Resistance specifications and test instructions are contained in the manufacturers' service manuals.

Other tests involve checking the solenoid vacuum valve and hoses for blockage or sticking. A typical test to see if the valve is sticking, figure 11-37, is to:

- Disconnect the vacuum hose at the EGR valve and connect a vacuum gauge to the hose.
- Start the engine and allow to idle.
- Disconnect the EGRVR solenoid harness connector from the EFR solenoid while monitoring the vacuum gauge.
- The EGR valve in this type of system typically requires 1.6 in-Hg (5.4 kPa) of vacuum to open. If the vacuum reading is greater than 1.6 in-Hg (5.4 kPa) after the solenoid is electrically disconnected, a mechanical EGRVR fault is possible.

A typical test to check for a blocked or restricted EGRVR is as follows, figure 11-38:

- Disconnect the vacuum hoses from the EGRVR.
- Plug the vacuum supply port.
- Use a hand vacuum pump to apply 10 to 15 in-Hg (34 to 51 kPa) of vacuum to the EGRVR vacuum source port.

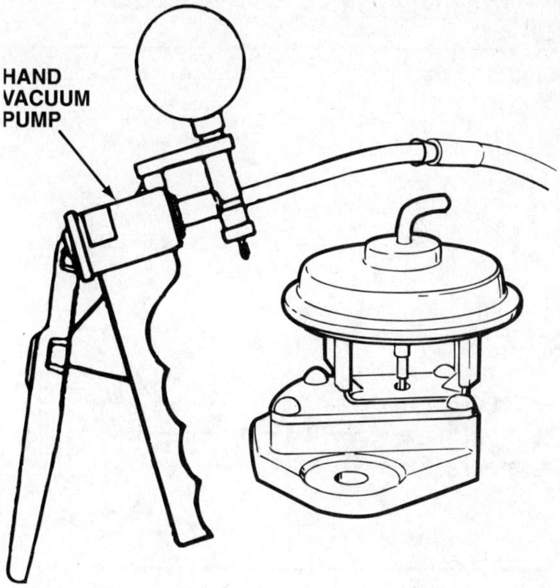

Figure 11-37. An EGRVR mechanical check means testing the vacuum signal to the EGR valve.

- The solenoid vent may be plugged if vacuum holds or is slow to vent.

DPFE Sensor

Many tests can be performed with the DPFE sensor depending on the type of problem and what DTCs, if any, are set. Types of DPFE sensor tests include:

- Checking the DPFE signal to the PCM. This can be accomplished using a DMM to measure the voltage on the circuit itself, or in some cases by monitoring the DPFE signal through the vehicle's data link connector (DLC) using a scan tool.
- Checking the integrity of the DPFE sensor's voltage reference (VREF) and ground circuits, figure 11–39.
- Checking for shorts, opens, or high resistance in the circuits between the DPFE and the PCM.

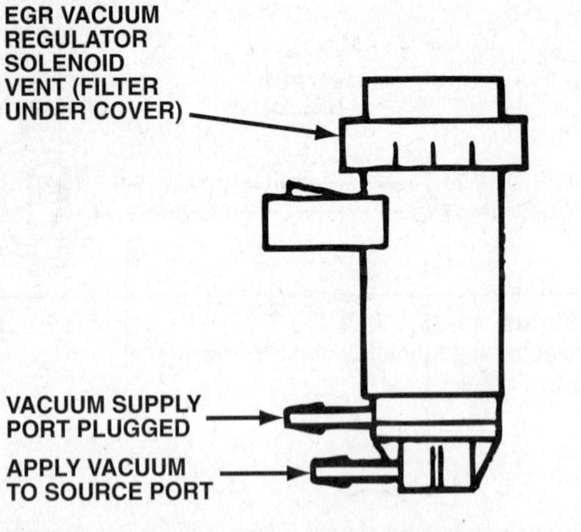

Figure 11-38. An EGRVR blockage check tests for a clogged filter under the vent cover.

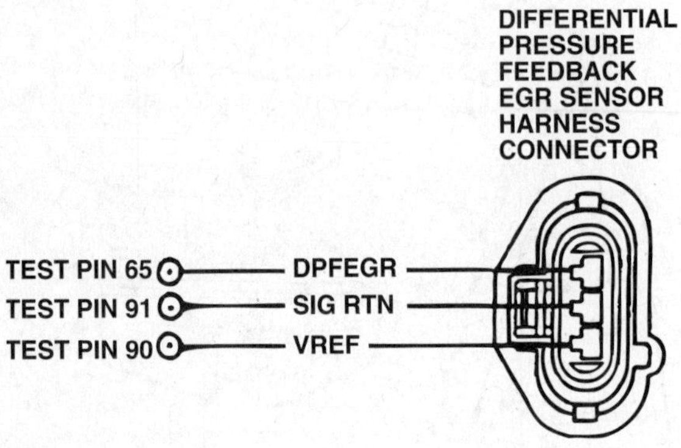

Figure 11-39. The DPFE sensor has reference voltage, ground return, and sensor signal wires like many other sensors.

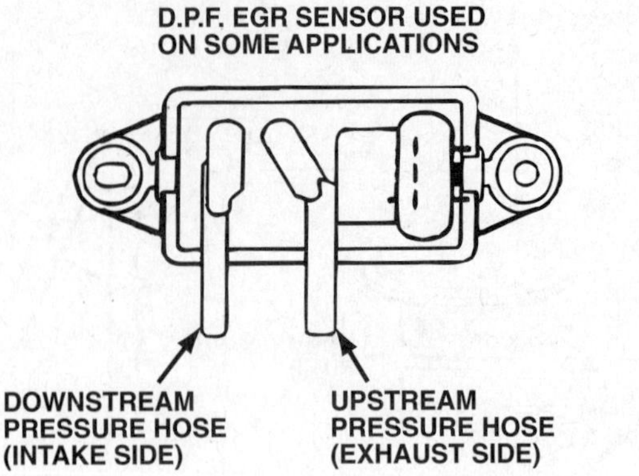

Figure 11-40. Test the DPF EGR sensor for leaks and blockage on both hose connections.

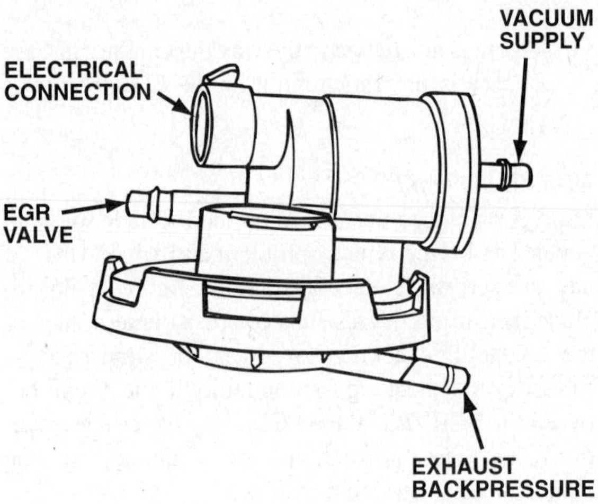

Figure 11-41. Vacuum and electrical checks are required to test an electronic EGR transducer.

- Checking for blockage or leaks in the downstream and upstream exhaust ports and passages, figure 11-40.

Backpressure Transducer System (Chrysler)

A typical backpressure transducer EGR system contains the following:

- EGR tube
- EGR valve
- Electronic EGR transducer
- Connecting hoses

The electronic EGR transducer, figure 11-41, contains an electrically operated solenoid and a backpressure transducer. Vacuum flows to the transducer when the PCM de-energizes (turns OFF) the solenoid. Exhaust system backpressure, when high enough, modulates a bleed valve in the transducer. With the solenoid de-energized (OFF) and the bleed valve closed, vacuum flows through the transducer to operate the EGR valve. De-energizing the solenoid, but not fully closing the transducer bleed hole (because of low exhaust backpressure), varies the vacuum applied to the EGR valve. Varying vacuum changes the amount of EGR flow.

OBD II System Monitor

The monitor activates when certain enable criteria are met (such as engine speed, temperature, rpm, and load). When the criteria are met, the EGR is turned OFF (solenoid energized) and the fuel trim is monitored through the oxygen sensor (O_2S) by the PCM. Turning the EGR off shifts the O_2S signal leaner. This lean shift should be indicated by the O_2S. Because the O_2S is used, the O_2S must pass its test before the EGR test is conducted. If the shift does not occur, the PCM will set a DTC, and after two consecutive trips it will illuminate the MIL.

Basic Testing

There are four main tests for this type of EGR system:

- EGR flow test
- EGR valve leakage test
- EGR valve control test
- Vacuum transducer test

EGR Flow Test

The following EGR flow test identifies if exhaust gas is flowing through the EGR valve. The test determines if the:

- EGR valve is operating
- EGR tube is plugged
- System passages in the intake or exhaust manifolds are plugged

This is NOT to be used as a complete EGR system test and is not accurate on integral backpressure transducer EGR valves because without backpressure, their vacuum motor diaphragms are vented to the atmosphere.

The engine must be running at normal operating temperature when performing the following:

1. Disconnect the hose at the vacuum motor fitting, figure 11-42, on top of the EGR valve vacuum motor.
2. Connect a hand-held vacuum pump to the vacuum motor fitting.
3. Slowly apply 5 inches of vacuum to the fitting.
4. While applying vacuum, engine speed should drop or the engine may stall. This indicates exhaust gas is flowing. If engine speed does not change, the valve may be defective, the EGR tube may be plugged, or the passages in the intake or exhaust manifolds may be plugged. If this is the case, continue as follows:
   - Remove the EGR valve from the engine.
   - Apply vacuum to the vacuum motor fitting and observe the stem on the EGR valve. If the stem is moving and not clogged with carbon, the EGR valve is operational. The problem is carbon clogging the EGR tube or passages.
   - Start the engine with the EGR valve removed from the engine. Check for both vacuum and exhaust from the EGR ports, tube, or passages.

EGR Valve Leakage Test

This test should be performed if the engine stalls or runs rough at idle. The test checks if the EGR valve is leaking, allowing exhaust gas flow at inappropriate times (such as at idle). The engine is OFF for the following procedure.

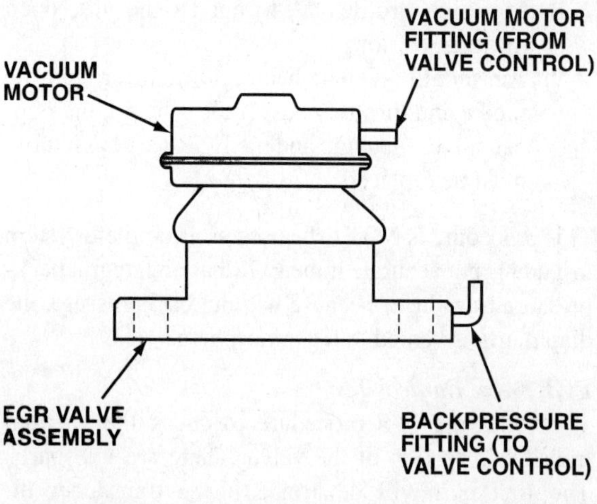

Figure 11-42. This EGR valve is tested on the car with a hand-operated vacuum pump.

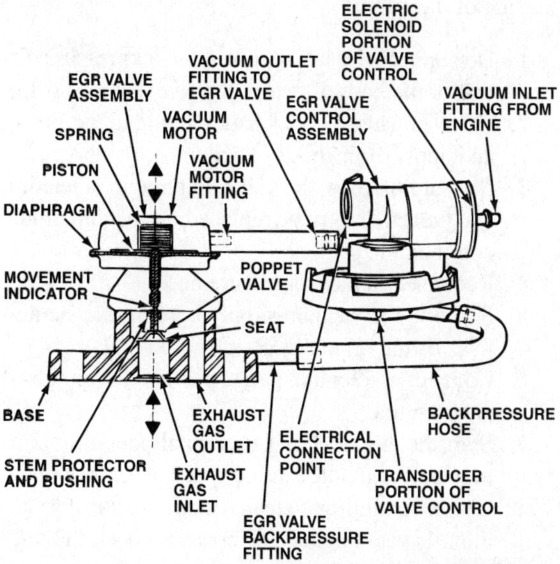

Figure 11-43. Apply compressed air to the backpressure fitting to check for poppet valve leakage.

1. Disconnect the hose from the vacuum motor fitting at the top of the EGR valve, figure 11-43.
2. Connect a hand-held vacuum pump to the fitting.
3. Apply 15 inches of vacuum.
4. Observe the vacuum gauge. If vacuum drops, the EGR valve diaphragm is ruptured, and the valve must be replaced. Continue with the procedure IF vacuum holds.
5. Disconnect the hose from the backpressure fitting, figure 11-43, on the EGR valve base.
6. Remove the air cleaner ductwork to expose the throttle plate.
7. Using a rubber-tipped air nozzle, apply 50 psi of compressed shop air to the backpressure fitting.

8. Move the throttle plate by hand to the wide-open throttle position.
9. Air should NOT be heard emitting through the intake and throttle body. If air is heard, the poppet valve is leaking and the EGR valve assembly must be replaced.

This procedure is NOT to be used as a complete system test and is not accurate if the EGR has an integral backpressure transducer because without backpressure, the diaphragm is vented to the atmosphere.

EGR Valve Control Test

The following is a procedure to check the vacuum transducer portion of the valve. There are two parts. The first part will determine if the transducer diaphragm at the backpressure side of the valve is leaking. The second part will determine if full intake manifold vacuum is flowing from the inlet to the outlet side of the valve.

1. Disconnect the hose from the backpressure fitting at the base of the EGR valve, figure 11-43.
2. Connect a hand-held vacuum pump to the fitting and apply 10 inches of vacuum.
3. If vacuum drops, the valve diaphragm is leaking and the valve must be replaced. If vacuum holds, continue the test.
4. Reconnect the backpressure hose.
5. Remove the vacuum supply hose at the vacuum inlet fitting on the EGR solenoid.
6. Connect a vacuum gauge to the disconnected vacuum line.
7. Start the engine, bring to normal operating temperature, and hold engine speed at 1,500 rpm.
8. Check for full and steady intake manifold vacuum. If vacuum is NOT present, check the vacuum line to the engine and repair as necessary before proceeding to the next step.
9. Reconnect the hose to the vacuum inlet fitting, disconnect the hose at the vacuum outlet fitting, and connect a vacuum gauge to the fitting.
10. Disconnect the electrical connector at the valve control to simulate an open circuit at the valve, which will activate the valve. This will set a DTC that will need to be erased after completing the test.
11. Run the engine at 1,500 rpm while checking for full manifold vacuum. In order for full vacuum to flow through the valve, exhaust backpressure must be present at the valve. It must also be high enough to hold the bleed valve in the transducer portion of the valve closed. Have an assistant briefly restrict the tailpipe to create backpressure. As backpressure builds, full vacuum should be observed on the gauge. If full vacuum is NOT present at the outlet fitting, but was present at the inlet fitting, replace the valve assembly.

This test is NOT to be used as a complete system test. Electrical operation of the valve should be checked using a scan tool. Refer to the manufacturer's service resources for specific instructions.

Linear EGR System (GM)

A linear EGR valve supplies recirculated exhaust gas to the engine independent of intake manifold vacuum. EGR flow is controlled through an orifice with a PCM-controlled pintle. The PCM controls the pintle position by monitoring the pintle-position feedback signal from the EGR position sensor. A linear EGR valve is typically activated when the engine is at normal operating temperature and at speeds above idle. The PCM uses information from the engine coolant temperature (ECT) sensor, throttle-position (TP) sensor, and mass air flow (MAF) sensor to control the EGR system.

OBD II System Monitor

The PCM tests the EGR system during deceleration by momentarily commanding the EGR valve to open while monitoring the manifold absolute pressure (MAP) signal. When the EGR valve is opened, the PCM should detect a proportional increase in the MAP signal. The PCM sets a DTC if the expected MAP signal change is not detected. Typically, there is only one EGR monitor test per ignition cycle. See figure 11-44 for a system electrical schematic.

Basic Testing

It is always best to follow the manufacturer's diagnostic procedure and pinpoint test based on a specific symptom or DTC. Study figure 11-45, a typical diagnostic procedure for insufficient EGR flow with an accompanying P0401 OBD II DTC.

EGR System Service

Removing and installing EGR system components is not difficult. Most are simply held in place by one or more fasteners. When replacing components:

1. Always use a new gasket when installing an EGR valve.
2. Tighten all mounting fasteners to specifications to prevent vacuum or exhaust leaks.
3. Make sure all EGR vacuum line connections are properly routed and tight. When replacing parts that have several vacuum line connections, it is a good idea to disconnect one line at a time and immediately reconnect it to the corresponding

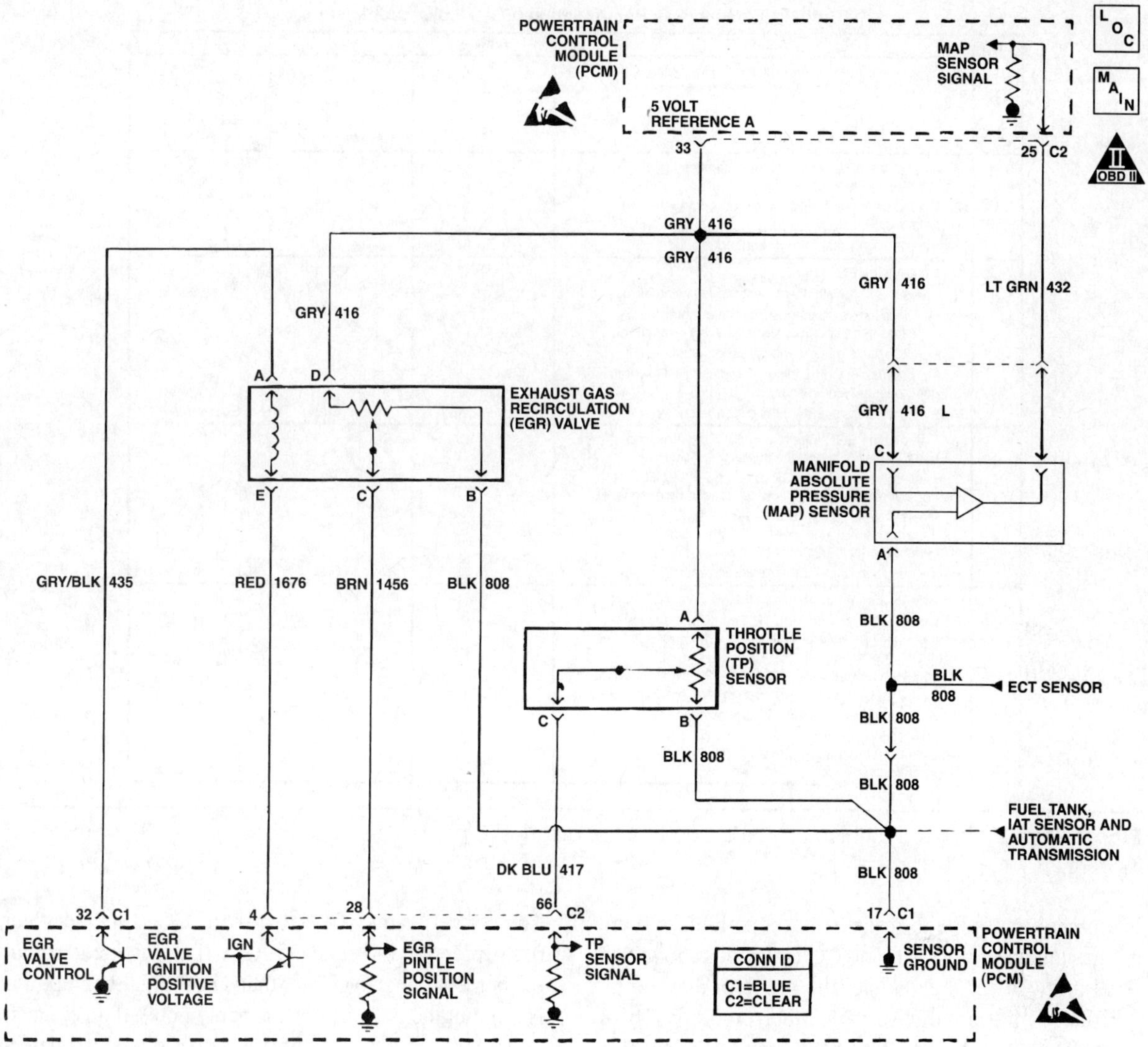

Figure 11-44. The GM Linear EGR system uses information from several sensors. (Courtesy of General Motors Corporation)

port on the new component or identify each by using a tag.

CATALYTIC CONVERTER SYSTEMS

Federal law requires that catalytic converters work efficiently to reduce emissions for at least 50,000 miles (80,000 km) with a properly maintained engine. Catalysts on vehicle models models 1995 and newer are warranted for 8 years or 80,000 miles. However, catalysts eventually suffer damage and the converter must be replaced. Converter failure is typically due to one of the following causes:

- Using fuels with lead, or too much sulfurous or phosphorus content
- Converter overheating from engine misfire or excessively rich air-fuel ratios
- Road hazards causing physical damage to the substrate

Catalyst Testing

Determining if the catalytic converter is operating efficiently is for the most part a matter of how new the vehicle is. Vehicles built prior to the era of OBD II must be tested manually; OBD II vehicles have a built-in catalyst efficiency monitor that checks catalyst efficiency.

4-Gas Analysis

The manual method of testing catalyst efficiency involves using a 3- or 4-gas analyzer. Chapter 2 of this *Shop Manual* discusses the use of this instrument. While a 2-gas analyzer will check HC and CO, the catalytic converter reduces these emissions to a level that

DTC P0401 Exhaust Gas Recirculation (EGR) Flow Insufficient

Step	Action	Value(s)	Yes	No
1	Was the Powertrain on Board Diagnostic System Check performed?	—	*Go to Step 2*	Go to *A Powertrain On Board Diagnostic (OBD) System Check*
2	1. Inspect the exhaust system for modification of original installed parts or leaks. 2. If a problem was found, repair exhaust system as necessary. Was a condition present that required repair?	—	*Go to Step 5*	*Go to Step 3*
3	1. Remove the EGR valve. 2. Visually and physically inspect the following items: • Pintle, valve passages and the adapter for excessive deposits or any kind of a restriction. • EGR valve gasket and pipes for leaks. 3. If a problem is found, clean or replace EGR system components as necessary. Was a condition present that required repair?	—	*Go to Step 5*	*Go to Step 4*
4	1. Remove the EGR inlet and outlet pipes from the exhaust manifold and the intake manifold. 2. Inspect the manifold EGR ports and the EGR inlet and outlet pipes for a blockage caused by excessive deposits, casting flashing or other damage. 3. If a problem is found, correct the condition as necessary. Was a condition present that required repair?	—	*Go to Step 5*	Go to Diagnostic Aids
5	1. Review and record the scan tool Fail Records data. 2. Clear DTC and monitor EGR Test Count display on the scan tool while operating the vehicle as specified in Diagnostic Aids. 3. Continue operating the vehicle EGR Test Count until 9 – 12 test samples have been taken. 4. Select scan tool Specific DTC information for DTC P0401. Note does test result indicate DTC P0401 Test Ran and Passed?	—	System OK	*Go to Step 2*

Figure 11-45. Follow the manufacturer's step-by-step diagnostic procedure when investigating EGR problems. (Courtesy of General Motors Corporation)

is almost impossible to measure at the tailpipe. When a 2-gas analyzer reads HC and CO emissions above the level allowed for the vehicle, the problem may be in the fuel, ignition, converter, or other emissions systems. Further testing will be necessary to isolate the cause of the high emissions and determine if the converter is at fault.

The 4-gas analyzer provides CO_2 and O_2 readings that can help indicate converter condition. At the ideal air-fuel ratio of 14.7 to 1, the CO_2 reading is high and the O_2 reading is low. In addition, the combined sum of CO and CO_2 is equal to approximately 14.7 percent at this ratio. As the air-fuel ratio leans out, O_2 increases and the combined sum of CO and CO_2 percentages decreases. With a richer mixture, O_2 decreases and the sum of CO and CO_2 percentages increases. The relationship is shown in figure 11-46.

The following are typical 4-gas readings with satisfactory combustion:

- CO_2—More than 10 percent (13–15 percent desired)
- O_2—1.0 to 2.0 percent
- HC and CO—Within specifications (may be low or even unmeasurable)

The converter may be defective if CO is greater than 0.5 percent and O_2 exceeds CO. If CO is greater than 0.5 percent but also greater than O_2, the air-fuel mixture is rich but the converter is probably good. If an exhaust system inspection or an infrared exhaust analysis determines that the converter is defective, it should be replaced. There are other methods of converter testing and to date, the EPA has not accepted any one method as reliable. It is therefore recommended that at least two methods be used prior to condemning a converter.

Testing a Converter by Temperature

One method of verifying proper catalytic converter operation is by measuring temperature. A converter is a catalyst that causes a chemical reaction to occur, which in the process generates heat. The amount of heat generated within the converter is directly related to how well the converter is working. As this chemical reaction occurs within the converter, the exhaust stream should pick up heat from the reaction. One method of determining whether a converter is working is by measuring the inlet and outlet temperature of the converter. A properly working converter should show about a 10 percent increase in temperature between the inlet and outlet of the converter. If the converter is not working at all then the inlet temperature will be greater than the outlet temperature. To test a converter by temperature readings, do the following:

1. Start the engine and operate at 2,500 rpm for at least two minutes. Make sure that closed loop operation has been achieved and that the coolant temperature is at operating temperature, figure 11-47.
2. Use a non-contact infrared pyrometer to measure both the inlet and outlet temperature of the converter while operating at 2,500 rpm, figure 11-48.
3. If the outlet temperature is 50 to 100°F higher than the inlet temperature, then the converter is working correctly. The outlet temperature should be about 10 percent higher than the inlet temperature.

NOTE: Some engines operate so efficiently that there may not be enough emissions in the exhaust stream to show an increase in converter temperature. If this is the case, it may be necessary to disable a cylinder to create a misfire. Do not operate the vehicle longer than 10 seconds with a cylinder disabled as the converter may overheat.

COMBINED CO AND CO_2 PERCENTAGE		APPROXIMATE AIR-FUEL RATIO (TO 1)
17.0		11.7
16.5		12.5
16.0	RICH	13.0
15.5		13.7
15.0		14.2
14.7	STOICHIOMETRIC	14.7
14.5		15.0
14.0	LEAN	15.5
13.5		16.0

Figure 11-46. The combined percentage of CO and CO_2 is closely related to the air-fuel ratio.

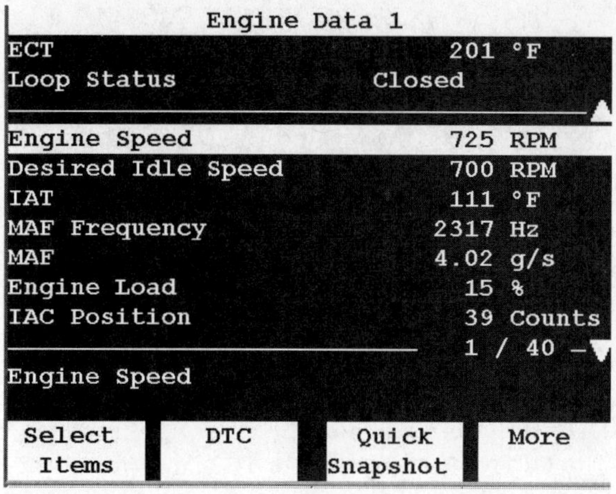

Figure 11-47. Make sure the vehicle is fully warmed up and operating in closed loop.

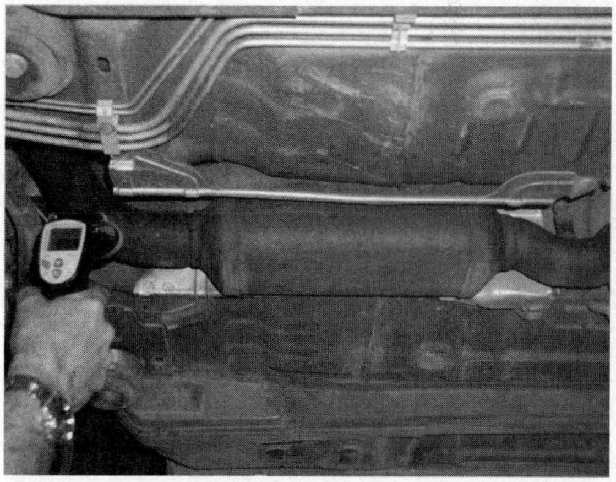

Figure 11-48. Measure the inlet and the outlet temperature of the converter.

Testing Exhaust Backpressure

A catalytic converter can fail and create excessive backpressure in the exhaust system. This can lead to reduced engine power and performance or in some cases an engine that will not even start. Clogged catalytic converters were more common when the beaded catalytic converter was used extensively; however, even the monolith-type converters can fail and create excessive exhaust backpressure. This is usually caused by overheating or physical damage. To test a vehicle for excessive backpressure, do the following:

1. Obtain an exhaust backpressure gauge, figure 11-49.
2. Remove the oxygen sensor in front of the catalytic converter and install the backpressure gauge, figure 11-50.
3. Start the engine and measure the backpressure at idle. Usually the maximum allowable backpressure is 1.25 psi (be sure to check the manufacturer's specifications).
4. Raise the engine speed to 2,500 rpm and measure the exhaust backpressure. It should not exceed 3 psi (check the manufacturer's specifications).
5. If either reading is above specifications, inspect the exhaust system before replacing the converter. If the exhaust system is OK, then replace the converter and perform the exhaust backpressure test again.

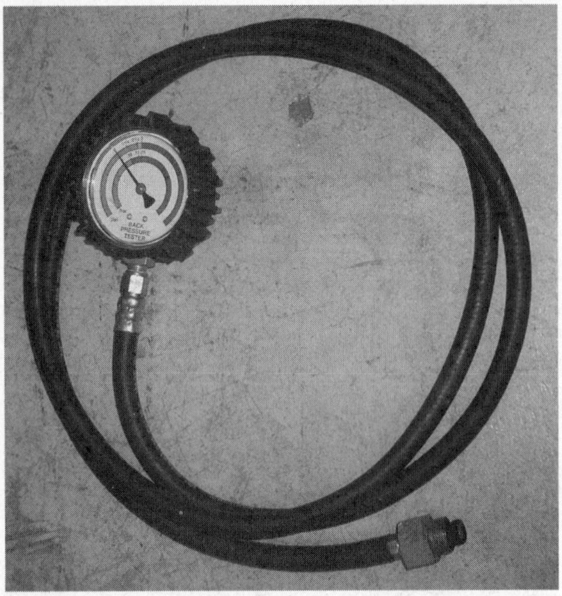

Figure 11-49. The exhaust backpressure gauge is used to measure the exhaust backpressure.

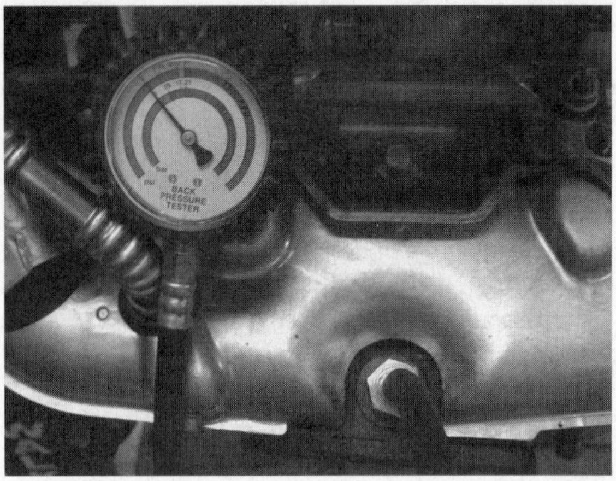

Figure 11-50. Remove an oxygen sensor in front of the converter and install the backpressure gauge.

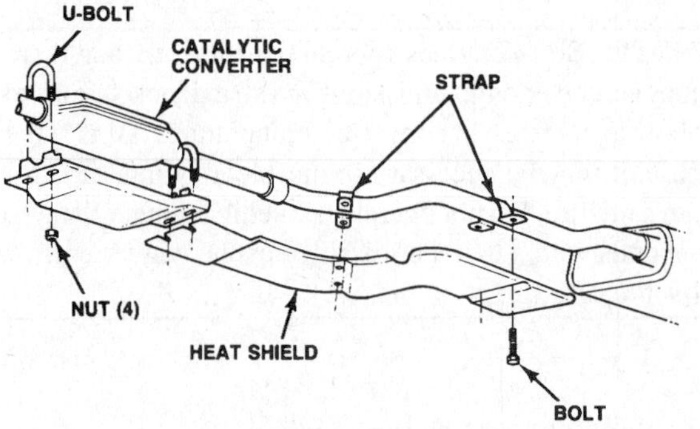

Figure 11-51. Note heat shield placement on this converter-equipped exhaust system.

OBD II System Monitor

The OBD II system monitor is similar for all domestic vehicles. The monitor is a function of the PCM that detects when catalytic converter efficiency deteriorates below a minimum level of effectiveness. The monitor relies mainly on front and rear heated oxygen sensors (HO_2S), which infer catalyst efficiency based on oxygen storage capacity and bonding. The front and rear HO_2S switching is counted under specific conditions for the purpose of calculating a before and after catalyst switching ratio. Once calculated, the ratio is compared against a threshold value. The catalyst fails if the switch ratio is greater than the threshold. In general, as catalyst efficiency decreases, the switch ratio of the HO_2S after the catalyst increases. The PCM activates the MIL when a fault is detected on two consecutive drive cycles (refer to manufacturer resources for the specific threshold).

Converter Service

Replacing a catalytic converter is essentially the same as replacing a muffler or resonator in the exhaust system. Converter installations differ slightly between vehicle makes and models; however, most utilize exhaust system flange-type or U-bolt connections. Typical examples are shown in figures 11-51 and 11-52. Listed below are the basic steps used to replace the converter on most vehicles:

1. Make sure the exhaust system and converter are cool enough to touch comfortably before starting work. This can prevent the possibility of serious burns.
2. Raise the vehicle off the ground to a comfortable working height.
3. Remove heat shields—generally retained by bolts or straps.
4. Disengage the air-injection tube from the converter (if applicable). The tube may be retained by bolts or clamps or it may be pressed into the housing.
5. Remove the catalytic converter flange fasteners at the front and rear. Some converters incorporate a short length of tubing at both ends. In this case, the connections with the exhaust tubing will have to be loosened.

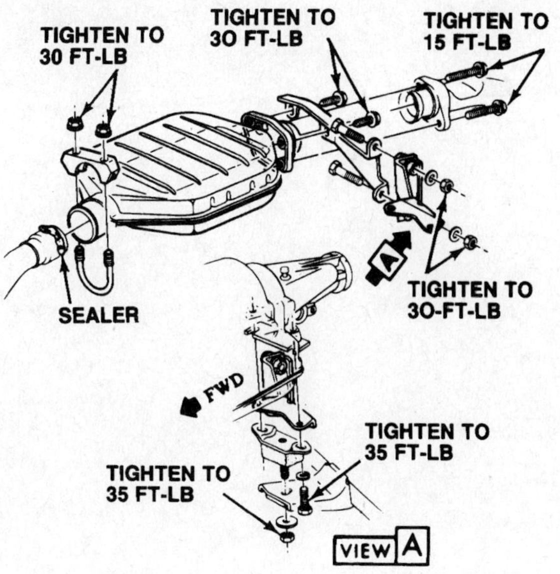

Figure 11-52. This GM catalytic converter uses both U-bolt and flange-type connections. (Courtesy of General Motors Corporation)

6. Separate the old converter unit from the exhaust system tubing and remove it from the vehicle.
7. Align the upper heat shield and position the new converter to the inlet pipe. New gaskets should be used where appropriate.
8. Loosely install the front attaching bolts and connect the outlet flange. Install the rear attaching fasteners loosely.
9. Reinstall the air-injection tube (if applicable). Align the converter with the remainder of the exhaust system.
10. Tighten all fasteners and install any remaining heat shields.
11. Lower the vehicle, start the engine, and check for exhaust system leaks.

12

Ignition System Inspection, Diagnosis, and Testing

OBJECTIVES

Upon completion and review of this chapter, you will be able to:

- Test ignition systems using an engine oscilloscope.
- Test and repair ignition primary triggering devices.
- Perform basic no-start diagnostics that are ignition related.

INTRODUCTION

Chapters 12 and 13 detail ignition and electronic control system services. This chapter contains general procedures for inspecting, diagnosing, and testing the ignition system circuits and components. Ignition system services, which include testing, removing, and replacing specific ignition system components, are detailed in Chapter 13. Chapter 14 describes removing, overhauling, and installing a distributor.

IGNITION SERVICE SAFETY

To prevent injury or damage to the ignition system, always follow these precautions during ignition system tests:

- Unless the procedure specifically states otherwise, switch the ignition off or disconnect the negative battery cable before separating any ignition system wiring connections. The high-voltage surge that results from making or breaking a connection may damage electronic components.
- Never touch, or short to ground, any exposed connections while the engine is cranking or running.
- Never short the primary circuit to ground without resistance; the high current that results damages circuits and components.
- Never create a secondary voltage arc near the battery or fuel system components. This may cause an explosion.
- Handle high-tension ignition cables carefully. Pulling or pinching cables causes internal damage that is difficult to detect.
- When testing the ignition system, follow the procedure specified by the manufacturer.

INSPECTION AND DIAGNOSIS

Ignition service begins with a visual inspection of the system components and circuitry. This is followed by a test drive to experience the symptoms and form a basic diagnosis.

Inspection

Check the following items during the inspection:

- Battery
- Secondary ignition components
- Primary ignition components
- Secondary ignition circuitry
- Primary ignition circuitry
- Related components

Remember, the battery must be at or near a full state of charge and the terminal connections must be clean and tight for *any* electrical system on the vehicle to function properly. The ignition system is no exception. A weak battery or high terminal resistance reduces available voltage and restricts current, which reduces ignition performance.

Check ignition components for secure mounting, correct installation, and good electrical connections. Inspect both the primary and secondary ignition components and their wiring. Check the entire length of wires and harnesses over for signs of damage. Verify the wires are properly routed and all connections are clean and tight.

Inspect any other components related to the ignition system. On a late-model vehicle, this includes all of the electronic engine sensors and actuators and their circuitry. Repair any problems found during the inspection before attempting to diagnose the ignition system.

Basic Diagnosis

If the engine does not start, perform a spark test to check for secondary voltage available from the coil. This is done by disconnecting a spark plug cable from the plug or removing the coil cable at the distributor cap, then attaching the cable to a spark plug simulator. A spark plug simulator is basically a plug without a side electrode and a grounding clamp attached to it, figure 12-1. Connect the secondary coil lead, or plug cable, to the simulator and clamp it firmly on a good ground, figure 12-2. Watch the plug simulator while cranking the engine. The simulator produces a bright blue are from the center electrode to its side each time the field collapses if secondary voltage is available from the coil.

When secondary voltage is available, the ignition system provides the current required to ignite the combustion charge. Therefore, the cause of the "no start condition" is most likely the mechanical, fuel, or engine control system, rather than an ignition system failure.

No secondary voltage available may be caused by a defective coil or some type of primary circuit failure. As explained later in this chapter, a series of tests is performed using a voltmeter, ohmmeter, and ammeter to isolate the source of the failure.

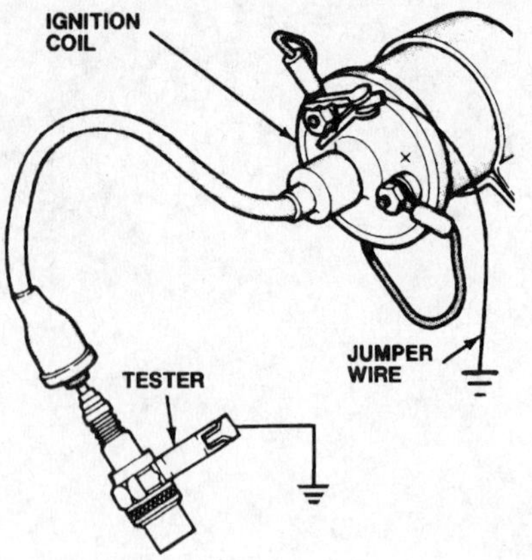

Figure 12-1. A spark plug simulator resembles a plug with a clamp attached to it.

Figure 12-2. Connect the plug simulator, crank the starter, and look for a bright blue arc across electrodes.

If the engine does start, drive the vehicle to experience the symptoms the customer is complaining about. The symptoms experienced during the test drive determine which areas of the ignition system to focus on during diagnosis, figure 12-3. When symptoms indicate a possible ignition failure, use an oscilloscope or engine analyzer to evaluate system performance and condition.

OSCILLOSCOPE DIAGNOSIS

Because an oscilloscope displays voltage changes during a period of time, it is an ideal instrument for examining the variable voltages produced in both the primary and secondary ignition circuits. All oscilloscopes operate on the same principle: A voltage trace, or waveform, is displayed on a viewing screen as a graph of voltage

CONDITION	CAUSE	CHECK
• Engine cranks normally but does not start and run	1. Open primary circuit 2. Coil grounded 3. No engine speed signal 4. Incorrect timing 5. Secondary voltage leak 6. Fouled spark plugs 7. No fuel 8. Mechanical failure	1. Circuit connections, coil, pickup coil, module, ignition switch. 2. Test and replace ignition coil. 3. Available voltage to the coil and CKP sensor signal. 4. Check and adjust, where applicable. 5. Coil tower, distributor cap, rotor, and high-tension cables. 6. Clean or replace. 7. Fuel delivery and volume. 8. Engine.
• Engine runs but misfires on one cylinder	9. Defective spark plug 10. Faulty plug wire 11. Defective distributor cap 12. Mechanical failure	9. Replace spark plugs. 10. Inspect and replace. 11. Inspect and replace. 12. Cylinder compression.
• Engine runs but misfires on different cylinders	13. Primary circuit resistance 14. Secondary voltage leak 15. Defective ignition coil 16. Fouled spark plugs 17. Fuel delivery problem 18. Mechanical failure	13. Circuit connections and continuity. 14. Coil tower, distributor cap, rotor, and cables. 15. Inspect, test, and replace as needed. 16. Inspect and replace as needed. 17. Fuel delivery and volume. 18. Engine.
• Engine runs but loses power	19. Incorrect timing 20. No timing advance 21. Fuel quality	19. Check and adjust. 20. Electronic control system. 21. Test and refuel.
• Engine backfires	22. Incorrect timing 23. Ignition crossfire 24. Exhaust leak 25. Fuel quality 26. Mechanical failure	22. Check and adjust. 23. Secondary wiring. 24. Inspect and repair as needed. 25. Test and refuel. 26. Engine.
• Engine knocks or pings	27. Incorrect timing 28. Too much timing advance 29. Lack of EGR 30. Fuel quality 31. Faulty knock sensor	27. Check and adjust. 28. Electronic control system. 29. Inspect and repair as needed. 30. Test and refuel. 31. Inspect and replace as needed.

Figure 12-3. Ignition system troubleshooting chart.

time. The vertical scale on the screen represents voltage and the horizontal scale represents time.

The voltage trace of an ignition event, or the firing of one spark plug, is divided into three periods or sections: dwell, firing, and open circuit, figure 12-4. Deviations from a normal pattern indicate a problem, so it is important to know what a good pattern for the system being serviced should look like. Although all ignition waveform patterns appear similar, there are variations between systems. Most scopes designed for automotive use display ignition traces in three different modes. Each mode is best used to isolate and identify particular kinds of malfunctions. The three basic patterns are:

- Superimposed
- Parade
- Raster

In a superimposed pattern, voltage traces for all cylinders are displayed on top of each other to form a single pattern on the scope screen, figure 12-5. This display provides a quick overall view of ignition system operation and is used to reveal certain major problems.

The parade pattern displays voltage traces for all cylinders one after the other across the screen from left to right in firing order sequence, figure 12-6. This allows easy comparison of voltage levels between cylinders. A parade display is useful for diagnosing problems in the secondary circuit.

A raster pattern shows the voltage traces for all cylinders stacked one above the other. The screen displays cylinders from bottom to top in firing order sequence, figure 12-7. This display is used to compare the time periods of the three sections of a voltage trace.

OSCILLOSCOPE SETTINGS AND CONNECTIONS

All oscilloscope manufacturers provide instructions for the use of their equipment. Although color codes and connector types vary, there are some basic similarities.

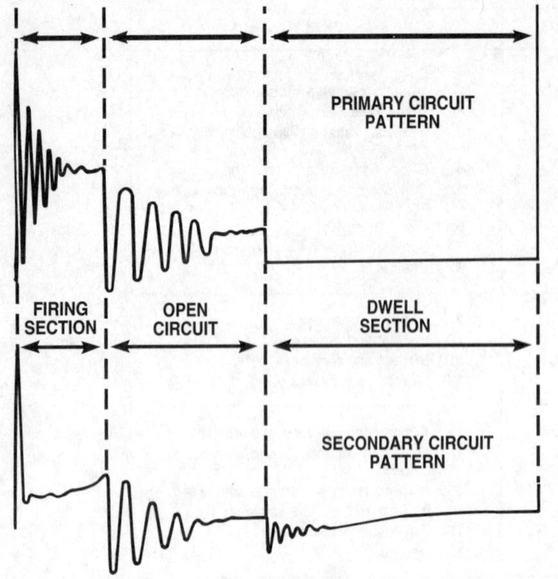

Figure 12-4. Ignition oscilloscope patterns, both primary and secondary, are divided into three distinct periods.

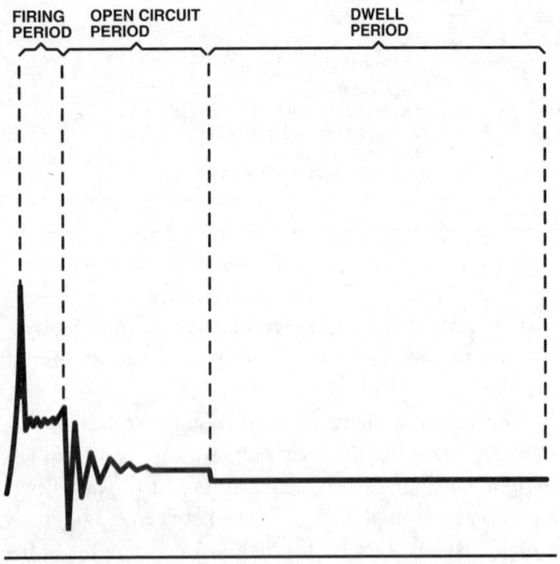

Figure 12-5. A superimposed pattern displays cylinder traces on top of each other.

Oscilloscopes installed in multifunction engine analyzer units are designed specifically for testing ignition systems and do not require any special considerations. Simply hook up the leads as instructed, then select primary or secondary and the type of display, superimposed, parade, or raster, on the scope control panel. Also select either the high or low voltage scale to be used.

The time base of the scope is automatically set to display one complete cycle of the ignition process. That is,

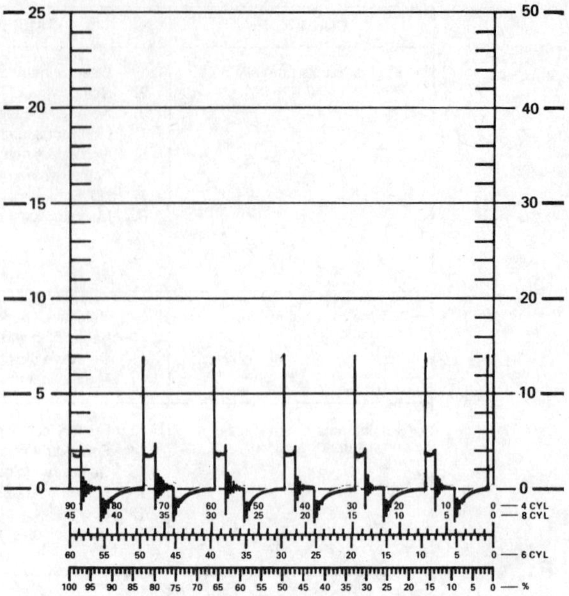

Figure 12-6. Parade pattern displays cylinder traces horizontally across the screen in firing order.

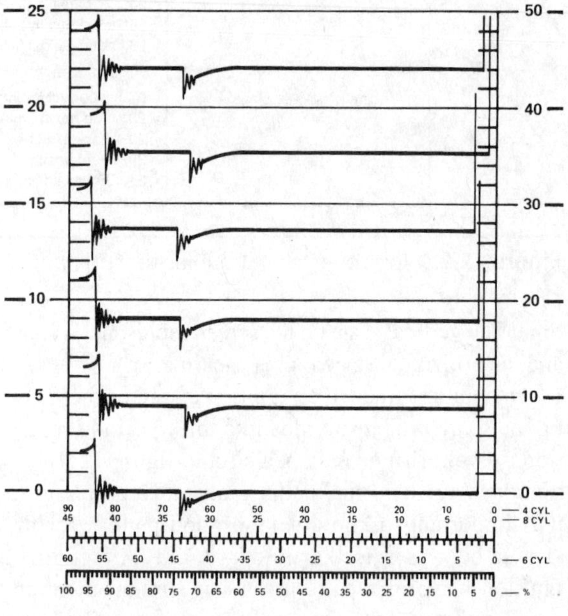

Figure 12-7. Raster pattern stacks cylinder traces vertically up the screen in firing order.

when connected to a 6-cylinder engine, the screen displays the firing sequence of the six cylinders, then starts again. As the engine speed changes, the scope maintains the same display, but the time represented changes.

Lab Scope Settings

It is possible to damage a lab scope if the sampled signal exceeds the capability of the scope. Always be cautious when measuring high secondary voltage to

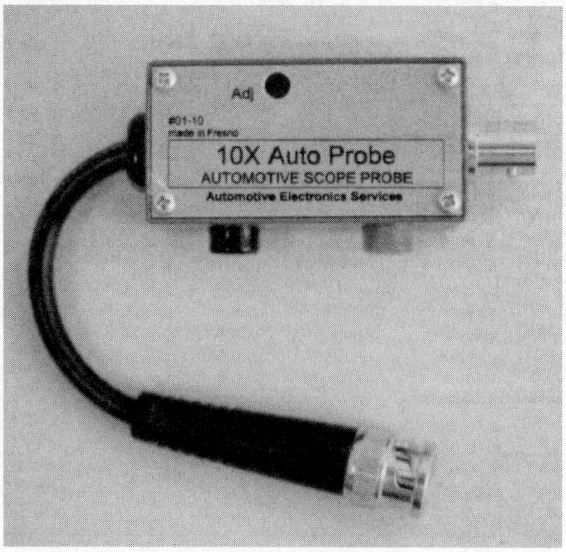

Figure 12-8. A 10X probe increases the display capability of the scope screen by a factor of 10.

prevent equipment damage. Know the limits of both AC and DC voltage that the scope can safely handle.

To display the signal, both the volts per division (VOLT/DIV) and time per division (TIME/DIV) must be selected on the scope control panel. The VOLT/DIV selection controls the height of the signal trace on the display screen. The lower the VOLT/DIV setting, the higher the displayed image. A higher VOLT/DIV setting results in a lower display. The VOLT/DIV setting determines how many volts are represented in each of the eight vertical grids or blocks on the display screen. As an example, consider a 5-volt square-wave signal. If the VOLT/DIV is set to 1 volt, the signal height changes five grids each time it switches on or off. However, the signal changes only one grid if the VOLT/DIV is set to 5 volts. The type of signal being sampled determines the VOLT/DIV setting to be used.

Ignition system voltage often exceeds the maximum voltage that a lab scope is capable of displaying. If the scope does not have a high-enough VOLT/DIV setting to properly display a higher voltage signal, a 10X probe may be used. These probes have a 10 to 1 attenuation, which reduces the signal by a factor of 10, figure 12-8. When sampling a 400-volt signal with a 10X probe, the signal displays on screen as 40 volts.

Sampling secondary ignition signals requires a special capacitive probe designed specifically for this purpose. These capacitive probes, or inductive clamps, generate a waveform trace based on fluctuations in the strength of the magnetic field and capture a secondary signal without connecting the scope directly to a high-voltage circuit. A capacitive probe for sampling the secondary ignition signals requires about a 1,000 to 1 attenuation.

Figure 12-9. Using an engine analyzer to sample ignition waveforms.

Before attaching the scope leads, adjust the VOLT/DIV switch so the screen displays both the highest and lowest voltage expected on the test circuit. Be aware, some scopes shunt excess voltage to ground to prevent damage to the instrument. If this happens while sampling primary ignition circuits, the collapse of the coil windings is disrupted, which stalls the engine.

Once a trace is visible on the scope screen, adjust the switch to fine tune and maximize the display. Most digital scopes built for automotive use have an automatic range finder that establishes initial settings based on the signal being sampled. Keep in mind, these features are designed simply to make using the scope easier and are not always the ideal settings for diagnosing a problem. Even with the most elaborate automatic ranging features, it is often necessary to make adjustments to the settings to maximize the display and get precise readings.

Scope Connections

When the scope is part of an engine analyzer, there is an assortment of cables, which are used to conduct a variety of tests, figure 12-9. Be sure to use the right cables, and to connect them correctly. Follow instructions from the equipment manufacturer. Modern engine analyzers often contain onscreen directions for conneting cables and conducting tests.

In general, to sample the primary ignition signal, the positive primary lead on the scope is connected to the primary negative terminal of the coil. The negative primary lead on the scope is connected to a good ground on the engine or battery.

To sample the secondary ignition, the secondary lead on the scope is connected to the coil wire, or clamped to the distributor cap if the coil is internal. This connection is made with an inductive clamp. An

Figure 12-10. Two inductive clamps used to pick up distributorless ignition system voltage signals.

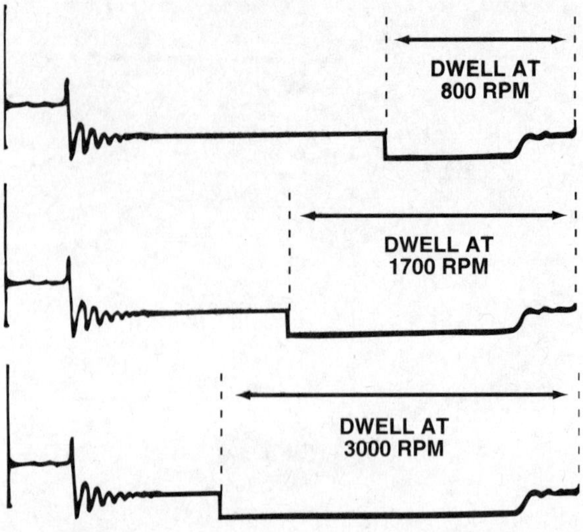

Figure 12-11. A dwell period that increases with engine speed is normal for some ignition systems.

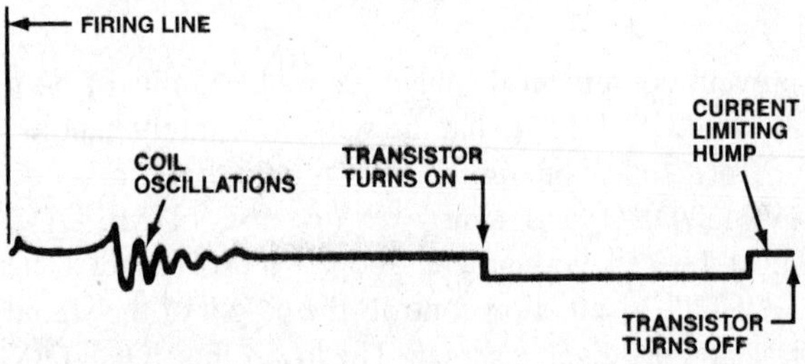

Figure 12-12. A current limiting transistor, when used, causes a small hump to appear on the trace at the end of the dwell section.

additional inductive clamp attaches to the number one spark plug lead to provide a triggering, or synchronization (sync), signal. This ensures that the waveform display begins with the first cylinder in the firing order.

On many late-model vehicles, especially those with distributorless or direct ignition systems, the coil terminals are not readily accessible for attaching the scope leads. Adapters and special test equipment are available for connecting a scope to these systems, figure 12-10. Follow instructions from the equipment manufacturer to connect these devices and test using procedures recommended by the vehicle manufacturer.

Ignition Waveforms

An oscilloscope is the ideal instrument for scrutinizing the variable voltages produced in the ignition circuits. As previously mentioned, all ignition system voltage signals, both primary and secondary, are divided into three periods: dwell, firing, and open circuit. Each period displays specific aspects of the ignition cycle.

Dwell Period

The dwell period is when the primary circuit is complete, and low-voltage current is building up the magnetic field of the coil. Dwell begins as the ignition module transistor switches on to complete the primary circuit and continues until the transistor switches off to begin the firing sequence. Dwell time varies for different systems, and some ignition systems also vary the dwell period according to engine speed or other operating conditions, figure 12-11.

A scope trace of the dwell period begins with a sharp drop of the signal as the coil driver completes the primary circuit to ground. This is followed by a relatively flat line which may have a slight ramp as current builds up in the secondary windings of the coil. When viewing, a secondary circuit waveform, the signal may ring, or display several diminishing oscillations, as the transistor switches to complete the primary circuit. Some ignition systems offer current limiting. This affects current in the primary circuit and produces a slight hump in the trace toward the end of the dwell period, figure 12-12. The exact shape and other characteristics of this current-limiting hump, appearing on both the primary and secondary trace, vary by system. Therefore, it is important to know the characteristics of the particular system being tested to form an accurate diagnosis. The dwell period ends with the abrupt upward stroke of the firing of the next cylinder.

Firing Period

The firing period, also called the spark line on a waveform trace, corresponds to the amount of time an electrical arc is bridging the spark plug gap and igniting the combustion charge. The firing period begins with a high inductive kick as the spark is established across the plug gap. Less voltage is required to maintain the arc once it is established, so the trace drops after the initial spike.

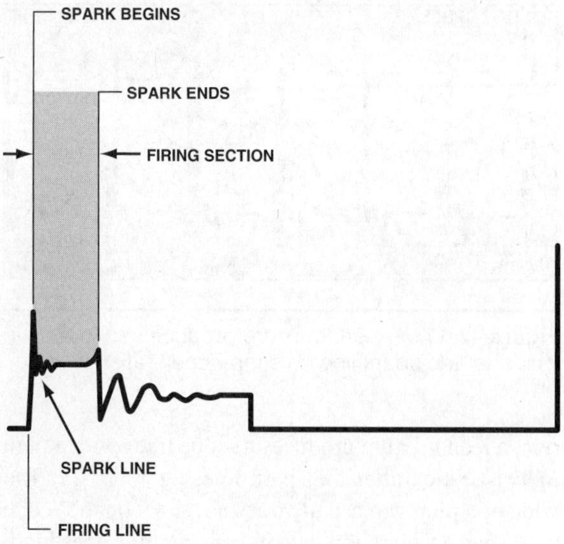

Figure 12-13. A typical primary firing period displays as a high spike followed by a spark line.

When sampling the primary circuit, the trace rings after the inductive kick in a series of diminishing oscillations which continue until the spark extinguishes, figure 12-13. Be aware, some systems suppress the ringing of the firing section, and an inductive kick followed by a relatively flat line is the normal pattern. Again, it is important to know the specific characteristics of the system being tested.

The trace drops following the inductive kick on a secondary pattern, as less voltage is needed to maintain an arc across the spark plug gap once it is established. On a good-running engine, the trace drops to less than one-quarter the height of the inductive kick, then continues as a near-horizontal line. A slight rise in this spark line is normal for some ignition systems, but too much rise indicates a fuel mixtures or combustion problem.

The firing period ends when there is no longer enough energy in the secondary windings of the coil to bridge the sparkplug gap. This opens the circuit to begin the final period of the ignition cycle.

Open Circuit Period

The open circuit period begins with a sudden drop of the trace as the spark plug stops firing. The trace continues as a relatively flat line until the transistor switches on to begin the dwell period for the next cylinder in the firing order. When the circuit opens, a considerable amount of energy remains in the secondary coil windings, even though there is not enough to bridge the plug gap. This remaining energy dissipates as heat and appears on both a primary and secondary waveform as ringing, figure 12-14. The primary transistor switches on the moment spark ends

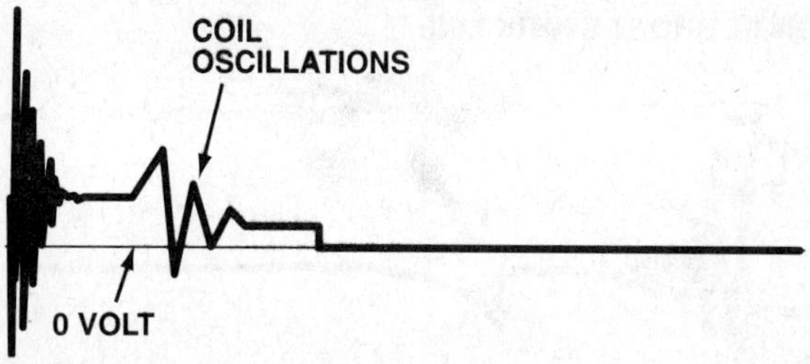

Figure 12-14. Diminishing oscillations appear on the trace as the coil dissipates residual voltage during the open circuit period.

on a Chrysler EIS systems, so there is no open circuit period. Look for a series of diminishing oscillations that flatten out to a near-horizontal line on an open circuit waveform.

Analyzing Ignition Waveforms

An ignition waveform contains a vast amount of information about circuit activity, mechanical condition, combustion efficiency, and fuel mixture. There are many variables which come into play, and a full explanation of ignition waveforms would fill an entire book. The information included here is simply to show how conditions outside the electronic circuitry can disrupt an ignition waveform. Electronic ignition patterns vary by system, so it is critical to know what is normal for the system being tested. Examples presented here are intended as a general reference only.

A number of ignition, fuel, and internal engine problems interfere with the firing period of a secondary ignition trace. Problems disrupting the firing trace are usually the result of high or low resistance either in the secondary circuit itself, or as the result of combustion problems within the cylinder.

Secondary ignition component failure due to wear is a common problem resulting in high circuit resistance. Whether caused by a damaged plug wire, distributor cap wear, or excessive spark plug gap, this additional resistance disrupts current and alters the scope trace. Higher voltage, which is required to over come this type of resistance, reduces current, and the energy in the coil dissipates more rapidly than normal. As a result, high open-circuit resistance produces a high firing spike followed by a high, short spark line, figure 12-15.

Other problems on the secondary circuit also increase resistance but have a different impact on the waveform. Corrosion on the cable terminals is a common problem that affects the spark line of a trace. Overcoming the resistance created by corrosion requires an increase in voltage to maintain a spark. Secondary terminal corrosion causes the spark line to start

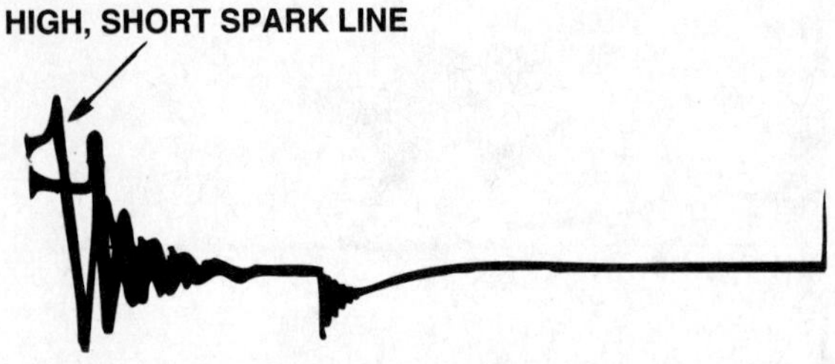

Figure 12-15. High open circuit resistance on one cylinder is quickly spotted on a superimposed display as a high, short spark line on the secondary trace.

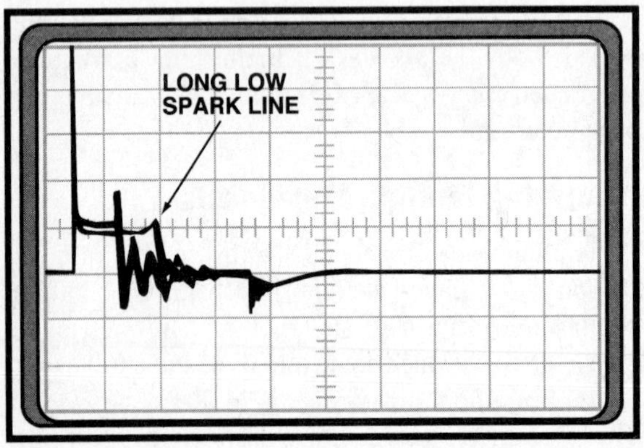

Figure 12-16. Low resistance on a cylinder produces a long, low spark line on the secondary trace.

high and slope downward sharply as voltage falls off. Poor connections on the primary circuits also add resistance that disrupts the trace. High primary resistance reduces the amount of energy the coil produces, which weakens the spark and results in a trace with a low inductive kick and short burn time.

Current always follows the path of least resistance, and low resistance in the secondary circuit causes current to bypass the spark plug gap. Carbon tracking in the distributor cap, poor secondary cable insulation, and fouled spark plugs are sources of low resistance that disrupt current. Low resistance reduces the voltage requirements and shows up on a trace as a low, long spark line, figure 12-16.

Fuel mixture problems often cause the trace of a spark line to jump erratically, and variations in the fuel mixture show up on the spark line as well. As fuel mixture in the plug gap varies, the voltage requirements to keep the plug firing also vary. Therefore, the spark line always has some variation. Turbulence in the cylinder and fuel mixture changes cause this variation. Diagnosing a fuel mixture problem is best done with an exhaust analyzer, rather than a scope.

Many problems are revealed on a scope only under snap acceleration, or engine load. On snap acceleration, a lean mixture produces a scope trace with a high inductive kick after the spark line, figure 12-17. Too wide of a plug gap usually does not show on the scope until snap acceleration, when the firing line goes high.

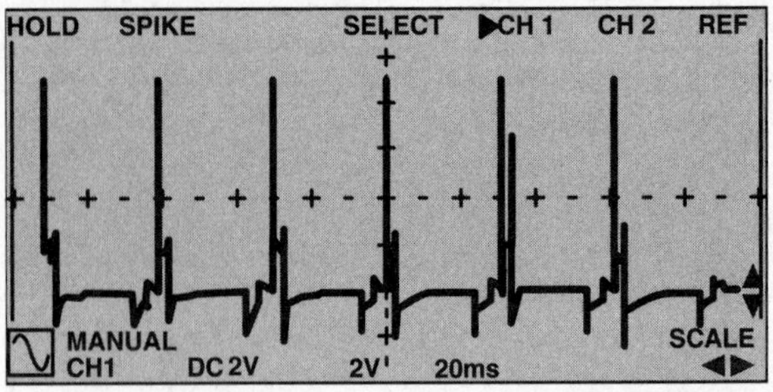

Figure 12-17. A lean mixture produces an inductive kick after the spark line on snap acceleration.

Primary Scope Patterns

Although the primary circuit is the low-voltage side of the ignition system, the signal may spike as high as 200 volts on some systems. Be familiar with the system tested and adjust the VOLTS/DIV setting accordingly if using a lab scope.

Firing Period

The firing period of the pattern corresponds to the amount of time the spark plug is firing. This is referred to as "burn time." Many engine analyzers record the burn time, which is a good indicator of the firing conditions. A long burn time is the result of low total resistance, such as that caused by low compression. A short burn time is the result of high resistance, such as that caused by high resistance in a plug wire. A normal display of the firing period appears with a high vertical spike, or firing line, as spark begins. This is followed by a relatively flat line until spark ends, figure 12-18.

Open Circuit Period

Once spark is extinguished, the open circuit, or intermediate, period begins. This section of the trace continues until an electronic signal from the ignition control module (ICM) or powertrain control module (PCM) switches on the primary circuit transistor. On some systems, such as the Chrysler EIS, there is no open circuit section because the primary transistor switches on the moment spark ends, figure 12-19.

An open circuit voltage trace displays the dissipation of energy remaining in the coil after plug firing is complete. The trace appears as a series of diminishing oscillations similar to those of the firing section, but considerably smaller. There should be at least three distinct coil oscillations on the trace, figure 12-20. Less than five oscillations indicates a defective coil primary winding. The end of the intermediate period is

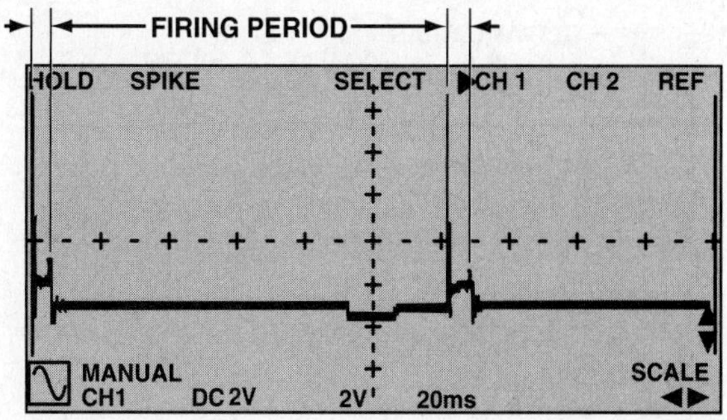

Figure 12-18. Look for an inductive kick and a series of diminishing oscillations in a primary firing period.

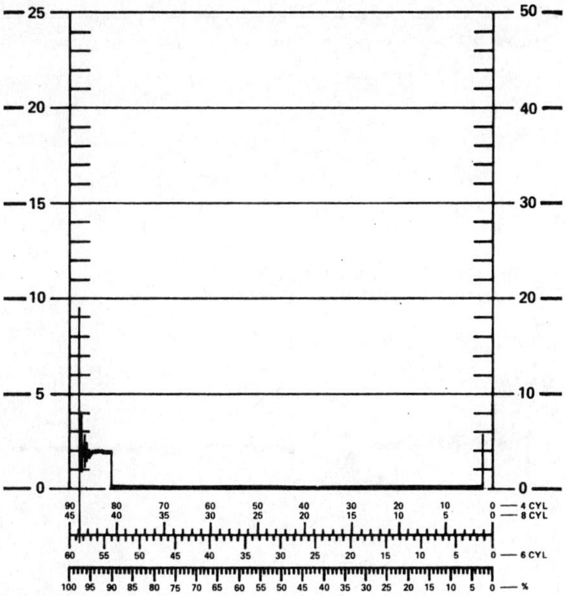

Figure 12-19. There is no open circuit period on some Chrysler systems because the primary transistor switches on as soon as the spark extinguishes.

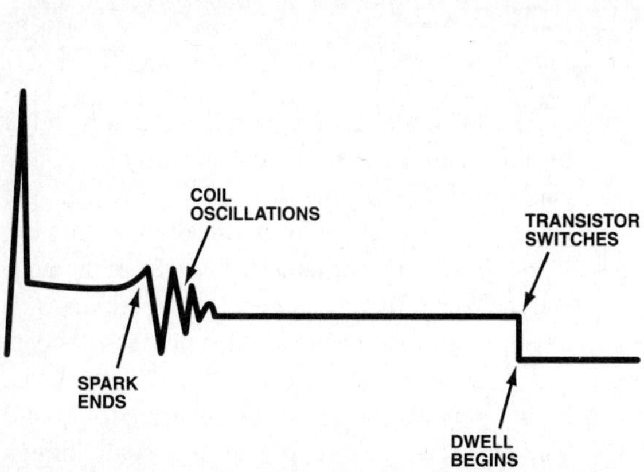

Figure 12-20. Look for at least three distinct coil oscillations in the primary open circuit period.

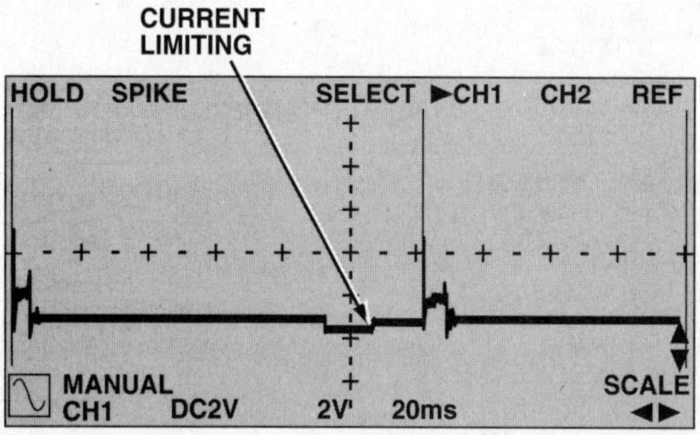

Figure 12-21. Current limiting produces a hump or oscillations toward the end of the dwell period on some systems.

marked by an abrupt drop. This is the point where the primary transistor turns on and dwell begins.

Dwell Period

The dwell period is the time the primary circuit is complete. During dwell, a low-voltage current is building up the magnetic field of the coil. Following the sharp drop of the transistor switching on, the trace appears as a relatively flat, horizontal line. The dwell period ends with the abrupt, upward stroke of the first firing oscillation of the next cylinder in the firing order. Small oscillations, or a hump, toward the end of the dwell period should be considered normal on some systems. If the control module contains a current limiting device, the hump indicates that the limiting circuit is activated, figure 12-21.

Use the raster pattern to compare the dwell period for each cylinder. Dwell sections should not vary by more than four to six degrees between cylinders. Dwell variations are generally caused by a worn distributor or timing chain. More important is any variation in the upward spike at the end of each trace. Variation at the end of the trace is variation in timing for each cylinder.

Secondary Scope Patterns

Secondary voltage traces also move left to right across the screen in firing, open circuit, and dwell sequence. Since normal firing voltage can exceed 40,000 volts, oscillations are displayed on an engine analyzer scope using a high-voltage scale graduated in kilovolt (kV), or thousand-volt, increments. With a lab scope, a capacitive pickup with a 10,000 to 1 attenuation is used to capture the entire waveform on the display screen and to protect the scope.

Firing Period

The firing period of the secondary pattern starts with a straight, vertical rise called the firing line or voltage spike. The peak of the spike indicates the amount of voltage required to create an arc across the spark plug air

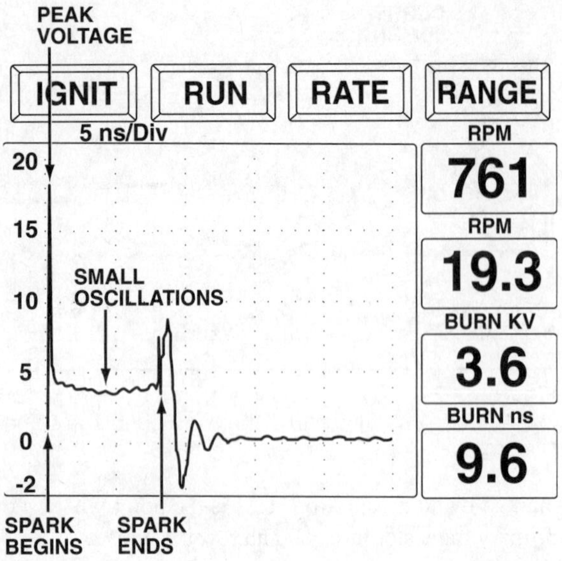

Figure 12-22. Look for small oscillations and a slight rise in the firing period on a secondary trace.

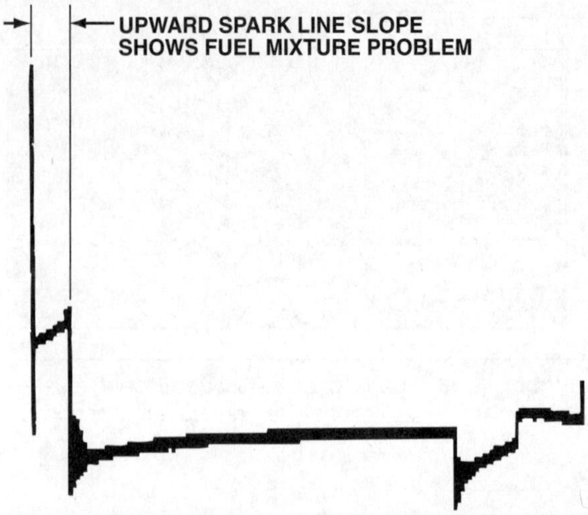

Figure 12-23. A secondary spark line that jumps or slopes upward indicates fuel mixture problems.

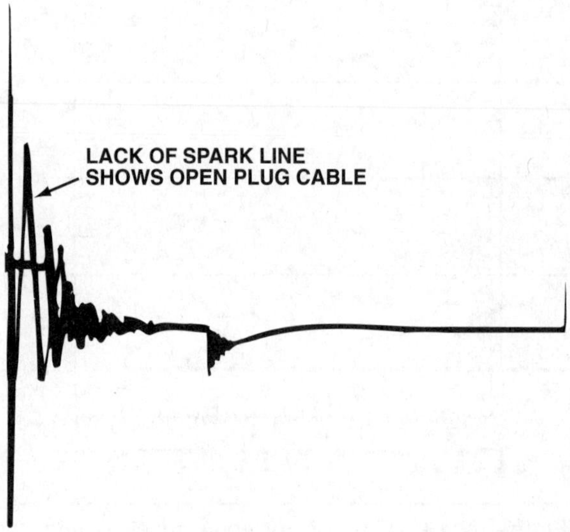

Figure 12-24. A large oscillation without a spark line on a secondary firing period indicates an open spark plug circuit.

gap. Once an arc is established, less voltage is required to maintain it. A normal trace drops to about one-quarter the height of the voltage spike, then continues as a horizontal spark line. A spark line represents continued current across the spark plug gap, and should be relatively flat and stable. However, a series of very small oscillations in the spark line is normal, as is a slight rise toward the end of the firing period, figure 12-22.

Damaged plug wires, worn distributor caps, and excessive plug gaps create high open-circuit resistance. High resistance appears as a high, short spark line. Corrosion on the cable terminals also creates high resistance, but corrosion causes the spark line to start high and slope sharply downward. Carbon tracking in the distributor cap, poor cable insulation, and fouled spark plugs are sources of low resistance. Low resistance shows up as a low, long, spark line. A spark line that jumps erratically or slopes upward or downward generally indicates a fuel mixture problem, figure 12-23. However, an upward sloping pattern may also be the result of a mechanical problem.

The firing period of a secondary trace reveals a number of ignition problems.

- A large firing oscillation with little or no spark line for one cylinder indicates an open circuit between the distributor rotor and the spark plug, figure 12-24. Use the parade display to isolate the faulty cylinder. Check for a disconnected or broken spark plug cable. This pattern can be created deliberately to check coil output on some systems, but causes component damage on others. Perform the check only when recommended by the manufacturer. To check, disconnect the cable from one of the spark plugs and start the engine. The top of the firing line of the open plug is the maximum available voltage from the coil.
- A parade pattern may show a short spark line still exists even when the spark plug cable is deliberately disconnected, figure 12-25. This indicates high voltage is causing a current drain to ground somewhere in the circuit. Current usually drains through the ignition cable insulation, the distributor cap, or the rotor. Carbon tracks tend to accumulate near the leakage over time.

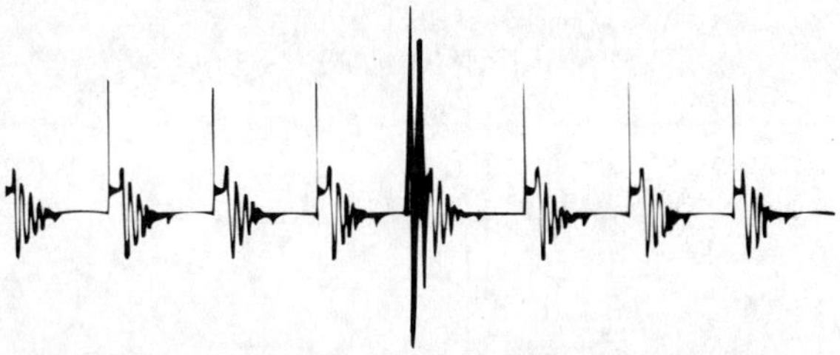

Figure 12-25. An open spark plug circuit displays a high peak and oscillations in a secondary parade pattern as high voltage leaks to ground.

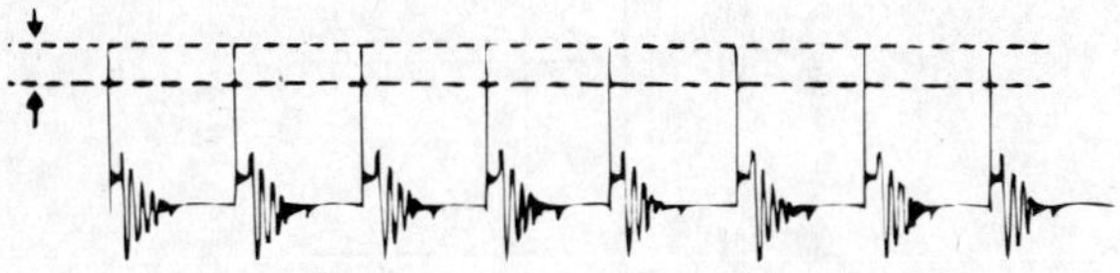

Figure 12-26. Look for fairly even firing line peaks when viewing a secondary parade pattern.

Figure 12-27. A high, short spark line on a secondary superimposed pattern indicates high resistance in one cylinder.

- There should be no more than a 20-percent difference between the highest and lowest firing spikes when comparing cylinders in a parade pattern, figure 12-26. Peak voltage variations are caused by fuel or electrical system problems. To separate the two, enrich the fuel mixture. If the spikes go down and engine speed increases, the problem is fuel related. If the spikes go down and engine speed remains unchanged, the plug gaps may be too great. If a single spike remains the same height, the cable to that plug may be damaged.
- One spark line that is higher and shorter than the rest indicates high resistance from an open circuit between the distributor cap and the spark plug, figure 12-27. A damaged or loose cable or a too-wide plug gap may be a fault. Corrosion on the cable terminals and in the distributor cap causes the spark line for one cylinder to start higher and angle downward more sharply than the others.

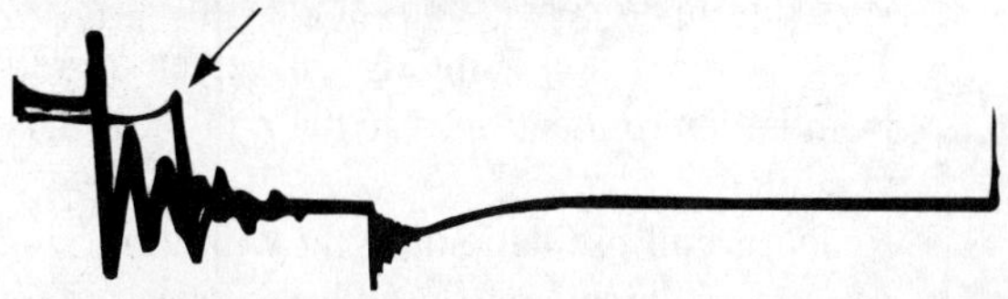

Figure 12-28. A low, long spark line on a secondary superimposed pattern indicates low resistance in one cylinder.

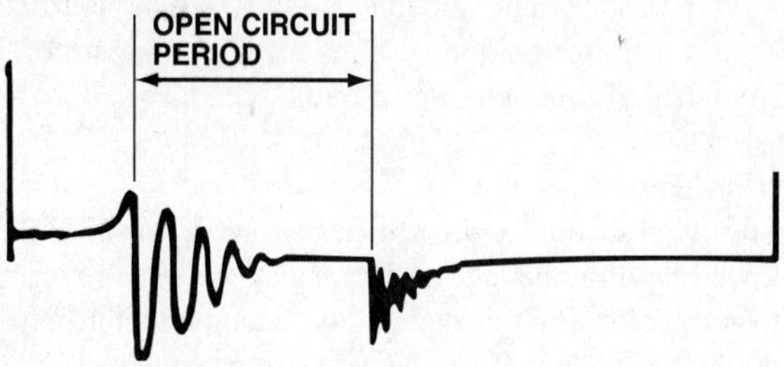

Figure 12-29. A secondary open circuit period displays diminishing oscillations.

- One spark line lower and longer than the rest indicates low resistance in the circuit between the distributor cap and the spark plug, figure 12-28. Check for carbon tracks in the distributor, poor cable insulation, or a fouled spark plug.
- A spark line that jumps erratically or slopes up rather than down is caused by an incorrect fuel mixture in that cylinder. The problem may be mechanical, such as sticking or worn valves, or caused by intake air leaks, or fuel-induction problems.

Open Circuit Period

A secondary open circuit trace indicates excess coil voltage being dissipated, and appears similar to a primary intermediate pattern. Look for a short vertical rise from the spark line followed by a series of diminishing oscillations, figure 12-29. Oscillations should be of relatively even width and taper down gradually to a near-horizontal line. The open circuit section ends with the module-on signal.

Use the open circuit period of the trace to check coil function. A faulty coil, primary circuit, or high-tension lead between the coil and distributor cap may cause a lack of oscillations. The firing period is used to detect a number of problems:

- Absent coil oscillations with a normal dwell section are the result of either a faulty coil or a high-resistance short to ground in the primary circuit.

- A pattern displayed upside-down indicates that coil polarity is reversed, usually because of reversed primary connections at the coil.
- The entire pattern jumping on screen results from an intermittent open in the coil secondary winding.
- Reduced coil oscillations along with a missing module-on signal are caused by an open circuit between the coil and the distributor cap.
- A variation in the firing signals on a raster pattern indicates timing differences among cylinders. The cause is generally worn ignition signal parts or a faulty module. Be aware, this is normal for those ignition systems using computer control of timing to regulate idle speed.

Dwell Period

The dwell section begins with a sharp vertical drop followed by diminishing oscillations as the trace returns to zero volts. The trace generally remains flat until the module switches off for the next firing sequence.

The length of the dwell section will vary for different systems, and may also be engine speed dependent. As with the primary trace, a slight rise, or hump, towards the end of the dwell period is normal if the control module has a current limiting device.

Dwell variation between cylinders is checked using the raster display, figure 12-30. A dwell variation of more than four to six degrees between cylinders indicates mechanical wear in the distributor, a loose timing chain or belt, or a faulty crankshaft position sensor. An ignition module is also capable of varying dwell. It is possible the dwell variation is normal, as the PCM controls timing to stabilize idle speed on some models.

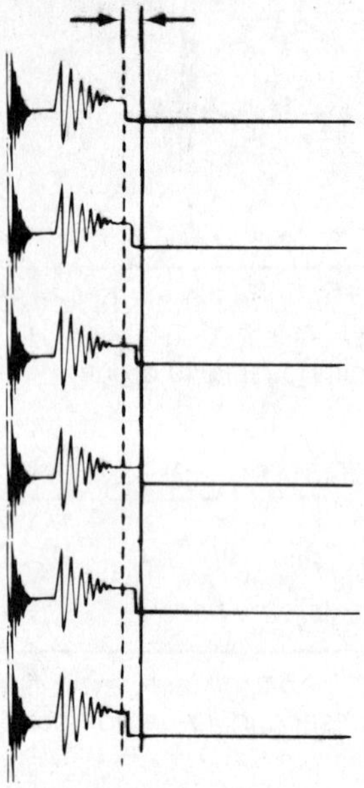

Figure 12-30. Use the raster display to check for too much dwell variation between cylinders.

COMPONENT TESTING

When oscilloscope testing reveals irregularities in the primary or secondary ignition systems, individual components and circuits are checked to isolate the source of the problem. Most ignition failures are the result of either high resistance, which impedes current, or low resistance, which allows too much current through the component or circuit. Component tests are performed using a digital multimeter (DMM), graphing multimeter (GMM), or lab scope.

Test specifications and procedures vary somewhat by manufacturer and model. Therefore, it is important to have accurate service information available for the systems being tested. This chapter contains general procedures which apply to most ignition systems. However, it is not to be used as a substitute for the factory service manual. Chapter 13 of this *Shop Manual* contains more detailed procedures for testing specific components.

Primary Circuit Testing

Primary circuit voltage has a direct effect on secondary circuit voltage. The loss of a single volt in the primary circuit can reduce secondary circuit voltage by as much as 10 kilovolts (10,000 volts). Common causes of primary circuit voltage loss include: high circuit resistance, insufficient source voltage from the battery, and low charging system output.

Sources of high primary resistance include:

- Loose, corroded, or damaged wiring connections
- An incorrect or defective coil
- A poor ground at the ignition module

Low source voltage can be caused by the following:

- Excessive starter motor current draw
- Low charging voltage
- A discharged battery

Resistance problems in the primary circuit are located by voltmeter testing. Begin with a check of available voltage at the battery and ignition coil, then perform voltage drop tests to locate the source of the problem.

Available Voltage

Two available voltage checks are performed to determine if the system is receiving enough voltage to op-

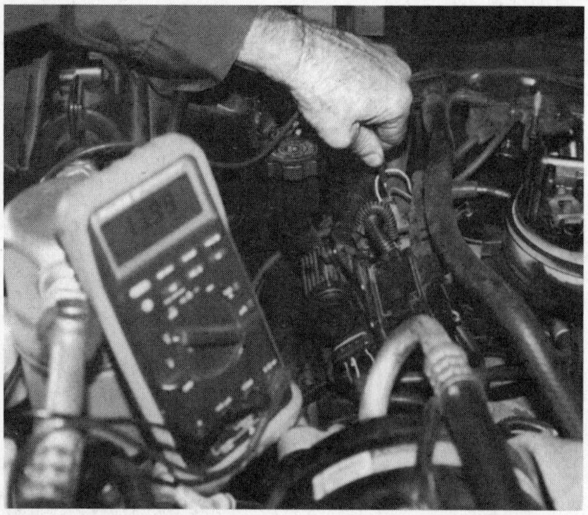

Figure 12-31. Checking for battery voltage available at the ignition coil.

erate. Because the ignition system depends on battery voltage, it is important to verify that the battery is at a sufficient state of charge. This is done by checking the voltage available from the battery while cranking the starter. To test, connect the positive voltmeter lead to the positive battery terminal, not to the cable connector, and connect the negative voltmeter lead to ground. Switch the ignition key and take a voltmeter reading while cranking the starter motor.

In general, a reading of 10 volts or more indicates that the battery is in good condition. Sources of low voltmeter readings include a voltage drop across the battery ground cable, excessive starter motor current draw, or incorrect charging system output. Refer to previous chapters in this *Shop Manual* as needed to repair the cause of low battery voltage while cranking before proceeding. After repairs are made, retest. If the battery is supplying sufficient voltage, continue by checking the available voltage at the coil.

Turn the ignition switch on without engaging the starter and note the voltmeter readings. Battery voltage should be available on all late-model systems, figure 12-31. Early electronic and breaker point ignition systems that use a ballast resistor normally have about 7 volts available to the coil. Next, crank the engine while monitoring the voltage. A scope or GMM is needed to properly evaluate the voltage drop to the coil. As the engine cranks or runs, the voltage drops because of the amperage when the coil primary is turned on. Any drop in voltage when the primary circuit is turned on is caused by resistance. A ballast resistor creates a large voltage drop because of the resistance of the resistor.

If voltmeter readings are not within specifications, check the primary circuit. Repair or replace any loose or damaged connections and repeat the test. If available voltage readings are still out of range, locate the source of the high resistance by conducting voltage drop tests along the primary circuit.

Voltage Drop

Voltage drop is the amount of voltage an electrical device normally consumes to perform its task. A small amount of voltage is always lost due to normal circuit resistance. However, excessive voltage drop is an indication of a high-resistance connection or failed component.

To check voltage drop, the circuit must be powered and under load. The circuit must also carry the maximum amount of current it is designed to handle under normal operating conditions. The amount of voltage drop considered acceptable varies by circuit. Low-current circuits that draw milliamps are affected by very small voltage drops, while the same amount of voltage drop has a negligible effect on a high-current circuit.

As explained in Chapter 2 of this *Shop Manual*, voltage drop along a circuit is measured by connecting the negative voltmeter lead to ground and probing at various points in the circuit with the positive meter lead. There are two methods of determining voltage drop: computed and direct measurements. Compute voltage drop by checking available voltage on both sides of a load. Then, subtract the voltage reading taken on the ground side of the load from the reading taken on the positive side. Direct voltage drop readings are taken by connecting the positive meter lead to the power side of a load and connecting the negative meter lead to the ground side of the component.

Manufacturers provide voltage drop specifications for various ignition components. High resistance on the control module ground circuit, often as a result of poor or corroded connections, is a common problem with electronic control systems. Measure the voltage drop across the ignition module ground as follows:

1. Connect the voltmeter positive lead to the negative, or distributor, coil terminal and connect the voltmeter negative lead to a good engine ground. Switch the ignition on and observe the voltmeter. Typically, a reading less than 0.2 volt indicates that the ground is in good condition, figure 12-32.
2. Crank or start the engine. Use a scope or GMM to measure the voltage during the dwell period; it should drop to about 0.5 volt. Be aware, voltage that does not drop below 1 volt is normal for some systems.

High voltage drop is often caused by a poor connection. Check and repair ground circuit wiring and connections, then repeat the test. If readings remain high, work

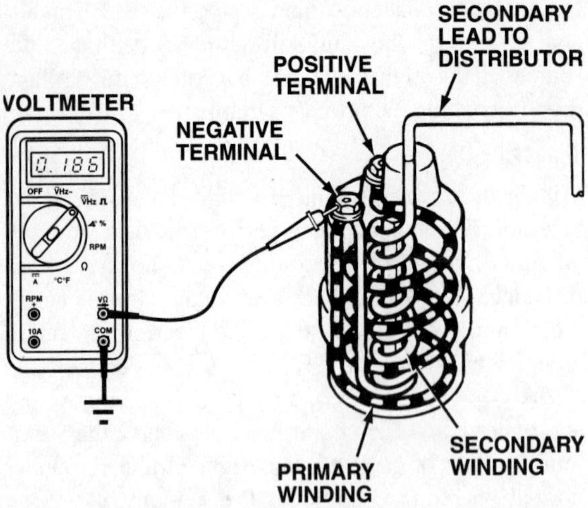

Figure 12-32. Checking voltage drop across the ignition module ground circuit.

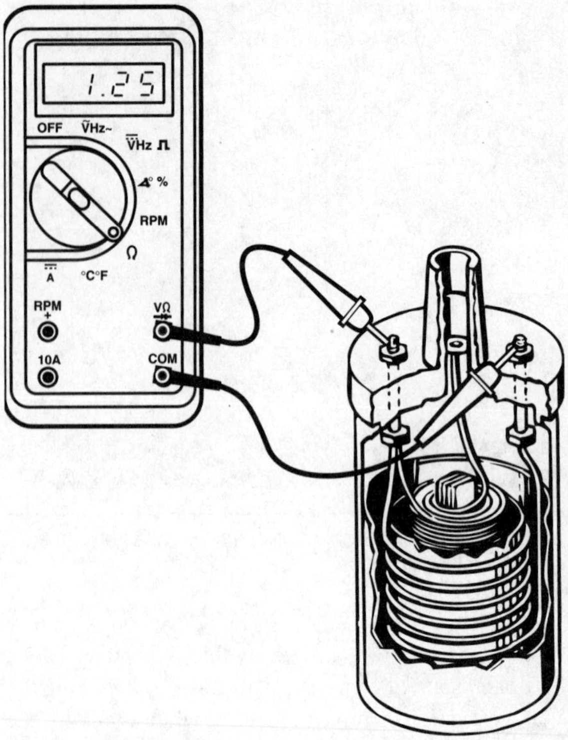

Figure 12-33. Checking primary coil winding resistance with an ohmmeter.

down the circuit taking readings to isolate the source of the high resistance. Voltage drop tests are performed on other ignition components and circuits in a similar fashion to locate high resistance areas. However, the amount of acceptable voltage drop varies by system and circuit. Accurate service specifications are essential for effective troubleshooting.

Ignition Coil Testing

For many years, the only test criteria vehicle manufacturers provided for checking an ignition coil were resistance specifications for the primary and secondary windings. Although checking winding resistance is quick and easy, it is not the most reliable method of testing a coil. To check winding resistance, simply disconnect the primary and secondary wires, take ohmmeter readings, then compare the results to specifications.

Set the ohmmeter to the lowest scale to check primary winding resistance on an oil-filled coil. Then, connect one ohmmeter lead to the positive, or battery, terminal of the coil, and connect the other meter lead to the negative, or distributor, terminal, figure 12-33. Take an ohmmeter reading and compare it to specifications, which are typically in the 0.50 ohm to 2.00 ohms range.

Secondary winding resistance on an oil-filled coil is measured with the ohmmeter set to the highest scale. To check, connect one ohmmeter lead to the coil tower secondary terminal, then touch the other meter lead to each of the primary terminals and note the meter readings, figure 12-34. Compare the lower of the two ohmmeter readings to specifications, which range anywhere from about 5,000 ohms to over 15,000 ohms depending upon the application.

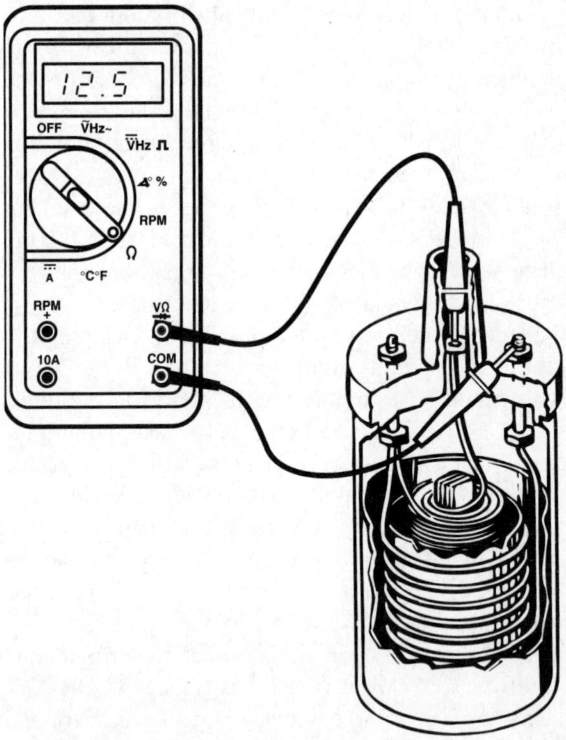

Figure 12-34. Checking secondary coil winding resistance with an ohmmeter.

A third ohmmeter check is made to test the coil windings for shorts. With the ohmmeter set on the lowest scale, connect one meter lead to either primary terminal and touch the other lead to the metal case of the coil. The meter should indicate infinite resistance. Any other reading indicates the coil windings are shorted to the case and the coil must be replaced.

Although checking winding resistance with an ohmmeter is a quick and easy way to evaluate an oil-filled coil, it is not an accurate method for checking the E-core and direct ignition coils used on modern engines. This is because an ohmmeter determines resistance by transmitting a low-voltage, low-current test signal through the circuit and measures the voltage drop of the test signal. A coil may be perfectly capable of carrying the test signal of the ohmmeter, but not the high-voltage, high-amperage signal produced in the windings during operation. Remember, current generates heat, and heat expands metal. It is fairly common for the windings of a coil to expand enough to create open or short circuits during operation, which often results in a misfire under load. This type of coil failure is impossible to detect with an ohmmeter.

Any leakage in the secondary circuit may cause a carbon track to form. Carbon tracking, the result of the circuit being bypassed, often appears on a rotor, cap, or the spark plug porcelain as a small black track. It is also possible for the ignition coil to carbon track inside between the primary and secondary windings. This type of coil damage does not show up in an ohmmeter test, but allows current to cross between the two windings. The only accurate way of determining the condition of a modern ignition coil is to measure coil current draw.

Coil Current Draw

Most manufacturers provide current draw specifications and test procedures for checking their ignition coils with an ammeter. However, the most accurate method of evaluating coil operation is to sample the signal on a lab scope or GMM using a current probe. This not only reveals the current draw, it also displays the build-up and release of energy in the windings. Both test methods, ammeter and current probe, are explained below.

Ammeter Testing

To test current draw with an intrusive ammeter, disconnect the positive, or battery, primary wire from the coil and connect it to the positive lead of the ammeter. Connect the negative lead of the ammeter to the positive terminal of the coil, figure 12-35. When using an inductive ammeter, fit the inductive pickup over the primary wire to the coil positive terminal leaving the wire connected.

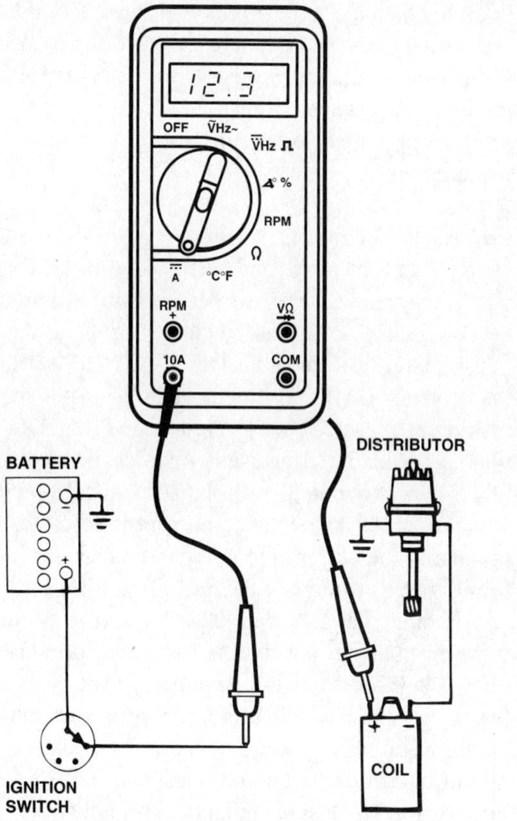

Figure 12-35. Intrusive ammeter connection for checking ignition coil current draw.

Some manufacturers recommend taking readings with the engine running, while others specify checking current draw while cranking the starter without starting the engine. Follow the recommended procedure for the vehicle being serviced. Observe the ammeter reading and compare it to specifications. When using a DMM for measuring amperage, it must have a MIN/MAX feature. A DMM averages the reading and a meter without MIN/MAX does not accurately display current because the current in the primary circuit is turned on for such a short time.

If there is no reading, the primary circuit is open. Repair the primary wiring and retest.

Higher-than-specified current draw is usually caused by an internal short circuit in the coil. However, installation of a coil not meeting the original equipment specifications of the vehicle being tested may also result in high current draw readings. In either case, the coil needs to be replaced. Make sure the replacement coil meets spedifications. Retest to confirm the repair.

Lower-than-specified current draw indicates a discharged battery, excessive resistance in the coil primary winding, loose or corroded primary connections, or high resistance in the primary wiring to the coil. Service the battery and primary connections as needed and perform voltage drop tests to locate high resistance in the primary wiring. Repair any problems found and retest. If no battery or circuit problems are found, replace the coil.

Current Probe Testing

Using a current probe and a lab scope or GMM is a quick, easy, and accurate way to measure current draw on any ignition coil. Set the current probe to the 100 mV per amp scale and adjust the VOLT/DIV setting of the scope or GMM so it is capable of displaying the maximum expected current. Adjust the TIME/DIV setting to capture the firing event. A typical firing event at idle occurs in about 20 milliseconds and delivers up to about 10 amps. Once the equipment is adjusted, simply clamp the current probe around either of the coil primary leads to capture the trace, figure 12-36.

Although the scope or GMM displays a current trace on both the positive and negative primary circuits, sampling from the negative circuit is the preferred method. Both circuits reveal peak current draw, but the negative, or ground side, provides a clearer picture of current build-up and release in the coil windings. However, on some models, especially those with direct ignition, it is difficult to access the negative circuit on each of the coils. Generally, these coils share a common power circuit accessible at a multi-plug connection of the harness.

There are basically two types of coil used on electronically controlled systems: those with current limiting and those without current limiting. Before attempting to analyze the current trace of an ignition coil, it is important to know which type is used.

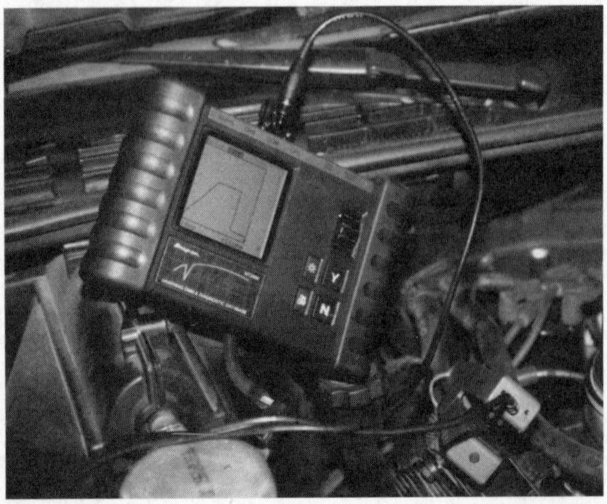

Figure 12-36. Checking coil current draw with a GMM and a current probe.

With current limiting, also known as peak and hold, current build-up in the coil windings stops once it reaches a predetermined level. Current is maintained at this level until the primary circuit opens to release the stored energy and fire the spark plug. A trace for this type of system quickly rises to peak control current, then continues as a horizontal line until the circuit opens and current drops immediately to zero, figure 12-37. Current remains at zero during the dwell period—that

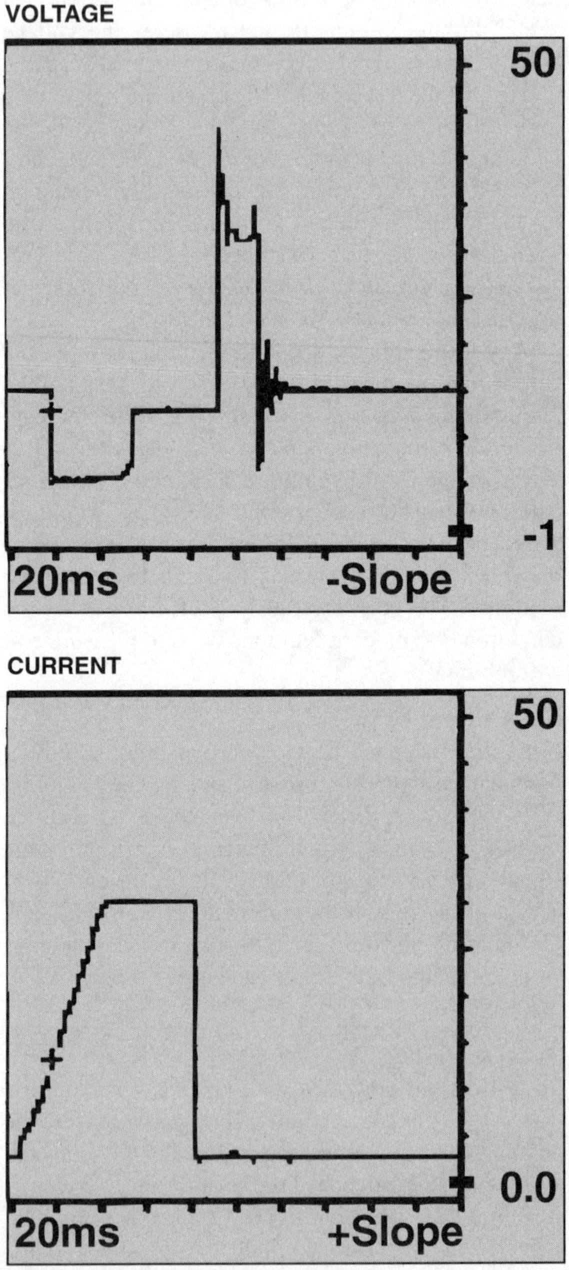

Figure 12-37. Typical primary ignition coil voltage and current trace with peak and hold control.

is, until it begins building up for the next cycle. Current limiting is an internal PCM or ICM, not an ignition coil, function. When current limiting is used, the PCM varies the dwell time in proportion to engine speed.

On systems without current limiting, the scope trace rises quickly to peak current, then drops immediately to zero as the circuit closes to fire the plug and begins the dwell section, figure 12-38. These systems control peak current with time. The dwell period in time remains constant regardless of engine speed, but changes when measured in dwell.

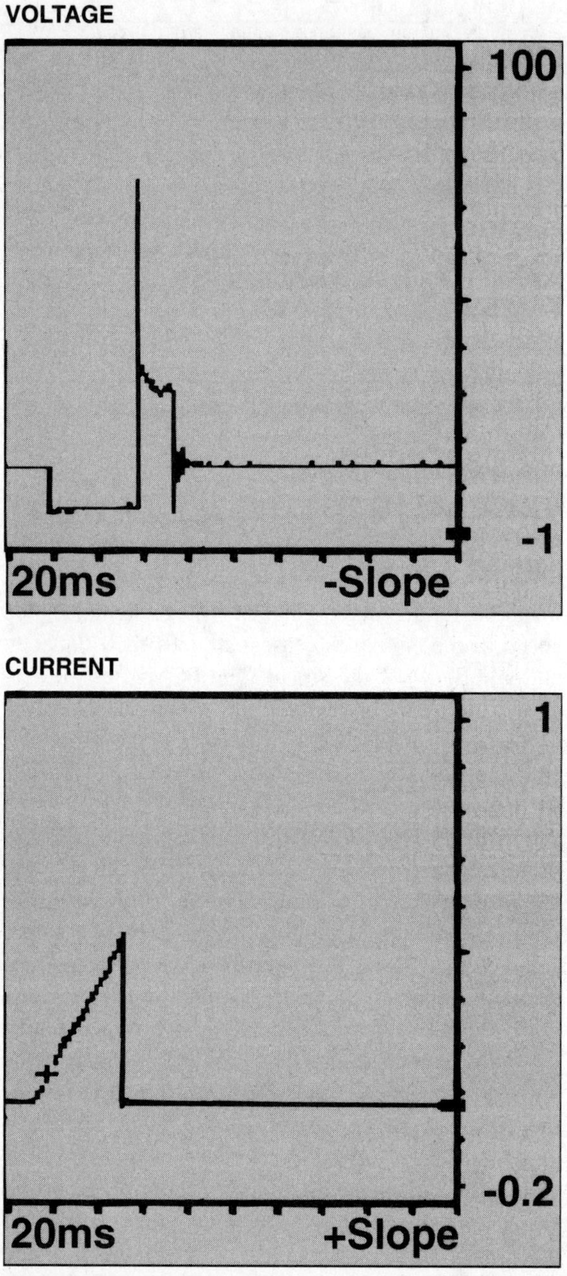

Figure 12-38. Typical primary ignition coil voltage and current trace without peak and hold control.

With either type of current control, look for a trace smoothly rising to the specified peak current and immediately dropping to zero as the primary circuit completes. Irregularities in the slope as current builds indicates resistance in the coil windings. A trace which does not reach the specified peak current indicates either high resistance in the coil or a power supply problem on the positive primary circuit. Ground circuit resistance is indicated by a voltage reading that does not drop close to zero volts. The trace continues as a horizontal line while the primary circuit is completed.

As previously mentioned, it is difficult or impossible to connect the current probe to the negative primary circuit on some models with direct ignition. These systems generally use a single positive circuit to supply all of the coils. Each coil has its own negative circuit. By connecting the current probe to the common positive circuit and adjusting the TIME/DIV setting of the scope or GMM, a current trace of all the coils is displayed in a parade pattern, figure 12-39. Current peaks should be even for all cylinders and at the level specified by the manufacturer.

Primary Ignition Triggering Devices

Electronic ignition systems rely on the signal of an engine speed sensor to determine when to open and close the primary circuit and fire the spark plugs. A number of different sensor designs are used to perform this task. The sensor, also referred to as a triggering device, may be located in the distributor or mounted on the engine block.

The primary circuit does not energize until the module receives a verifiable engine speed (rpm) signal from the crankshaft position (CKP) sensor. The CKP sensor provides the ignition module or powertrain control module (PCM) information on crankshaft speed and location. Some systems use an additional camshaft position (CMP) sensor for more precise timing of the ignition firing sequence, figure 12-40. A CMP sensor is used to provide the synchronization signal on distributorless ignition and sequential fuel-injection systems,

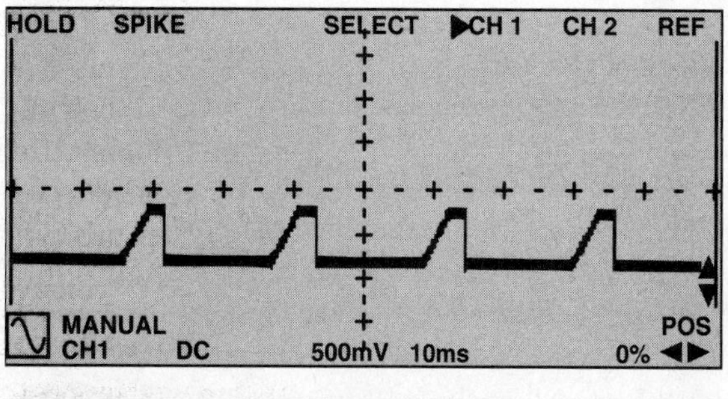

Figure 12-39. Parade display of direct ignition coil current taken on the positive circuit.

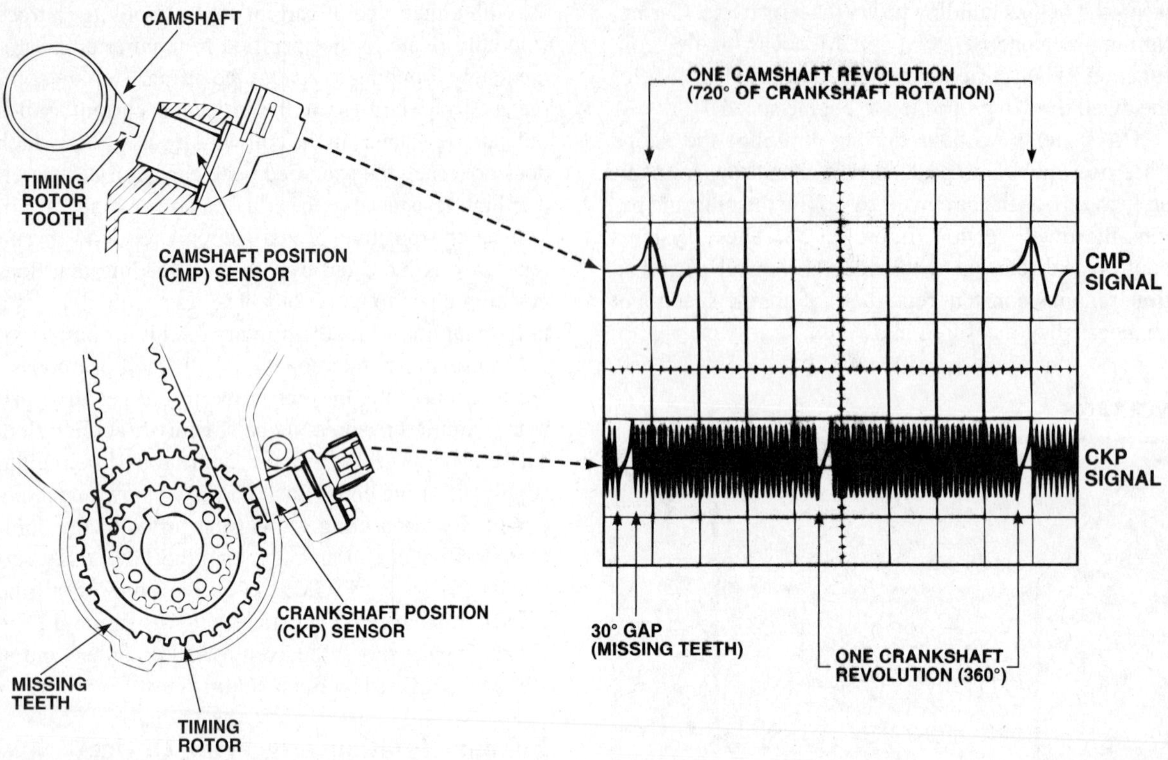

Figure 12-40. Some systems use both a crankshaft position (CKP) sensor and camshaft position (CMP) sensor to control ignition coil firing.

and for misfire detection on OBD II vehicles. There are four basic designs for automotive speed sensors:

- Magnetic pickup
- Hall Effect switch
- Optical sensor
- Magneto resistive sensor

The most accurate way to diagnose an engine speed sensor problem is to monitor the signal on a lab scope or GMM. A magnetic pickup is a sine wave–producing analog device, while Hall Effect switches and optical sensors are digital and produce a digital or square waveform. For all types, the ignition module or PCM uses the frequency of the signal, or how fast it repeats, to determine the speed of the sensed component.

Magnetic Pickup

Whether called a magnetic pickup coil, variable reluctance sensor, or permanent magnet generator, this type of sensor generates an analog alternating current (AC) voltage signal. These sensors are self-contained and do not require an applied voltage to produce a signal. Since these sensors produce their own signal, it is possible to get a waveform with correct amplitude and frequency, even though there are problems on the chassis ground circuits.

The pickup coil is wound around a permanent magnet, whose field expands and collapse as the teeth of a rotating trigger wheel pass by it, figure 12-41. The magnetic field generates an AC voltage signal. As the speed of the engine increases, so does the frequency and amplitude of the signal. The PCM uses the frequency of the signal to determine rotational speed.

Pickup coils have a two-wire circuit, a positive lead and a negative lead, connecting them to the PCM. For best results, connect both scope probes directly to the sensor leads as close to the sensor as possible. Backprobe a connector to connect the scope. Avoid piercing the wire insulation. A good trace generally sweeps up the positive slope and drops on the negative slope, figure 12-42. Check for inverted probe or wiring harness connections if the vertical spike is on the positive slope of the waveform, figure 12-43. On some models, the PCM cannot accurately monitor the signal if the harness connections are inverted, or reversed. This may be the source of a driveability problem. The shape of the trace peak varies for different sensor designs, but most signals must reach a minimum amplitude, or peak voltage, before the PCM recognizes them. Look for uniformity in the trace cycles. Remember too, the PCM watches for the signal to switch from positive to negative and negative to positive. This should occur at

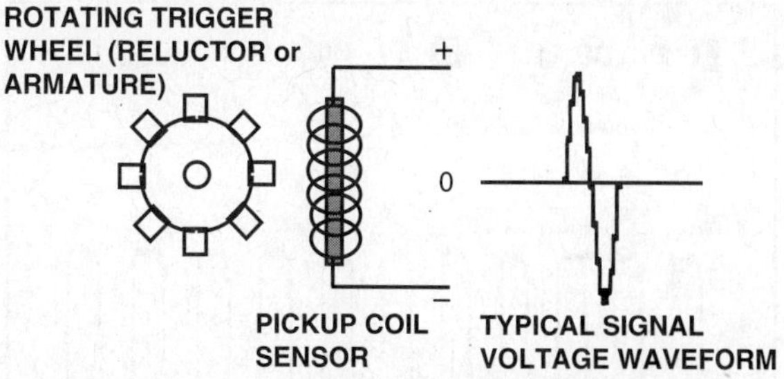

Figure 12-41. A pickup coil is self-powered and generates an analog AC signal as the magnetic field expands and collapses.

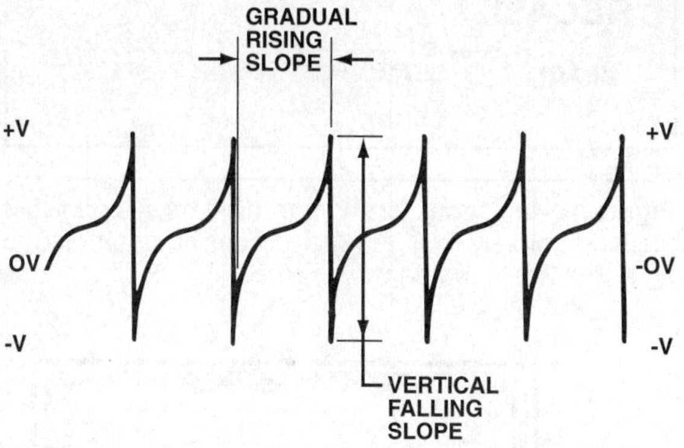

Figure 12-42. A pickup coil generates an AC pattern that ramps up and drops sharply.

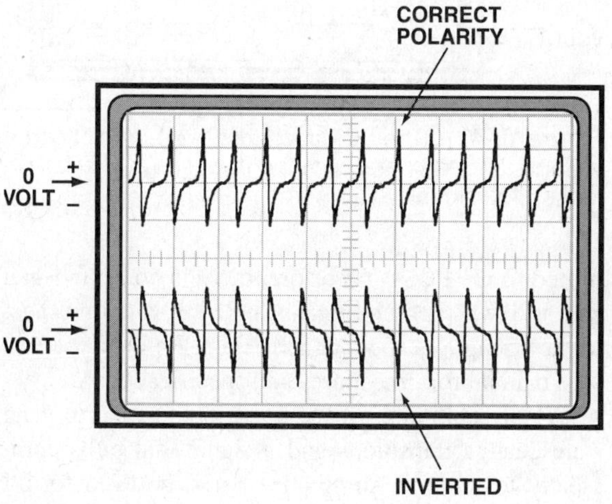

Figure 12-43. Reversed pickup coil polarity, which causes driveability problems, produces an inverted trace.

the 0-volt baseline. If not, check for poor, or resistive, connections on the PCM and chassis grounds.

The resistance of a pickup coil can be checked with an ohmmeter because all pickup coils operate at a specific resistance. However, specifications and test points vary between manufacturers and models. On some vacuum advance distributors, vacuum is applied while testing the pickup coil.

In general, measurements are taken by connecting the ohmmeter leads to opposite sides of the coil. If the reading is not within specifications, replace the coil. To check for a grounded coil, connect one ohmmeter lead to ground and touch the other ohmmeter lead alternately to each of the pickup coil connectors. The ohmmeter should show infinite resistance at all test points. If not, replace the pickup coil.

Air Gap Check and Adjustment

On some magnetic pickups, the air gap between the pickup coil and trigger wheel is adjustable. Air gap has no effect on the dwell period; dwell is determined by the control module. However, the air gap must be set to a specific clearance when a new pickup unit is installed. Although the air gap should not change during service, it should be checked before performing troubleshooting tests. Always use a brass, or other nonmagnetic, feeler gauge to check and adjust air gap. The pickup coil magnet attracts a steel feeler gauge and cause an inaccurate adjustment. Air gap differs between manufacturers, as well as between models. Therefore, the correct specifications are necessary when making adjustments.

Hall Effect Switch

A Hall Effect switch uses a microchip to switch a transistor on and off and generate a digital signal transmitted to the PCM. As a shutter wheel passes between the Hall Effect element and a permanent magnet, the magnetic field expands and collapses to produce an analog voltage signal. The Hall element contains an electronic logic gate that converts the analog signal into a digital one. This digital signal is used to trigger transistor switching, figure 12-44. The transistor transmits a digital square waveform at a variable frequency to the PCM.

Hall Effect switches produce a square waveform, whose frequency varies in proportion to the rotational speed of the shutter wheel. A Hall Effect switch does not generate its own voltage, and power must be provided for the deivce to work.

Hall Effect operation requires a three-wire circuit—power input, signal output, and ground. The Hall element receives an input voltage from either the ignition switch or the PCM to power it. Activity from the magnetic field and shutter blade opens and closes a switch to ground on the input signal. Be aware, output signal voltage is not always the same as input signal voltage. Therefore, accurate specifications are needed for testing.

A Hall Effect switch creates a square waveform as the transistor switches on and off. To get a good scope trace, attach the positive scope probe to the transistor

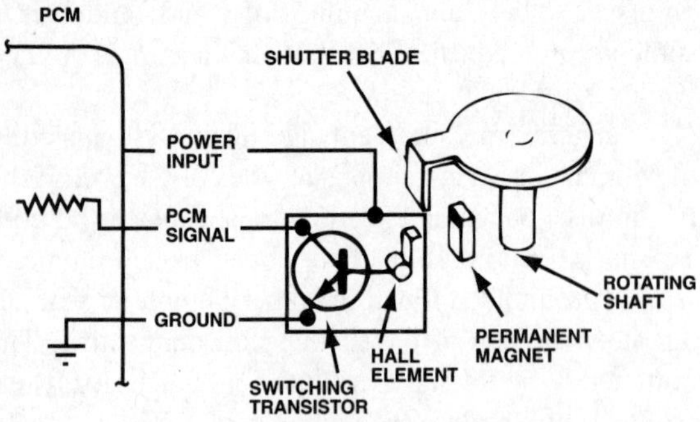

Figure 12-44. A Hall Effect switch uses the analog signal of a magnetic field expanding and collapsing to switch a transistor and generate a digital signal to the PCM.

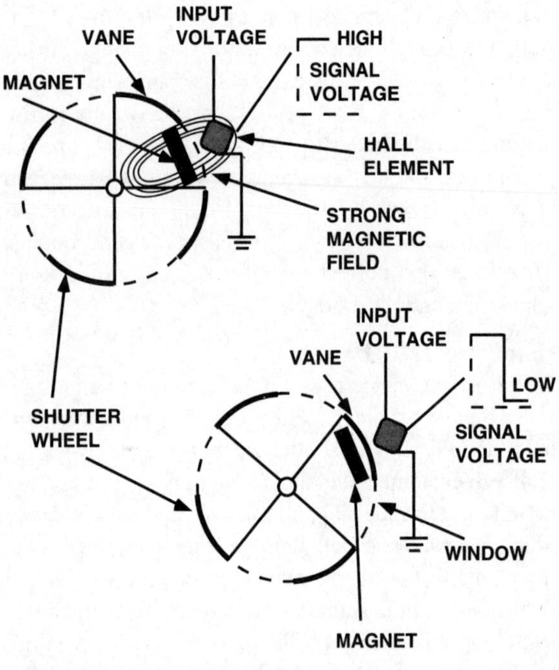

Figure 12-45. The signal of a Hall Effect switches high as the shutter wheel changes the magnetic field from strong to weak.

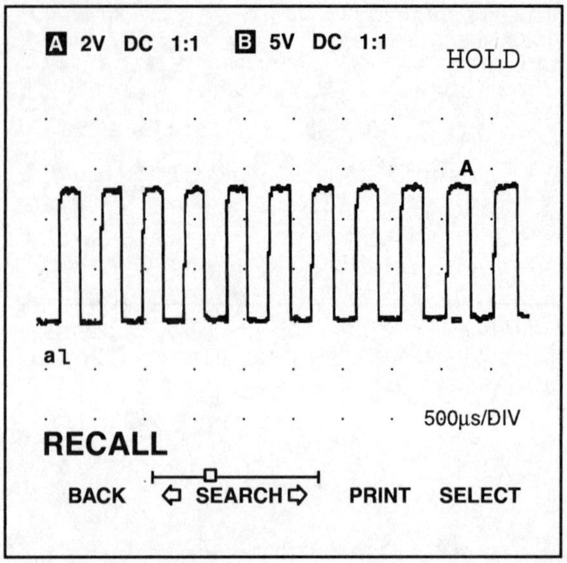

Figure 12-46. Look for sharp transistors, a signal that pulls completely to ground, and consistent amplitude on a Hall Effect switch waveform.

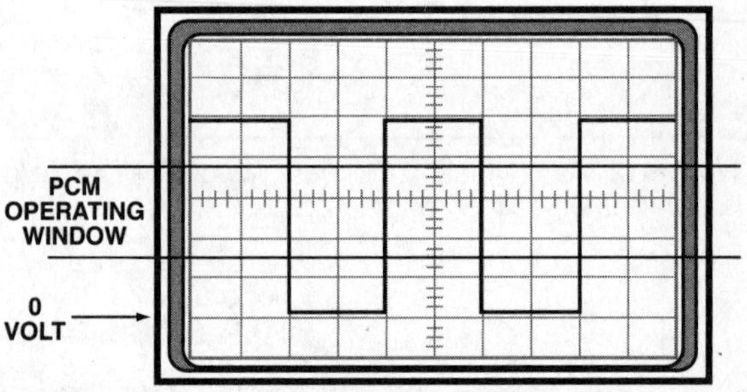

Figure 12-47. The PCM monitors the midpoint of the voltage range to determine the frequency of a Hall Effect switch signal.

output signal and connect the negative probe to the sensor ground. For optimum results, connect both probe leads as close as possible to the sensor. When a shutter vane is between the sensing element and the permanent magnet, the magnetic field is weak because the shutter blade disrupts it. When there is an open shutter blade window between the element and magnet, the magnetic field is strong. Each time a shutter window opens or closes, the signal generated by the Hall element changes state. That is, it switches on or off, figure 12-45. Depending upon the application, this signal may be amplified or inverted before it is transmitted to the PCM. Therefore, there is no established rule as to whether the signal is high or low with respect to the strength of the magnetic field. Refer to the service manual for procedures and specifications.

When examining a scope trace, look for sharp, clean state change transitions and a signal that pulls completely to ground. Amplitude should be even for all waveforms, the pattern should be consistent, and peaks should be at the specified voltage level, figure 12-46. The shape and position of the slots on the shutter wheel determine the shape and duty cycle of the waveform. Some Hall Effect switch patterns have a slight rounding at the top corners of the trace. This is normal for these units and does not indicate a problem on the circuit. Keep in mind, the PCM looks for switching at the midpoint of the voltage range, not at the top or bottom, figure 12-47. However, rounding at the bottom corners

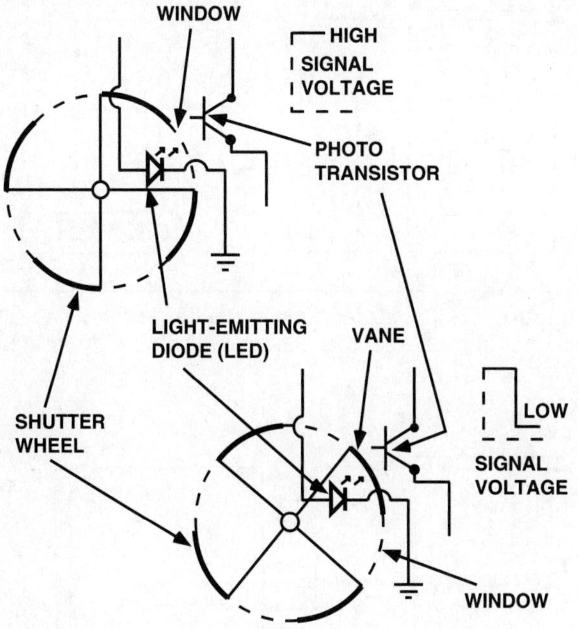

Figure 12-48. Optical sensors use an LED and a light-sensitive transistor to transmit a variable-frequency digital square waveform to the PCM.

of the trace is not normal and does indicate a problem, even though the PCM does not process this portion of the signal. Rounded corners at the bottom of the trace, or a signal not completely dropping to zero, indicates high resistance on the ground circuit. High ground circuit resistance is often the result of a poor connection, making it difficult for the signal to completely ground. Hall Effect sensors produce an output signal by grounding the signal voltage supplied. This voltage must be applied to the sensor by the PCM or other component. Disconnect the sensor to check for the correct bias voltage on the signal wire. Also, check for the correct voltage on the power circuit to the Hall element. A problem here causes problems on the signal circuit.

As a quick check, a logic probe may be used to verify the voltage signal is switching from high to low.

Optical Sensor

An optical sensor uses a light-emitting diode (LED), a shutter wheel, and a phototransistor to produce a digital signal that changes frequency in proportion to rotational speed, figure 12-48. As with a Hall Effect switch, an optical sensor requires an external power source and uses a three-wire circuit. One wire carries power to operate the LED, one is the signal generated by the transistor, and the third provides a common ground path.

Signal voltage, which is usually 5 volts, switches on and off as the rotating shutter passes between the phototransistor and LED to open and close the ground circuit. When the shutter allows light to shine on the phototransistor, the base of the transistor switches and the signal voltage changes state. When the reflector plate blocks the light to the phototransistor, the base of the transistor switches again and signal voltage changes as well.

Optical sensors are more expensive to manufacture and more delicate than a magnetic pickup or Hall Effect switch. Therefore, they are the least common of the three types. However, Chrysler, General Motors, and a few import manufacturers use optical sensors on some of their distributor-type ignition systems. Optical sensors are typically used as vehicle speed sensors and engine speed sensors because their high-speed data rate is more accurate for high rpm applications than other sensor designs.

When viewed on a scope, an optical sensor produces a square waveform similar to that produced by a Hall Effect switch. Optical sensors ground the applied signal voltage to produce an output signal. This applied, or bias, voltage is generally provided to the sensor by the PCM. Disconnect the sensor to check for the correct bias voltage on the signal wire.

Magneto Resistive Sensor

Many of the late-model systems use a magneto resistive sensor, which is a magnetic pickup with a square wave output. This sensor has the same signal as a Hall Effect switch or optical sensor. A magneto resistive sensor uses a magnetic pickup to produce an analog signal, which is converted to a digital signal inside the sensor.

As with the Hall Effect sensor, three wires are required—power input, signal output, and ground. However, a bias voltage is not required on the signal wire, which provides the full on/off signal. Many magneto resistive sensors are referred to as Hall Effect sensors because of the three wires and square wave signal. A visual inspection of the sensor confirms the type of sensor.

Secondary Circuit Testing

Problems in the secondary ignition circuitry are usually easily detected on the oscilloscope trace. Typical failures are the result of high or low resistance in the high-tension cables, distributor cap and rotor wear or damage, and worn or damaged spark plugs. Caps, rotors, and plugs are inspected and replaced as needed when irregularities appear on the secondary scope trace. High-tension cables are checked with an ohmmeter to confirm their continuity.

Excessive resistance in the secondary circuit causes a number of driveability problems, such as engine misfires, higher burn voltages, and shorter burn times. Damaged high-tension cables are often the cause of

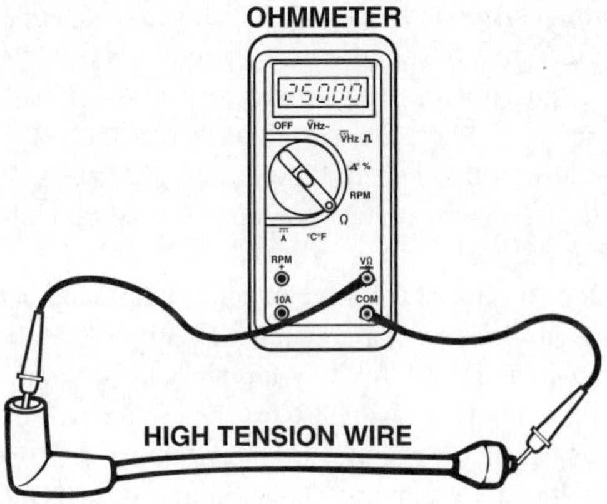

Figure 12-49. A spark plug cable should register about 4,000 ohms per foot of resistance on an ohmmeter.

excess secondary resistance. High cable resistance often results from:

- Loose or corroded connections at the distributor cap terminal or spark plug
- Damage to the cable conductor from heat, vibration, or mishandling
- A broken terminal at either end of the cable

Inspect the cables for obvious signs of damage and replace as needed. If damage is not visibly apparent, check the resistance of the cables with an ohmmeter. A typical television-radio suppression (TVRS) cable should measure about 4,000 ohms per foot. Test resistance by removing the high-tension cable from the engine and attaching the ohmmeter leads to each end of the cable, figure 12-49. If meter readings are out of specification, replace the cable.

Replace damaged secondary parts as needed, then check operation on the oscilloscope to confirm the repair.

ELECTRICAL NOISE

Electrical noise is a loosely defined term used to describe abnormal voltage levels on one circuit resulting from electrical activity on a second circuit. The two circuits share no common connection. The interference, or noise, is induced through the insulation of the wiring.

An electromagnetic field is created whenever current travels through a conductor, and the strength of the field changes when the current stops and starts. Each change in field strength creates an electromagnetic signal wave. The frequency of the wave increases in proportion to the speed of the current changes. When the signal wave is strong enough, it may be induced into an adjacent circuit.

In this phenomenon, known as electromagnetic interference (EMI), the signal wave is generated by the

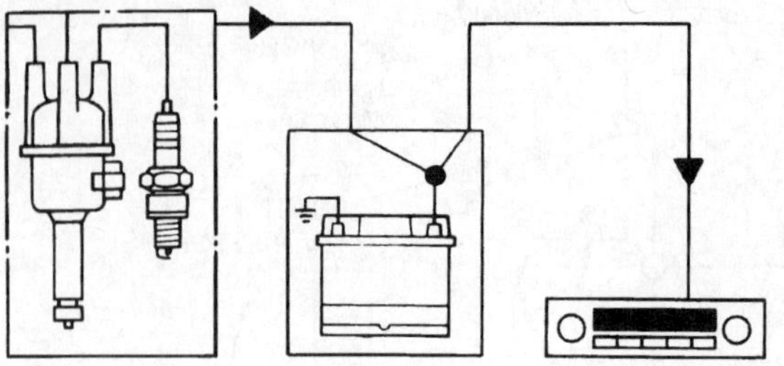

Figure 12-50. A conductive couple forms when a signal from the source is transmitted directly to the receiver.

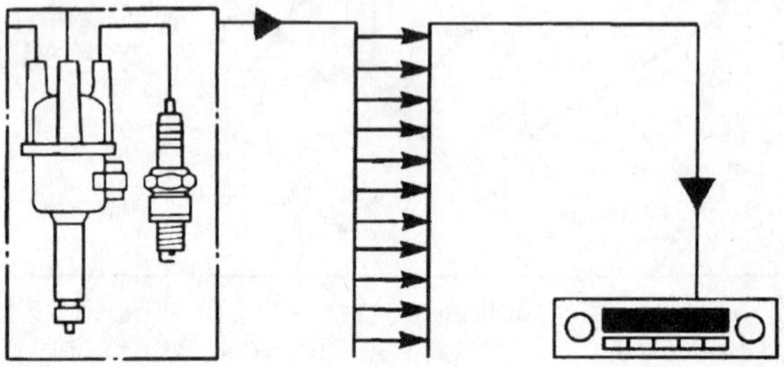

Figure 12-51. A capacitive couple forms when an electrostatic field forms between the source and receiver.

source and transferred into the receiver. EMI can be transmitted in four ways:

- Conductive coupling
- Capacitive coupling
- Inductive coupling
- Electromagnetic radiation

A conductive couple occurs when interference is directly transmitted from the source to the receiver, figure 12-50. Capacitive coupling takes place when applied voltage creates an electrostatic field between the source and receiver, figure 12-51. An inductive couple forms when applied voltage creates an electromagnetic field between the source and receiver, figure 12-52. Electromagnetic raditaion occurs when the receiver acts as an antenna and picks up signal waves generated by the source, figure 12-53.

In automotive applications, electrical noise may be classified into two categories: radio frequency interference (RFI) and electromagnetic interference (EMI). Although RFI is a form of EMI, radio frequency signals are much stronger and should be considered separately.

Radio Frequency Interference

Radio frequency interference produces a signal wave strong enough to interfere with radio and television

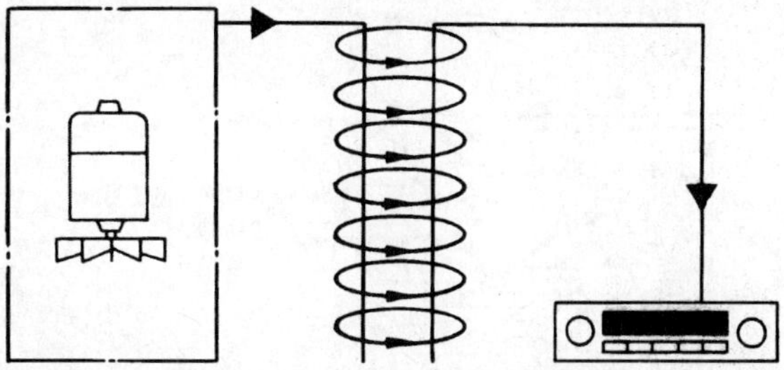

Figure 12-52. Inductive coupling occurs as a magnetic field forms and collapses between the source and receiver.

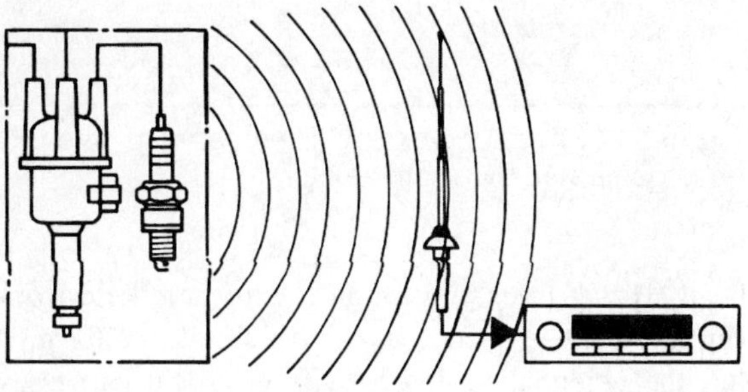

Figure 12-53. Electromagnetic radiation is when the receiver acts as an antenna to pick up signals generated by the source.

broadcast transmissions. Secondary ignition circuits are capable of producing waves at the rate of thousands of cycles per second, or kilohertz (kHz). Signals of this magnitude fall into the range of commercial broadcast signals. In addition to disrupting broadcast signals, RFI transmissions also create serious problems on the low-power circuits of an automotive computer control system.

To prevent RFI from affecting commercial broadcasts, vehicle manufacturers are required to use television-radio-suppression (TVRS) cables on the secondary ignition system. These cables have a high-resistance, nonmetallic conductor core, such as carbon, linen, or graphite-impregnated fiberglass wrapped into several layers of insulating material, figure 12-54. Due to their high resistance, TVRS cables raise the voltage applied to the spark plugs. This offers the additional benefits of improved ignition characterisitcs and prevents the formation of some plug deposits. Higher voltage at the spark plug permits the combustion of leaner air-fuel mixtures, and also reduces the amount of current present after ignition is complete. This reduction of current extends the life of the spark plugs and the distributor cap terminals.

In general, RFI transmission is of little concern because vehicle manufacturers have taken steps to eliminate the problem. However, replacement ignition cables that do not meet original equipment specifications, or are incorrectly installed, may cause unwanted signals to be induced into other circuits.

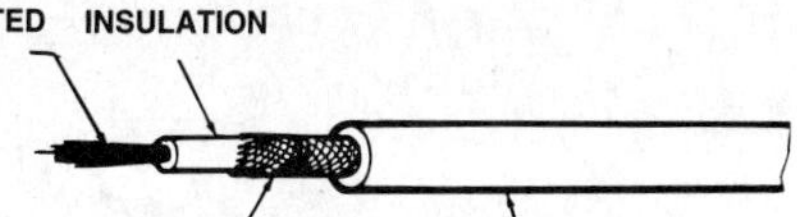

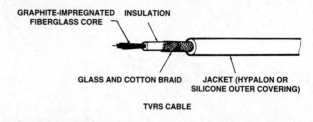

Figure 12-54. A TVRS cable has a nonmetallic, high-resistance core surrounded by several layers of insulation.

A common RFI problem is cross firing. Cross fire occurs when voltage from one cable is induced into an adjacent cable, causing a spark plug to fire out of sequence. To prevent cross fire, route cables so they do not touch, and avoid running them parallel. This parallel rule also applies to any primary wiring and harness in close proximity to the ignition cables. When conductors cross each other at an angle, the magnetic field is disrupted and induced voltage is prevented.

Electromagnetic Interference

The revolving commutator in motors, the natural fluctuations in the alternator output circuit, electromechanical voltage regulators, and vibrating horn contacts all produce EMI transmissions. In addition, a wave is created whenever a relay or switch opens or closes a circuit, and when a solenoid is activated. Static electric charges developed from the friction of tires contacting the road, drive belts turning pulleys, axles and driveshafts rotating, and clutch and brake applications can also produce EMI transmissions.

Manufacturers are aware of this problem and take steps to suppress EMI on suspect circuits. Common suppression methods include the use of capacitors, suppression filters, ground straps, and shields.

A capacitor is a device which absorbs and stores voltage. When installed at the switching point of a circuit, a capacitor absorbs changing voltage and prevents EMI transmission. Manufacturers often install capacitors on the primary circuit at the ICM, figure 12-55, and on the output terminal of the alternator, figure 12-56. Suppression filters, which also absorb unwanted voltage signals, are often used to suppress the natural voltage fluctuations produced by the commutator of an electric motor, figure 12-57. Most replacement parts do not include the capacitor or suppression filter. When replacing parts, it is important to transfer any capacitive suppression device from the old part to the new one to prevent EMI.

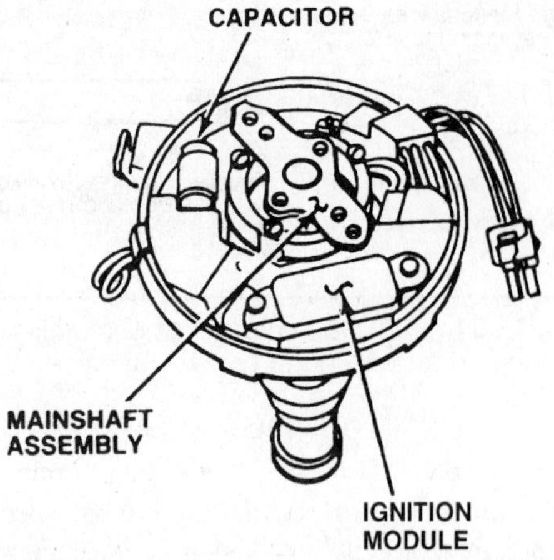

Figure 12-55. A capacitor at the ICM absorbs EMI produced by the primary circuit switching on and off.

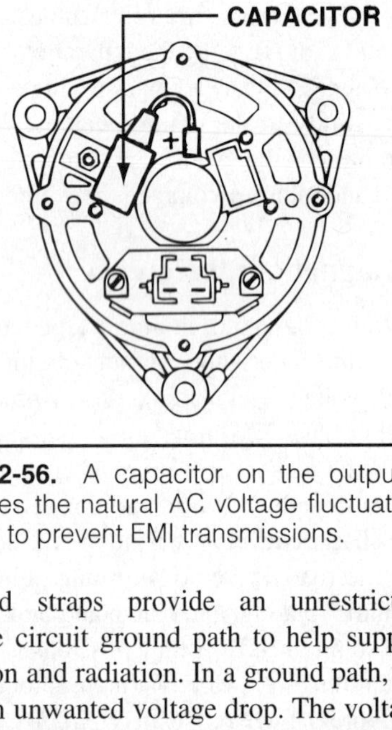

Figure 12-56. A capacitor on the output terminal suppresses the natural AC voltage fluctuations of an alternator to prevent EMI transmissions.

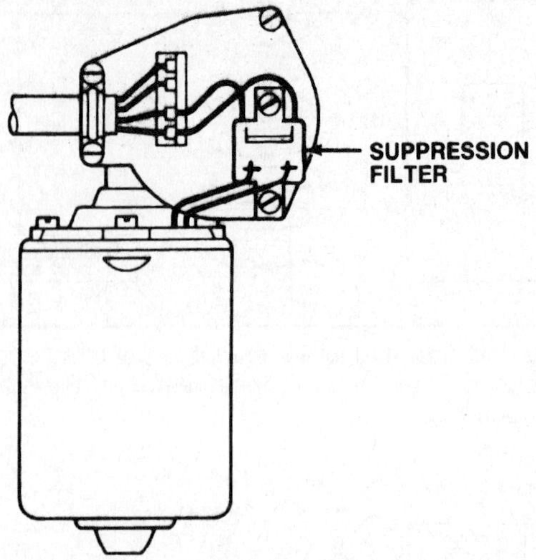

Figure 12-57. Suppression filters absorb interference generated by electric motors.

Ground straps provide an unrestricted, low-resistance circuit ground path to help suppress EMI conduction and radiation. In a ground path, resistance creates an unwanted voltage drop. The voltage difference across the resistance is a potential EMI source. Ground straps are typically installed between body parts and bushing-mounted components to ensure good conductivity. Some late-model vehicles have a ground strap connecting the hood to the fender or bulkhead. These straps prevent an electrostatic field from forming between body panels. If a field does form, it has the potential to couple with electrical harnesses mounted to the sheet metal.

A simple metal shield surrounding a component capable of generating EMI signals effectively blocks any wave transmission. Several manufacturers use a metal shield on their distributors to control RFI radiation. In addition, the metal housing of an onboard computer acts as a shield to prevent EMI waves from entering the unit and causing internal damage.

Any time a suppression device is removed when servicing a vehicle, it is important to reinstall it on assembly. Problems related to EMI transmissions are difficult to diagnose because they are often random and unpredictable. Use a DMM, scan tool, or lab scope to monitor suspect circuits. Then, watch for abnormal circuit activity while activating switches, relays, solenoids, and motors. Also, use a ground strap to make sure your body is grounded when handling components containing a microprocessor. Be aware, simply sliding across a velour seat can generate enough static electricity to damage delicate computer circuits.

13

Ignition System and Distributor Service

OBJECTIVES

Upon completion and review of this chapter, you will be able to:

- Test and replace/repair ignition system control modules.
- Test and replace ignition coils.
- Perform spark plug service per manufacturer's specifications.
- Test and replace as required spark plug wires.
- Inspect and replace distributor caps and rotors as needed.
- Perform distributor overhaul procedures as required.

INTRODUCTION

This chapter addresses testing, repairing, and replacing individual components of the ignition system. Refer to the appropriate sections of this chapter to repair ignition failures revealed by the diagnostic and test procedures in Chapter 12.

Keep in mind, most modern ignition systems contain delicate and expensive electronic components that are easily destroyed by a voltage surge. The high-voltage surge that results from simply making or breaking a connection may cause damage. Always observe the safety precautions outlined in Chapter 12 of this *Shop Manual* to avoid damage and personal injury while working on the ignition system.

The ignition, fuel injection, and emission control operations are electronically linked together by the engine management system on modern vehicles. Every ignition system service includes accessing the onboard diagnostic program to check for and clear diagnostic trouble code (DTC) records. Ignition circuit and component problems commonly set a misfire DTC on OBD II vehicles, which must be cleared before emissions testing. When a DTC is found, follow the troubleshooting procedure recommended by the manufacturer exactly, without skipping steps, to locate the source.

PRIMARY SUPPLY CIRCUIT SERVICE

Failures in the primary supply circuit, which includes all the wiring and devices that connect the positive battery terminal to the positive ignition coil terminal, result in no or low available voltage readings at the coil. Although exactly how the power is delivered to the coil varies significantly by make and model, all ignition systems receive power through the ignition

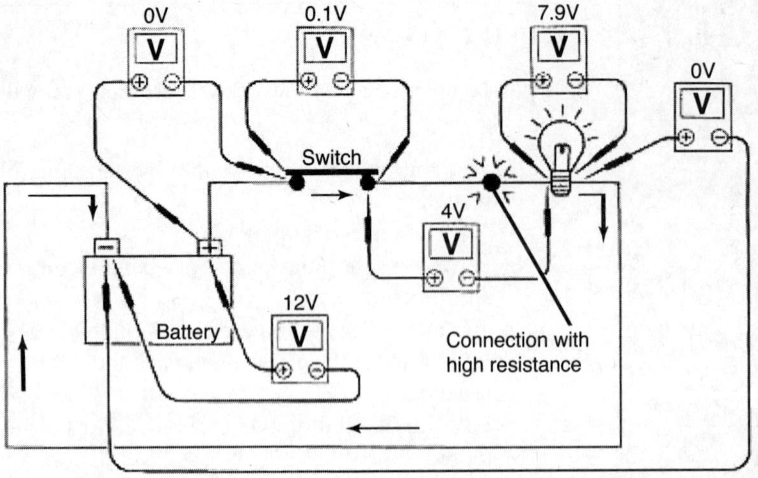

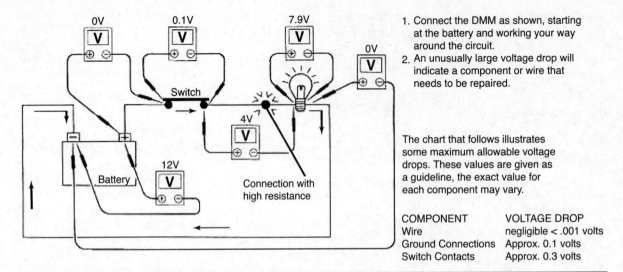

1. Connect the DMM as shown, starting at the battery and working your way around the circuit.
2. An unusually large voltage drop will indicate a component or wire that needs to be repaired.

The chart that follows illustrates some maximum allowable voltage drops. These values are given as a guideline, the exact value for each component may vary.

COMPONENT	VOLTAGE DROP
Wire	negligible < .001 volts
Ground Connections	Approx. 0.1 volts
Switch Contacts	Approx. 0.3 volts

Figure 13-1. Checking voltage drops in the primary ignition circuit is the same as checking voltage drops in lighting circuits. Measure across each connection switch. (Courtesy of Nissan North America, Inc.)

switch. Therefore, the ignition is linked directly to the starting and charging systems. Be aware, low available voltage to the coil may be a symptom of a starting or charging circuit failure.

Locate an accurate circuit diagram before attempting to troubleshoot the circuit. Low available voltage is a result of high circuit resistance and no voltage indicates an open circuit. Check for high resistance and open circuits by voltage drop testing. Almost any nonresistive electrical component can be voltage drop tested with a voltmeter, even if no specifications are available. Nonresistive electrical components of the ignition primary circuit include connectors, relays, and switches. With the engine running, work down the circuit from the coil to the ignition switch taking readings at each circuit connection and component, figure 13-1. Make repairs as needed if any high resistance is found.

Testing a Connector

The procedure for voltage drop testing an individual, single-wire connector is the same as for a multi-pin connector. Measure voltage drop across each circuit of a multi-pin connector. A fairly common problem is moisture collecting inside the connector shell, which shorts the individual conductors together. Some short circuits have high resistance, which results in some, but not all, of the current following the short. The remaining current travels the normal circuit path. Use a voltmeter to measure the voltage drop through the connector with the circuit powered and operating. To compute voltage drop across a connector:

1. Attach the positive voltmeter lead to the power-side circuit wire going into the connector and the negative meter lead to ground. Note the reading.

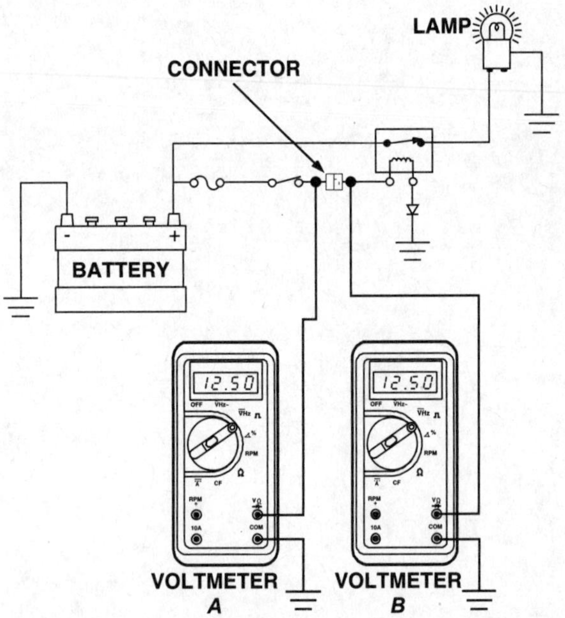

Figure 13-2. There should be negligible voltage drop across a connector, so readings are the same on both sides.

2. Next, hook the positive voltmeter lead to the load side or ground side circuit wire coming out of the connector, and the negative meter lead to ground. While taking a reading, wiggle or shake the connector and watch for any voltage change. Loose connector contacts cause the voltage to vary when the wire is moved.
3. The voltage readings on both sides of the connector should be the same, figure 13-2. Any difference represents the voltage loss of the connector.

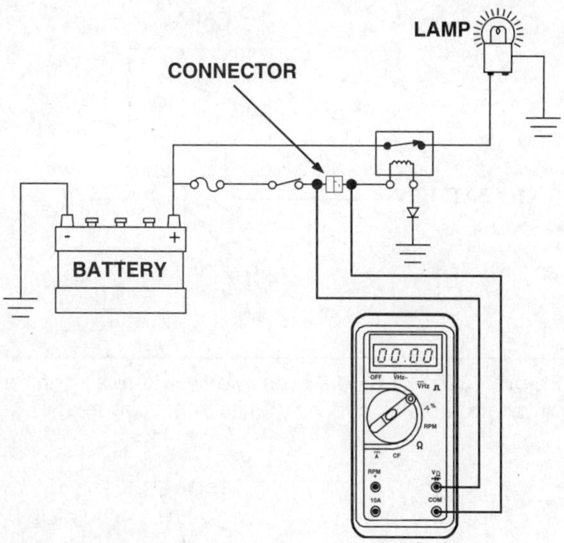

Figure 13-3. A direct voltage drop reading across a connector should be zero.

Voltage drop across a connector is directly measured by connecting the positive voltmeter lead to the power side of the connector and the common meter lead to the load side of the connector, figure 13-3. A direct reading across the connector should be zero volts. Any voltage measured indicates resistance in the connection.

Unless the contacts are loose, corroded, burned, or shorted to other contacts in a multi-pin connector, there should not be any voltage drop across a connector. If there is, the connector may need replacement, especially if the contacts are corroded or burned. Clean, tighten, and retest loose connections.

Testing a Relay

A relay, which is a mechanical device that uses a low-current circuit to open and close a switch and control higher current on another circuit, installs somewhere in the primary ignition circuit on a number of systems. The PCM switches the relay control circuit to energize and deenergize the relay coil. An electromagnet in the coil closes the power circuit contacts to allow a high-current signal through the power circuit. Relays may be switched on either the power or ground side of the circuit. Both types may be used at different locations on the same vehicle. Two common relay failures are defective coil windings and insufficient contact between the points, both of which increase resistance and create a voltage drop across the circuit.

To test the power side of a relay for circuit resistance, use a voltmeter to measure voltage drop. Voltage drop across a relay should be measured with the load drawing current. To compute voltage drop across a relay:

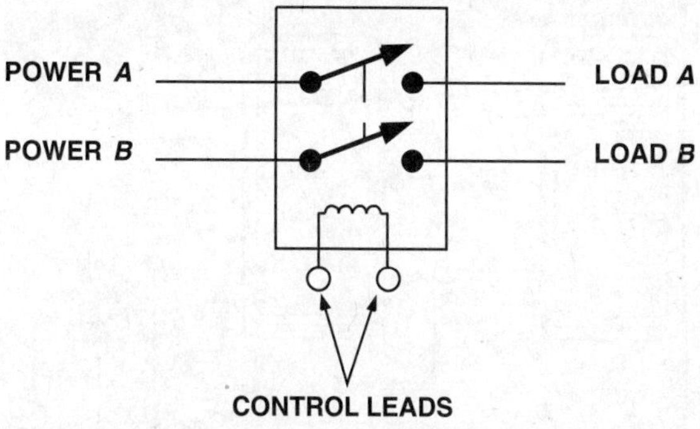

Figure 13-4. A single control circuit operates more than one load circuit on some relays.

1. Inspect the relay to determine which wires are the primary, or control, leads and which are the secondary, or load, leads. Some relays have one primary and multiple secondary circuits, figure 13-4. Refer to a pin chart or specifications in the service manual to accurately identify the power side of the relay.
2. Connect the positive voltmeter lead to the circuit wire entering the relay on the power side and the common meter lead to a good chassis or engine ground. Activate the relay by powering the primary, or control, circuit and note the voltmeter reading.
3. Leave the common meter lead grounded and connect the positive voltmeter lead to the circuit wire, leaving the relay on the load or ground side. With the relay activated, note the voltmeter reading.
4. The voltage readings on both sides of the relay should be the same, figure 13-5. The voltage drop is the difference.

Voltage drop across a relay is directly measured by connecting the positive voltmeter lead to the power-side secondary lead and the negative voltmeter lead to the load-side secondary lead. The voltage measured with the relay activated is the voltage drop of the relay, figure 13-6.

On most relays, there should be virtually no voltage loss across the relay contacts, unless they are corroded or burned. Although a voltage drop of 0.1 to 0.2 volt is acceptable on relays that are used to carry starter motor current, it is too much for all other relays. Any excessive voltage loss in a relay means it needs replacement.

Testing a Diode

When a relay is used to regulate current through electronic control circuits, a diode is often wired in parallel with the coil, figure 13-7. A diode makes a good one

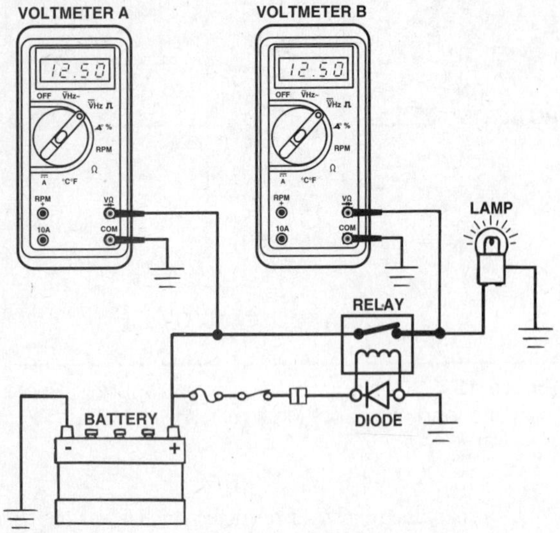

Figure 13-5. Voltage readings should be the same on both sides of a relay load circuit.

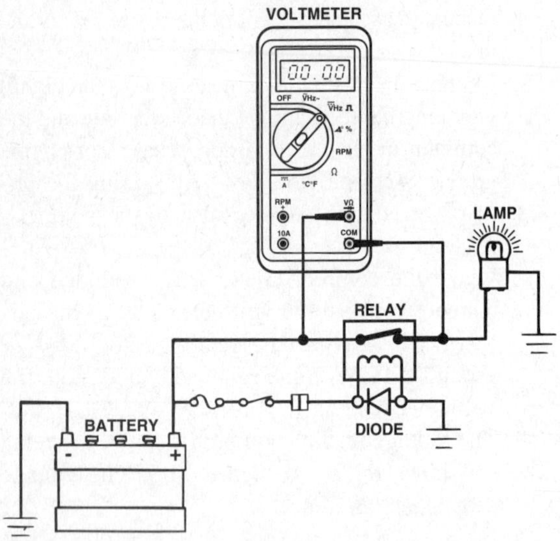

Figure 13-6. Look for a zero meter reading when measuring voltage drop directly across a relay load circuit.

way electrical *check* valve because it offers little or no resistance to current in one direction and is an open circuit to current in the other direction. The diode prevents voltage spikes generated by the coil from entering the electronic circuits.

A diode in parallel with a coil cannot be checked with a DMM. Instead, either an analog ohmmeter or a test lamp is used to check a parallel diode. A good diode has a higher ohms reading in one direction when the leads are reversed. With a test lamp, power the circuit through the test lamp and reverse the leads. The lamp should be brighter in one direction.

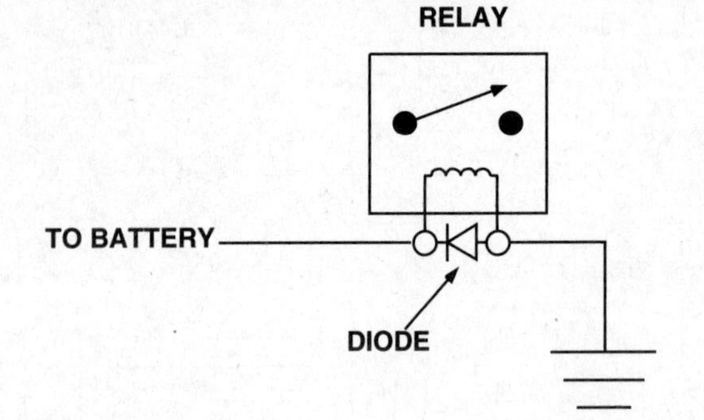

Figure 13-7. A diode in parallel with a relay control circuit protects electronic circuits from voltage spikes.

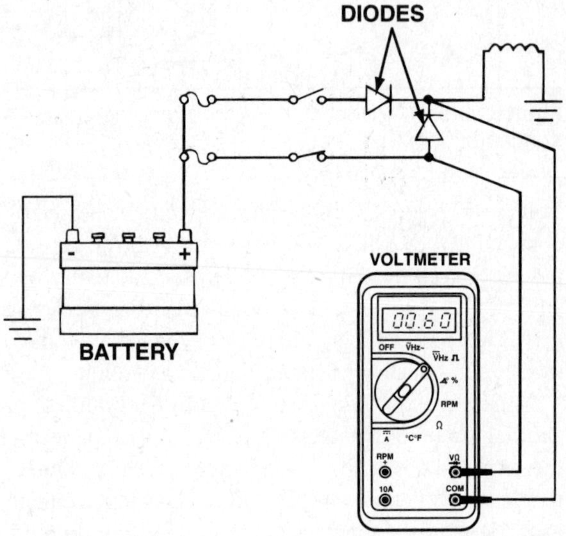

Figure 13-8. Taking a direct voltage drop reading across a diode.

A diode in series with a coil is checked by voltage drop testing. Take two voltmeter readings to compute voltage drop: one between the input, or power, side of the diode and ground, and one between the output, or ground, side of the diode and ground. Take readings with the circuit powered. Direct voltage drop readings are taken by connecting the voltmeter across the diode and powering the relay, figure 13-8. With either method, voltage drop across the diode should be about 0.6 volt.

Testing a Switch

To test any switch for circuit resistance, use a voltmeter to measure voltage drop through poles of the switch. Refer to the service manual of the vehicle being repaired to determine how current is routed through the switch in the different positions when test-

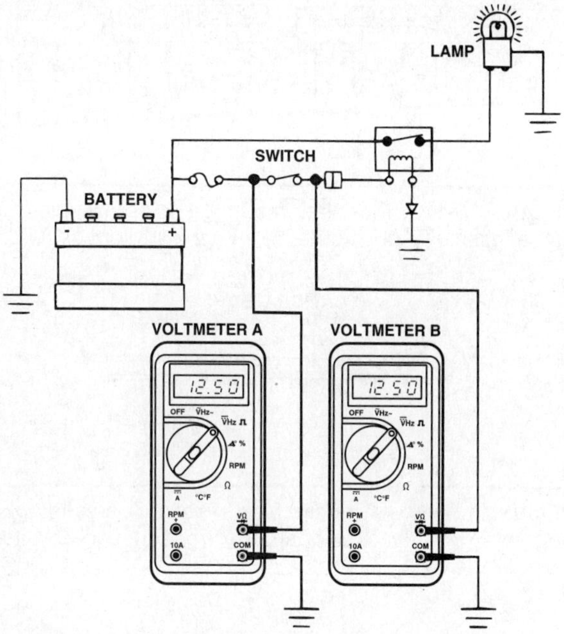

Figure 13-9. Voltmeter readings taken on either side of a switch should be equal.

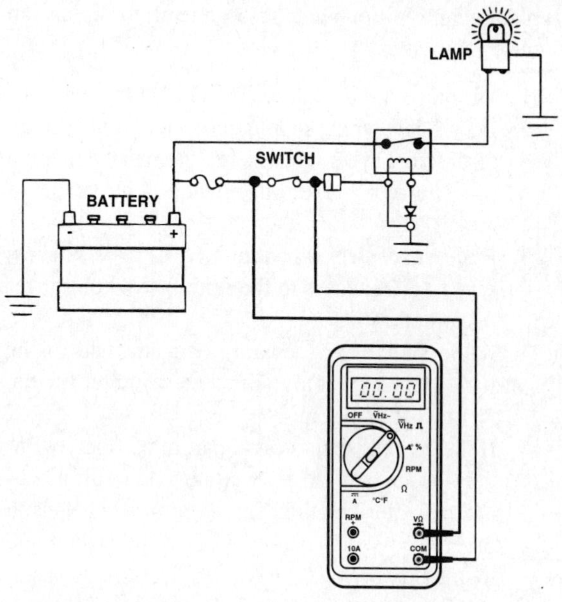

Figure 13-10. Expect a zero reading when measuring voltage drop directly across a switch.

ing an ignition switch. Voltage drop readings are taken with the circuit under power and operating. To compute voltage drop across a switch:

1. Connect the positive voltmeter lead to the power-side wire and the common meter lead to a good ground. Activate the switch and note the reading.
2. Leave the common meter lead grounded and connect the positive voltmeter lead to the wire leaving the switch on the load-side. Activate the switch and note the voltmeter reading, figure 13-9.
3. The voltage readings on both sides of the switch should be the same. Any difference is the voltage drop of the switch.

Measure voltage drop directly by connecting the voltmeter across the switch poles and activating the switch, figure 13-10. With either method, there should not be any voltage dropped in the switch. Voltage drop on a switch is usually caused by switch contacts that are corroded, pitted, or burned. If a switch has a voltage drop, replace it.

Ignition Switch

The ignition switch is typically tested without removing it from the vehicle by disconnecting the wiring harness multi-plug and checking continuity through the switch in all positions. Readings are generally taken with an ohmmeter. Locate the pin chart in the service manual for testing the ignition switch, figure 13-11. Replace the switch if continuity is other than specified.

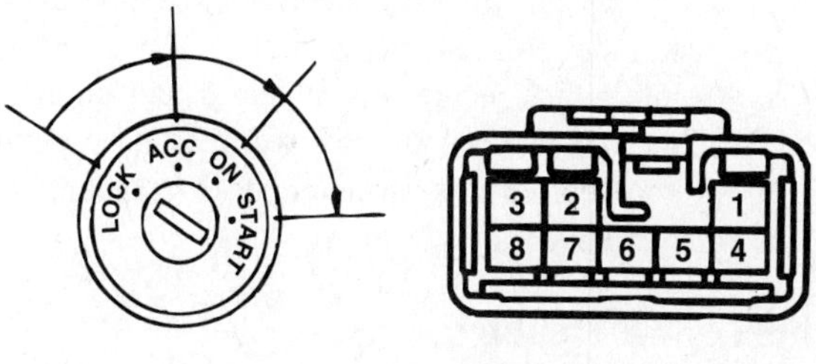

SWITCH POSITION	TESTER CONNECTION	SPECIFIED CONDITION
LOCK	-	NO CONTINUITY
ACC	5 - 7	CONTINUITY
ON	4 - 5 - 7 2 - 3	CONTINUITY
START	4 - 7 - 8 1 - 2 - 3	CONTINUITY

Figure 13-11. Pin chart for testing a 1998 Toyota Celica ignition switch. (Provided courtesy of Toyota Motor Sales U.S.A., Inc.)

An ignition switch assembles to the ignition lock tumbler, which mounts onto the steering column. On some models, the switch is an integral part of the lock assembly, while on others the electrical switch is a separate component. With either type, ignition switches are not repairable and must be replaced when defective.

Ignition Switch Replacement

Ignition switch replacement procedures vary by make and model. Refer to the factory service manual when replacing an ignition switch. In general, to replace an ignition switch:

1. Disconnect the negative battery cable.
2. On models with a supplemental restraint system, disable the system following the service manual procedure to prevent accidental deployment of the air bag.
3. Remove all steering column cover panels necessary to gain access to the ignition switch for replacement.
4. If the switch is a separate part attached to the lock cylinder, remove the fasteners and lift the switch off the lock.
5. If the switch is an integral part of the lock cylinder, loosen and move, or remove, any tilt mechanisms, interlock linkages, column brackets, and other accessories that connect to the switch and lock assembly.
6. Remove the lock assembly from the steering column. The lock assembly is often secured to the steering column with tamperproof bolts, which are removed with a small chisel and hammer.
7. Install the replacement switch.
8. Connect and secure any linkages and components removed or loosened to access the switch.
9. Fit the steering column cover panels and connect the negative battery cable.

Fuses and Fusible Links

The ignition switch power circuit is protected by a fuse, a fusible link, or both. Check these circuit protection devices if there is no voltage to the ignition switch with the key off. Fuses and fusible links carry a specified current without damage, but if this predetermined rating is exceeded, they open, or blow, to prevent circuit damage. Installing a new part does not solve the problem, as whatever caused the first fuse or link to blow quickly does the same to the replacement. Always determine and correct the cause of the blown device before installing a replacement.

Available voltage at the ignition switch that is lower than battery voltage with the key on indicates high circuit resistance between these two points. Locate the source by voltage drop testing. Repair as needed. If system voltage is available to, but not through, the ignition switch, the problem lies in the switch itself or one of the parallel circuits.

Fuses

Although fuses are available in a range of standard sizes and types, most domestic and import manufacturers use only blade-type fuses in late-model applications, figure 13-12. A miniature version, as well as a larger "maxi" version, of the standard blade-type fuse have gained in popularity in recent years. All fuses are rated by current capacity and the internal filament burns through to open the circuit when current rating is exceeded.

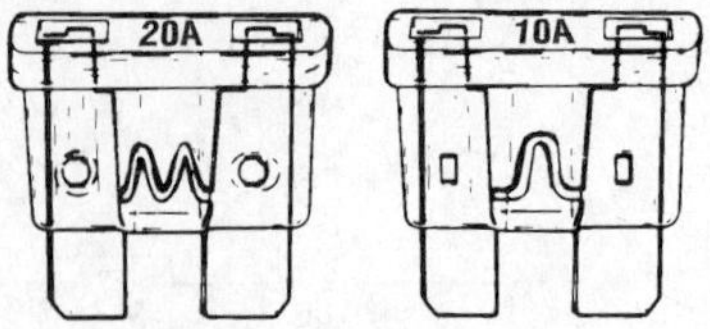

Figure 13-12. Most late-model vehicles use blade-type fuses of various current ratings to protect circuits.

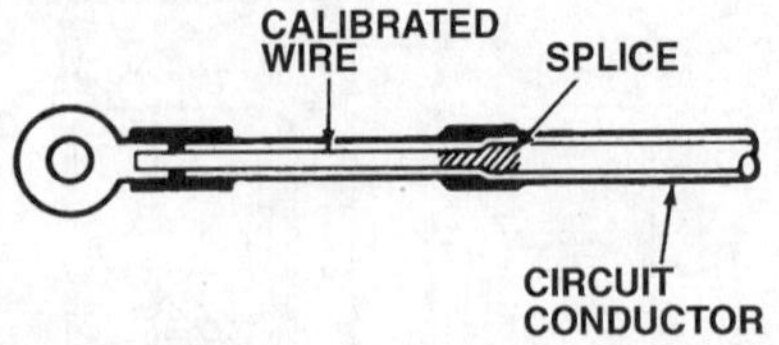

Figure 13-13. The calibrated wire of a fusible link opens to prevent circuit damage if current exceeds its amperage rating.

A fuse generally is checked while still in place, using a test lamp. If the lamp does not light on either side of the fuse, turn the ignition switch on and recheck. A voltmeter is used in a similar manner. A burned, or blown, fuse has a hot side and a dead side. Because a fuse is meant to burn out when current is too high, it should never be replaced with one whose rating is higher than the original equipment. A fuse with too high a rating causes circuit or component damage if an overload occurs.

Fusible Links

A fusible link is generally used as a backup to fused circuits to prevent harness damage in the event of a fire or major electrical malfunction. Fusible links are made of a special smaller wire and are part of the main wiring harness, figure 13-13. They react the same as a fuse, but have higher current ratings. Each fusible link protects several circuits that have their own fuses with a lower rating. A fusible link usually is connected in series with another accessory fuse. Some late-model vehicles have replaced this type of fusible link with a blade-type maxifuse that plugs into a relay or fuse box in the engine compartment.

When replacing a blown wire-type fusible link, be sure the replacement part is the correct size and properly rated for the circuit load. There are five different sizes, and each has its own current rating. To make the

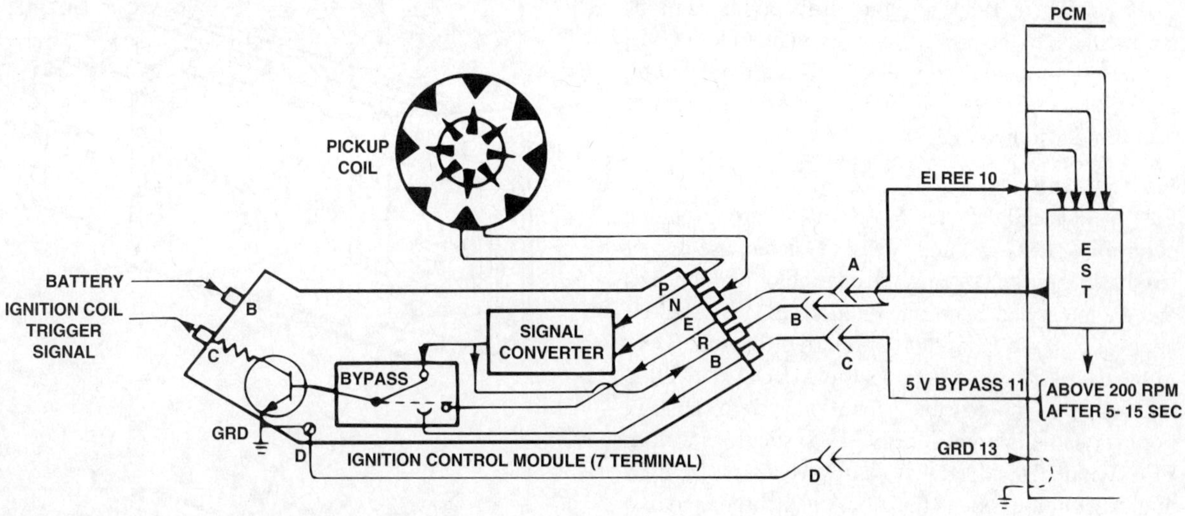

Figure 13-14. The ignition control module on this General Motors system electronically opens and closes the coil primary circuit. (Courtesy of General Motors Corporation)

repair, cut out the fusible link at the nearest connector on each side. Attach each end of the replacement wire, which must be the same length as the section removed, to the standard wire and solder the connection with rosin-core solder. Refer to Chapter 1 of this *Shop Manual* as needed for wire-splicing procedures. Allow the splice to cool, then wrap the connection with tape to insulate it. Never use acid-core solder on an electrical connection, as acid reacts to the electrical current, causing electrolysis that destroys the wire.

IGNITION CONTROL MODULE SERVICE

In general, an electronic ignition system functions as follows: The ignition control module (ICM) processes an input voltage from the distributor signal generator to complete the primary coil circuit and fire the spark plugs. Some signals are generated by a pickup coil or Hall Effect switch mounted in the distributor housing. On these systems, the ICM, which is also referred to as the igniter, generally mounts on or near the distributor. Direct, or distributorless, ignition systems (DIS) use a crankshaft position (CKP) sensor, as well as a camshaft position (CMP) sensor on some models, to generate an input signal to the ICM. The ICM generates the output signal to complete the primary circuit and initiate firing, figure 13-14. The ICM is incorporated into the powertrain control module (PCM) on some systems. On other DIS systems, the ICM is a separate component that mounts in the engine compartment. The ICM may bolt to a bracket on the engine, fasten to a fender panel or the bulkhead, or form the base of a coil pack.

Regardless of the ICM design or location, voltage drop tests are performed to locate open or grounded distributor signal generator circuits. Signals available at the ICM provide information on engine speed, timing advance, and cylinder firing. Monitoring these signals on a multi-trace lab scope provides a glimpse of how well the electronic control system is working. The ICM input signals must be correctly synchronized in order to fire the correct spark plug at the right time in the compression cycle. Ignition timing electronically adjusts to compensate for changes in engine speed and load.

Testing

Methods to determine whether an ICM is good or defective vary between manufacturers and systems. Procedures outlined by the vehicle manufacturer must be followed. In general, eliminate all other possibilities before condemning the module. Verify that power is available to the module, and check for voltage drop across connections on the power and ground circuits. Since modules provide the primary circuit ground connection, simply cleaning and tightening the ground circuit connections often solves the problem. Many modules ground through the fasteners that attach the unit to the engine or distributor. Voltage drop on an electronic ground circuit should never exceed 0.01 volt. Always check the ground circuit before replacing a component.

Chrysler Systems

Early Chrysler electronic ignition systems have a separate ICM that generally bolts to the engine compartment bulkhead, but the ignition control functions are incorporated into the PCM on all late-model applications.

Beginning in 1984, all ignition functions were incorporated into the powertrain control module (PCM). Ignition timing advance is electronically regulated by the PCM. Access the onboard diagnostic program to check and test the system.

Ford Systems

Late-model Ford vehicles, those with electronic engine control IV (EEC-IV) and EEC-V systems, use one of two basic ignition systems: the distributor ignition (DI) and the integrated electronic ignition (EI). On EEC-IV applications, the EI system is referred to as the electronic distributorless ignition system (EDIS) and DI is referred to as the distributor ignition system (DIS) in Ford service literature. All EEC-IV systems, as well as EEC-V models with a DI system, have a separate ICM unit. All ignition control functions are incorporated into the PCM on EEC-V models with an EI system.

Ford has very specific self-test programs built into the PCM on all models with electronic control systems. Accessing the self-test program is part of the diagnostic procedure. Results of a self-test determine which pinpoint tests in the Powertrain Control/Emissions Diagnostic Manual are conducted to isolate the source of the failure. Individual components, such as the ICM, should be tested only as and where instructed by the pinpoint tests.

Regardless of the type of engine management and ignition system, any Ford ICM functions in a similar fashion and provides valuable information for diagnosing ignition and driveability problems. Ford provides specifications and instructions for performing voltmeter checks of the ICM in the pinpoint test procedures. Monitoring the ICM signals on a multi-trace lab scope reveals how well the ignition and control system are operating. Refer to the appropriate service manual to identify the correct circuits for testing.

A typical Ford ICM is a small rectangular unit with either two six-pin connectors, one at either end, or a single twelve-pin connector, figure 13-15. In addition to battery power from the ignition switch, the harness connectors carry input signals from engine sensors and the PCM, feedback signals to the PCM, and output signals to the ignition coil or coils. The ICM size, location, configuration, and number of circuits used vary by model and year. The assembly attaches to a bracket or other mounting surface on the engine with screws. The ICM mounting screws provide the ground circuit connection for both the ICM and PCM, so make sure they are clean and tight. The ICM receives battery voltage from the ignition switch and a parallel circuit delivers voltage to the positive primary coil terminals. A ground circuit connects the ICM to the PCM.

The ICM receives a profile ignition pickup (PIP) signal, generated by either a CKP sensor or a CMP sensor depending upon application, which provides base timing information. On EI models, a magnetic pickup coil type CKP sensor provides input to the ICM through a two-wire, one positive and one negative, circuit. The output of the CKP sensor provides the PIP signal, which is transmitted to the PCM on another circuit. A missing tooth on the CKP sensor tone ring provides a crankshaft position reference that is clearly visible on the scope screen, figure 13-16. The frequency and amplitude of the analog signal increase in proportion to engine speed. Some, but not all, distributorless systems use an additional CMP sensor signal to synchronize ignition timing and sequential fuel injection (SFI). The CMP signal is also used for misfire detection on OBD II systems. The CMP sensor is a

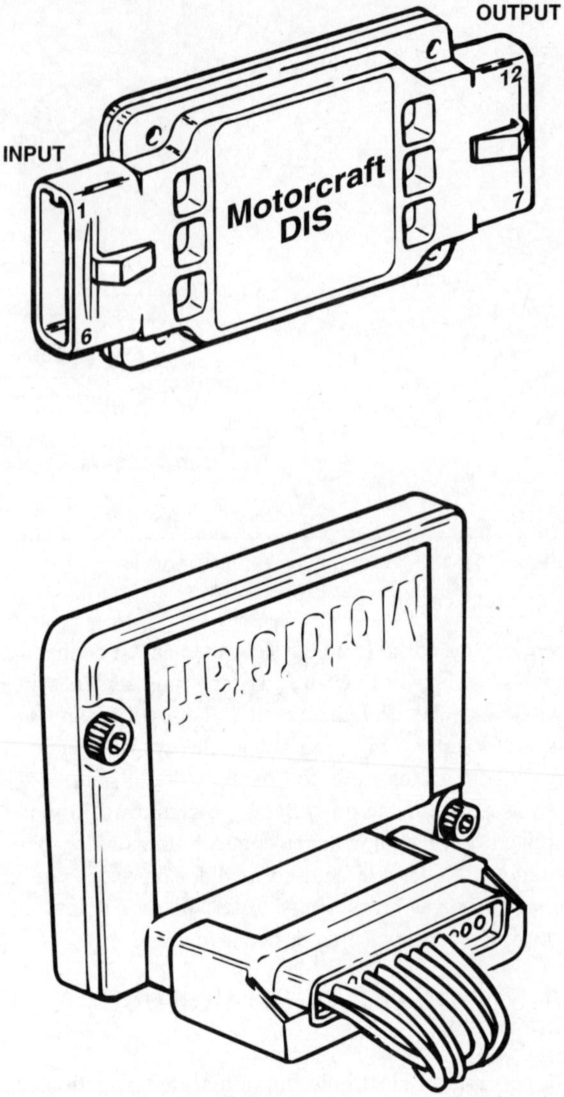

Figure 13-15. There are several versions of the Ford stand-alone ICM. (Courtesy of Ford Motor Company)

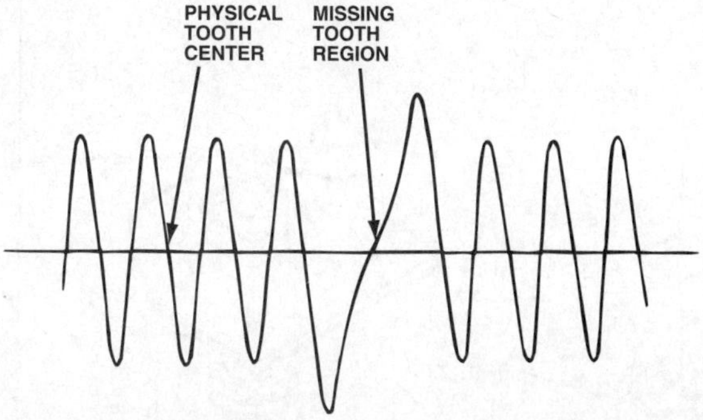

Figure 13-16. A missing tooth provides a crankshaft position signature on this Ford pickup coil signal.

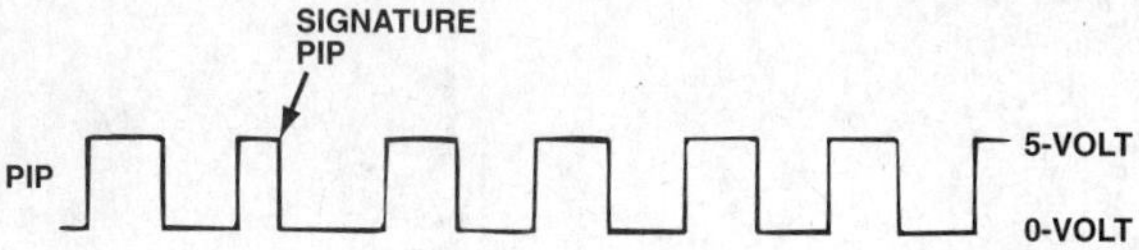

Figure 13-17. Position signature on the PIP signal from a Ford Hall Effect CMP sensor.

Hall Effect switch that provides what Ford calls the cylinder identification (CID) signal to the ICM. Look for a digital square wave that increases in frequency with engine speed and pulls completely to ground.

On late-model DI systems, the distributor houses a Hall Effect CMP sensor that provides a PIP signal to the ICM on a single-wire circuit, figure 13-17. A parallel circuit carries the CMP sensor signal to the PCM. One of the shutter wheel windows is smaller than the other openings, so for each camshaft revolution the pulse width of one cycle in the waveform is different from the rest. This signature waveform provides the timing synchronization reference and the engine will not start if it is absent. Be aware, the same distributor may be used with two different shutter wheels: one with a signature and one without. The signature is used to time SFI, and if installed on a non-SFI engine causes incorrect ignition timing. The frequency of the digital signal increases in proportion to engine speed.

On all systems, the ICM receives an additional input signal from the PCM that actually triggers the opening and closing of the negative primary coil circuit. This is referred to as the spark output (SPOUT) signal on older systems and the spark angle word (SAW) signal on newer models. The difference between the two is in how the computer processes the information; both are a digital square wave signal.

The ICM output circuits connect to the negative primary coil terminals and to the PCM. The number of primary coil circuits varies by system, but, as explained in Chapter 12 of this *Shop Manual,* all are quickly checked using a current probe. The ICM outputs diagnostic feedback information for the self-test program to the PCM on what Ford calls the ignition diagnostic monitor (IDM) circuit. A tachometer signal may be taken off the IDM circuit on some models, and is provided by a separate circuit on others.

General Motors Systems

Control of dwell and timing of the primary circuit to generate secondary current is a PCM function on all current General Motors domestic vehicles. Although some models may have a separate ICM, it is used only to switch the primary current off and on. The following applies to older ignition systems, both distributor and direct, that use a separate ICM to complete the primary ignition circuit.

On systems with the ICM assembly in the distributor, the module is a separate component that is replaceable, figure 13-18. General Motors distributorless ignition systems have a separate ICM unit that assembles onto the base plate below the coil pack, figure 13-19. Early systems have the CKP signal to the ICM, with the PCM controlling only the timing, while later systems have the CKP signal to the PCM, with the PCM controlling both timing and dwell. Test procedures are the same for both types. The integrated direct ignition system used on the Quad Four engine also uses a separate ICM that bolts to the cover shield and connects to the coils with a short harness, figure 13-20.

Although there are several variations of each module design, testing is similar for all of them. Certain signals should be available at select pins of the ICM, but the correct pin must be identified. Check the manual for the vehicle being serviced to identify ICM pin assignments. Most General Motors systems have the following signals at the ICM:

- Battery
- Coil primary
- Crankshaft position
- Reference
- Bypass
- Electronic spark timing
- Ground

The module receives battery voltage through the ignition switch. This power terminal, which is generally marked either "+" or "B+," should have at least 12 volts available with the key on. The ground circuit connects the ICM to the PCM, but does not provide the ground path for the ignition module. The ICM grounds through the mounting screws that attach it to the distributor base plate, coil-pack base, or cover shield. Make sure these connections are clean, tight, and offer no resistance.

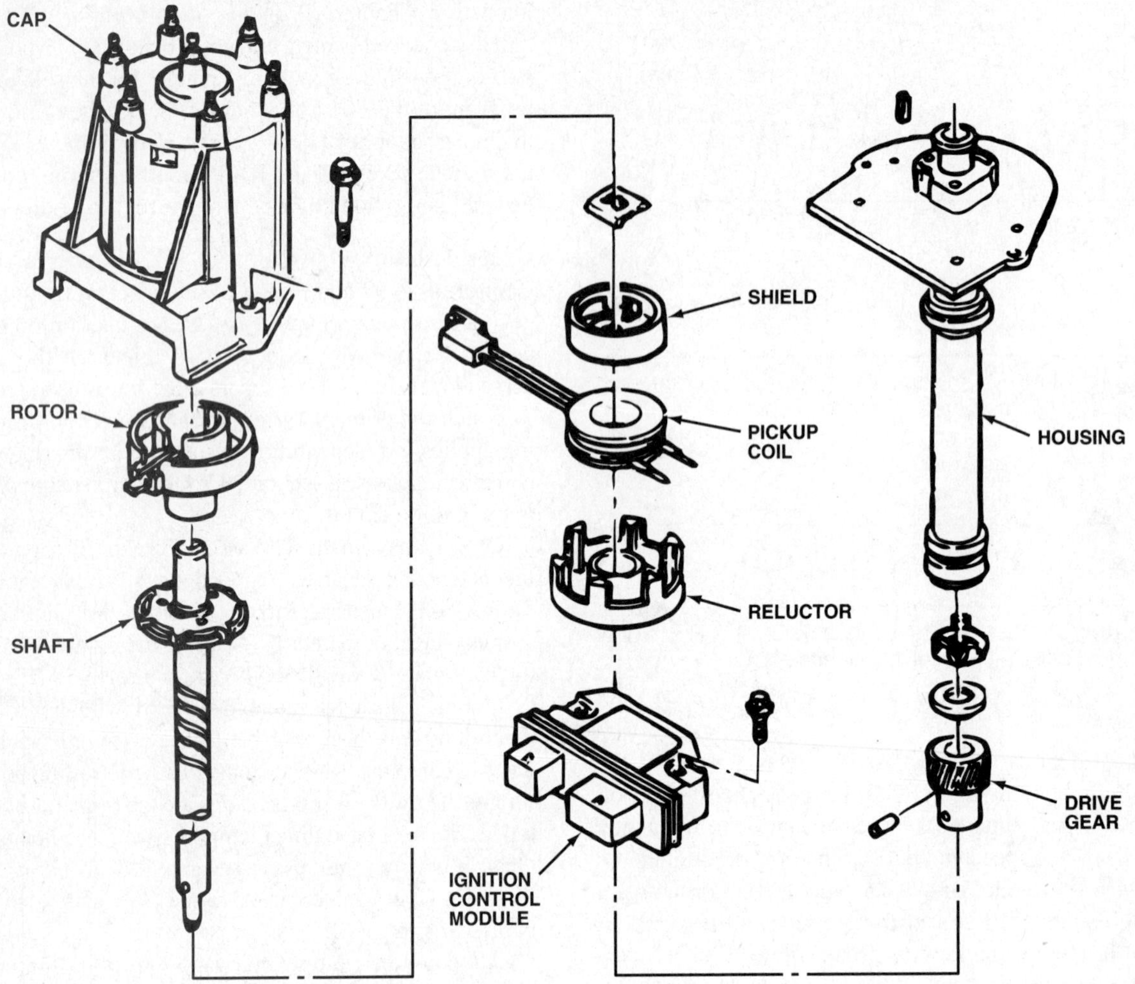

Figure 13-18. Some General Motors systems have a separate ICM installed inside the distributor. (Courtesy of General Motors Corporation)

Figure 13-19. The ICM bolts to the coil pack base plate on early direct ignition systems.

The ICM connects to the negative primary coil on what is commonly referred to as the TACH terminal on earlier models. This terminal is marked "C" or "TACH" on some applications, and is identified only by a number on others. This circuit is sampled with a current probe, as explained in Chapter 12 of this *Shop Manual,* to check coil operation.

Two circuits, often labeled "P" and "N," are the positive and negative circuits from the magnetic pickup coil used as a CKP sensor. Check these circuits on a scope and look for an alternating current (ac) pattern that varies in frequency and amplitude in proportion to engine speed. Some systems have a signature wave that is used to identify the number one cylinder, figure 13-21. The CKP sensor either assembles into the distributor, bolts to the engine, or attaches to the base of the coil pack assembly.

The reference circuit, labeled "R" on some models, carries a 5-volt reference voltage transmitted by the

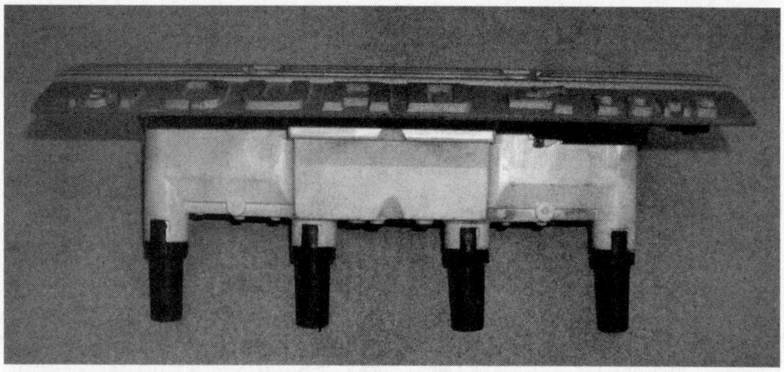

Figure 13-20. A Quad-Four ICM bolts to the coil assembly cover.

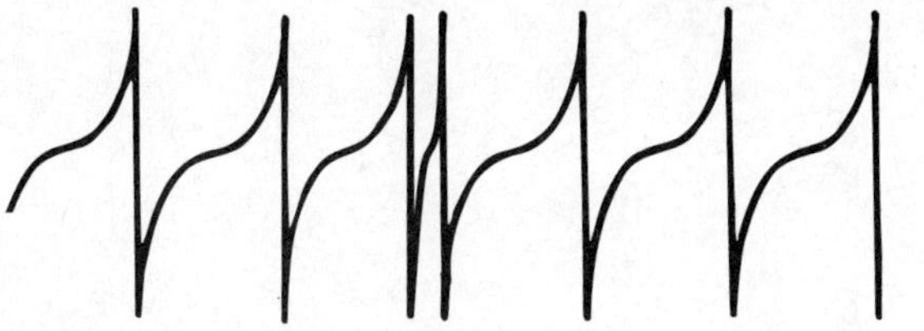

Figure 13-21. A CKP sensor often generates a signature wave to identify the number one cylinder.

PCM. The ICM internally switches the reference signal on and off to ground as the transistor in the module switches. On a scope, the reference circuit produces a variable frequency square waveform, figure 13-22. Each cycle, or pulse, of the waveform coincides with the crankshaft position of one cylinder for ignition firing.

The bypass and electronic spark timing (EST) terminals, often labeled "B" and "E," are the circuits the PCM uses to control timing advance. Once the engine starts, or exceeds a minimum rpm, the PCM applies a 5 volt signal to the ICM on the bypass circuit. The PCM, rather than the ICM, is controlling the opening and closing of the primary circuit when this bypass voltage signal is present. The EST circuit carries a variable frequency digital signal from the PCM to the ICM that switches the transistor on and off when the system is under PCM control.

The ground circuit provides a ground path between the PCM and ICM, but is not the ground source for the ICM. As previously mentioned, the ICM is grounded through the mounting fasteners.

Import Systems

Like their domestic counterparts, the PCM regulates the opening and closing of the primary coil circuit directly on most late-model import ignition systems. When the PCM does all of the control, a simplified ICM is used only as a switch for the primary circuit. This type of ICM is often called a power transistor, or

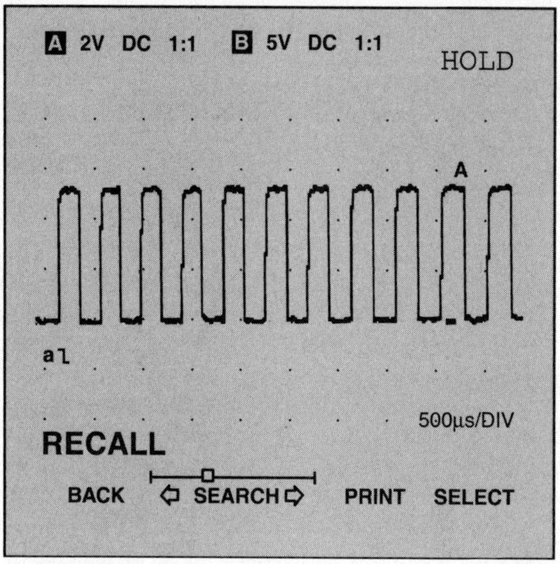

Figure 13-22. The reference circuit of the General Motors ICM cycles on and off with the base timing for each cylinder.

minimum function. An ICM that controls dwell, but the PCM controls timing, typically has more than three wires. Commonly called an igniter, this type of ICM may be located in the engine compartment or inside the distributor, figure 13-23.

The latest style is the individual direct ignition (IDI) system, which uses an individual coil for each spark plug. These "coil-over-plug" systems may be controlled by either a remote minimum-function ICM or an ICM built into each coil. The PCM transmits a control signal for timing and dwell to each spark plug. Some systems have a positive-switching ICM control signal, while others go negative, or ground, to turn on the primary circuit. The signal, which may include a signature wave, is easily viewed on a scope or graphing multimeter (GMM), figure 13-24.

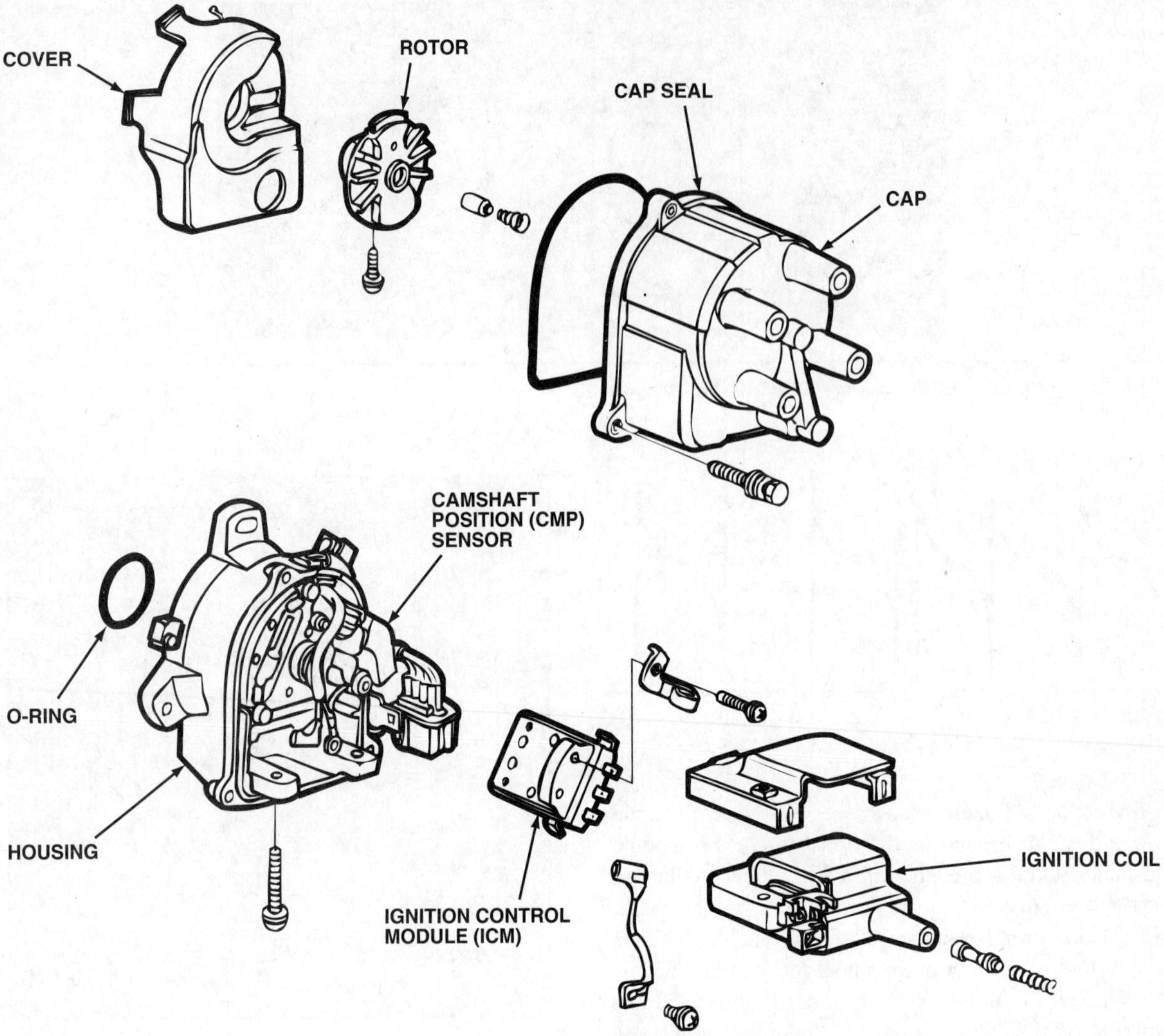

Figure 13-23. The ICM, along with the coil, assembles into this Honda Accord distributor. (Courtesy of American Honda Motor Co., Inc.)

A number of igniter designs are used and the number of circuits, location and mounting, test procedures, and specifications vary considerable by year, make, and model. Have accurate service information available for the system being repaired. In general, any ICM must have voltage applied, usually battery voltage from the ignition switch, and a nonresistive ground connection. The ICM mounting fasteners often complete the ground circuit. Since the ICM opens and closes the negative primary coil circuit, one of the terminals should carry a signal that switches on and off ground while the engine is running or cranking. If not, replace the ICM.

Replacement

Replacing an ICM that mounts in the distributor requires partial disassembly of the distributor. Often, it is easier to simply remove the distributor from the engine, disassemble it, and replace the ICM on the bench. Distributor removal, disassembly, and assembly techniques are detailed later in this chapter.

To replace a Ford ICM, carefully disconnect the harness multi-plugs and remove the mounting screws. A thin coat of silicone grease or heat sink compound is applied to the base of the ICM to increase heat transfer from the unit to the mounting bracket. Avoid getting grease in the mounting bolt holes or on the screws. Remember, both the ICM and PCM ground through the ICM fasteners. Any grease on the threads acts as an insulator and creates resistance in the ground circuit.

On General Motors systems with the ICM at the base of the coil pack, remove the coil assembly from the engine to replace the module. The coils bolt onto the module. With a direct ignition system, the coil

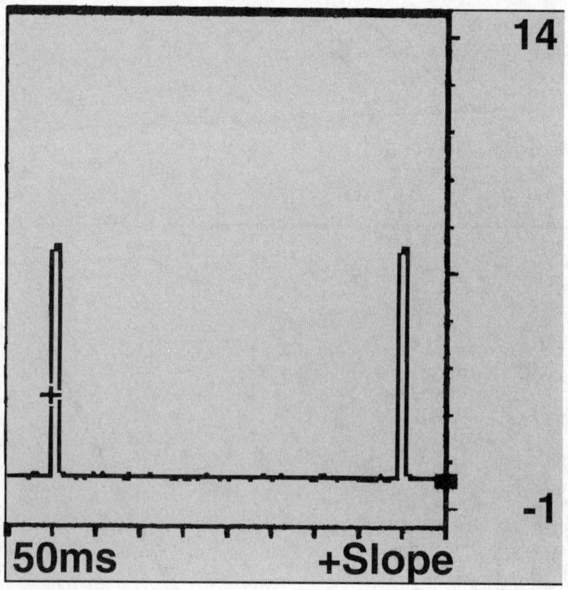

Figure 13-24. A positive switching ICM control signal and signature wave captured on a GMM.

housing assembly is removed from the engine as a unit and disassembled on the bench.

SPEED AND REFERENCE SENSOR SERVICE

The PCM or ICM uses the signals of the speed and reference sensors to determine the rotational speed and position of the crankshaft, camshaft, or both. The signal may be an analog ac voltage generated by a magnetic pickup coil, or the digital square wave produced by a Hall Effect switch or optical sensor. Regardless of design, the PCM uses the frequency of the sensor signal to determine the speed of the rotating component. Speed and reference sensors are tested with a digital multimeter (DMM). Operation of the sensor is observed by monitoring the signal on a lab scope or GMM.

Pickup Coil

Whether called a magnetic pickup coil, variable reluctance sensor, or permanent magnet generator, this type of sensor generates voltage as an analog signal. These sensors are self-contained and do not require an applied voltage to produce a signal. However, a slight bias voltage is applied to one coil lead on most systems. Since these sensors produce their own voltage, it is possible to get a waveform with correct amplitude and frequency, even though there are problems on the chassis ground circuits.

The pickup coil is wound around a permanent magnet, whose field expands and collapses as the teeth of a rotating trigger wheel, or tone ring, pass by it, figure 13-25. The magnetic field generates an AC voltage signal and, as the speed of the engine increases, so does the frequency and amplitude of the signal. Two wires, a positive lead and a negative lead, connect the sensor to the PCM. With the addition of a circuit built into this type of sensor, it becomes a magneto-resistive sensor. The analog signal is rectified to a variable-frequency digital square wave signal on these sensors.

For best results, attach scope probes and DMM leads directly to the sensor leads as close to the sensor as possible. Back-probe the harness connector to take readings.

A scope trace of the sensor generally sweeps up the positive slope and drops on the negative slope. Check for inverted probe or wiring harness connections if the vertical spike is on the positive slope of the waveform of most engines, figure 13-26. Be aware, this inverted pattern is normal for some manufacturers, such as Honda and Isuzu. The shape of the trace peak varies for different sensor designs, but most signals must reach a minimum amplitude before the PCM recognizes them. Look for uniformity in the trace cycles broken only by the signature of a missing tooth where applicable. The PCM monitors for the signal to switch from positive to negative and negative to positive, which should occur at the 0-volt baseline. If not, check PCM and chassis grounds.

Manufacturers generally provide resistance specifications for checking internal pickup coil circuits with a DMM. Sensor output voltage can also be checked using a DMM. Magnetic sensor output voltage is related to engine speed. Typically, it ranges between about 500 millivolts at cranking speed to approximately 100 volts at high rpm. When diagnosing an engine that cranks but does not run, the sensor output should exceed the minimum specified output while cranking. A signal that switches slowly typically cannot register on a DMM. An example would be a CMP sensor, which cycles once, or produces one sine wave, for each two revolutions of the crankshaft.

Pickup coil sensors may be mounted in the distributor or on the engine. Both CKP and CMP sensors mount at various locations on the engine. The tone ring for a CKP sensor may be on the crankshaft pulley, flywheel, or the crankshaft. Typically, a sensor has a short harness with a two-pin connector and the assembly attaches to the engine with one or two bolts. Note the harness routing before removing the sensor and install the replacement in the same manner.

Pickup sensors are not adjustable. However, they must be properly aligned and installed with a precise

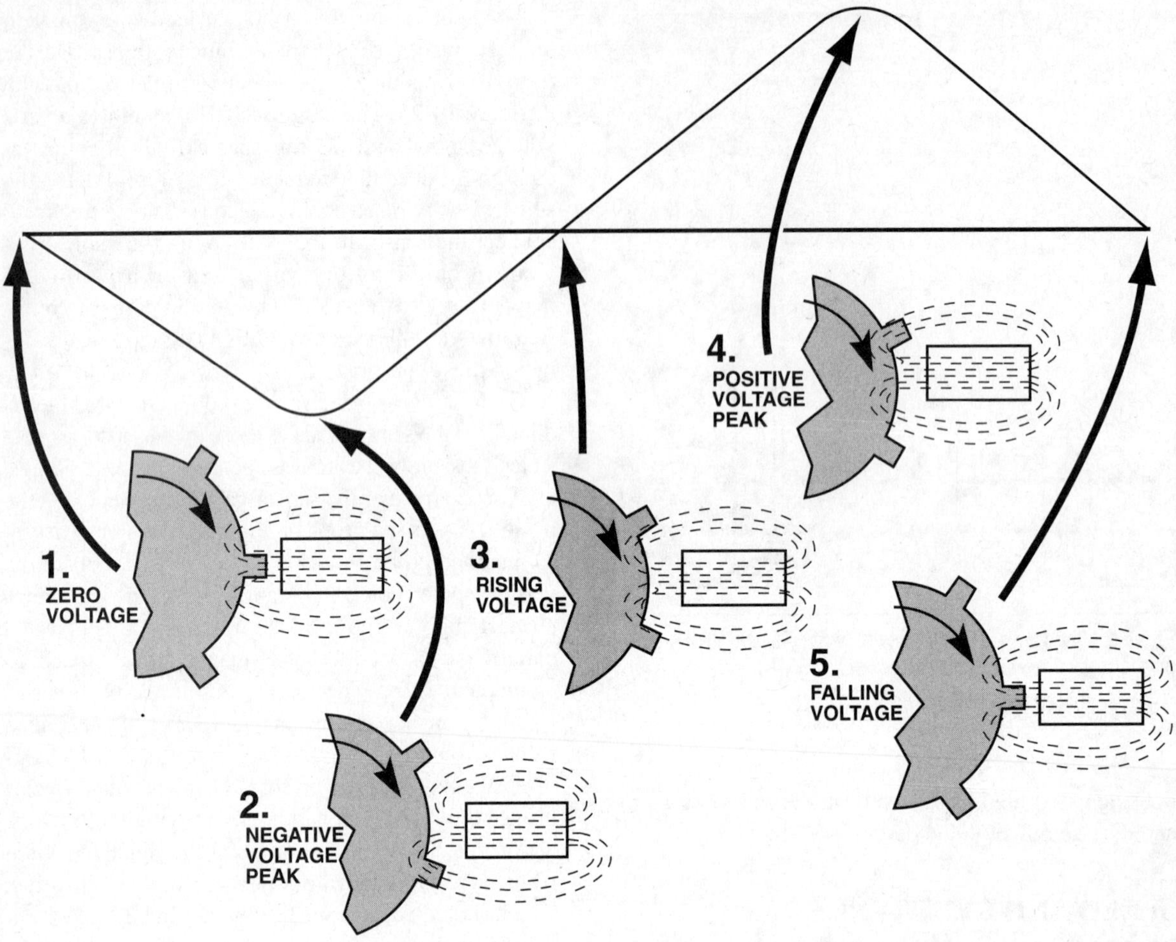

Figure 13-25. The teeth of the trigger wheel disrupt the magnetic field of a pickup coil to produce an AC signal.

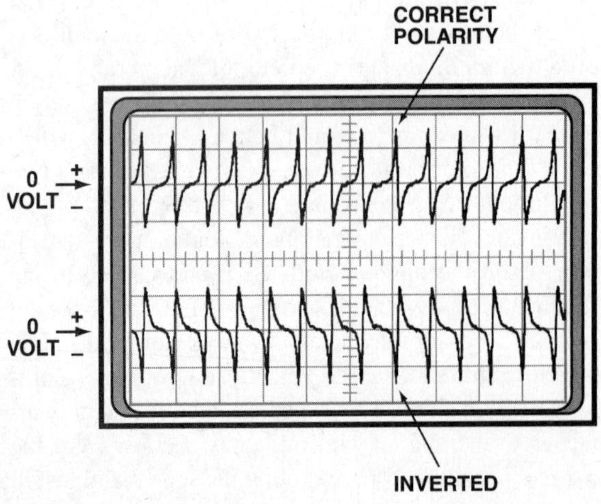

Figure 13-26. Reversed, or inverted, polarity on a pickup coil may cause driveability problems.

gap between the teeth and sensor to function properly. Misalignment, incorrect gap, damage to the tone rings, or accumulated debris may be the cause of an engine that sags, stumbles, hesitates, or diesels. Sensor failure often prevents the engine from starting, or causes it to die immediately after startup. Sensor damage due to road debris or vibration may cause engine misfire, stumble, or intermittent no-start conditions.

Hall Effect Switch

A Hall Effect switch uses a microchip to switch a transistor on and off and generate a digital signal that is transmitted to the PCM. As a shutter wheel or the magnet passes between the Hall element and a permanent magnet, the magnetic field expands and collapses to produce an analog signal. The Hall element contains a logic gate that converts the signal into a digital signal, which triggers transistor switching, figure 13-27. Hall Effect switches produce a square waveform, whose frequency varies in proportion to the rotational speed of the shutter wheel.

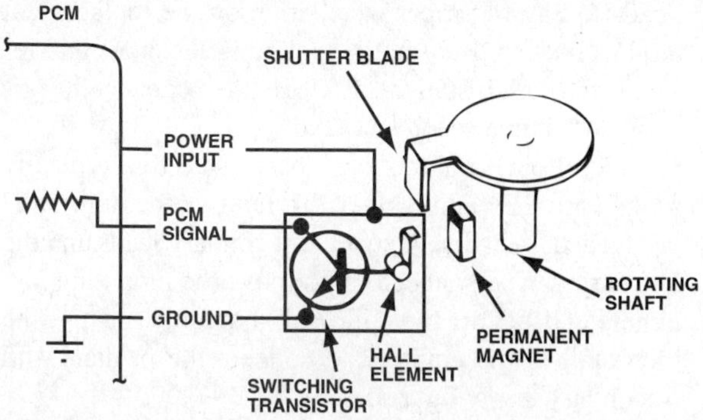

Figure 13-27. A Hall Effect switch consists of the Hall element, a permanent magnet, and the shutter wheel.

Applications for Hall Effect switches are similar to those for magnetic pickup sensors. However, since a Hall Effect switch does not generate its own voltage, power must be provided for the device to work. Therefore, Hall Effect operation requires a three-wire circuit—power input, signal output, and ground.

The Hall element receives an input voltage from either the ignition switch or the PCM to power it. Activity of the magnetic field and shutter blade opens and closes a switch to ground on the input signal. Be aware, output signal voltage is not always the same as input signal voltage. Therefore, accurate specifications for the unit being tested are required.

Hall Effect switches seem to function better on a higher voltage signal than other electronic components, and are often an exception to the 5-volt reference rule. Keep this in mind when checking supply voltage, the unit may require 7, 9, or even 12 volts to operate properly. On engines that use more than one Hall Effect switch, such as crankshaft and camshaft sensors, a common circuit generally supplies the power.

The most effective way to check a Hall Effect switch is to monitor the signal on a scope. Look for sharp, clean state-change transitions and a signal that pulls completely to ground. Amplitude should be even for all waveforms, the pattern should be consistent, and peaks should be at the specified voltage level, figure 13-28. The shape and position of the slots on the shutter wheel determine the shape and duty cycle of the waveform. Some Hall Effect switch patterns have a slight rounding at the top corners of the trace that can generally be overlooked. Remember, the PCM looks for switching at the midpoint of the voltage range, not at the top or bottom. However, rounding at the bottom corners of the trace is often caused by high resistance on the ground circuit. Also, check for the correct voltage on the power circuit to the Hall element. A problem here causes problems on the signal circuit.

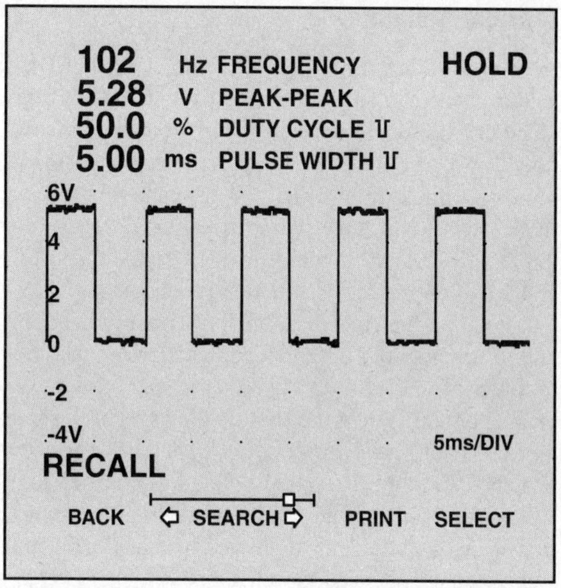

Figure 13-28. Look for clean, sharp transitions and a zero-volt baseline on a Hall Effect switch trace.

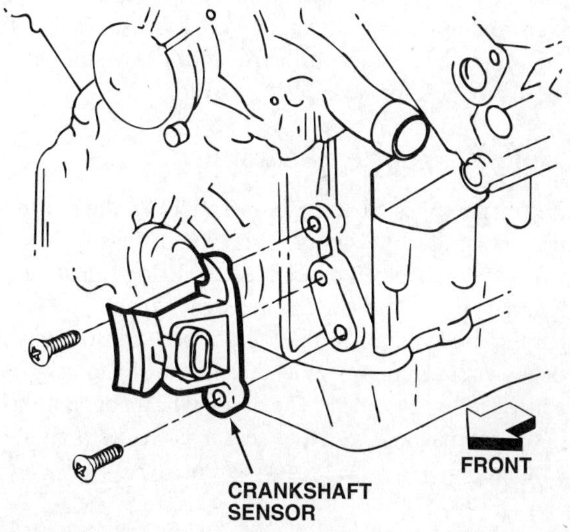

Figure 13-29. This General Motors Hall Effect CKP sensor bolts to the front of the engine. (Courtesy of General Motors Corporation)

When used as a CKP or CMP sensor, Hall Effect switches generally install in the distributor. Some Hall Effect sensors mount to the engine at the crankshaft or camshaft. These are replaced by removing the mounting fasteners and separating the wiring harness connector, figure 13-29.

Optical Sensor

An optical sensor uses a light-emitting diode (LED), a shutter wheel, and a phototransistor to produce a digital signal that changes frequency in proportion to rotational speed. Like a Hall Effect switch, an optical sensor requires an external power source and uses a three-wire circuit. One wire carries power to operate the LED, one is the signal generated by the transistor, and the third provides a common ground path.

Signal voltage, which is usually 5 volts, switches on and off as the rotating shutter passes between the phototransistor and LED. This opens and closes the ground circuit. When the shutter allows light to shine on the phototransistor, the base of the transistor switches, causing the signal voltage to change state. When the reflector plate blocks the light to the phototransistor, the base of the transistor switches again, and signal voltage changes as well.

Optical sensors are more expensive to manufacture and more delicate than a magnetic pickup or Hall Effect switch. Therefore, when used, they generally assemble into the distributor to protect them. Typically, optical sensors are used as engine speed sensors because their high-speed data rate is more accurate for high rpm applications than any other sensor designs. When viewed on a scope, the waveforms are similar to those produced by a Hall Effect switch.

Magneto Resistive Sensor

The magneto resistive sensor is a later variation of the magnetic inductive sensor. This type of sensor uses a magnetic pickup coil to produce an analog signal, and internal circuitry to convert the signal to a digital signal that is delivered to the PCM. These sensors have three wires: ignition, ground, and the digital signal generated by the sensor. The ac signal cannot be monitored. A magneto resistive sensor produces a signal similar to that of a Hall Effect sensor.

IGNITION COIL SERVICE

Chapter 12 of this *Shop Manual* explains how to check an ignition coil by measuring coil current draw with an ammeter or current probe. An ohmmeter is used to measure the resistance of the primary and secondary windings. If current draw or winding resistance is out of specifications, replace the coil.

Oil-filled and some E-core coils attach to a bracket that bolts onto the engine. Other E-core coils install in the distributor. To replace a remotely mounted coil, disconnect the secondary lead from the coil tower and the primary wires from the terminals. Label the primary leads to ensure proper installation on the replacement coil as needed. Loosen the mounting fasteners and remove the coil. Install the new coil, then connect the primary wires and secondary lead.

A coil pack mounts on a base plate that typically bolts onto the engine block. To remove the unit, label as needed, then disconnect the secondary leads and the primary wiring connectors. Remove the mounting fasteners and lift off the coil pack assembly. Secure the new unit to the engine and connect the primary and secondary leads, figure 13-30.

Direct ignition coils fit onto the spark plug and are often secured by small bolts or screws, figure 13-31. A two-pin connector provides the primary circuit connections. On some designs, each coil is serviced individually. On others, the coils for all of the cylinders remove from the engine as an assembly, figure 13-32. The assembly is torn down on the bench to service the individual coils on some, but not all, designs.

IGNITION CABLE SERVICE

Inspect high-tension ignition leads and measure resistance with an ohmmeter as described in Chapter 12 of this *Shop Manual.* When replacing cables, make sure the new leads meet original equipment specifications. Most ignition systems use 7-mm cables. However, a few use larger 8-mm cables. The larger cables provide additional dielectric resistance in a system where secondary voltage exceeds 40 kV. Use the proper-size replacement cables; they are not interchangeable.

Ignition cables generally push-fit into the distributor cap or DIS coil. Twist and pull up on the boot to remove

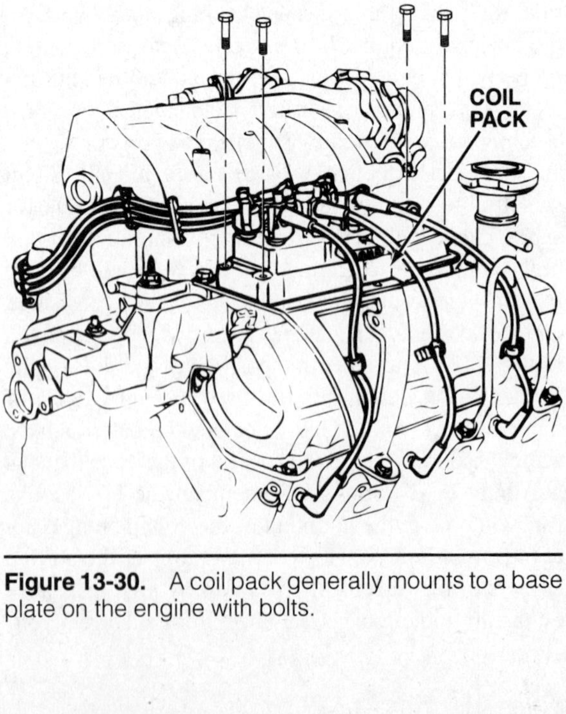

Figure 13-30. A coil pack generally mounts to a base plate on the engine with bolts.

the cable from the cap. Fit the new cable to the cap so the terminal seats firmly on the tower, figure 13-33. Fit the rubber boot seal over the tower, or DIS terminal, and squeeze it to remove any trapped air.

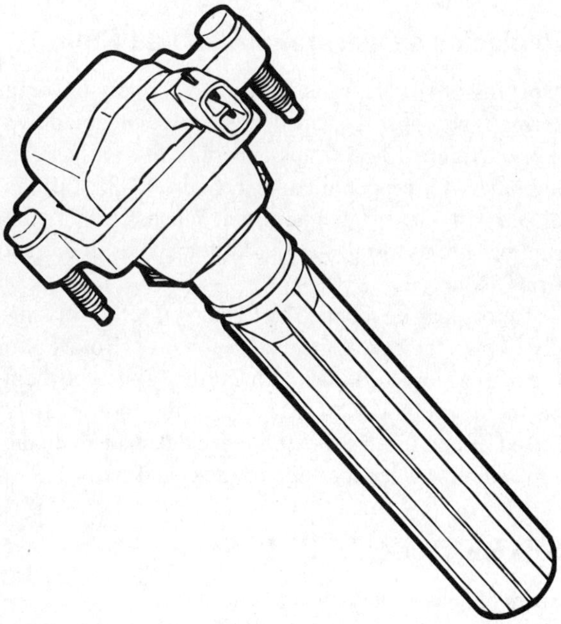

Figure 13-31. Two bolts secure this Chrysler direct ignition coil to the engine. (Courtesy of DaimlerChrysler Corporation)

The cables used on some Chrysler engines have locking terminals that form the distributor contact terminal within the cap. To replace this type of cable, remove the distributor cap and compress the terminal ends with pliers, figure 13-34. Some distributor caps use a male ignition cable terminal that looks much like a spark plug. The cable end snaps onto the terminal instead of fitting down inside the cap tower.

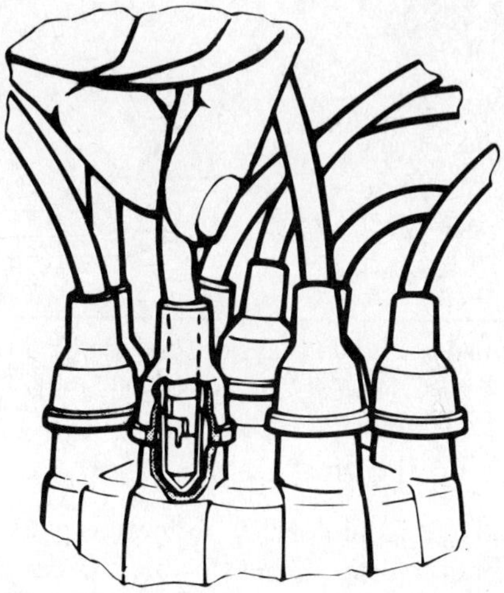

Figure 13-33. Make sure the cable end completely seats in the distributor cap.

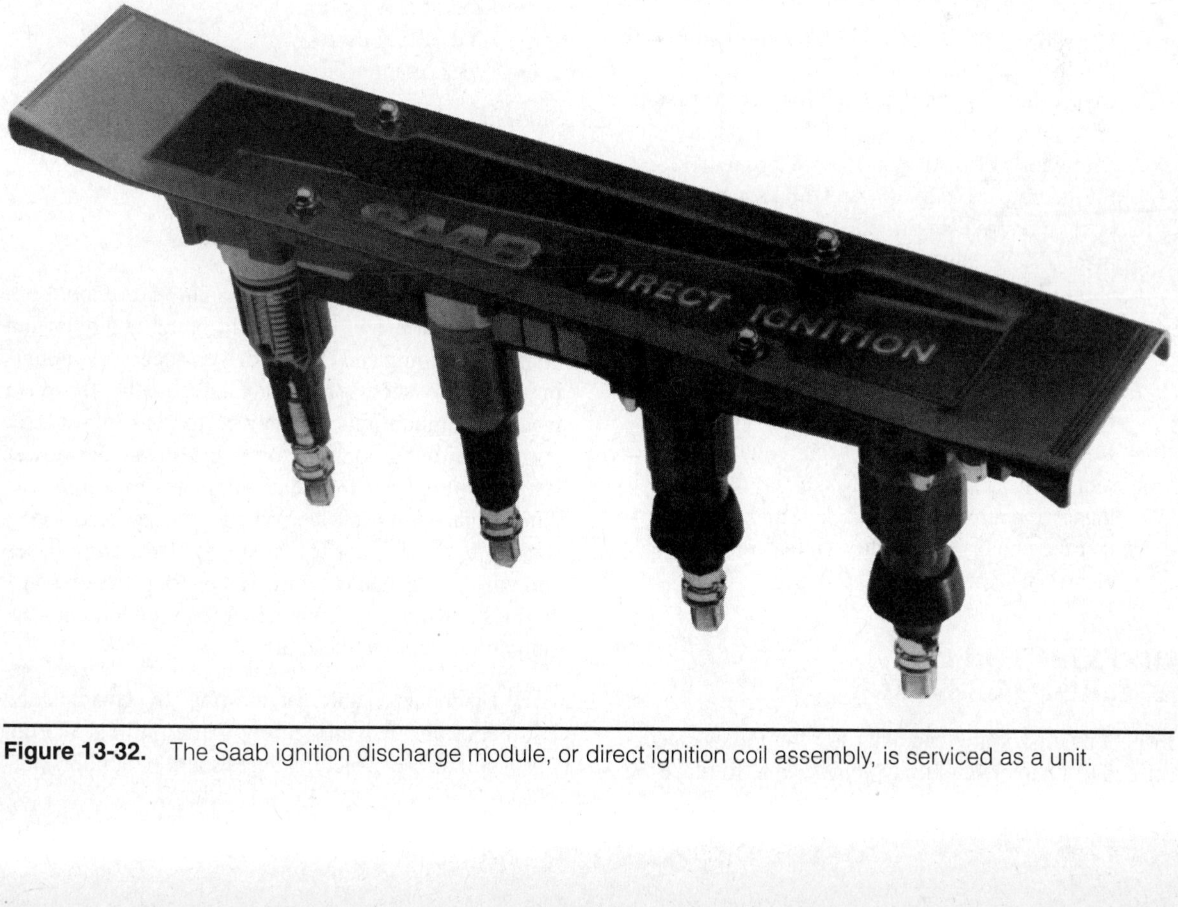

Figure 13-32. The Saab ignition discharge module, or direct ignition coil assembly, is serviced as a unit.

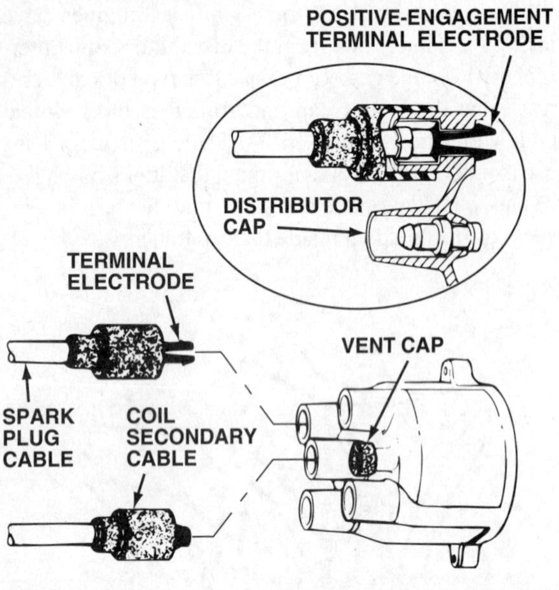

Figure 13-34. Some Chrysler spark plug cable ends lock to the cap and pliers are needed to remove them.

Replacing Ignition Cables

To install new ignition cables, follow this procedure to avoid confusion and possible incorrect installation:

1. If the distributor cap uses locking terminals, remove the cap. If it is a traditional cap, leave it on the distributor and begin at cylinder number one and work in firing order sequence.
2. Disconnect only one cable from a spark plug and the distributor cap at a time.
3. Check the cap tower terminal for dirt, corrosion, or damage. Replace the cap if there is a heavy accumulation or damage.
4. Fit the new cable to the cap while squeezing the boot to remove any trapped air when it seats in place. Make sure the terminal is firmly seated and the rubber boot seals over the tower.
5. Connect the other end of the cable to the spark plug. Make sure the terminal and boot fit properly onto the plug terminal and the cable is routed as the original.
6. Repeat this procedure for each remaining cable, including the coil cable. Make sure that the cables are secured in their brackets and cannot contact the exhaust manifold or interfere with other electrical wiring.

DISTRIBUTOR CAP AND ROTOR SERVICE

Inspect the distributor cap and rotor for damage as described in Chapter 12 of this *Shop Manual.* Replace the parts if any sign of wear or damage is found. Never clean cap electrodes or the rotor terminal by filing. If electrodes are burned or damaged, replace the part. Filing changes the rotor air gap and increases secondary circuit resistance.

Replacing a Distributor Cap and Rotor

Most distributor caps are replaced by simply undoing spring-type clips, L-shaped lug hooks, or holddown screws, figure 13-35. Caps generally have a locating lug or slot for proper alignment, figure 13-36. Lift the old cap off and install a new one. When replacing the cap, be sure to install the spark plug cables in correct firing sequence.

Rotors may be retained by holddown screws or simply fit onto the end of the distributor shaft. Rotors also use one or more locating lugs to correctly position them on the distributor shaft. Make sure the rotor is fully seated. If not, it can strike the cap and damage both the cap and rotor when the engine cranks and starts.

SPARK PLUG SERVICE

Spark plugs are routinely replaced at specific service intervals as recommended by the vehicle manufacturer. However, these intervals are guidelines and actual spark plug service life will vary. Spark plug life depends upon:

- Engine design
- Type of driving done
- Kind of fuel used
- Types of emission control devices used

Since the spark plugs are the final component in all secondary circuits, the remainder of the circuit cannot perform properly if they are not in good condition.

Spark Plug Removal

Spark plug access is limited on many late-model engines due to a maze of air conditioning and emission control plumbing and engine-driven accessory mountings. Engine accessories may have to be loosened from their mountings and moved to get to the plugs. Air conditioning compressors, air pumps, and power steering pumps are frequent candidates for relocation during spark plug service. When moving these accessories, be careful not to damage or strain any hoses and wiring attached to them. The spark plugs on some engines are most easily reached from underneath the vehicle. To remove the spark plugs:

1. Disconnect cables at the plug by grasping the boot and twisting gently while pulling. Do not pull on the cable. Insulated spark plug pliers pro-

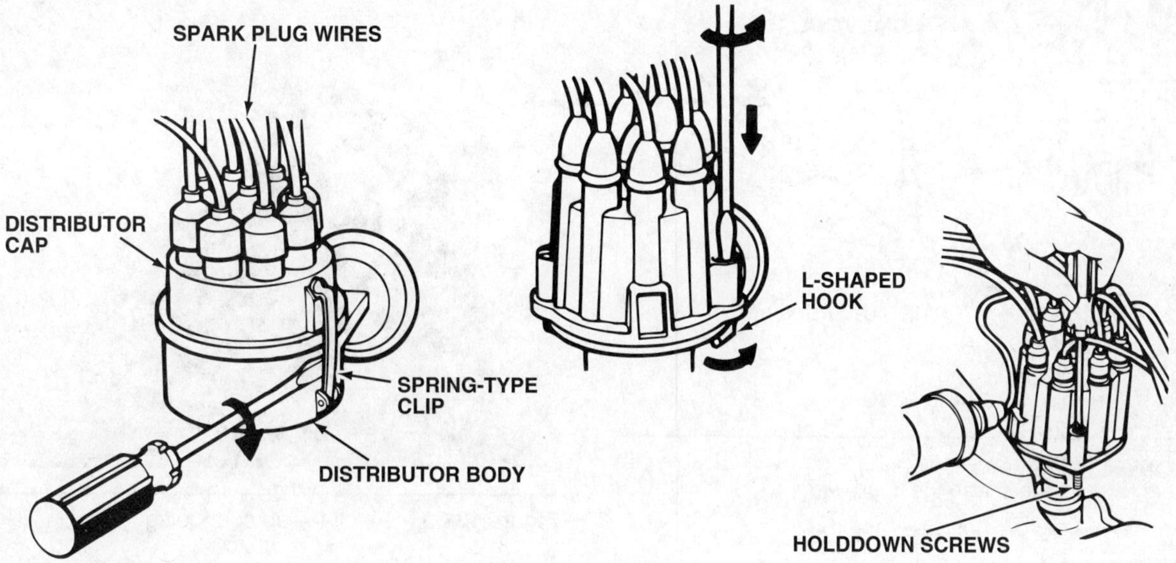

Figure 13-35. A number of methods are used to attach the cap to the distributor housing.

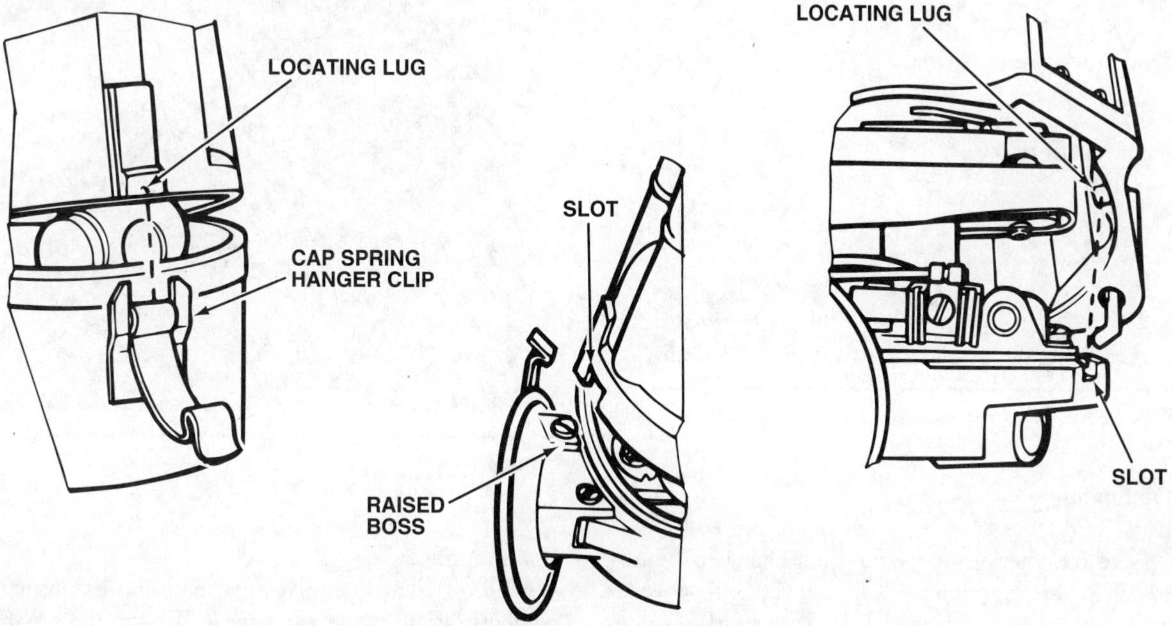

Figure 13-36. Most caps have a locating lug and install on the distributor in only one way.

vide a better grip and are recommended when working near hot manifolds, figure 13-37.

2. Loosen each plug one or two turns with a spark plug socket, then clear dirt away from around the plugs with compressed air.
3. Remove the plugs and keep them in cylinder number order for inspection.
4. When removing plugs with a gasket, be sure the old gasket comes out with the plug.

Spark Plug Diagnosis

Examining the firing ends of the spark plugs reveals a good deal about general engine conditions and plug operation. The insulator nose of a used plug should have a light brown-to-grayish color, and there should be very little electrode wear, figure 13-38. These conditions indicate the correct plug heat range and a healthy engine. Some common spark plug conditions that indicate problems follow.

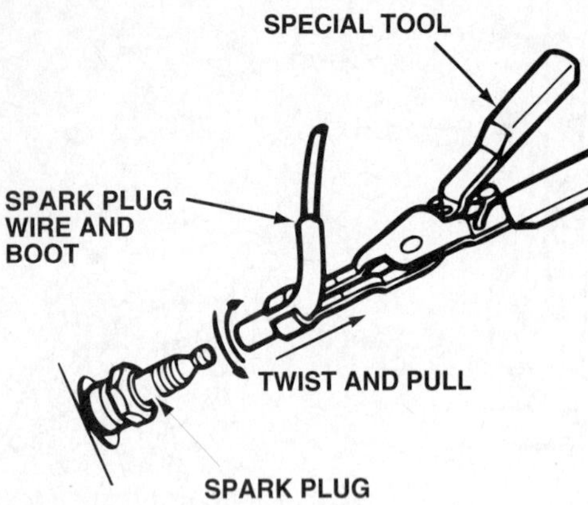

Figure 13-37. Twist and pull the boot with a cable pliers to separate it from the spark plug.

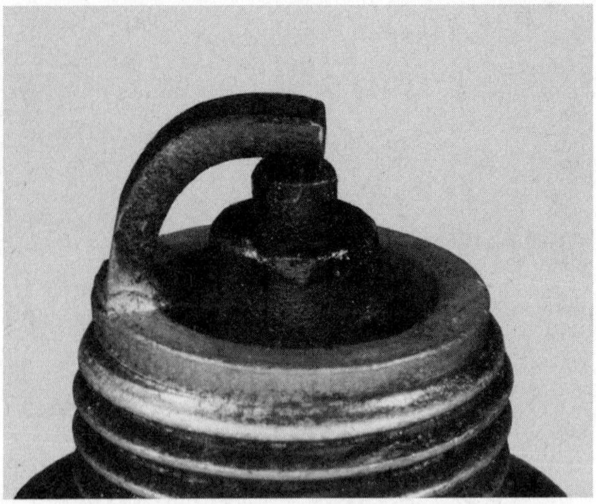

Figure 13-39. An oil-fouled spark plug.

Figure 13-38. Normal used spark plug.

Figure 13-40. A carbon-fouled spark plug.

Oil Fouling

Dark, wet deposits on the plug tip are caused by excessive oil entering the combustion chamber, figure 13-39. Piston ring, cylinder wall, and valve guide wear are likely causes in a high-mileage engine. Also, a defective PCV valve can draw oil vapor from the crankcase into the intake and oil-foul the plugs.

Carbon Fouling

Soft, black, sooty deposits on the plug end indicate carbon fouling, figure 13-40. Carbon results from a plug that is operating too cold. Check for spark plugs with an incorrect heat range, an overly rich air-fuel mixture, weak ignition, inoperative manifold heat control valve, retarded timing, low compression, and faulty plug wires or distributor cap. Carbon fouling may also result from overloading due to excessive stop-and-go driving.

Ash Fouling

Certain oil or fuel additives that burn during normal combustion can create ash deposits, figure 13-41. Ash deposits are light brownish-white accumulations that form on and around the electrode. Normally, ash deposits are nonconductive, but large amounts may cause misfiring.

Splash Fouling

Small dark patches visible on the insulator indicate splash fouling, figure 13-42. Deposits breaking loose from pistons and valves and splashing against the hot plug insulator cause splash fouling. The condition often occurs after engine servicing that restores engine power and higher combustion temperatures. Splash-fouled plugs are generally cleaned and reinstalled.

Figure 13-41. An ash-fouled spark plug.

Figure 13-42. A splash-fouled spark plug.

Gap Bridging

Gap bridging is usually due to conditions similar to those described for splash fouling, figure 13-43. The difference is that deposits form a bridge across the electrodes and cause a short that prevents firing. This condition is common in engines with poor oil control.

Insulator Glazing

Shiny, yellow, or tan deposits are a sign of insulator glazing, figure 13-44. Frequent hard acceleration with a resulting rise in plug temperature causes glazing. The high temperature melts normal plug deposits and fuses them into a conductive coating that causes misfiring. Severe glazing cannot be removed and the plugs must be replaced.

Figure 13-43. A spark plug with the gap bridged.

Figure 13-44. Insulator glazing on a spark plug.

Overheating

Spark plug overheating is indicated by a clean, white insulator tip, excessive electrode wear, or both, figure 13-45. The insulator may also be blistered. Incorrect spark plug heat range, incorrect tightening torque, overadvanced timing, a defective cooling system, or a lean air-fuel mixture may be the cause of overheating.

Detonation

Detonation causes increased heat and pressure in the combustion chamber that exerts extreme loads on engine parts. Fractured or broken spark plug insulators are a sign of detonation, figure 13-46. Overadvanced timing, a lean fuel-air mixture, low gasoline octane,

Figure 13-45. Overheating causes excessive spark plug electrode wear.

Figure 13-47. A preignition damaged spark plug.

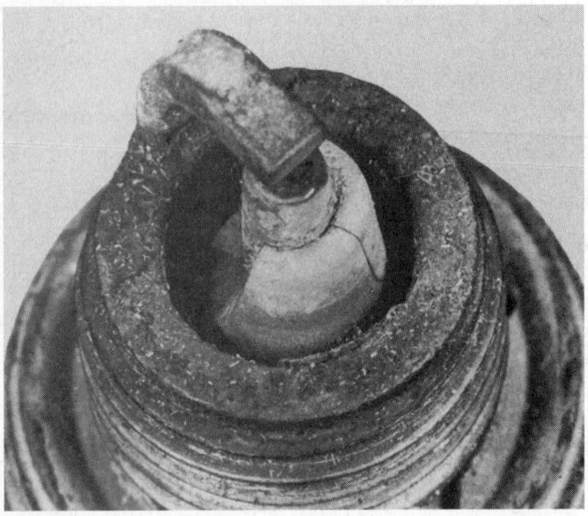

Figure 13-46. A detonation-damaged spark plug.

Figure 13-48. An excessively worn spark plug.

and engine lugging are contributing factors. An EGR valve that fails to open is also a cause of detonation.

Preignition

Preignition, which is the air-fuel charge igniting before the plug fires, causes severe damage to the spark plug electrodes, figure 13-47. Combustion chamber hot spots or deposits that hold enough heat to prematurely ignite the air-fuel charge are common causes of preignition. Other sources of preignition include crossfiring between plug cables, a plug heat range much too hot for the engine, and loose spark plugs.

Even if the color of the insulator and deposits is normal, rounded and worn electrodes indicate that a plug should be replaced, figure 13-48. These plugs are simply worn out. The voltage required to arc across the gap has increased and continues to do so with additional use. Misfiring under load is a clue to worn-out plugs. Such plugs also contribute to poor gas mileage, loss of power, and increased emissions.

Another problem that may be seen on a spark plug is carbon tracking down the insulator. Tracks appear as dark gray lines running the length of the insulator. This problem has become more common as newer spark plug designs tend to have shorter, and smooth-sided, insulators.

Spark Plug Installation

Spark plugs, both new and used, must be correctly gapped before they are installed. Although a wide variety of gapping tools is available, a round wire feeler

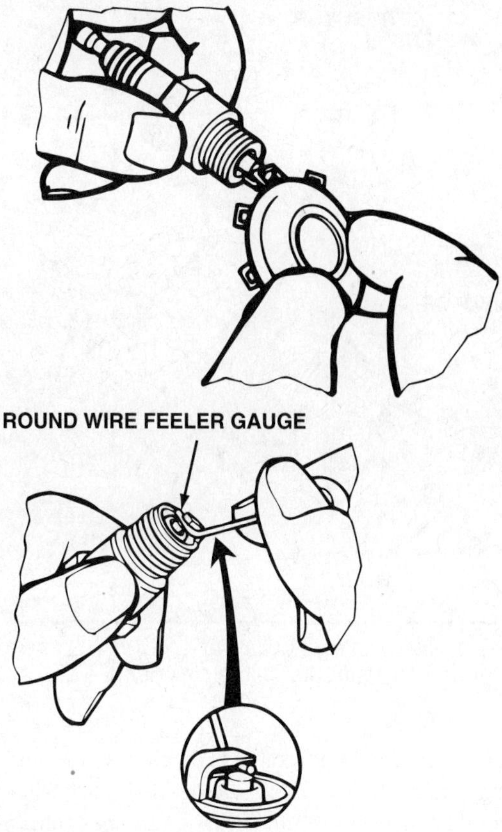

Figure 13-49. Checking spark plug gap with a feeler gauge.

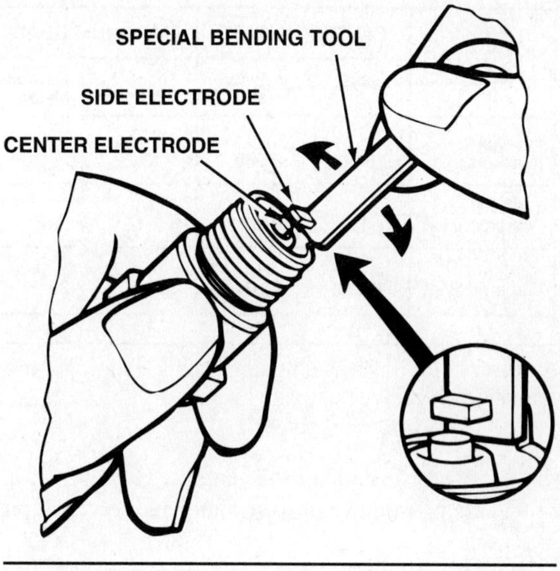

Figure 13-50. Adjust spark plug gap by carefully bending the side electrode.

gauge is the most efficient for used plugs, figure 13-49. Adjust the gap by carefully bending the ground electrode, figure 13-50.

- Do not assume that new plugs are correctly gapped.
- Do not make gap adjustments by tapping the electrode on a workbench or other solid object. This causes internal plug damage.

Cleaning the threaded plug holes in the cylinder head with a thread chaser ensures easy spark plug installation. With aluminum heads, use the tool carefully to avoid damaging the threads.

Some manufacturers recommend using an antiseize compound or thread lubricant on the plug threads. Use thread lubricant only when specified by the manufacturer. Antiseize compound is commonly used when installing plugs in aluminum cylinder heads. Be sure to use the specific compound recommended by the manufacturer, as not all are compatible with aluminum. Whenever thread lubricant or antiseize is used, reduce the tightening torque slightly.

Once the plug gap has been properly set, install plugs as follows:

1. Wipe any dirt and grease from the cylinder head plug seats.
2. Check that any gaskets used on the plugs are in good condition and installed.
3. Fit the plugs in the engine and run them in by hand.
4. Tighten the plugs to specification with a torque wrench, figure 13-51.

DISTRIBUTOR SERVICE

This section contains general procedures to remove, disassemble, test, repair, and install the distributor of a modern electronic ignition system. Procedures are similar for all makes and models, and these instructions apply to most distributor repair services. Always check the vehicle service manual for exact procedures and specifications. Overhaul procedures for five common domestic distributor models are presented as photographic sequences at the end of this chapter.

A modern distributor is an engine-driven switch assembly. The assembly typically contains two switches: a primary switch that provides an engine speed or position signal to the ignition control module (ICM), and a secondary switch that routes secondary voltage to the spark plugs. The ICM is part of the assembly on some distributors. Some distributor services, such as replacing the cap and rotor, are generally performed without removing the distributor from the engine. On some models, it is possible to replace the pickup coil, Hall Effect switch, or ICM with

PLUG TYPE	CAST-IRON HEAD		ALUMINUM HEAD	
	Foot-Pounds	Newton-Meters	Foot-Pounds	Newton-Meters
14-MM Gasketed	25-30	34-40	13-22	18-30
14-MM Tapered Seat	7-15	9-20	7-15	9-20
18-MM Tapered Seat	15-20	20-27	15-20	20-27

Figure 13-51. Always tighten spark plugs to the specified torque value.

the distributor installed in the engine. However, it is often easier to remove the distributor and service it on the bench.

With the exception of V8-powered General Motors light trucks, the only domestically produced vehicles in current production that use a distributor are those with import engines, such as the 3.0 L Mitsubishi V-6 in Chrysler minivans, the Chevrolet Metro built by Suzuki, and the Nissan-powered Mercury Villager. Distributors are also used on some late-model vehicles from BMW, Honda, Nissan, Mercedes-Benz, Mitsubishi, Saab, Toyota, and Volvo. Although designs vary considerably, all distributors are serviced in a similar fashion.

Some late-model distributors, such as those used by BMW and Mercedes-Benz, are simply a secondary switch driven off the camshaft. The rotor bolts to a flange at the end of the camshaft and the cap bolts onto the camshaft cover. An O-ring seals the cap to the cover. Opening and closing the primary coil circuit is controlled by the PCM. The PCM receives speed and position inputs from sensors mounted elsewhere on the engine. A periodic inspection of the cap, rotor, and O-ring, which are replaced on an as-needed basis, is the only distributor service required.

DISTRIBUTOR REMOVAL AND INSPECTION

The principles of distributor removal are the same for all systems, although the exact details will vary somewhat according to manufacturer and engine. Many distributors require loosening or relocating other engine accessories, such as the air cleaner or air intake ducts, in order to remove them. It may be helpful to tag all electrical leads and vacuum lines that are disconnected during distributor removal. Also, be sure to observe all electrical safety precautions and shop safety regulations during distributor removal and in-

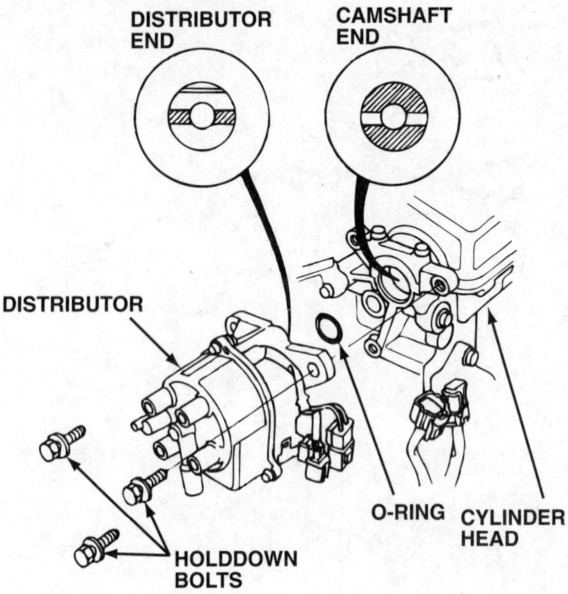

Figure 13-52. Three bolts attach this Honda camshaft-driven distributor to the cylinder head.

stallation. It is always advisable to disconnect the negative battery cable to prevent accidental arcing when servicing the distributor. If not, make sure the ignition is switched off.

Removal

The distributors used on some late-model engines are driven directly off the end of the camshaft. An offset drive dog attached to the distributor shaft engages a machined slot in the camshaft, figure 13-52. Cylinder positioning has no effect on distributor removal or installation because the drive connects to the camshaft in only one position. These distributors usually have two or three slotted ears that fit over studs in the cylinder head and are retained by nuts. Remove the distributor simply by removing the nuts and lifting the assembly from the engine.

Most distributors install into a machined bore in the engine block and are gear-driven. A helical, or bevel, gear on the distributor gear engages a spiral gear on the camshaft or auxiliary shaft in the engine, figure 13-53. Many distributors are held in the engine with a hold-down clamp and a bolt, figure 13-54. The clamp rides over a mounting flange on the distributor base. A machined slot on the distributor housing fits over a threaded stud on the engine block on other designs. A nut secures the distributor assembly to the engine. With either type, once the mounting fasteners are removed, the distributor should lift off the engine fairly easily. Be sure to mark the position of the distributor

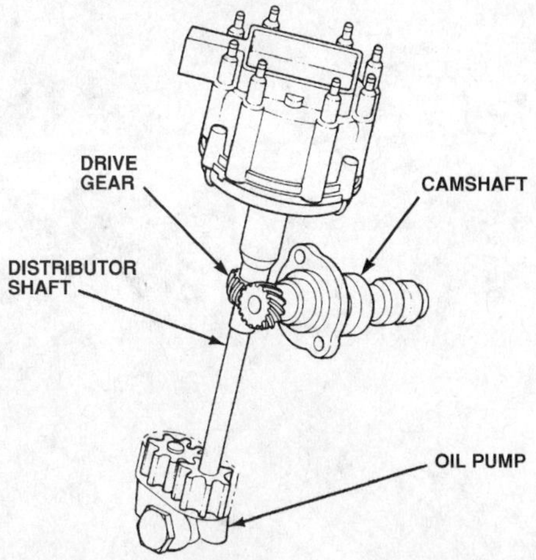

Figure 13-53. Distributor and oil pump drive are taken off the same camshaft gear on many designs.

Figure 13-55. Mark the distributor position on the engine before loosening fasteners.

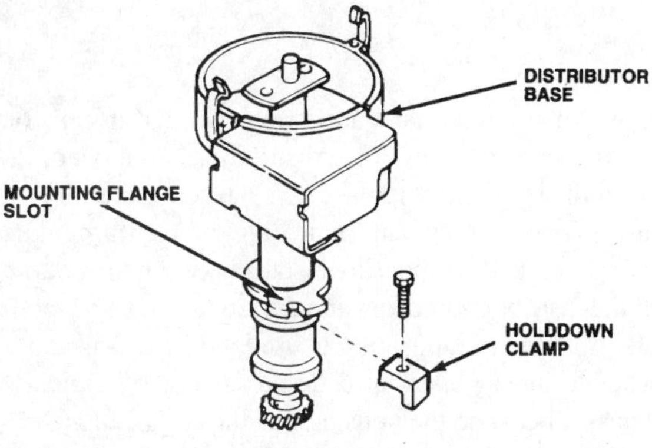

Figure 13-54. Many distributors are attached to the engine with a holddown clamp and bolt.

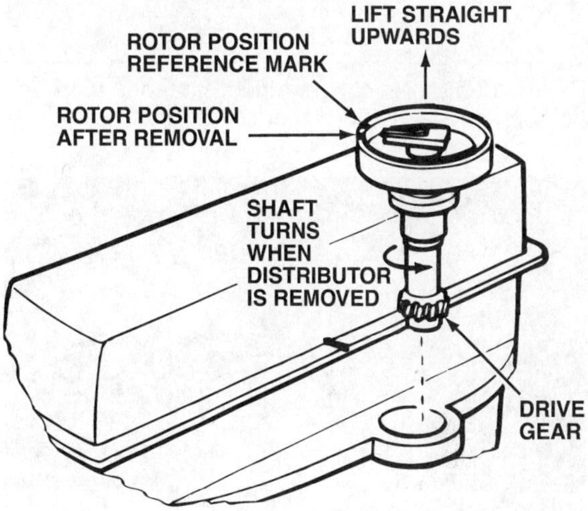

Figure 13-56. Make rotor and engine position reference marks on the distributor housing for easy assembly.

on the engine before loosening the holddown fasteners, figure 13-55. To remove a typical distributor:

1. Disconnect and remove the distributor cap. On most engines, the ignition cables do not need to be disconnected from the cap; simply position the cap and wires out of the way.
2. Using a wrench on the crankshaft pulley bolt, rotate the engine by hand so that the number one cylinder is up to top dead center on the compression stroke, or the distributor rotor points straight ahead or up.
3. Mark a reference point on the distributor housing in line with the position of the rotor tip. Also mark reference points on the distributor housing and engine so the distributor can be installed in the same position, figure 13-56.
4. Disconnect the primary circuit wiring to the distributor. Disconnect any vacuum lines attached to the distributor as well.
5. Loosen and remove the distributor holddown fastener along with the holddown bracket, if used. Offset "distributor" wrenches make loosening difficult-to-reach fasteners easy, figure 13-57.
6. Twist the distributor housing gently to loosen it from the engine, then remove the distributor by pulling it up and out of the engine block bore, figure 13-58. Do not use force to remove the distributor; it should come out easily. If the distributor is stuck in the bore, it is likely the result of a dry and hardened O-ring seal. Apply a little penetrating oil to loosen it.

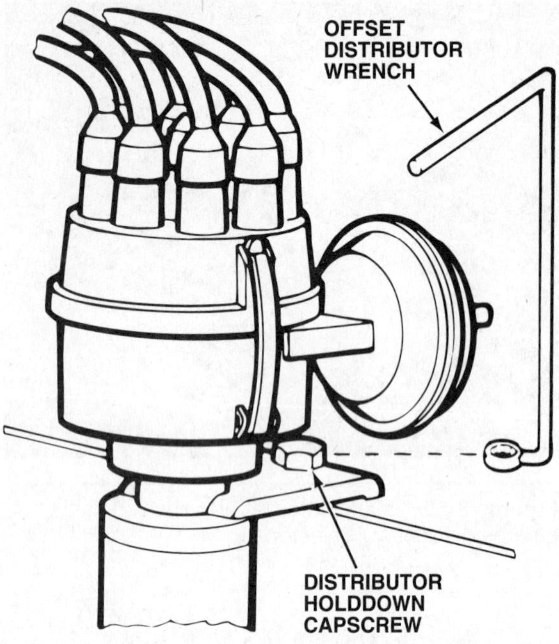

Figure 13-57. An offset wrench is needed to loosen holddown fasteners on some distributors.

Figure 13-58. With a slight twist to break the seal, most distributors slip easily from the bore.

7. The shaft of a gear-driven distributor rotates slightly as the assembly is removed and the gears disengage. Note how far the rotor moves off the reference mark as the distributor is removed. Install the distributor with the rotor in this position.
8. After removing the distributor, cover the engine block bore with a shop rag to keep dirt and contaminants out of the crankcase.

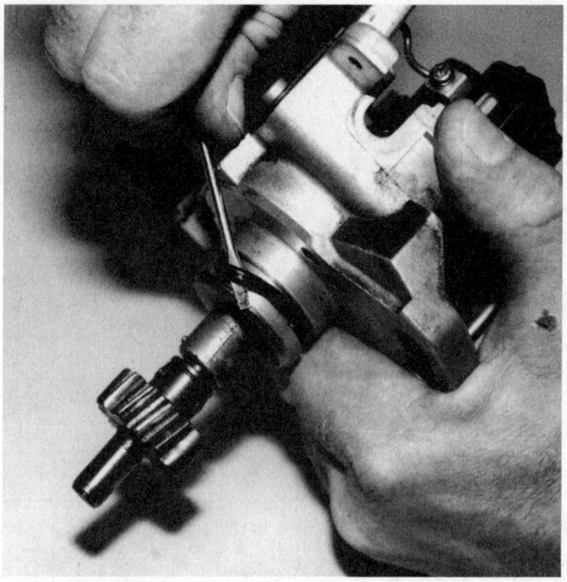

Figure 13-59. Use a screwdriver or pick to remove an O-ring.

Many distributors have an O-ring seal that installs on the housing. Whenever the distributor is removed, inspect the O-ring and replace it as needed. Carefully remove O-rings to prevent scratching the housing, figure 13-59. A metal sealing ring installs between the bottom of the distributor housing and the engine on some models. If a metal sealing ring is used, be sure to install it when fitting the distributor. Some older distributors use a gasket between the housing and the engine block.

Some engines, such as older small-block Ford V8s, use an intermediate shaft between the oil pump and the distributor. This shaft should remain in the engine when the distributor is removed. However, sludge buildup inside the engine may cause the shaft to stick in the distributor drive gear and disengage the oil pump as the distributor is removed. If the shaft comes all the way out with the distributor, there is no problem because it is easily reinstalled. But, a shaft that comes part way out and then drops off into the engine while the distributor is being removed causes a serious problem. The shaft generally falls into the front of the oil pan or timing cover, which must be removed to retrieve it.

To avoid dislodging an intermediate shaft, slowly and carefully lift the distributor just enough to reach under the housing to the bottom of the drive gear. If the intermediate shaft is coming out with the distributor, grasp it and remove it along with the distributor. Do not attempt to push the shaft back in place because it has already pulled free of the oil pump. To install an intermediate shaft, hold it firmly with a special grip-

Figure 13-60. Inspect the drive gear or dog for signs of wear and damage.

ping tool or long-nosed snapring pliers and lower it into the engine. Rotate the shaft to engage the drive dog with the oil pump, then release the gripping tool. Make sure the shaft is completely seated before installing the distributor.

Some other distributor installations have unique features that require special handling and procedures as well. Always check the service manual for the vehicle being repaired for information and specifications.

Inspection

Once the distributor is removed, use solvent and a brush to clean all oil, grease, dirt, and rust from the distributor shaft and housing. Take care to keep solvent away from the bushings, electronic components, and vacuum advance unit, if used. Dry the distributor with low-pressure compressed air and perform a preliminary inspection. Look for:

- Binding or excessive shaft movement, check end play and side play
- Worn or chipped drive gear teeth, or dog lugs, figure 13-60
- Loose or damaged electrical leads and terminal connections
- Loose or damaged component mountings
- Signs of inadequate lubrication

If the distributor is to be serviced, install it in a bench vise with protective, or soft, jaws. If protective jaws are not available, wrap the distributor housing with shop rags to protect it. Avoid overtightening the vise, as the aluminum distributor housing is easily distorted and damaged.

If shaft end play and side play feel excessive, check for wear using a dial indicator. Attach the dial indicator base to the distributor housing or the vise. The housing is preferable as it eliminates false readings from the housing moving in the vise. Position the indicator plunger on the top end of the shaft, slightly preload the plunger, and zero the dial face. Then, push up on the base of the shaft and take an indicator reading. This is the end play; compare results to specifications. Too much end play is often the result of a worn thrust washer, which usually installs under the drive gear.

To measure side play, move the indicator plunger to the side of the shaft, preload the plunger, and zero the dial face. Next, firmly grasp the distributor housing and rock the shaft back and forth. Side play is displayed on the dial indicator; compare readings to specifications. Excessive side play is the result of worn bushings or bearings, or a bent shaft. To check for a bent shaft, leave the dial indicator in position and watch the dial face while rotating the shaft through one complete revolution. If the dial indicator shows excessive runout, the shaft is bent and must be replaced.

Check the movement of the advance mechanism on older-model distributors that use mechanical weights or a vacuum diaphragm to advance the spark timing. To check mechanical systems, simply hold the shaft and twist the advance mechanism. The weights should move freely without binding and fully extend to stops, then return smoothly under spring force when released. If not, expect to find damaged bearings or bushings, or built-up dirt and corrosion between parts on disassembly. Use a hand vacuum pump to check a vacuum advance unit. Apply vacuum while watching the movement of the breaker plate. Again, watch for smooth and full movement during application and release.

Continuity of electronic devices installed in the distributor, such as a pickup coil, Hall Effect switch, optical sensor, or ignition control module, is checked with an ohmmeter before disassembly. Test points and specifications vary by model. Refer to the appropriate service manual for exact procedures.

Once the distributor has been inspected, it is possible to determine what repairs are needed to return it to service. With late-model distributors, it is often more economically feasible to replace the entire unit, rather than overhaul it. In fact, individual components are not always available to rebuild a distributor. Always consider the options and follow the course of repair that best suits the needs of the customer.

DISTRIBUTOR DISASSEMBLY AND ASSEMBLY

Distributor designs vary not only by manufacturer and engine, but often by model and year as well. Therefore, it is important to have accurate service instructions and specifications available. The procedures presented here are general and apply to most distributor overhauls. Photographic sequences that detail the overhaul of several popular domestic distributors are included following the general guidelines. Whenever a distributor is to be overhauled, read through the entire procedure in the service manual before beginning the teardown. Also be sure to have any required special tools and anticipated replacement parts available before disassembly.

Disassembly

Follow the service manual procedures and disassemble the distributor into component pieces, figure 13-61. This usually involves removing the drive gear, which is attached to the shaft with a retainer pin or snapring. Before removing the drive gear, mark it and the shaft to ensure they are in the same position on assembly. When driving out the retainer pin, make sure the distributor shaft is firmly supported to prevent damage, figure 13-62. The drive gear is an integral part of the shaft on some models. Other components, such as base plates, electronic devices, and advance mechanisms, attach to the shaft or housing with small bolts, snaprings, circlips, springs, or other fasteners. Keep track of where the fasteners install, as fitting them in the wrong location on assembly may interfere with shaft rotation or advance movement.

Most distributor shafts are supported by bushings that press-fit into the housing. However, some models use bearings rather than bushings. If the bushings or bearings require replacement, the best way to remove them is with a bushing puller. A brass drift and a hammer may be used to drive the bushings or bearings out of the housing on some models. When using a drift, take care to avoid scratching or nicking the housing bore. Also, work slowly from side to side when driving out a bushing to avoid cocking it in the bore.

Once the distributor is disassembled, clean the inside of the distributor and all the moving parts, except vacuum units, with solvent and a small brush. Dry the clean parts with low-pressure compressed air. Make sure that all dirt and solvent residue are removed, especially from electrical connections. Wipe dirt off with a clean, lint-free cloth or swab with electrical contact cleaner. Carefully inspect all components for signs of damage or wear and replace parts as needed.

Assembly

Refer to the service manual instructions to correctly assemble the distributor. Always replace any O-rings or gaskets that seal the housing to the engine. Lightly lubricate all moving parts during assembly. Keep in mind, proper lubrication is essential, but too much lubrication causes problems. Distributor rotation sprays excess lubricant around the inside of the distributor where it collects dirt and causes short circuits.

Use the correct-size bearing or bushing driver to install new bearings or bushings into the distributor housing. Make sure the bearings or bushings install straight in the bore and are fully seated. With bushings,

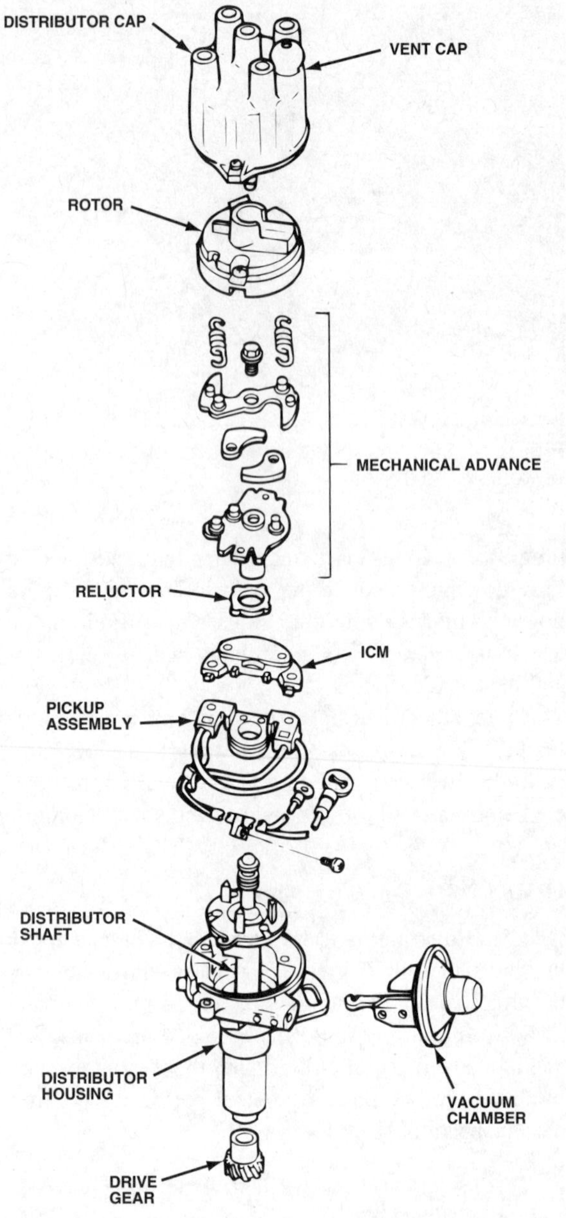

Figure 13-61. An exploded view diagram is a handy reference when overhauling a distributor.

Figure 13-62. Support the distributor shaft to drive out a gear retainer pin.

Figure 13-63. Turn the shaft to align the rotor with the mark made earlier on the distributor.

it may be necessary to ream the inside diameter to provide adequate shaft clearance once they are installed. Be sure to thoroughly clean bushing chips and shavings from the assembly after reaming.

Many solid-state distributors require that a silicone lubricant or heat sink compound be applied to specific locations within the distributor. Typical points are on the ignition module base, primary connections, rotor contact tip, and distributor connections. The silicone lubricant prevents corrosion and moisture that might interfere with voltage signals. A thin film of heat sink compound is generally applied to the mounting base of the module to help dissipate heat. Avoid applying excess lubricant or heat sink compound.

Once the distributor is assembled, check shaft rotation, end play, and side play. Also check the movement of any advance mechanisms used as previously described. Perform a few quick ohmmeter checks to verify the continuity of replacement electronic parts.

DISTRIBUTOR INSTALLATION

After the distributor has been serviced, follow these general directions to install it on the engine:

1. Make sure the negative battery cable is disconnected or the ignition is switched off to prevent accidental arcing.
2. Check that the number one cylinder of the engine is at top dead center of the compression stroke and in firing position.
3. Rotate the distributor shaft to align the rotor with the reference mark on the housing, figure 13-63. Turn an additional amount, as noted during removal, to compensate for gear engagement.
4. Align the distributor housing mark with the reference point on the engine.
5. If an O-ring or gasket is used, make sure it is in position. Lightly lubricate O-rings with motor oil to ease installation.
6. Insert the distributor into the engine, turn the rotor slightly as needed to engage the drive gears, and push lightly to seat the distributor.
7. Make sure the distributor is fully seated in the engine and engaged with the camshaft and oil pump drive.
8. Check that the rotor and reference mark are aligned. If out of position, note the amount. Then raise the distributor up, reposition the rotor the same amount, and lower it back into place.
9. When installing a distributor that attaches directly to the camshaft end, align the tangs on the distributor shaft with the slots in the end of the camshaft. Seat the distributor against the cylinder head and install the fasteners.
10. Fit the distributor holddown clamp, fasteners, and any other items used to hold the distributor in the engine. Draw the mounting fasteners up snugly, but do not fully tighten them.
11. Connect the distributor primary wiring connectors. Also attach the vacuum lines to the advance unit, if used.
12. Install the distributor cap and any other items removed during disassembly. Then, connect the negative battery cable.

After installing the distributor, start and run the engine. Check base ignition timing per the manufacturer's procedures. Adjust timing as needed, where applicable, then fully tighten the distributor holddown fastener.

DISTRIBUTOR OVERHAUL PROCEDURES

The following pages contain photographic procedures to disassemble, overhaul, and reassemble several common domestic solid-state distributors. Included are the:

- Delco-Remy 8-cylinder HEI model
- Delco-Remy 4-cylinder HEI-EST model
- Motorcraft Universal TFI-I model
- Chrysler 4-cylinder Hall Effect model
- Chrysler 6-cylinder optical distributor

Chapter 13 of the accompanying *Classroom Manual* has additional information and illustrations of these distributors. Refer to the *Classroom Manual* as needed when overhauling a distributor. Using techniques previously explained in this chapter, inspect and test distributor components at appropriate times during the overhaul procedure.

Keep in mind, all of the distributors discussed here were used for a number of years on an assortment of models. As a result, there may be some slight variations between the examples shown and the actual distributor being serviced. Remember, these photographic sequences are for instructional purposes only and are not to be substituted for the factory service manual procedures.

DELCO-REMY 8-CYLINDER HEI DISTRIBUTOR OVERHAUL PROCEDURE

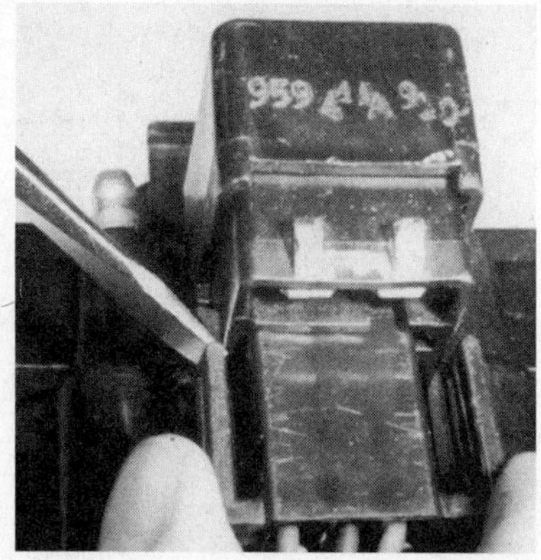

1. Carefully pry the module connector from the cap terminal with screwdriver. Release the four latches and remove the cap and coil from the distributor.

2. Remove the three screws from the coil cover, then remove the four coil mounting screws. Note the position of the ground leads, then lift off the coil.

3. Remove the rubber seal, carbon button, and spring. If the seal is brittle, replace it. The button and spring are the secondary coil lead; replace them if worn or broken.

4. Remove the two attachment screws and lift off the rotor. Note the location of the pickup coil leads, tag them if needed, then disconnect them from the module.

5. Disconnect the harness connector from the terminals on the opposite end of the module and slip the primary lead grommet from the distributor housing.

6. Remove the two mounting screws and lift the module off the distributor. Wipe the silicone grease from the module and mounting base. Apply fresh grease on assembly.

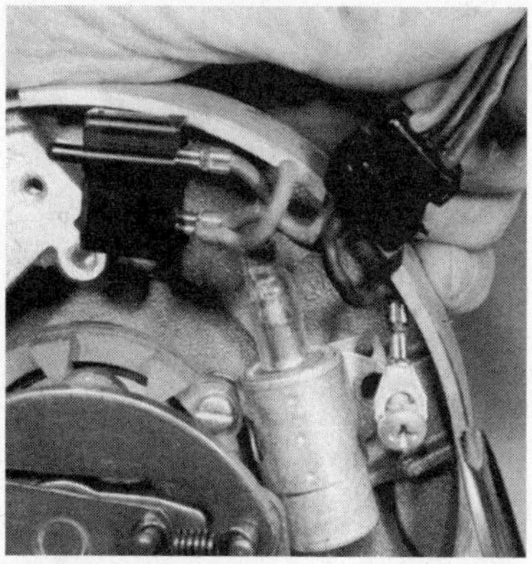

7. Remove the screw securing the RFI capacitor and lift the capacitor and primary lead harness from the distributor housing.

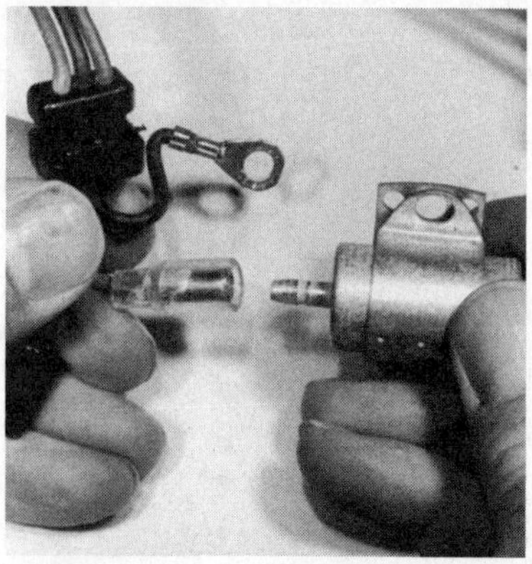

8. Disconnect the capacitor from the primary lead. Inspect and test the capacitor and primary lead for wear, continuity, and short circuits. Replace worn or damaged parts.

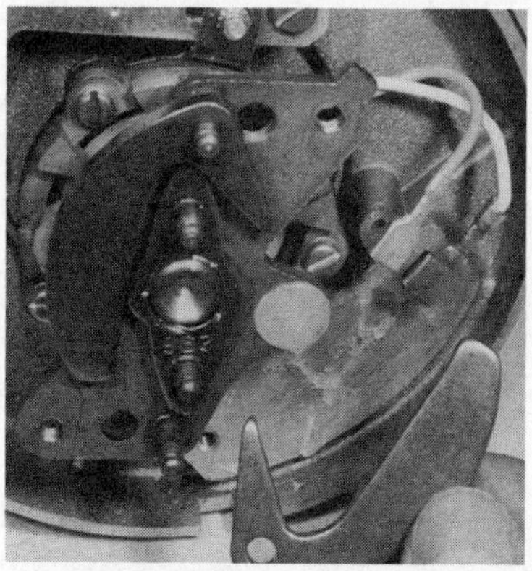

9. Make position marks on the advance springs, pins, and weights for ease of assembly. Then, remove the two springs, weight retainer, and weights.

DELCO-REMY 8-CYLINDER HEI DISTRIBUTOR OVERHAUL PROCEDURE

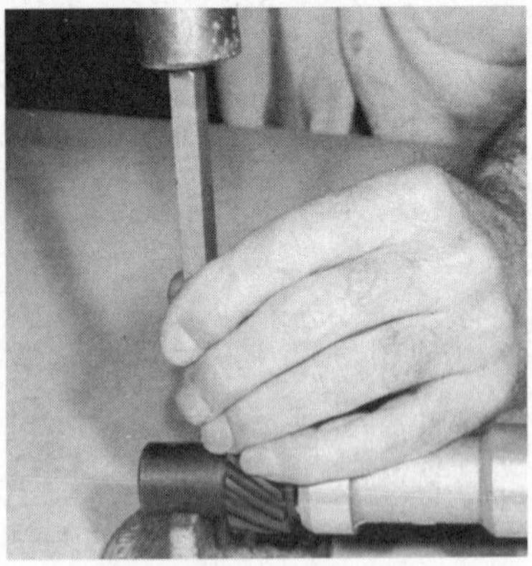

10. The shaft must be removed to remove the pickup coil and pole piece. Support the gear as shown and drive out the roll pin to separate the gear and shaft.

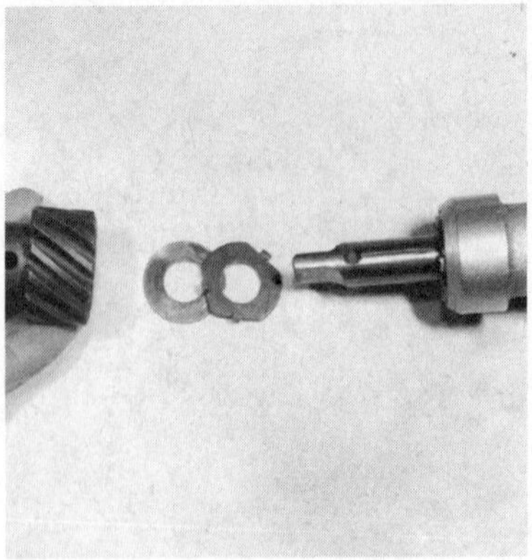

11. Remove the drive gear, shim, and tabbed washer. Note the positions for assembly. The dimple toward the bottom of the gear must align with the rotor tip when assembled.

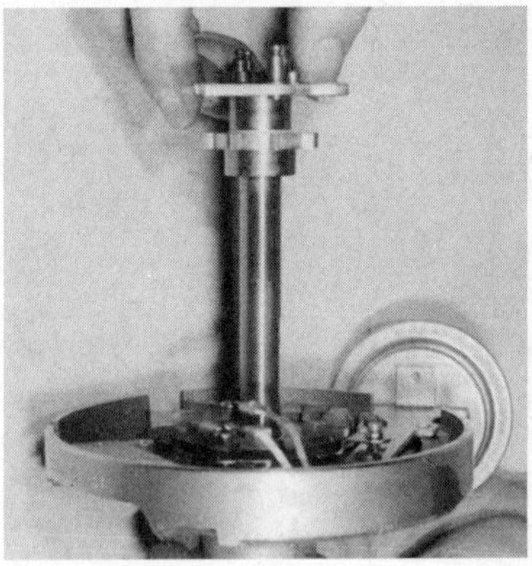

12. Carefully remove the shaft, trigger wheel, and weight base from the distributor to avoid damaging the bushings. Separate the trigger wheel from the shaft.

13. Remove the three screws attaching the pole piece to the pickup coil. Remove the pole piece along with the rubber gasket installed beneath it.

14. Slip the pickup coil from the retainer and lift it off the distributor. During operation, the retainer is rotated by vacuum to advance timing.

15. The pickup coil retainer is fastened to the bushing at the top of the housing with a waved washer. Carefully pry the washer from its slot.

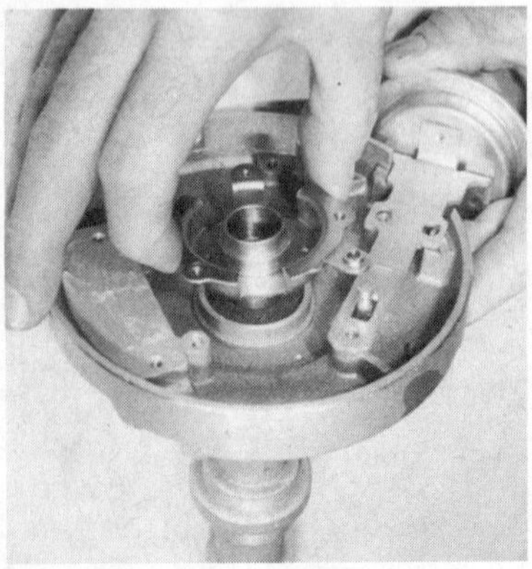

16. Separate the vacuum advance link and lift off the pickup coil retainer along with the felt washer beneath it. Remove the two mounting screws and the vacuum unit.

17. To assemble, lubricate or replace the felt washer beneath the pickup coil retainer and apply a thin coat of silicone grease to the module mounting base.

18. Fit the vacuum unit and assemble the pickup coil into the housing. Tighten the vacuum unit screws and make sure the wave washer seats in the bushing to hold the pickup coil retainer.

DELCO-REMY 8-CYLINDER HEI DISTRIBUTOR OVERHAUL PROCEDURE

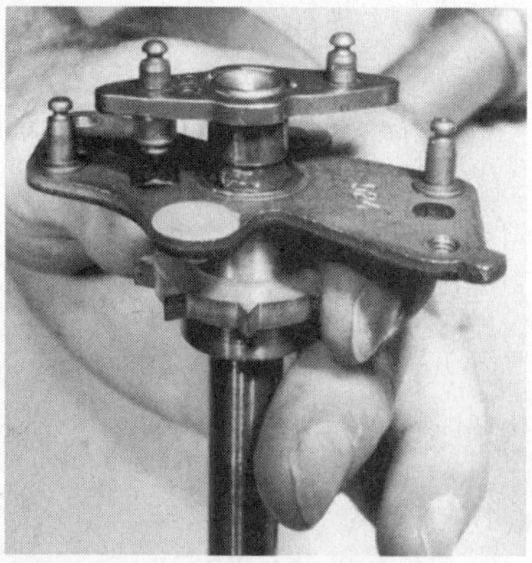

19. The trigger wheel assembles onto the distributor shaft. Make sure parts are clean and lubricate the bushings with fresh motor oil. Avoid using too much oil.

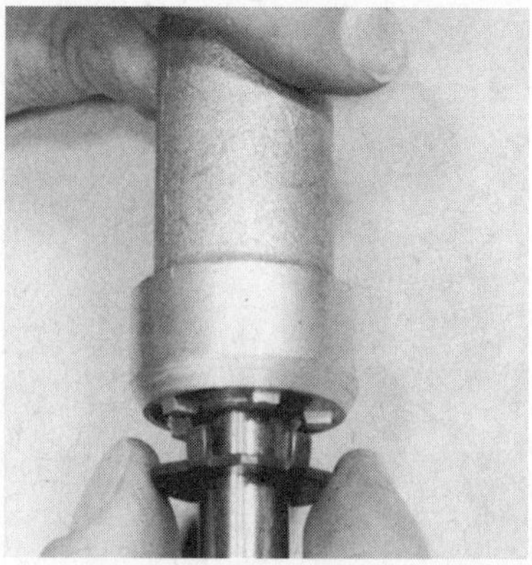

20. Carefully insert the distributor shaft and trigger wheel assembly into the housing to avoid hitting the bushings. Assemble the washer, shim, and gear onto the shaft.

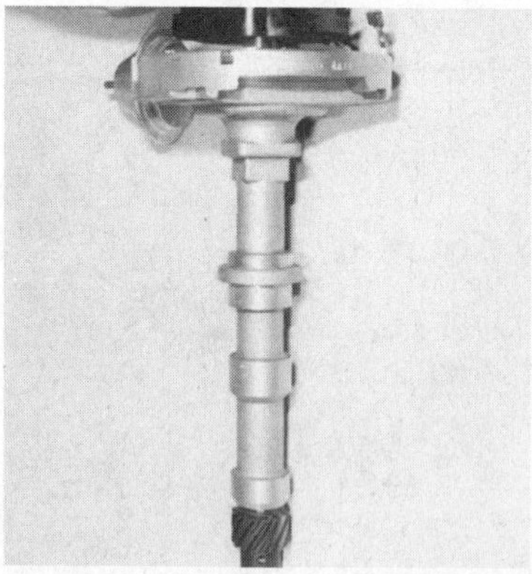

21. The drive gear is correctly installed if the dimple on the gear aligns with the rotor tip. If so, drive the roll pin through the gear and shaft to secure it.

22. Assemble the advance weights in their original positions onto the shaft base plate. Fit the retainer over the pins and shaft, then install the springs in their original positions.

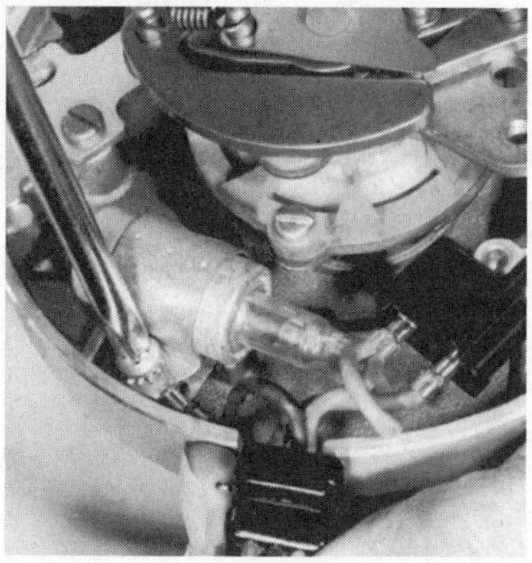

23. Connect the primary harness to the capacitor and fit the harness grommet into the housing slot. Install the capacitor in the distributor and tighten the screw.

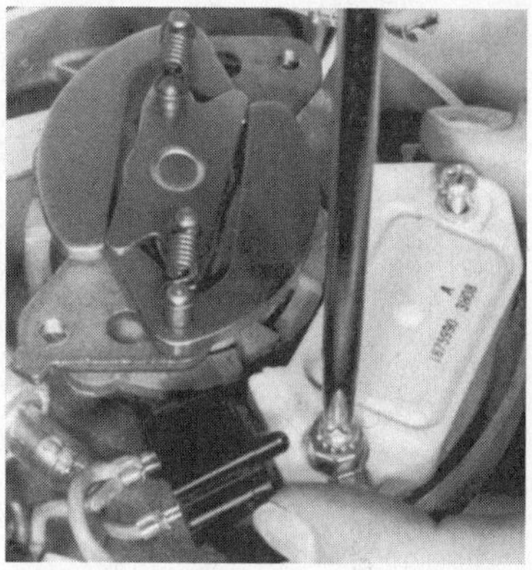

24. Apply heat sink compound to the module, then attach it on the mounting base with two screws. Remember, these are ground connections and they must be clean and tight.

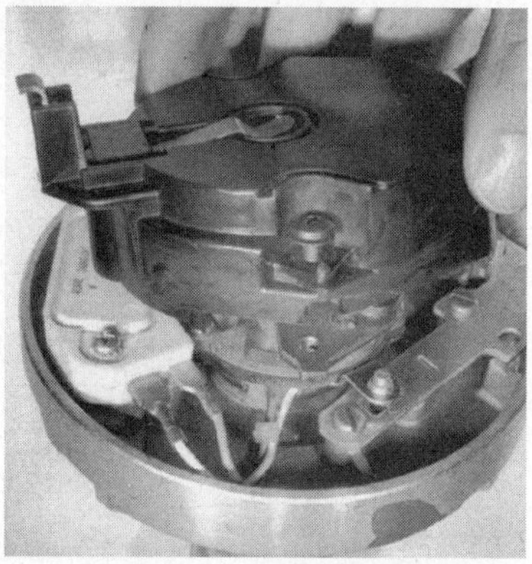

25. Attach the primary connector and pickup coil leads to the module, then check shaft rotation and advance movement. Fit the rotor and tighten two screws.

26. Fit the carbon button and spring into the cap and install the rubber seal. Then, fit the coil into the cap.

27. Install the coil mounting screws and the coil cover. Place the cap on the distributor and secure it with the holddown latches.

DELCO-REMY 4-CYLINDER HEI-EST DISTRIBUTOR OVERHAUL PROCEDURE

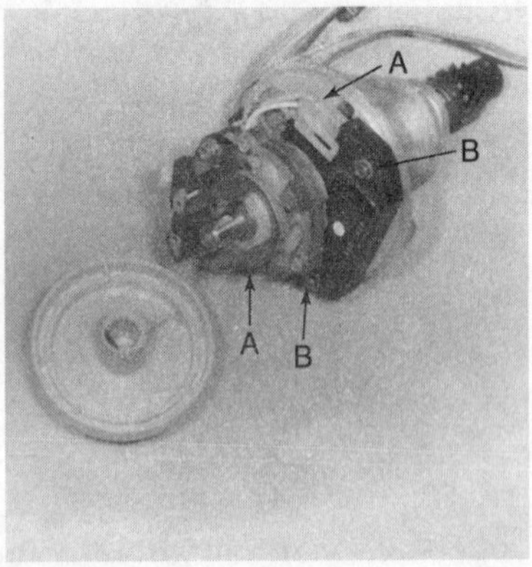

1. Pull the rotor straight off the shaft. Unclip and disconnect the leads on both sides of the HEI-EST module (A), then remove the module screws (B).

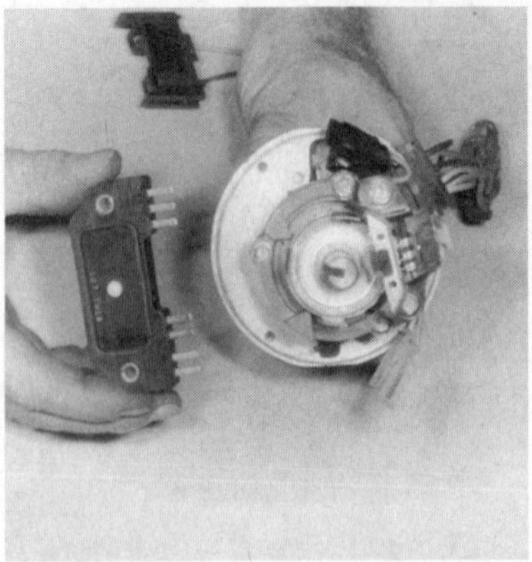

2. Lift the HEI-EST module from the distributor base. Clean the base of any silicone lubricant left by the module.

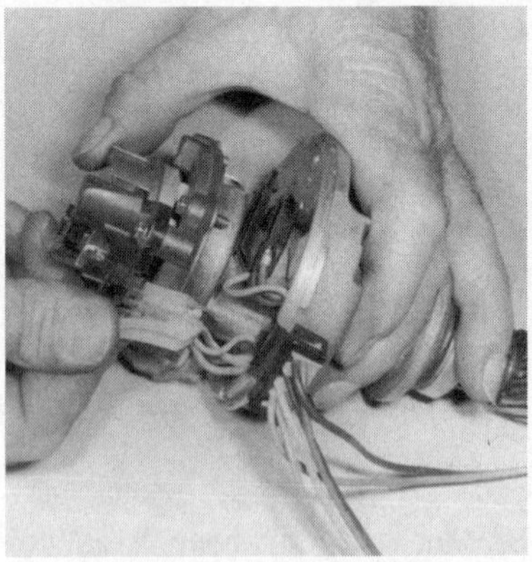

3. This Hall Effect switch provides the EST signal. Carefully unclip and remove the electrical connector from the Hall Effect switch.

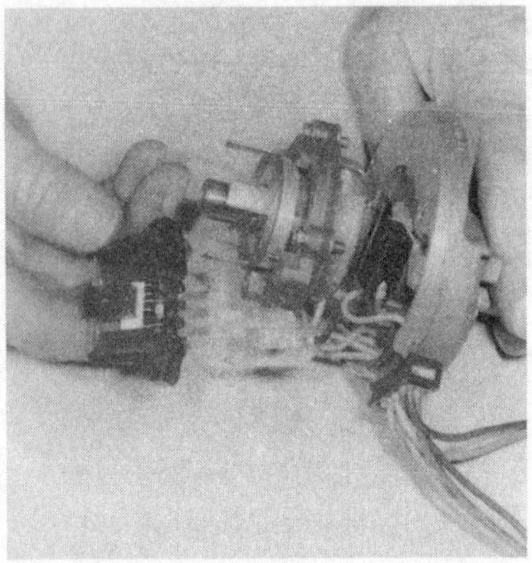

4. Remove the two screws holding the Hall Effect switch to the pickup coil assembly and lift the Hall Effect switch from the distributor.

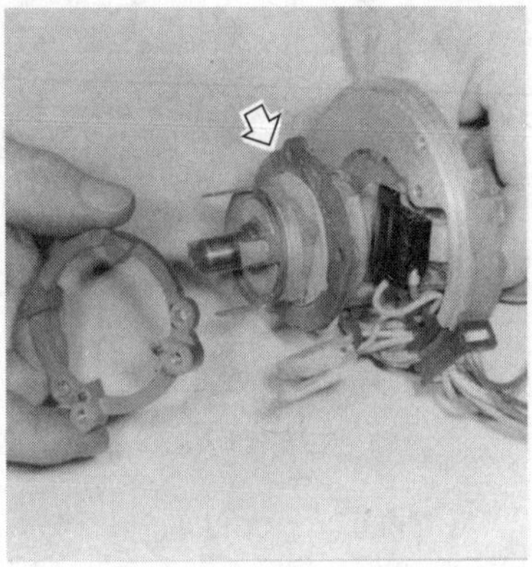

5. Remove the three screws holding the pole piece and magnet assembly. Remove the pole piece first, then remove the magnet (arrow).

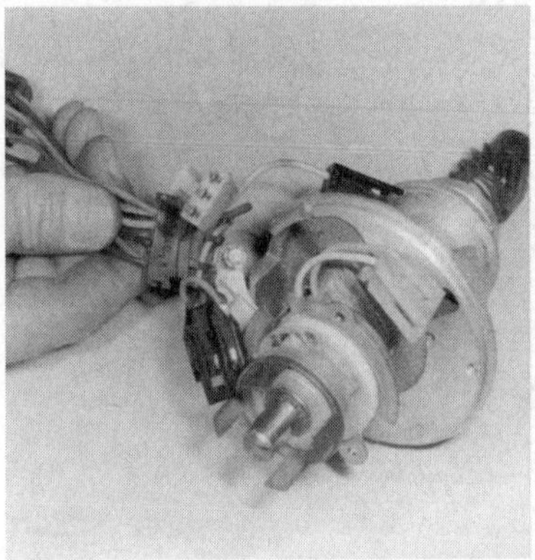

6. Remove the screw holding the wiring harness to the distributor housing and lift off the wiring harness.

DELCO-REMY 4-CYLINDER HEI-EST DISTRIBUTOR OVERHAUL PROCEDURE

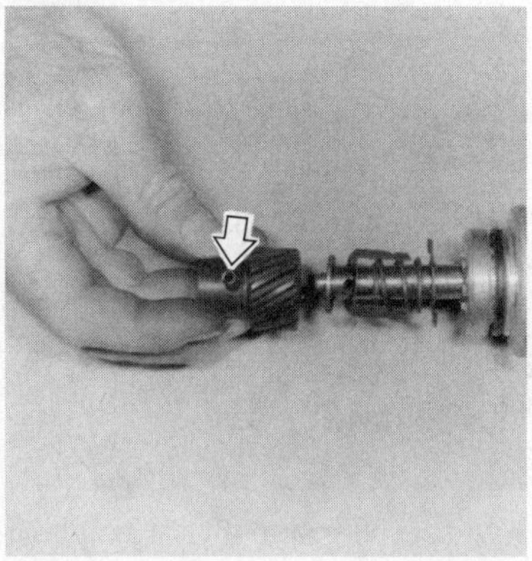

7. Drive out the roll pin (arrow) with a suitable punch to free the drive gear. Remove the drive gear, small washer, spring assembly, and large washer.

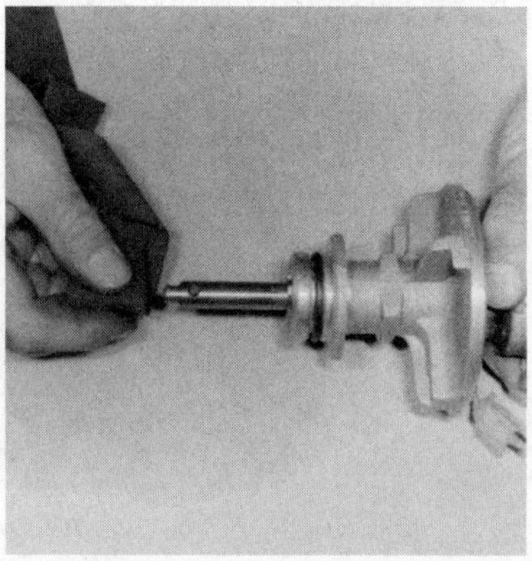

8. Deburr the distributor shaft with crocus cloth to prevent damage to the housing bushings when the shaft is removed.

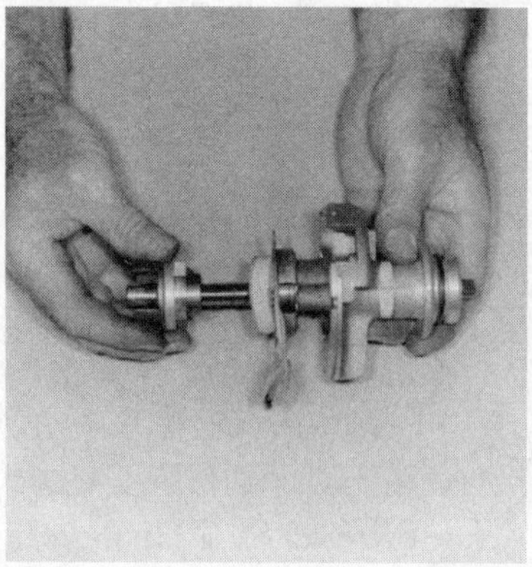

9. Carefully withdraw the shaft along with the attached shutter blade and trigger wheel from the distributor housing.

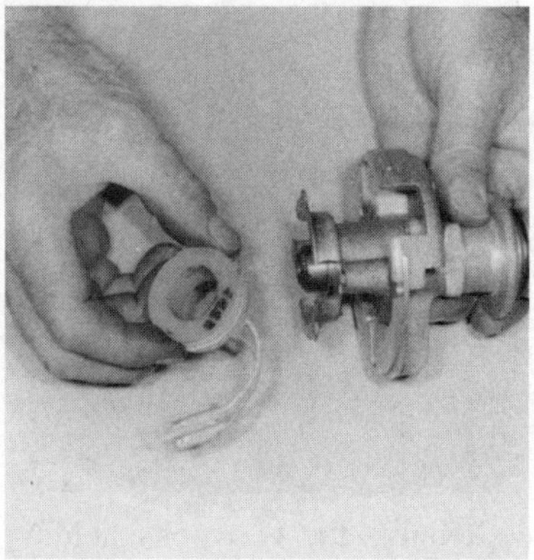

10. Remove the magnetic pulse pickup coil from the cup on the distributor base.

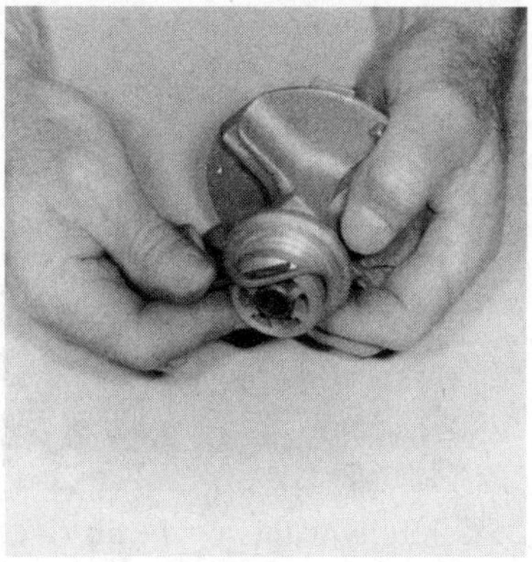

11. Use a small screwdriver as shown to remove the O-ring on the housing base. Install a new O-ring.

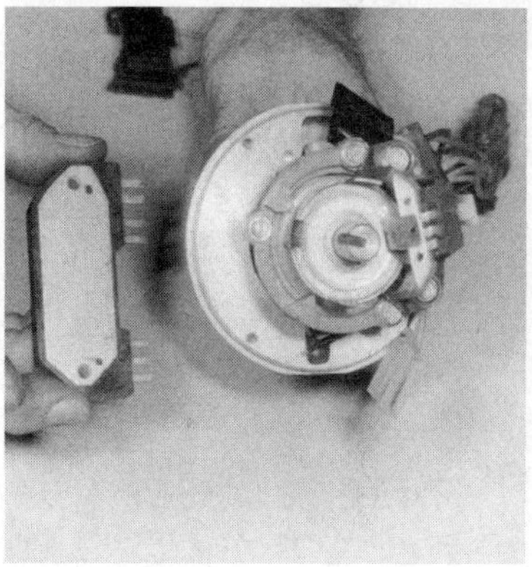

12. Assemble the distributor in reverse order. Be sure to apply a thin coat of silicone lubricant to the module base before installing it on the distributor base.

MOTORCRAFT UNIVERSAL TFI-I DISTRIBUTOR OVERHAUL PROCEDURE

1. Remove and set aside the cap and rotor. Then, make reference marks on the distributor shaft and drive gear for correct positioning on assembly.

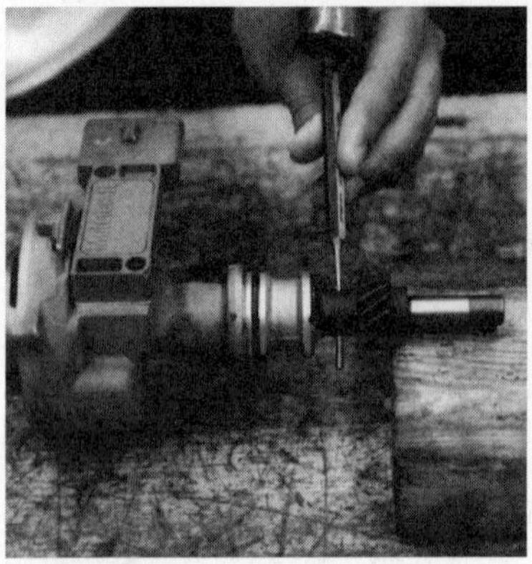

2. Support the distributor shaft and gear and use a pin punch and hammer to drive the roll pin completely out.

3. A special gear puller is available to remove the gear from the shaft. Remove the gear with a suitable puller.

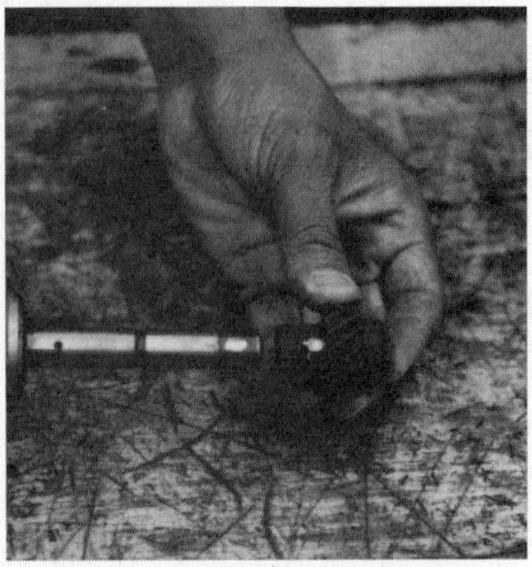

4. Once the gear is removed, clean any built-up varnish and debris from the end of the shaft to prevent scoring the bushings when the shaft is removed.

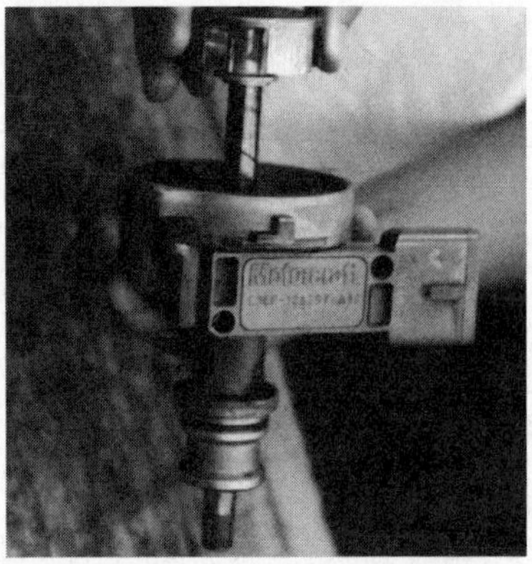

5. Carefully withdraw the distributor shaft from the distributor housing. Penetrating oil and a few light hammer taps may be needed to free the shaft.

6. Remove the two bolts holding the module in place. Then, slide the module down to disconnect the terminals and remove it from the housing.

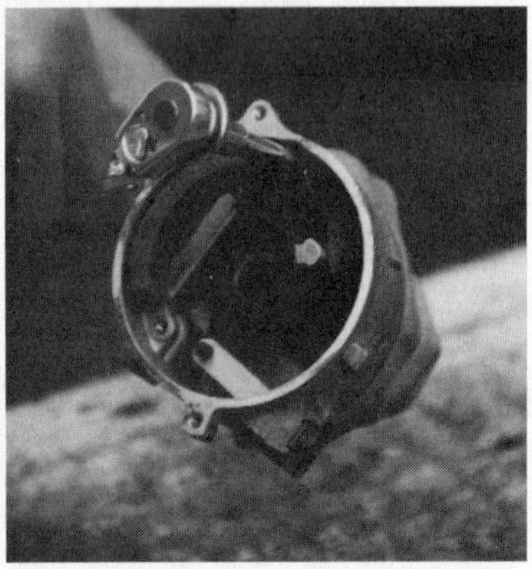

7. Remove the two bolts that attach the magnet portion of the stator assembly to the distributor housing.

8. Remove the single bolt that attaches the stator bracket to the distributor housing.

9. Carefully pry up the stator bracket to free it from the pin on the pickup assembly, then slide the bracket out of the housing.

MOTORCRAFT UNIVERSAL TFI-I DISTRIBUTOR OVERHAUL PROCEDURE

10. Lift the stator assembly, which includes the pickup and the magnet, out of the distributor housing.

11. Fit the new stator assembly into the housing so that the magnet bolt holes align and the pickup pin is correctly positioned.

12. Fit the bracket into the housing and onto the pin. Then, install and tighten the bracket screw and the two magnet screws.

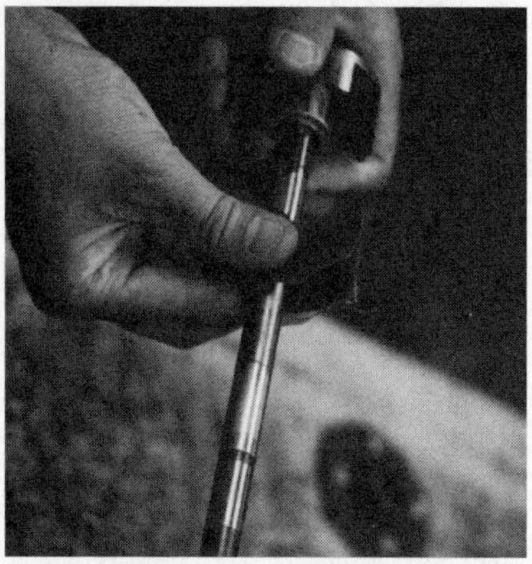

13. Polish the load-bearing area of the distributor shaft with crocus cloth to remove any nicks, burrs, or scratches. Then, slip the shaft into the housing.

14. Carefully position the drive gear so the reference marks and the pin bores are aligned. Use a press to seat the gear on the shaft.

15. Press the gear on slowly and carefully until the bore on the gear aligns with that of the shaft. When correctly aligned, the pin should start by hand.

16. Install the roll pin using a punch and hammer. Drive the pin in until it is flush with the gear.

17. Apply the silicone grease supplied with the new part on the back of the module in an even coat.

18. Fit the new module into the connector terminals and onto the housing. Install the two module screws and the cap and rotor to complete the overhaul

CHRYSLER 4-CYLINDER HALL EFFECT DISTRIBUTOR OVERHAUL PROCEDURE

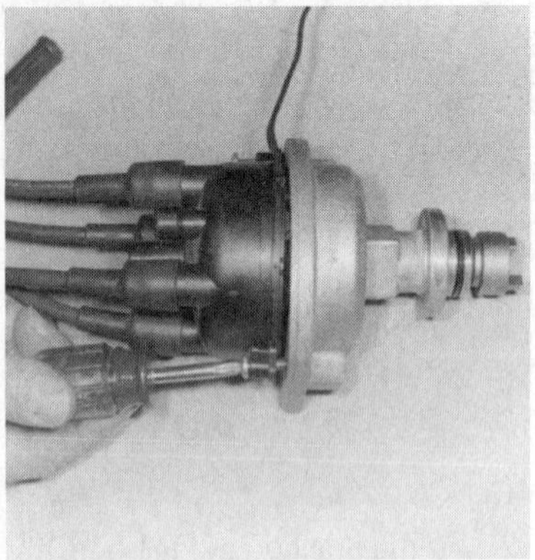

1. Loosen the two screws holding the distributor cap to the base. Lift the cap straight off to prevent rotor damage.

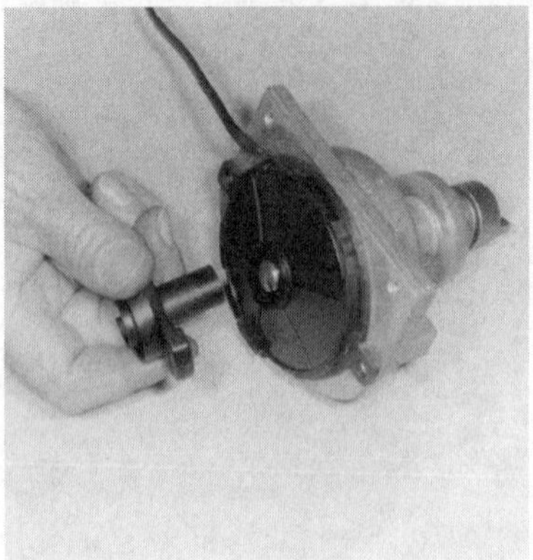

2. The rotor has an internal tab that fits into a slot in the shaft; pull the rotor straight off the distributor shaft.

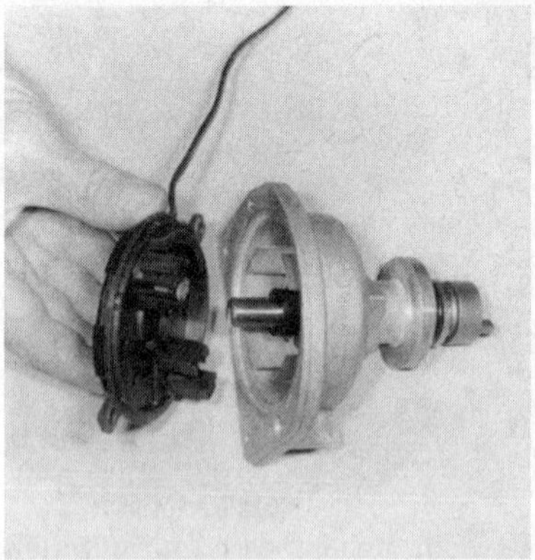

3. Lift off the Hall Effect pickup assembly. This unit is held in place by the distributor cap screws and is loose when the cap is removed.

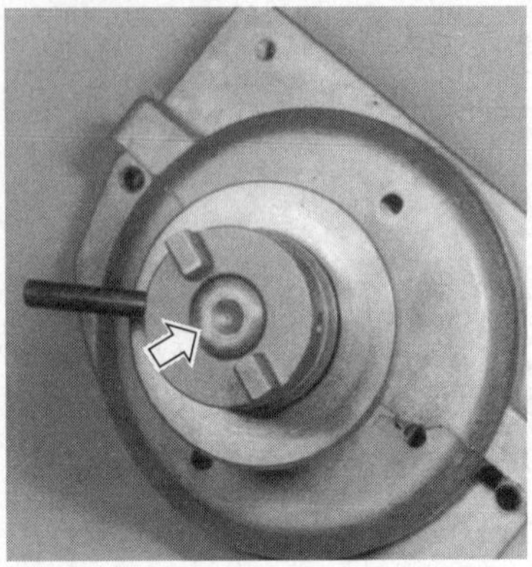

4. Mark the base of the shaft and the drive dog (arrow) to provide an alignment reference for assembly.

5. Support the drive end of the distributor. Then, carefully drive the roll pin out of the drive and shaft with a punch and hammer.

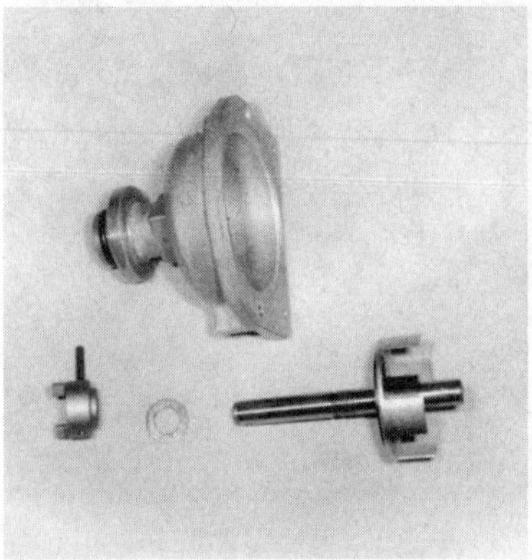

6. Remove the drive dog from the shaft, deburr the end of the shaft with crocus cloth, then carefully pull the shaft from the housing.

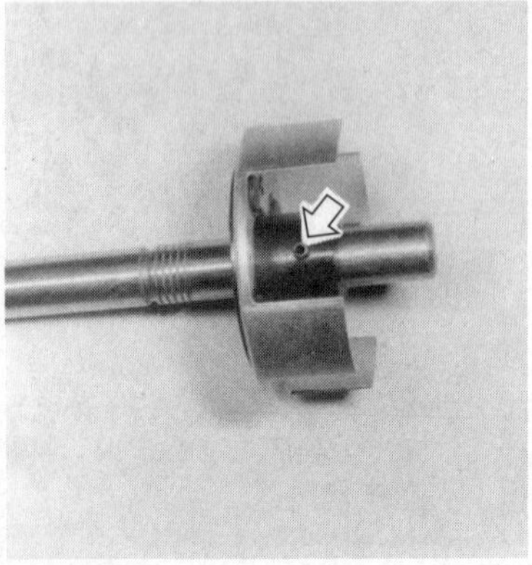

7. A small roll pin (arrow) fastens the shutter blade to the distributor shaft. Remove and install with a small punch and hammer as needed.

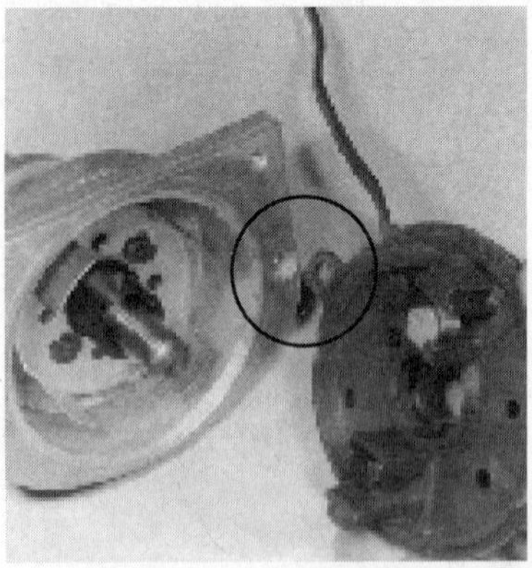

8. A tab on the Hall Effect pickup fits a recess in the housing to align the assembly. Make sure the pickup is fully seated.

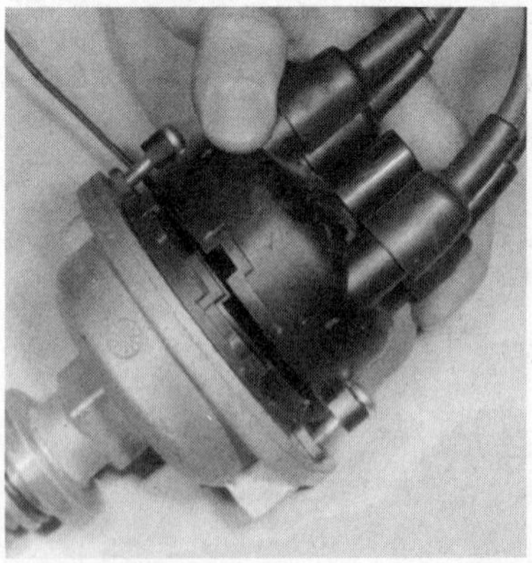

9. Fit the cap onto the Hall Effect pickup and distributor assembly. Make sure everything is aligned, then tighten the two screws to complete the assembly.

CHRYSLER 6-CYLINDER OPTICAL DISTRIBUTOR OVERHAUL PROCEDURE

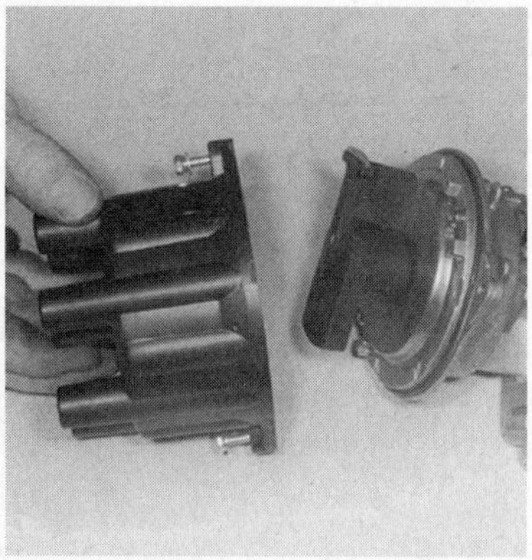

1. Loosen the two screws holding the distributor cap to the housing and remove the cap.

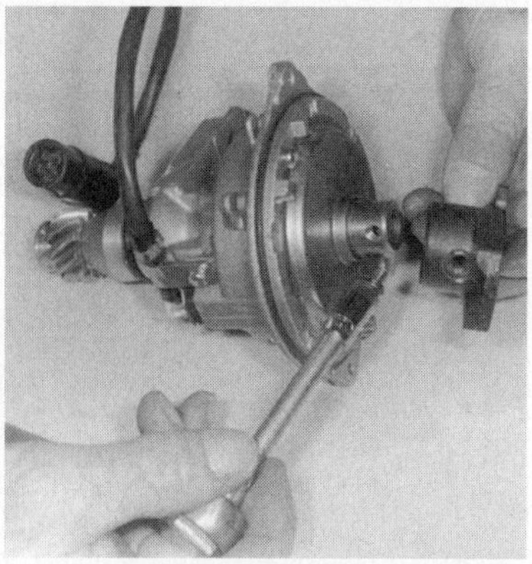

2. A screw underneath the rotor holds the rotor to the distributor shaft. Remove the screw and slip the rotor off the shaft.

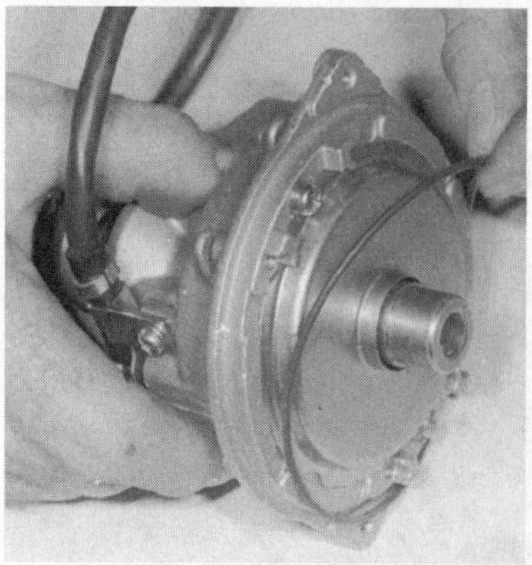

3. Carefully remove the O-ring that seals the cap to the distributor housing. Reuse the O-ring if it is not damaged or deteriorated.

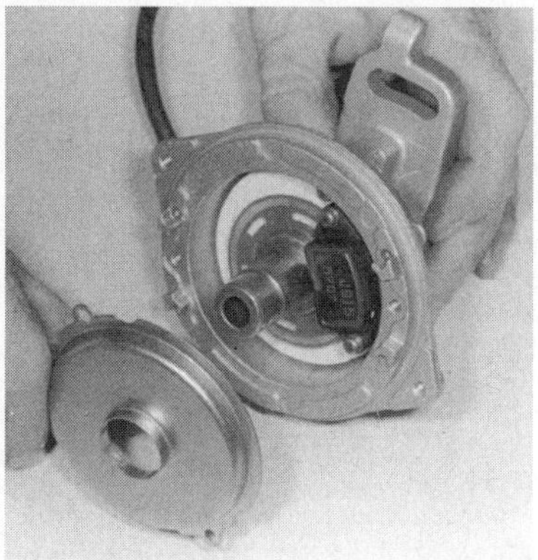

4. Remove the dust cap from the housing. This cap reduces electromagnetic interference (EMI) to prevent contamination of the sensor signal.

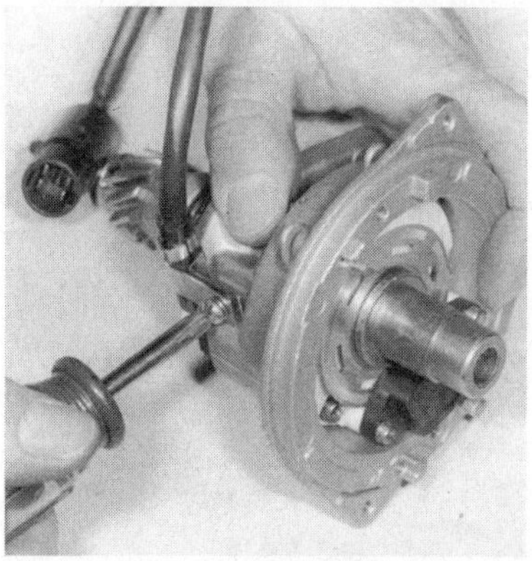

5. A screw attaches the wiring harness to the housing. Remove the screw, then disconnect the harness from the optical sensor.

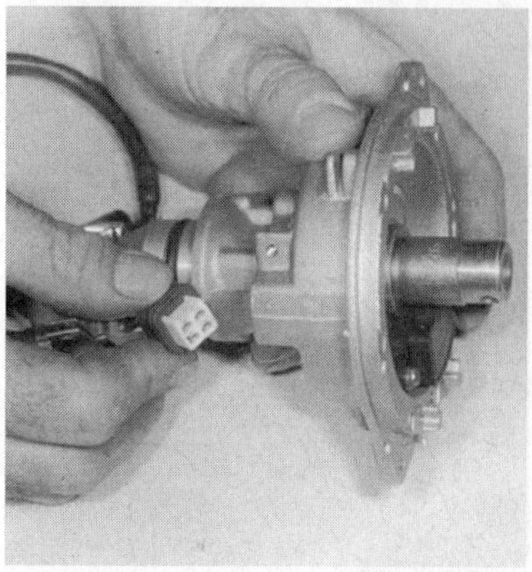

6. Gently work the wiring harness connector plug and weatherproof seal free and remove it from the base of the distributor.

7. Note that the screws holding the optical module are sealed with epoxy (A); do not try to remove them. Remove the screw in the center of the distributor shaft (B).

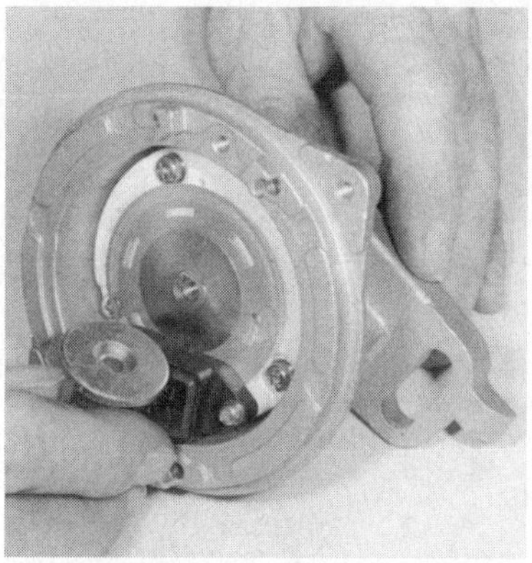

8. Once the screw is removed, slip the upper bushing of the end of the shaft to expose the disc and spacer assembly.

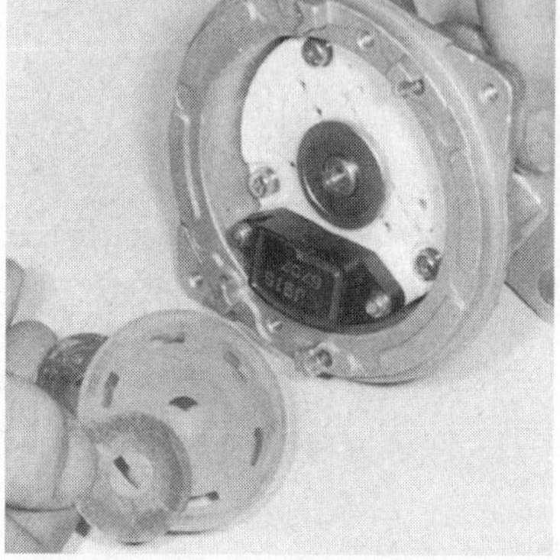

9. Note the position of the disc and spacer for assembly alignment. Then, carefully remove the disc and spacer assembly from the housing.

CHRYSLER 6-CYLINDER OPTICAL DISTRIBUTOR OVERHAUL PROCEDURE

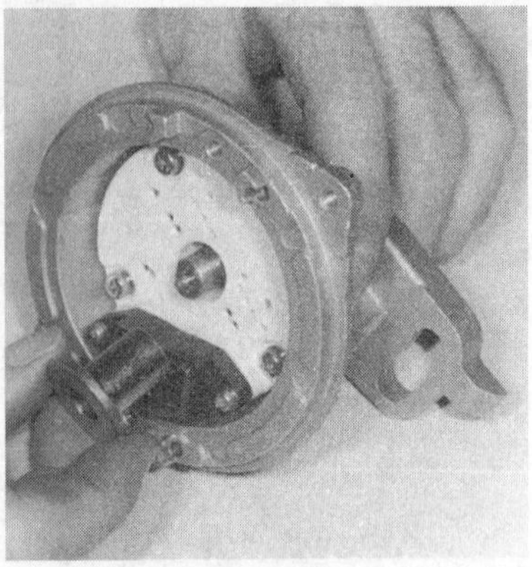

10. Pull the lower bushing out of the housing. Machined flats on the shaft and bushing align the assembly and prevent bushing rotation.

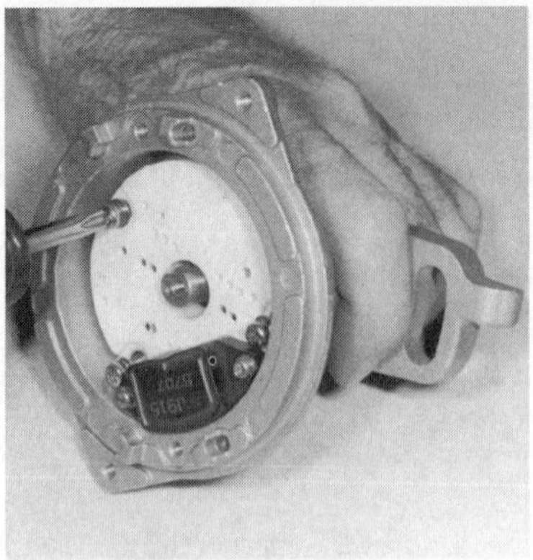

11. Three screws attach the base of the optical module assembly to the distributor housing. Remove to replace the module assembly if needed.

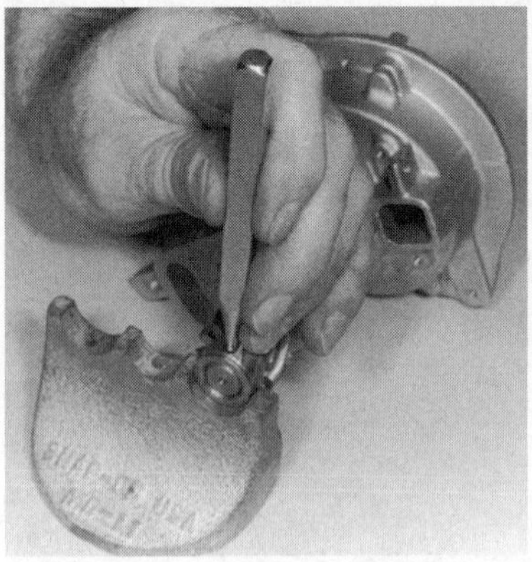

12. Support the distributor housing and drive the roll pin from the gear and shaft with a punch and hammer if shaft removal is required.

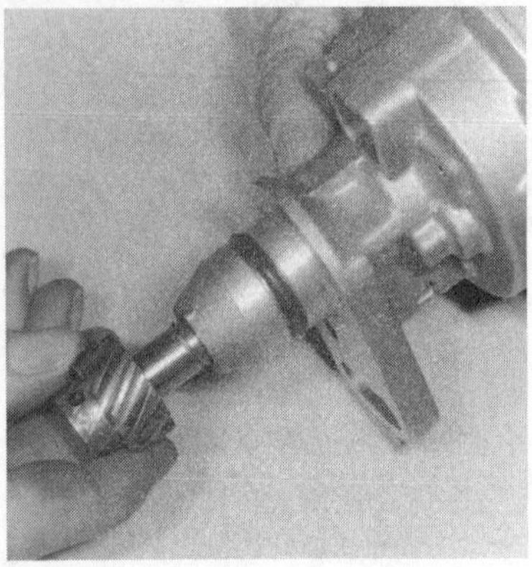

13. Slide the drive gear from the end of the shaft. There is no need to deburr the end of the shaft before removing it because the housing contains no bushings.

14. Remove the two screws that hold the shaft and bearing assembly in the housing, then slip the shaft and bearing out of the housing.

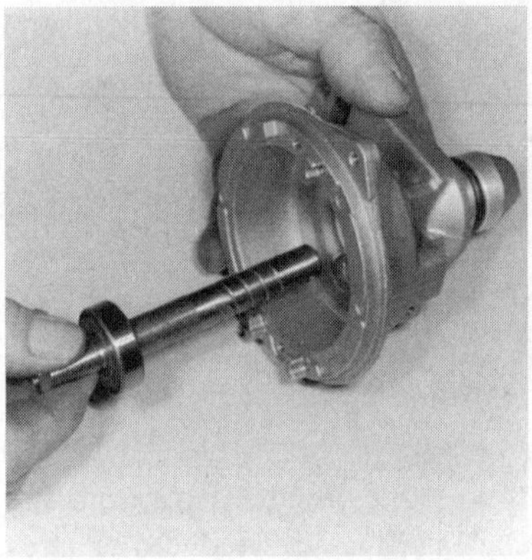

15. Replace the O-ring on the drive gear end of the housing, then assemble the distributor in reverse order.